T0318988

Sustainable Food Waste-to-Energy Systems

Sustainable Food Waste-to-Energy Systems

Edited by

Thomas A. Trabold

Callie W. Babbitt

Academic Press is an imprint of Elsevier
125 London Wall, London EC2Y 5AS, United Kingdom
525 B Street, Suite 1650, San Diego, CA 92101, United States
50 Hampshire Street, 5th Floor, Cambridge, MA 02139, United States
The Boulevard, Langford Lane, Kidlington, Oxford OX5 1GB, United Kingdom

Library of Congress Cataloging-in-Publication Data
A catalog record for this book is available from the Library of Congress

British Library Cataloguing-in-Publication Data
A catalogue record for this book is available from the British Library

ISBN 978-0-12-811157-4

For information on all Academic Press publications
visit our website at https://www.elsevier.com/books-and-journals

Publisher: Joe Hayton
Acquisition Editor: Raquel Zano
Editorial Project Manager: Mariana L. Kuhl
Production Project Manager: R.Vijay Bharath
Cover Designer: Mark Rogers

Typeset by SPi Global, India

Dedication

TAT: I dedicate this work to my mother Beatrice Trabold and my late father John Trabold. I also extend special thanks to my wife Nancy and children Lily, William, and Josef for their constant love and support.

CWB: I dedicate this work to my daughter Nora, with the hope that sustainability advances can ensure a livable planet for her and for future generations.

Contents

Contributors

Numbers in parenthesis indicate the pages on which the authors' contributions begin.

William R. Armington (259), Golisano Institute for Sustainability, Rochester Institute of Technology, Rochester, NY, United States

Callie W. Babbitt (1,177,259), Golisano Institute for Sustainability, Rochester Institute of Technology, Rochester, NY, United States

Roger B. Chen (259), Golisano Institute for Sustainability, Rochester Institute of Technology, Rochester, NY, United States

Bruce I. Dvorak (111), Department of Civil Engineering and Department of Biological Systems Engineering, University of Nebraska-Lincoln, Lincoln, NE, United States

Jacqueline H. Ebner (177), College of Liberal Arts, Rochester Institute of Technology, Rochester, NY, United States

Matthew J. Franchetti (203), Department of Mechanical, Industrial and Manufacturing Engineering, The University of Toledo, Toledo, OH, United States

Somik Ghose (203), Department of Mechanical, Industrial and Manufacturing Engineering, The University of Toledo, Toledo, OH, United States

Serpil Guran (141,159), Rutgers University EcoComplex "Clean Energy Innovation Center", Bordentown, NJ, United States

Swati Hegde (11,69,177), Golisano Institute for Sustainability, Rochester Institute of Technology, Rochester, NY, United States

Rodrigo A. Labatut (47), Department of Hydraulic and Environmental Engineering, Pontifical Catholic University of Chile, Santiago, Chile; Bioprocess Analytics LLC, Gansevoort, NY, United States

Vineet Nair (29), Golisano Institute for Sustainability, Rochester Institute of Technology, Rochester, NY, United States

Jennifer L. Pronto (47), Bioprocess Analytics LLC, Gansevoort, NY, United States

Kirti Richa (231), Energy Systems Division, Argonne National Laboratory, Lemont, IL, United States

Erinn G. Ryen (231), Wells College, Aurora, NY, United States

Jeyamkondan Subbiah (111), Department of Biological Systems Engineering and Department of Food Science and Technology, University of Nebraska-Lincoln, Lincoln, NE, United States

Thomas A. Trabold (1,11,29,69,89,177), Golisano Institute for Sustainability, Rochester Institute of Technology, Rochester, NY, United States

Shwe Sin Win (11,89,177), Golisano Institute for Sustainability, Rochester Institute of Technology, Rochester, NY, United States

Rami M.M. Ziara (111), Department of Civil Engineering, University of Nebraska-Lincoln, Lincoln, NE, United States

Acknowledgment

This work would have been impossible without the support of the Golisano Institute for Sustainability at Rochester Institute of Technology, which since its founding in 2006 has actively promoted technology development that comprehends the fundamental tenets of environmental, economic, and social sustainability. We acknowledge the New York State Pollution Prevention Institute (NYSP2I) and its Sustainable Food Program that has been a leading voice in our region promoting best practices for diverting food waste from landfills, and identifying and developing alternative methods that convert these materials into value-added products. The National Science Foundation (NSF) is also acknowledged for providing funding for "Managing Energy, Water, and Information Flows for Sustainability across the Advanced Food Ecosystem" (CBET-1639391), part of the *Innovations at the Nexus of Food, Energy, and Water Systems (INFEWS)* program. NYSP2I and NSF have provided support for a large cohort of MS and PhD students who will become the next generation of thought leaders in sustainable food systems.

Chapter 1

Introduction

Thomas A. Trabold and Callie W. Babbitt

Golisano Institute for Sustainability, Rochester Institute of Technology, Rochester, NY, United States

Food is, of course, essential for organisms to live, thrive, and propagate, but the relationship between human beings and the food they consume has evolved in many ways over the centuries. The Haudenosaunee (Iroquois) people, who widely inhabited our region of New York State before the arrival of European settlers, had a deep spiritual relationship with food. The "Three Sisters" of corn, squash, and beans were presented as gifts directly from the Creator and considered essential to sustaining community life (Lewandowski, 1987). In most regions of the globe, food has been assigned great significance in representing cultural, religious, ethnic, and national identity. But today, few people in the developed economies of the world have direct, personal involvement in the enterprise of producing food, and therefore wasting it does not seem particularly arrogant or vulgar. Whereas the Iroquois associated their very existence with the ability to plant and nurture certain staple crops, in modern times we discard half-eaten apples or sandwiches without even a hint of introspection.

As the global community begins to seriously consider the implications of human-induced climate change and a rising population expected to exceed 11 billion by the end of the century (United Nations, 2015), it is important to address the problem of food waste and the very real sustainability threat it represents. While much of the developed world tries to decide how to stop wasting so much food and the precious natural resources contained therein, other much larger populations in developing countries suffer from severe food insecurity. But even in the United States, it has been estimated that one in seven people face some food insecurity throughout the year (ReFED, 2016). Such disparities between the "haves" and "have-nots" have many political and social justice implications that are beyond the scope of this book. Our objective is to approach this problem from a strictly technical standpoint, and in this context the imbalance between food supply and demand indicates that significant inefficiency exists in the global food system.

According to a recently published study, the impacts of food waste on the U.S. economy are staggering, accounting for 21% of fresh water use, 19% of fertilizer use, 18% of cropland, and over 21% of landfill volume, all at an annual cost of $218 billion, more than half of which is borne directly by consumers (ReFED, 2016). To develop a viable societal strategy for addressing the food waste problem, it is instructive to first consider some of the trends responsible for changing attitudes toward the agricultural and food system in the United States and other developed economies.

- Food is plentiful

Before the organized practice of agriculture, the lifestyle offered by hunter-gatherers was able to support a global population of only about 4 million (Tilman et al., 2002). The "Green Revolution" spurred by synthetic fertilizers, new strains of disease-resistant crops, and heavy water input has provided broad abundance of food, at least in the more developed regions of the world, that has reduced hunger and enabled urban centers to expand. In parallel with the upward trend in per capita food supply (Fig. 1.1), the caloric intake of relatively wealthy populations has been on a steady upward trajectory, to the point that previously rare ailments such as diabetes and obesity are now common across all socioeconomic groups (e.g., Geiss et al., 2014).

- Food is inexpensive

At the beginning of the 20th century, Americans spent between 40% and 45% of their limited income on food (Chao and Utgoff, 2006). As agricultural became more heavily mechanized and crop yields improved, the unit cost of food production dropped significantly, while at the same time average inflation-adjusted personal incomes increased appreciably. The result is that average expenditures on food have dropped from nearly 23% of disposable personal income in 1929 to around 11% since 2000 (Fig. 1.2). The current spending on food is dwarfed by the costs of both housing and transportation, and is comparable to what the average American now spends on insurance and pension payments (Bureau of Labor Statistics, 2016).

Sustainable Food Waste-to-Energy Systems. https://doi.org/10.1016/B978-0-12-811157-4.00001-2

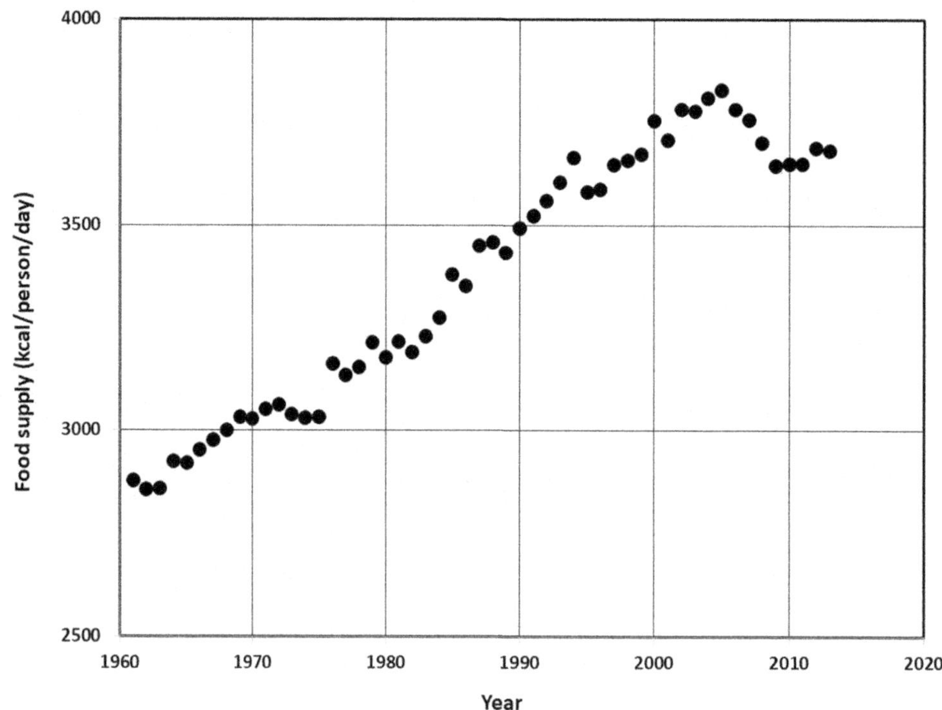

FIG. 1.1 Growth in U.S. food supply in kcal/person/day. *Source: Roser and Ritchie, 2017; primary data from FAO (https://ourworldindata.org/food-per-person/).*

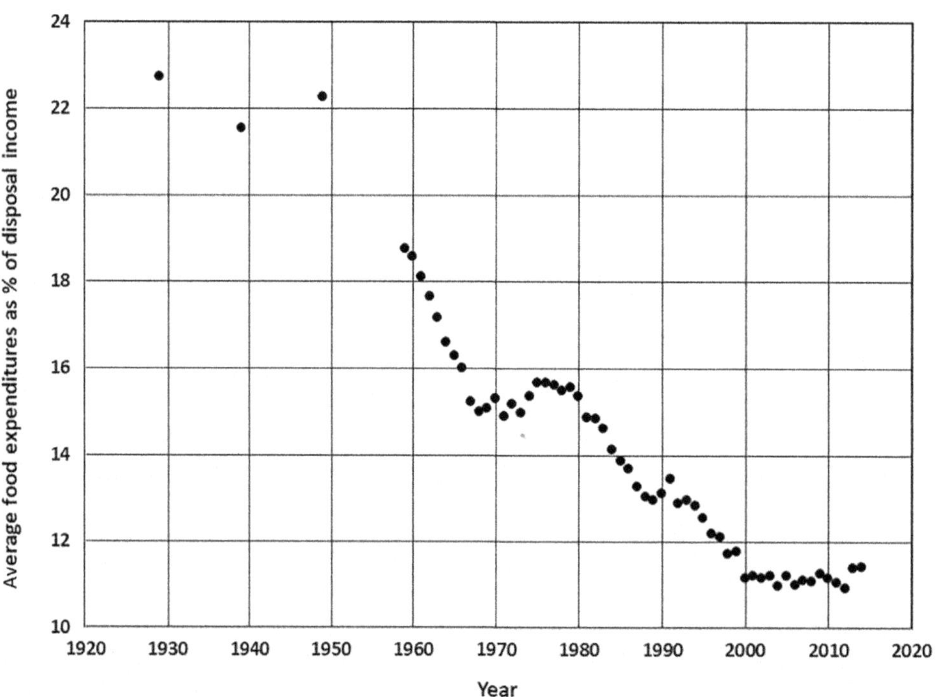

FIG. 1.2 Average U.S. food expenditures as a percentage of disposal personal income. *Source: Calculated by the Economic Research Service of the U.S. Department of Agriculture, from various data sets from the U.S. Census Bureau and the Bureau of Labor Statistics. https://www.ers.usda.gov/data-products/food-expenditures.aspx.*

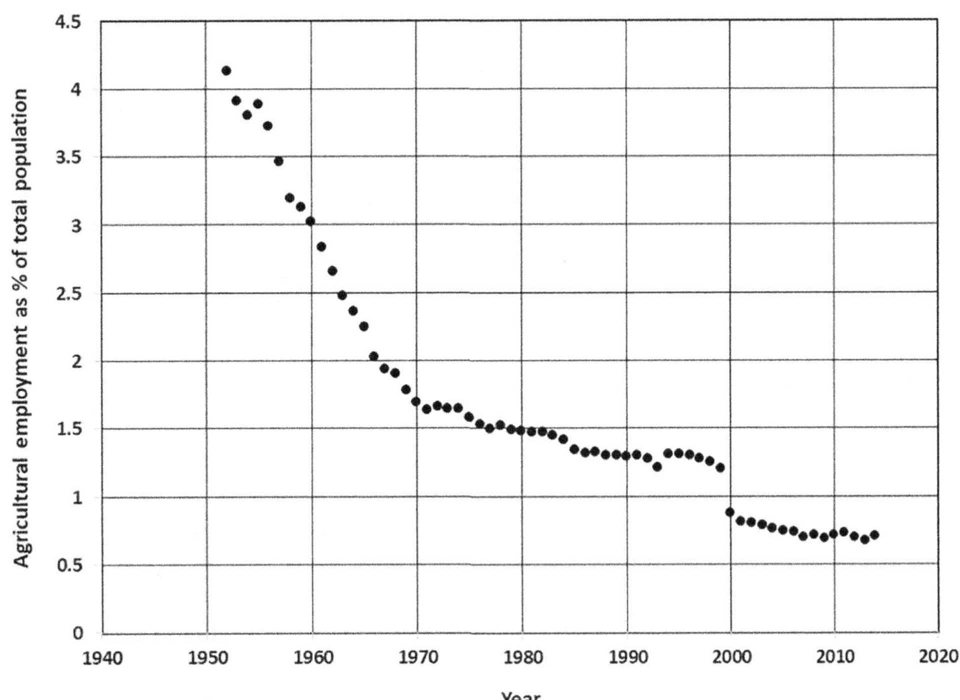

FIG. 1.3 Agricultural employment as a percentage of total U.S. population. *Source: FAOSTAT tool of the Food and Agriculture Operation (FAO) of the United Nations. http://www.fao.org/faostat/en/#data.*

- Food production requires few people

In the early 1800s, agriculture accounted for nearly 80% of the U.S. labor force. The number of agricultural workers peaked around 12 million in 1910, which at the time comprised roughly 30% of the labor force and 13% of the total population (Sullivan, 1996). With advances in technology and mechanization, family farms have to a great extent been displaced by large industrial operations, and now the agricultural system employs fewer than 3% of the available labor force and less than 1% of the total population (Fig. 1.3). This trend has important social implications in that few people are now directly connected to farms or know someone who is, and thus have limited knowledge of the processes associated with food production and distribution, and of their underlying contributions to national economic activity and security.

- Food is no longer only a local resource

The imported share of total average food consumption in the United States was less than 8% in the early 1980s and stabilized around 11% by 2001 (Jerardo, 2003). However, various trade agreements signed since then, most notably the North American Free Trade Agreement (NAFTA) among the United States, Canada, and Mexico, has resulted in an increased fraction of imported foods in American diets and a steady increase in per capita expenditures on food imports (Fig. 1.4). According to estimates from the U.S. Department of Agriculture, in 2013 the import share of U.S. food consumption was 19% based on volume and 20% based on value.[1] In addition to food products originating outside the United States, another significant consideration is the distance domestically produced food travels from the point of origin to the point of consumption (so-called food miles). Some recent studies have illustrated that greenhouse gas impacts from food transport are far less than impacts associated with the production phase (i.e., meats vs. vegetables; Weber and Matthews, 2008). However, food imports and transport are usually associated with increased use of packaging and can increase the rate of food waste from spoilage and damage during transport, or from rejection of consumer-ready products imported from countries with lower safety standards (Brooks et al., 2009a). Also, the global "cold-chain" provided by refrigeration technology has increased consumers' food choices but has also contributed to food waste across the farm-to-fork spectrum (Heard and Miller, 2016).

The aggregated impact of these major trends in the U.S. food system is illustrated in Fig. 1.5, with all data normalized to year 2000. In each case, second-order polynomials were fitted to the raw data in Figs. 1.1–1.4 to provide a simplified picture

1. https://www.ers.usda.gov/topics/international-markets-trade/us-agricultural-trade/import-share-of-consumption/#data.

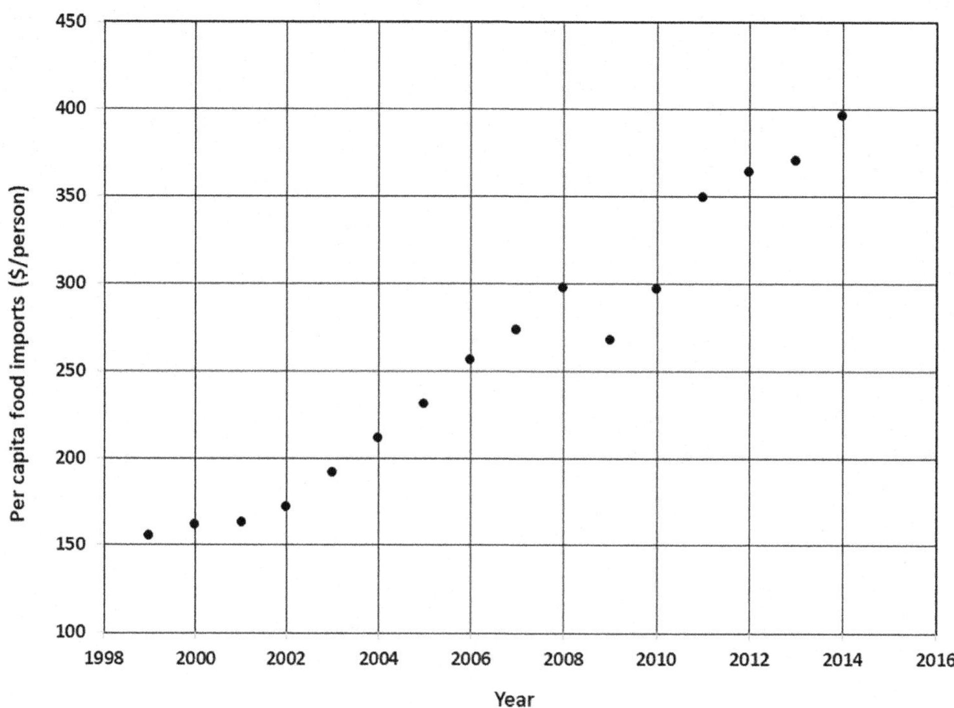

FIG. 1.4 U.S. per capita expenditures on imported food, including animal products, plant products, and beverages. Note the distinct drop in 2009 following the global financial crisis. *Sources: Unites States Department of Agriculture, www.fas.usda.gov/gats; Brooks, N.L., Regmi, A., Jerardo, A., 2009. US food import patterns, 1998–2007. USDA, Economic Research Service.*

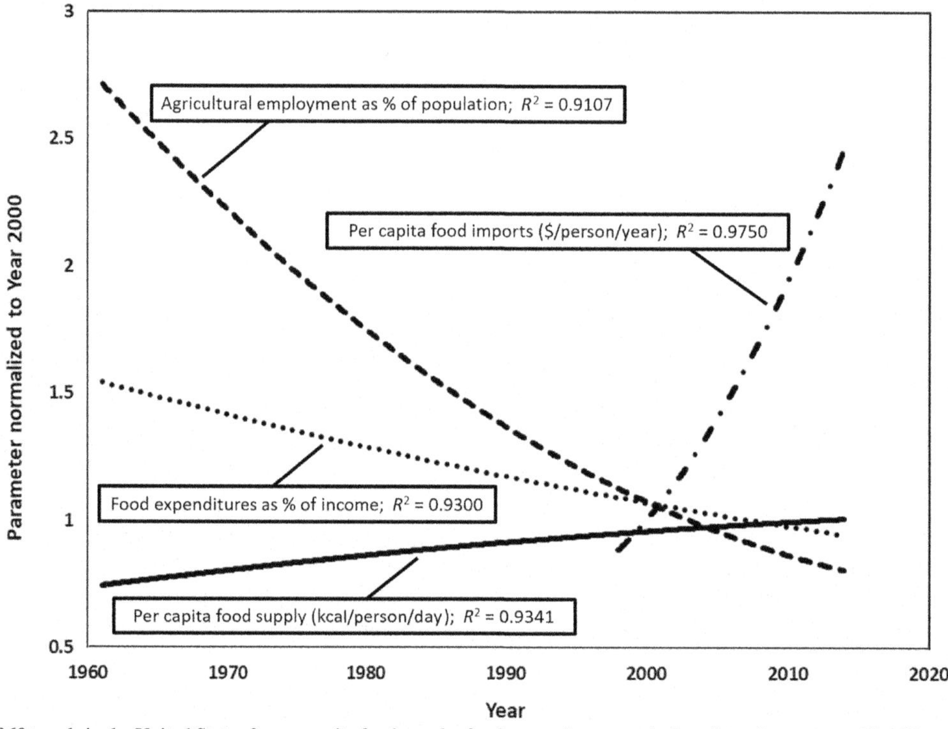

FIG. 1.5 Post-1960 trends in the United States for per capita food supply, food expenditures, agricultural employment, and food imports, all normalized to year 2000. Lines represent second-order polynomial fits to raw data presented in Figs. 1.1–1.4, respectively, with associated correlation coefficients indicated. Note that not all curves pass through a value of "1" in year 2000 because of the fitting of polynomial regression curves to the normalized data in Figs. 1.1–1.4.

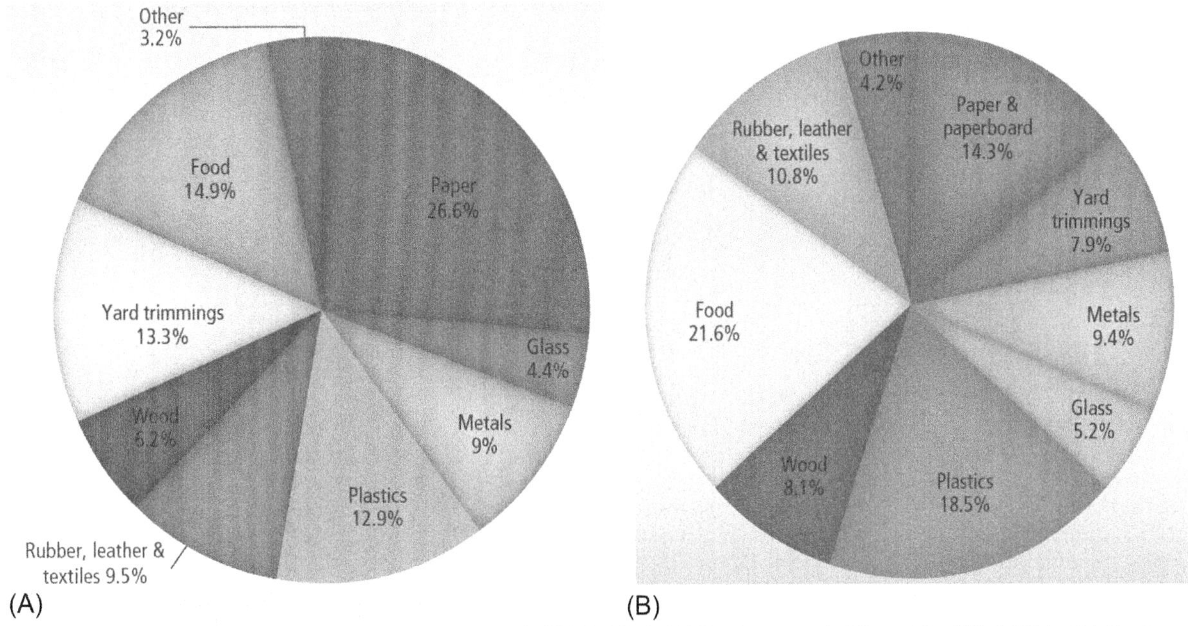

FIG. 1.6 Material proportions of U.S. municipal solid waste (MSW) in 2014: (A) Total generated; (B) total landfilled (EPA, U.S. Environmental Protection Agency, 2016).

of the trends that emerged after 1960. As the per capita food supply has steadily increased, the cost of food as a percentage of disposal income has declined. Also, while food has become more plentiful and inexpensive, the fraction of the U.S. population that is directly involved in its production has dropped to a very large degree. All these changes have occurred in the last 50 years while the level of per capita food imports has increased, with a significant jump observed since the beginning of this century. Even one not trained in sociology or human psychology might hypothesize that the trends illustrated in Fig. 1.5, which all relate in some way to the perceived societal value of food, would have an impact on the amount of food waste created, and the embodied energy and water content that is lost (Cuéllar and Webber, 2010).

Direct historical data for total U.S. food waste is not readily available, but we do have a reasonably good understanding of the fraction of food currently produced that is not consumed by humans. Kantor et al. (1997) estimated that in 1995 the fraction of U.S. food waste was 27% of food produced, but this estimate did not include losses occurring on farms or in food processing operations. Several more recent studies estimate the food waste fraction, considering all contributions from agricultural residues to postconsumer waste, at between 30% and 40% (e.g., Hall et al., 2009; Gunders, 2012; Buzby et al., 2014; ReFED, 2016). However, there is still a fair amount of uncertainty as many studies have based these estimates on the same outdated data (Parfitt et al., 2010). Other sources of data are available for the total volume of food waste that is generated (14.9% of all municipal solid waste, MSW) and then sent to landfill (21.6%); Fig. 1.6 (EPA, U.S. Environmental Protection Agency, 2016). Applying the former food waste fraction to the annual data for per capita MSW (Fig. 1.7) would imply that the generation of food waste steadily increased from 1960 until around 2000, when the level of per capita MSW appears to have leveled off.

Another factor that makes the problem of food waste so difficult to address, beyond the societal trends outlined above, is that the characteristics of the waste change significantly across the farm-to-fork spectrum. Crop residues generated in agricultural operations are generally homogeneous and concentrated in a relatively small number of locations. As these raw food materials are transported to consumer-facing businesses (grocery stores, restaurants, institutional food services, etc.) or to industrial operations for further processing, they are often mixed with other food materials and thus the waste produced in these stages becomes more heterogeneous and geographically dispersed. Another critically important factor is that at the food processing and consumer-facing business stages, packaging is often added to the material flow, further complicating the handling of waste. As these food materials proceed to the final consumption stage, any resulting waste is usually combined with all other types of waste typically generated at the scale of an individual household, and thus relatively small amounts of food waste are disposed of at many individual locations (over 116 million households in the United States as of 2015[2]). While technologies exist to sustainably convert food waste to energy and other value-added products, as will be

2. https://www.census.gov/quickfacts/table/PST045216/00.

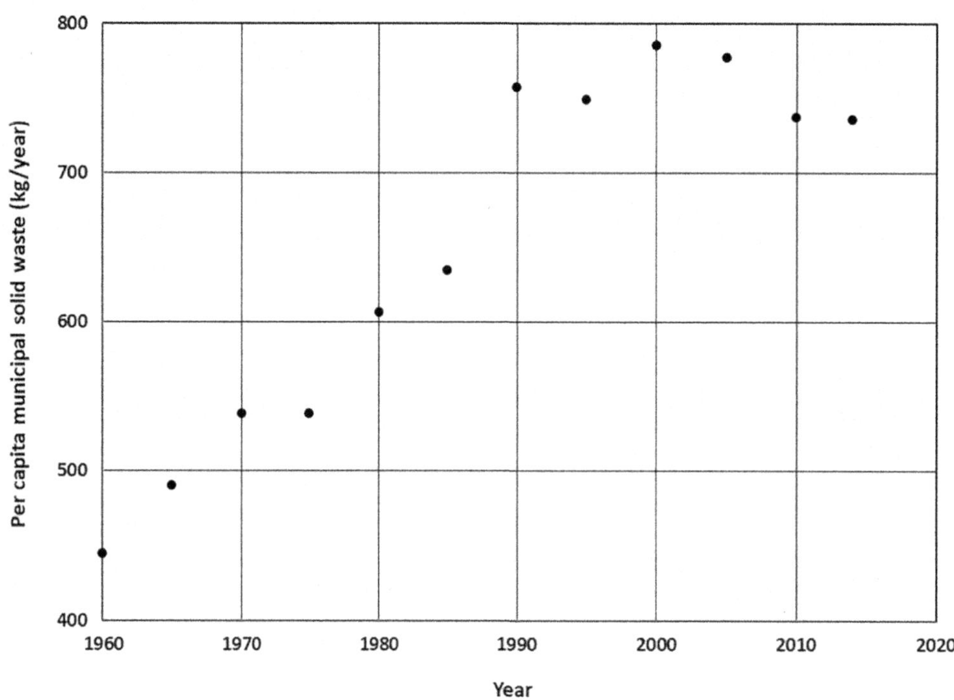

FIG. 1.7 U.S. per capita municipal solid waste (MSW) generation (EPA, U.S. Environmental Protection Agency, 2016).

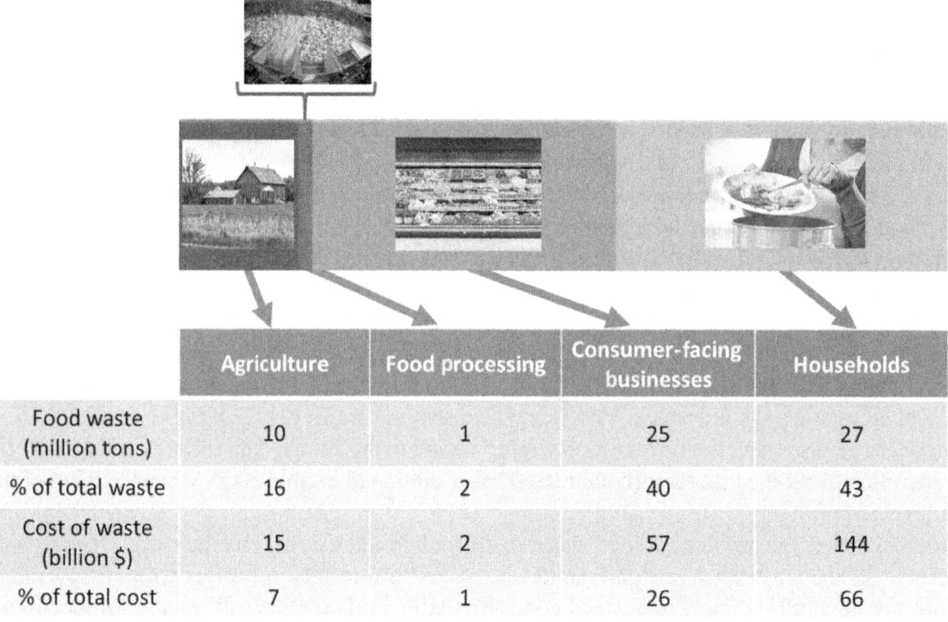

	Agriculture	Food processing	Consumer-facing businesses	Households
Food waste (million tons)	10	1	25	27
% of total waste	16	2	40	43
Cost of waste (billion $)	15	2	57	144
% of total cost	7	1	26	66

FIG. 1.8 U.S. food waste across the farm-to-fork spectrum (ReFED, 2016).

discussed throughout this book, the logistics of collecting and transporting all this material is an important aspect of the problem that cannot be overlooked.

According to the comprehensive ReFED report (ReFED, 2016), the United States currently produces 63 million tons of food waste per year, with the breakdown by food sector stage shown in Fig. 1.8. Agriculture produces an estimated 10 million tons of waste per year, but this is a relatively small impact in terms of total waste produced and its associated cost. Food processing wastes are an even smaller contribution, because manufacturers have already implemented many automation and efficiency improvements, and the waste generated is of reasonably consistent composition and often suitable for alternative uses, such as animal feed. The most significant opportunities for reducing food waste, in terms

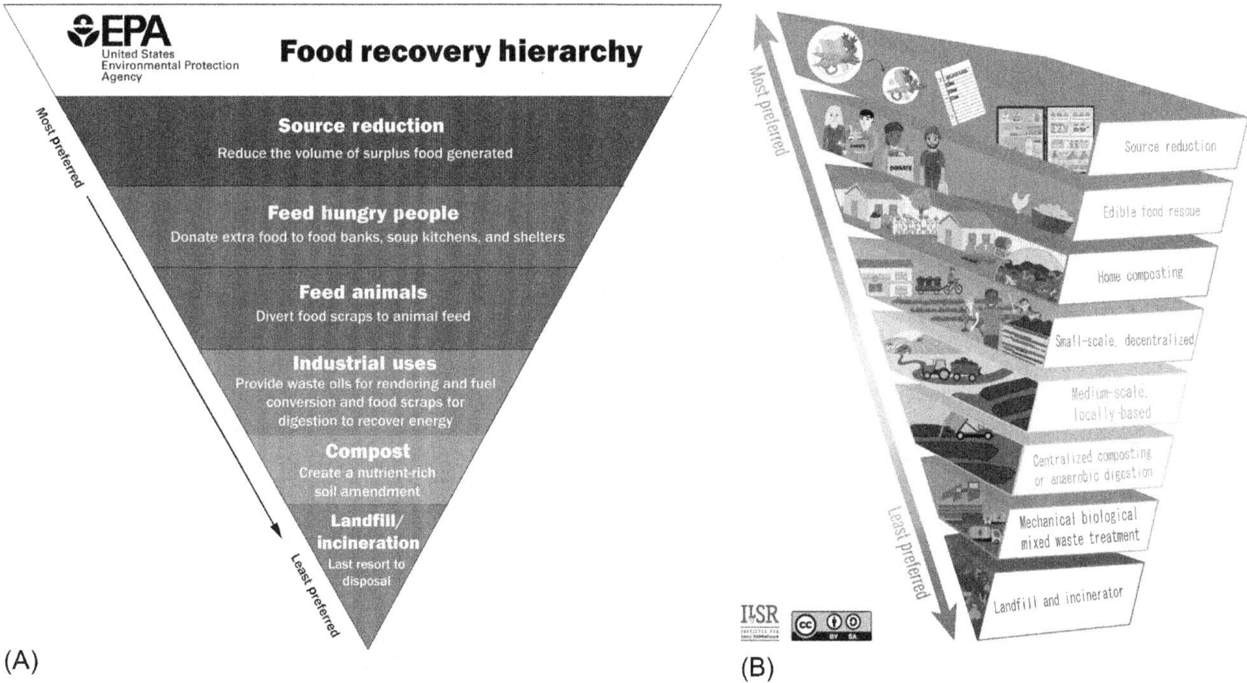

FIG. 1.9 (A) Food recovery hierarchy (U.S. Environmental Protection Agency). (B) Expanded food recovery hierarchy that distinguishes between centralized and decentralized conversion technologies (Institute for Local Self-Reliance https://ilsr.org/, a national nonprofit organization working to strengthen local economies, and redirect waste into local recycling, composting, and reuse industries. It is reprinted here with permission.).

of both landfill diversion and cost savings, are at the consumer-facing business and household stages, which combined account for 83% of the food waste and 92% of the associated cost. In Chapter 2, a detailed analysis is provided of the specific sources of waste at each stage of the farm-to-fork spectrum, the associated physical and chemical characteristics, and also the important effect of packaging. Food and packaging must be considered as an integrated "system" as we explore alternatives for diverting food waste from landfills, including sustainable energy production.

Recognizing the extent of the food waste problem and the need to define and prioritize alternative uses, the U.S. Environmental Protection Agency (and many organizations in other countries, such as WRAP in the United Kingdom) has developed a food recovery hierarchy to provide general guidelines for waste generators in selecting among different pathways for diverting waste from landfills (Fig. 1.9A). Certainly, the top priority is to simply reduce waste at its source through improved process efficiency, better sales projections, meal planning, etc. Even if all reasonable steps are taken to limit waste generation in the first place, it is inevitable that some fraction of food items produced at farms, manufactured in food processing plants, and sold by consumer-facing businesses would still be available on the market without sufficient consumer demand. In these cases, the highest value would be placed on donations to feed hungry people, but there are limits on the quantity and types of food that can be handled in this manner. The second option is to feed animals, and here the opportunities are more diverse, because many "lost" food materials that are generated during processing (e.g., vegetable skins or fruit pomace) would not be fed to humans, but serve quite well as feed for livestock. For excess food materials that cannot be fed to humans or animals, the remaining options are "industrial uses" comprised mostly of methods to recover energy, composting, and lastly disposal methods including landfilling, incineration, and waste water treatment. In Chapter 3, these various conventional methods as presented in the food recovery hierarchy are explored in more detail to understand the current state of the art of food waste handling in the United States, most of which also applies to other developed economies.

The middle portion of this book considers energy recovery methods that could be among the "industrial uses" identified in Fig. 1.9A, including technologies that are currently applied at industrial scale and others that are still largely at the research and development phase.

- *Anaerobic digestion (AD)* (Chapter 4) is a series of biochemical process in which microorganisms degrade organic matter in the absence of oxygen to produce a "biogas" that is rich in methane (CH_4) and thus has the potential to generate thermal and/or electrical energy. The conventional use of AD has been to convert animal manure, mostly on dairy farms, but has more recently been used to treat combined manure and food waste in so-called codigestion operations (Holm-Nielsen et al., 2009; Ebner et al., 2015, 2016).

- *Fermentation* (Chapter 5) is another biochemical process whereby yeasts and bacteria convert organic molecules into acids, solvents, or alcohol-based fuels (Ebner et al., 2014; Hegde et al., 2018). The most common commercial systems convert the sugars in corn to produce ethanol, blended with gasoline for transportation fuel (Wang et al., 2012).
- *Transesterification* (Chapter 6) is the acid- or base-catalyzed reaction of triglycerides in oils (including those extracted from energy crops and waste cooking oil) and an alcohol into a mono-alkyl ester, commonly known as biodiesel (Meher et al., 2006; Leung et al., 2010).
- *Bioelectrochemical systems* (Chapter 7), including microbial fuel cells and microbial electrolysis cells, use bacteria to convert organic and inorganic matter and directly produce electrical current or hydrogen (Logan et al., 2006; Logan, 2010).
- *Gasification and pyrolysis* (Chapter 8) are thermochemical conversion methods in which biomass is processed at high temperature with less than the stoichiometric level of oxygen required for full combustion or incineration. Products can include hydrogen-rich syngas, liquid bio-oil and biochar, a stable form of solid carbon (Ahmed and Gupta, 2010).
- *Hydrothermal liquefaction* (Chapter 9) converts wet biomass by depolymerization under conditions of moderate temperature and high pressure into a high energy density bio-oil product (Toor et al., 2011).

Important considerations in the deployment of any food waste-to-energy technology are the associated environmental and economic benefits relative to other available options. For example, prior studies have suggested that corn-based ethanol can have a larger carbon footprint than conventional transportation fuels depending on all the upstream emissions resulting from growing and processing the feedstock, operating industrial-scale fermentation and distillation processes, and blending with gasoline (e.g., Pimentel and Patzek, 2005; Wang et al., 2007). It is therefore necessary to consider the life cycle of the entire system to quantify its relative environmental impact (Xu et al., 2015), as described in Chapter 10. Similarly, food waste-to-energy system economic analysis must consider all the benefits and costs associated with the primary food waste feedstocks (e.g., is the waste obtained at no cost, or is a "tipping" fee assessed) and all the coproducts generated that may have economic value. For example, in the case of AD, electricity may be the main intended "product," but other potential value-added outputs are waste heat from the engine-generator set, cow bedding from the solid (undigested) fraction of the effluent stream, and fertilizer from the liquid fraction. As outlined in Chapter 11, economic analysis also must consider the system deployment approach. Whereas large, centralized facilities generally have lower capital and operation and maintenance (O&M) costs per unit of waste processed, the waste generator may not extract financial benefit from the material which contains their invested resources. Conversely, smaller distributed systems may cost more per unit of waste processed, but individual waste generators or communities of generators may retain more of the benefits, and transportation costs can be reduced or eliminated altogether. A more recently developed food recovery hierarchy scheme assigns a higher value to local and community-based composting and AD (Fig. 1.9B).

Beyond understanding methodologies for quantifying environmental and economic performance of food waste-to-energy systems, it is important to also consider the existing policy and regulatory framework that influences what technologies progress from lab-scale demonstration through the "valley of death" to commercial viability. In Chapter 12, we explore policy initiatives in the United States and other developed economies affecting both food waste disposal (such as commercial landfill bans recently legislated in several states) and incentives for producing and marketing renewable resources for electrical and thermal energy, as well as transportation fuel. Finally, in Chapter 13, we assess how deployment of food waste-to-energy systems can truly satisfy the "triple bottom-line" of economic, environmental, and social sustainability. This assessment includes such considerations as logistics of food waste transport and energy distribution, the important influences of community perception and engagement, and the combination of multiple feedstocks, conversion technologies, and coproducts in an integrated "biorefinery" architecture.

Our objective in developing this book has been specifically to describe available pathways for conversion of food waste to sustainable energy products. However, it is important to note that there are a number of existing publications that provide a wealth of information relevant to productive food waste utilization, including treatment methods and potential uses of food waste (Arvanitoyannis, 2010), treatment of solid and liquid food wastes and the associated environmental impacts (Kosseva and Webb, 2013), sustainable energy from biomass resources (de Jong and van Ommen, 2014), and processing technologies for value-added products from food waste (Galanakis, 2015). It is also important to emphasize from the outset that using food waste as a primary feedstock for energy production may not be environmentally favorable relative to incumbent technologies (based upon life cycle assessment), and also that energy is not necessarily the most economically or socially sustainable product. Notable advances have been made in recent years to convert food waste to a wide variety of value-added products, including bioplastics, specialty chemicals, and food supplements (Lin et al., 2013), but assessment of these technologies is beyond the scope of the present work. Whatever the desired outcome may be from valorizing food waste, achieving true system-level sustainability requires that a delicate balance is achieved among environmental, economic, and social considerations, which likely has significant geographic variability and may change over time. Additionally, although this overview of sustainable food waste-to-energy systems focuses largely on data and case studies from the United States, it is expected that the findings will have broader applicability, especially in rapidly developing Asian countries which are expected to dominate global food waste production in the foreseeable future (Adhikari et al., 2006).

REFERENCES

Adhikari, B.K., Barrington, S., Martinez, J., 2006. Predicted growth of world urban food waste and methane production. Waste Manag. Res. 24 (5), 421–433.

Ahmed, I.I., Gupta, A.K., 2010. Pyrolysis and gasification of food waste: syngas characteristics and char gasification kinetics. Appl. Energy 87 (1), 101–108.

Arvanitoyannis, I.S., 2010. Waste Management for the Food Industries. Academic Press, New York.

Brooks, N., Buzby, J.C., Regmi, A., 2009a. Globalization and evolving preferences drive US food import growth. J. Food Distrib. Res. 40 (1), 39–46.

Bureau of Labor Statistics, U.S. Department of Labor, 2016. News release. Consumer spending – 2015. August 30, 2016.

Buzby, J.C., H.F. Wells and J. Hyman, 2014. The estimated amount, value, and calories of postharvest food losses at the retail and consumer levels in the United States. EIB-121. U.S. Department of Agriculture, Economic Research Service, February 2014.

Chao, E.L. and Utgoff, K.P., 2006. 100 Years of U.S. consumer spending—data for the nation, New York City, and Boston. Report 991. U.S. Department of Labor.

Cuéllar, A.D., Webber, M.E., 2010. Wasted food, wasted energy: the embedded energy in food waste in the United States. Environ. Sci. Technol. 44 (16), 6464–6469.

de Jong, W., van Ommen, J.R., 2014. Biomass as a Sustainable Energy Source for the Future: Fundamentals of Conversion Processes. John Wiley & Sons, Hoboken, NJ.

Ebner, J., Babbitt, C., Winer, M., Hilton, B., Williamson, A., 2014. Life cycle greenhouse gas (GHG) impacts of a novel process for converting food waste to ethanol and co-products. Appl. Energy 130, 86–93.

Ebner, J.H., Labatut, R.A., Rankin, M.J., Pronto, J.L., Gooch, C.A., Williamson, A.A., Trabold, T.A., 2015. Lifecycle greenhouse gas analysis of an anaerobic codigestion facility processing dairy manure and industrial food waste. Environ. Sci. Technol. 49 (18), 11199–11208.

Ebner, J.H., Labatut, R.A., Lodge, J.S., Williamson, A.A., Trabold, T.A., 2016. Anaerobic co-digestion of commercial food waste and dairy manure: characterizing biochemical parameters and synergistic effects. Waste Manag. 52, 286–294.

EPA, U.S. Environmental Protection Agency, 2016. Advancing sustainable materials management: 2014 fact sheet. Report EPA530-R-17-01. November 2016.

Galanakis, C.M. (Ed.), 2015. Food Waste Recovery: Processing Technologies and Industrial Techniques. Academic Press. London.

Geiss, L.S., Wang, J., Cheng, Y.J., Thompson, T.J., Barker, L., Li, Y., Albright, A.L., Gregg, E.W., 2014. Prevalence and incidence trends for diagnosed diabetes among adults aged 20 to 79 years, United States, 1980–2012. JAMA 312 (12), 1218–1226.

Gunders, D., 2012. Wasted: how America is losing up to 40 percent of its food from farm to fork to landfill. NRDC Issue Paper. August 2012 IP:12-06-B.

Hall, K.D., Guo, J., Dore, M., Chow, C.C., 2009. The progressive increase of food waste in America and its environmental impact. PLoS One 4 (11).

Heard, B.R., Miller, S.A., 2016. Critical research needed to examine the environmental impacts of expanded refrigeration on the food system. Environ. Sci. Technol. 50 (22), 12060–12071.

Hegde, S., Lodge, J.S., Trabold, T.A., 2018. Characteristics of food processing wastes and their use in sustainable alcohol production. Renew. Sustain. Energy Rev. 81, 510–523.

Holm-Nielsen, J.B., Al Seadi, T., Oleskowicz-Popiel, P., 2009. The future of anaerobic digestion and biogas utilization. Bioresour. Technol. 100 (22), 5478–5484.

Jerardo, A., 2003. Import share of US food consumption stable at 11 percent. USDA, Economic Research Service.

Kantor, L.S., Lipton, K., Manchester, A., Oliveira, V., 1997. Estimating and addressing America's food losses. Food Rev. 20 (1), 2–12.

Kosseva, M., Webb, C., 2013. Food Industry Wastes: Assessment and Recuperation of Commodities. Academic Press, London.

Leung, D.Y., Wu, X., Leung, M.K.H., 2010. A review on biodiesel production using catalyzed transesterification. Appl. Energy 87 (4), 1083–1095.

Lewandowski, S., 1987. Diohe'ko, the three sisters in Seneca life: implications for a native agriculture in the finger lakes region of New York state. Agric. Hum. Values 4 (2–3), 76–93.

Lin, C.S.K., Pfaltzgraff, L.A., Herrero-Davila, L., Mubofu, E.B., Abderrahim, S., Clark, J.H., Koutinas, A.A., Kopsahelis, N., Stamatelatou, K., Dickson, F., Thankappan, S., 2013. Food waste as a valuable resource for the production of chemicals, materials and fuels. Current situation and global perspective. Energ. Environ. Sci. 6 (2), 426–464.

Logan, B.E., 2010. Scaling up microbial fuel cells and other bioelectrochemical systems. Appl. Microbiol. Biotechnol. 85 (6), 1665–1671.

Logan, B.E., Hamelers, B., Rozendal, R., Schröder, U., Keller, J., Freguia, S., Aelterman, P., Verstraete, W., Rabaey, K., 2006. Microbial fuel cells: methodology and technology. Environ. Sci. Technol. 40 (17), 5181–5192.

Meher, L.C., Sagar, D.V., Naik, S.N., 2006. Technical aspects of biodiesel production by transesterification—a review. Renew. Sustain. Energy Rev. 10 (3), 248–268.

Parfitt, J., Barthel, M., MacNaughton, S., 2010. Food waste within food supply chains: quantification and potential for change to 2050. Philos. Trans. Royal Soc. B Biol. Sci. 365 (1554), 3065–3081.

Pimentel, D., Patzek, T.W., 2005. Ethanol production using corn, switchgrass, and wood; biodiesel production using soybean and sunflower. Nat. Resour. Res. 14 (1), 65–76.

ReFED, Rethink Food Waste through Economics and Data, 2016. A roadmap to reduce U.S. food waste by 20 percent.

Roser, M., Ritchie, H., 2017. Food per Person. Published online at OurWorldInData.org. Retrieved from: https://ourworldindata.org/food-per-person [Online Resource].

Sullivan, R., 1996. Trends in the agricultural labor force. Chapter 12, In: Simon, J. (Ed.), The State of Humanity. Wiley-Blackwell, Oxford and Cambridge, MA.

Tilman, D., Cassman, K.G., Matson, P.A., Naylor, R., Polasky, S., 2002. Agricultural sustainability and intensive production practices. Nature 418 (6898), 671–677.

Toor, S.S., Rosendahl, L., Rudolf, A., 2011. Hydrothermal liquefaction of biomass: a review of subcritical water technologies. Energy 36 (5), 2328–2342.

United Nations, 2015. World Population Prospects—the 2015 revision. ESA/P/WP.241. Department of Economic and Social Affairs, New York.

Wang, M., Han, J., Dunn, J.B., Cai, H., Elgowainy, A., 2012. Well-to-wheels energy use and greenhouse gas emissions of ethanol from corn, sugarcane and cellulosic biomass for US use. Environ. Res. Lett. 7 (4), 045905.

Wang, M., Wu, M., Huo, H., 2007. Life-cycle energy and greenhouse gas emission impacts of different corn ethanol plant types. Environ. Res. Lett. 2 (2) 024001.

Weber, C., Matthews, H., 2008. Food-miles and the relative climate impacts of food choices in the United States. Environ. Sci. Technol. 42 (10).

Xu, C., Shi, W., Hong, J., Zhang, F., Chen, W., 2015. Life cycle assessment of food waste-based biogas generation. Renew. Sustain. Energy Rev. 49, 169–177.

FURTHER READING

Brooks, N.L., Regmi, A. and Jerardo, A., 2009b. US food import patterns, 1998–2007. USDA, Economic Research Service.

Brundtland, G.H., et al., 1987. Our Common Future. Oxford University Press, New York.

Chapter 2

Waste Resources in the Food Supply Chain

Thomas A. Trabold, Shwe Sin Win and Swati Hegde

Golisano Institute for Sustainability, Rochester Institute of Technology, Rochester, NY, United States

2.1 INTRODUCTION

As introduced in Chapter 1, food waste is a major global problem that has significant environmental, economic, and social sustainability implications. It is widely recognized that many of the conventional practices for handling food waste resources (Chapter 3) are inadequate in their ability to recover embodied energy and water. To begin to transition the global food supply chain (FSC) to alternative waste management technologies, including the waste-to-energy systems described in Chapters 4–9, it is necessary to first clearly define what is meant by food "waste" and "loss," and then quantify how much waste occurs at each stage of the FSC: agriculture, food processing, consumer-facing businesses, and households.

Part of the challenge surrounding quantification of food waste is that there is no universal definition of what is actually considered "waste," and as these materials move through the supply chain, they become increasingly heterogeneous and geographically dispersed. Huge volumes of waste are created in agricultural and food processing operations, but in many cases these waste streams are of fairly uniform composition and generated at a relatively small number of physical locations. At the other extreme, when food waste is generated at millions of individual residences, it is often a mixture of many different types of materials and usually combined with the rest of the solid waste typically produced in the normal function of operating a household. A significant complication is the introduction of packaging at the food processing stage, and this second material phase needs to be comprehended in developing potential waste-to-energy (WtE) solutions for converting waste generated downstream at consumer-facing businesses and households.

As discussed in a number of recent publications (e.g., Ebner, 2016; Hall, 2016; Derqui et al., 2016; Bellemare et al., 2017; Corrado et al., 2017), there is a wide variety of definitions that have been proposed for food waste materials, and proper quantification of resources available for alternative utilization pathways requires that these definitions are clearly established from the outset. The definitions proposed by the Food and Agriculture Organization of the United Nations (FAO, 2013) are summarized as follows:

- "**Food loss** refers to a decrease in mass (dry matter) or nutritional value (quality) of food that was originally intended for human consumption. These losses are mainly caused by inefficiencies in the food supply chains, such as poor infrastructure and logistics, lack of technology, insufficient skills, knowledge and management capacity of supply chain actors, and lack of access to markets. In addition, natural disasters play a role."
- "**Food waste** refers to food appropriate for human consumption being discarded, whether or not after it is kept beyond its expiry date or left to spoil. Often this is because food has spoiled but it can be for other reasons such as oversupply due to markets, or individual consumer shopping/eating habits."
- "**Food wastage** refers to any food lost by deterioration or waste. Thus, the term "wastage" encompasses both food loss and food waste."

In an earlier publication from the same organization (Gustavsson et al., 2011), somewhat different definitions were recommended:

> *"Food losses refer to the decrease in edible food mass throughout the part of the supply chain that specifically leads to edible food for human consumption. Food losses take place at production, post-harvest and processing stages in the food supply chain (Parfitt et al., 2010). Food losses occurring at the end of the food chain (retail and final consumption) are rather called "food waste", which relates to retailers' and consumers' behavior."*

Sustainable Food Waste-to-Energy Systems. https://doi.org/10.1016/B978-0-12-811157-4.00002-4

Several studies also attempted to distinguish between FSC losses that are "planned" or "unavoidable," from those that are "unplanned" or "avoidable." For example, Quested and Johnson (2009) offered these definitions:

- *Avoidable Food Waste*: "Food and drink thrown away that was, at some point prior to disposal, edible (e.g. slice of bread, apples, meat)."
- *Possibly Avoidable*: "Food and drink that some people eat and others do not (e.g. bread crusts), or that can be eaten when a food is prepared in one way but not in another (e.g. potato skins)."
- *Unavoidable Food Waste*: "Waste arising from food or drink preparation that is not, and has not been, edible under normal circumstances (e.g. meat bones, egg shells, pineapple skin, tea bags)."

Although these definitions appear to be tailored for losses generated during the consumption phase, they may also apply to upstream FSC stages as well. For example, in the case of potatoes (described in detail in connection with Fig. 2.4), the skins are often generated as waste in the process of manufacturing frozen French fries or as a prepared food item in a grocery store or restaurant, and thus may be considered a "possibly avoidable" waste resource across the food supply chain.

For the purpose of defining the universe of food materials available for waste-to-energy conversion, it is reasonable to be as broad as possible, and identify any potential feedstocks that could be used as feedstock, even if currently they are diverted to beneficial use. For this purpose, we recommend the definition proposed by Stenmarck et al. (2016) for The European Union's FUSIONS program, focused on reducing food waste through social innovation:

> *"Food waste: Fractions of food and inedible parts of food removed from the food supply chain to be recovered or disposed (including composted, crops ploughed in/not harvested, anaerobic digestion, bioenergy production, co-generation, incineration, disposal to sewer, landfill or discarded to sea)."*

Thus, in the discussion that follows, we make no distinction between food "waste" and "loss" and use the terms interchangeably to represent the mass of material that leaves the food supply chain for any reason prior to human consumption.

Once the geographic scale is appropriately constrained, it is important to acquire as much data as possible to adequately quantify total annual waste volumes and the generation locations, as well as significant temporal variations (e.g., due to production schedule, seasonal demand, etc.), waste phase (solid, liquid, packaged; Chapter 3), and physical/chemical characteristics that will dictate the "best" WtE conversion pathway.

2.2 GLOBAL PERSPECTIVE

There are a number of relevant studies that have considered the quantification of food loss and waste on a global scale (e.g., Parfitt et al., 2010; Lipinski et al., 2013; Ghosh et al., 2016). Although it is clear that consistent data are seriously lacking, especially in less economically developed countries, several general trends have been widely reported. First, the distribution of waste among different stages of the FSC varies greatly across different global regions (Fig. 2.1). In developing countries, the majority of waste occurs toward the agriculture end of the supply chain, due to inefficiencies in harvesting and immediate product handling, and the lack of suitable infrastructure for storage and refrigeration. In more affluent countries, the major source of food waste is at the consumption phase, often because of the wide disparity between the amount of food needed for healthy living and what is actually procured, some of which spoils before it can be consumed. As presented in Table 2.1, consumption losses range from 61% of total food waste in North America and Oceania to only 5% in sub-Saharan Africa, while combined production, handling, and storage account for 76% of waste in sub-Saharan Africa and 23% in North America and Oceania. The widely cited study of Gustavsson et al. (2011) considered different food waste dynamics among medium- and high-income countries in three regions (Europe including Russia; United States, Canada, Australia, and New Zealand; China, Japan, South Korea) and low-income countries in four regions (sub-Saharan Africa; North Africa, Central Asia and Western Asia; South and Southeastern Asia; Latin America), and identified these key differences:

- Medium- and high-income country food wastes result primarily from:
  - Deficient quality, including aesthetic defects.
  - Scraps generated during food processing, including transportation losses.
  - Poor environmental conditions during display in retail facilities, which accounts for over 50% of fruit and vegetable waste.
  - Lack of proper planning and communication in food service operations.
  - Use of "best by" dates that encourage consumers to discard food that is still edible.

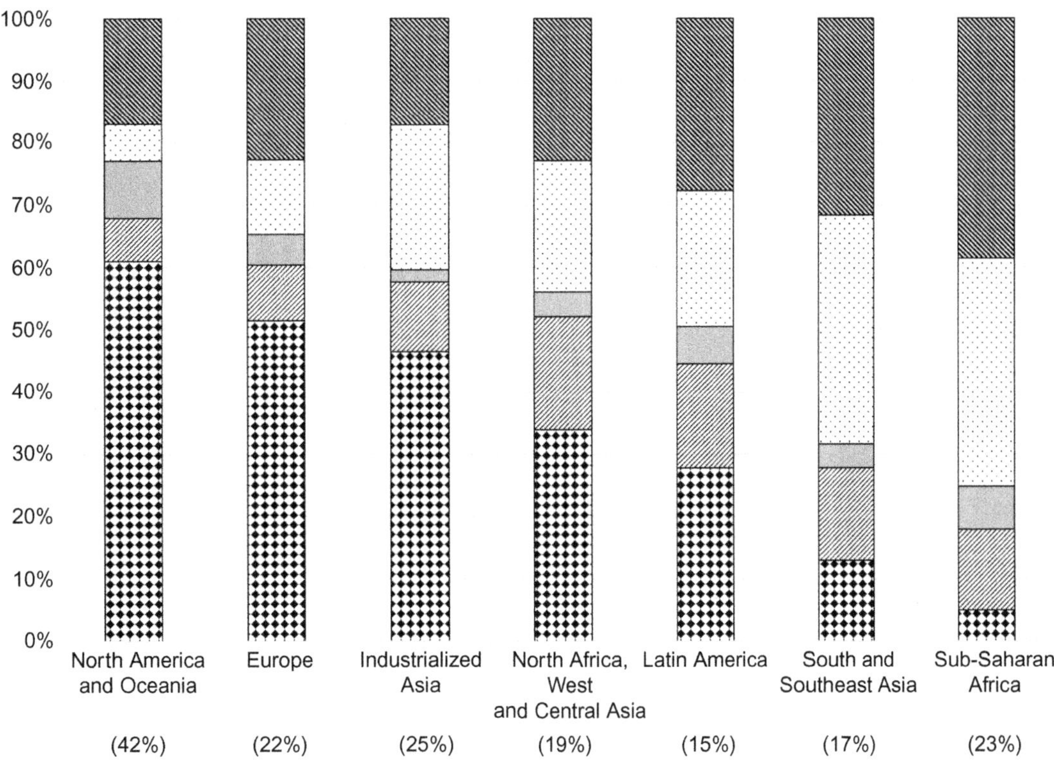

FIG. 2.1 Food waste across the FSC in different global regions, based on percent of kcal lost (from 2009 data reported by Lipinski et al., 2013). Percentages below region names indicate share of total available food that is wasted.

TABLE 2.1 Comparison Between Agriculture and Consumption Waste in Different Global Regions (Lipinski et al., 2013)

Region	Consumption Losses (% of Total Waste)	Agriculture[a] Losses (% of Total Waste)
North America and Oceania	61	23
Europe	52	35
Industrialized Asia	46	40
North Africa, West and Central Asia	34	44
Latin America	28	50
South and Southeast Asia	13	69
Sub-Saharan Africa	5	76

[a]*Combined losses from production, handling, and storage.*

- Low-income country food wastes result primarily from:
  - Poor storage facilities which result in rodent and insect infestation, especially in warm, humid climates.
  - Poor infrastructure and transportation, combined with a lack of refrigeration.
  - Inadequate market facilities that are unsanitary and also lack refrigeration facilities.
  - Poor packaging that jeopardizes food products moving from the agriculture phase to processing and/or retail.

It is interesting to note that despite the widely different distributions of waste in these various regions, the data in Fig. 2.1 indicate that the total amount of waste as a percentage of available food resources is similar in Europe, Industrialized Asia, and sub-Saharan Africa (22%–25%). However, North America and Oceania have by far the least inefficient food systems (i.e., 42% of available food is wasted), and more than twice the per capita rate of food waste (on a kcal basis) of any other

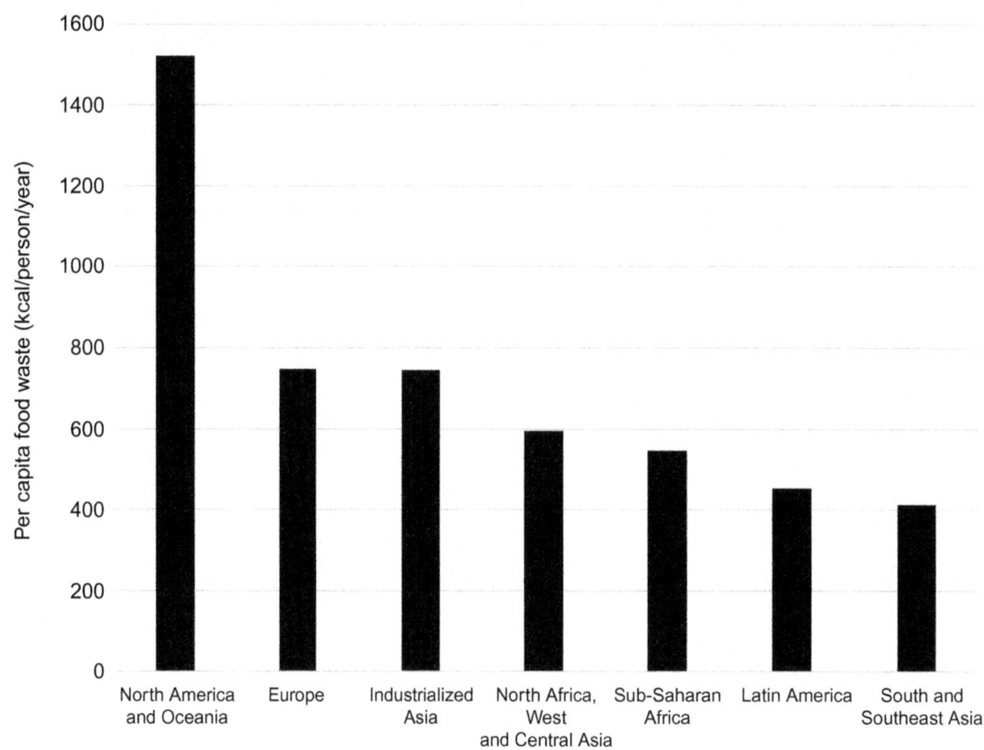

FIG. 2.2 Per capita food waste for different global regions, based on 2011 data (Lipinski et al., 2013).

region (Fig. 2.2). The relative economic impacts of food loss in different global regions also play an important role. For example, in the United States food accounts for about 11% of the average household budget (Fig. 1.2; USDA, 2016), while in some developing countries (particularly in poor rural regions) a significant fraction of income is spent on food (Barrett and Dorosh, 1996).

2.3 NATIONAL PERSPECTIVES

The global food waste overview in Section 2.2 provides a macroscale picture of where opportunities may exist in different regions for upcycling food waste resources, including application of waste-to-energy conversion technologies. However, it is necessary to delve into country-level data to develop a better perspective of what options offer the best pathways for maximizing economic and environmental benefits of food waste utilization. As is the case in many areas of energy research, larger and more economically advanced countries tend to have available more reliable primary data, but as described later there are a number of published studies that focus on analysis of food waste resources in a variety of specific countries and regions. As reported by Parfitt et al. (2010), there are major gaps in the availability of accurate country-scale food waste data, with much of the information for developing countries acquired many years ago, and information for rapidly expanding countries (including the BRIC nations of Brazil, Russia, India, and China) largely absent altogether.

Thi et al. (2015) conducted a study of food waste trends in developing countries and found a strong correlation between per capita gross national income (GNI) and the food waste generation rate when few or no "zero waste" policies have been implemented. For example, with a per capita GNI of $1570 (US dollars, based on 2009–2013 World Bank data) India had a per capita food waste generation rate of 0.06 kg/day (Ranjith, 2012; Manipadma, 2013). Conversely, Brazil (the most affluent developing country studied, with per capita GNI of $11,690) had a per capita food waste generation rate of 0.17 kg/day (Corsten et al., 2012). Among developed countries, the United States had the highest food waste generation rate of 0.52 kg/person/day. However, it was also observed that many developed countries such as Germany, Singapore, Sweden, and Denmark have more advanced waste utilization strategies and thus the per capita food waste generated falls below the general trend that represents data for all the developing countries considered, but also Taiwan, the United Kingdom, and the United States. For example, with nearly the same per capita GNI as the United States, Singapore's food waste generation per person is 23% lower (0.40 kg/day; NEA, 2013).

Because of the now global recognition that food waste is a major environmental, economic, and social challenge, there has recently been a growing number of published studies on region- and country-specific food waste characterizations and assessments of potential alternative utilization strategies. Since the publication of the work of Thi et al. (2015) that presented food waste data for 25 developed and developing countries, there has been a number of new studies, especially focused on rapidly expanding Asian economies. Ong et al. (2017) recently explored trends in food waste valorization in India, Thailand, Singapore, Malaysia, and Indonesia. They found that there is significant potential in waste-to-energy conversion and upcycling to value-added food products and "green" chemicals, but also that legislation and public perception play key roles in supporting these new industries. Minten et al. (2016) studied the potato sector in Bangladesh, India, and China and reported that product transportation contributes significantly to losses across the FSC and recommended investment in improved cold storage facilities to alleviate this problem. The research of Gokarn and Kuthambalayan (2017) identified challenges inhibiting the reduction of losses in the Indian food supply chain, which accounts for 20% of gross domestic product (GDP) and 50% of the population's employment. They recommended that the greatest impact could be achieved by addressing challenges related to food characteristics (perishability, quality variation, seasonality, and bulkiness), supply chain uncertainty, market infrastructure, and food policy and regulation. In a focused study of food waste in the hotel sector of Jaipur City, India, Gandhi et al. (2017) showed there is significant potential for biogas production and economic return through improved collection efficiency and waste utilization. Several recent studies from China have also characterized food waste in particular FSC sectors or cities. For example, Wang et al. (2017) surveyed 195 restaurants in four cities and reported that waste produced per meal depends on many factors, with more waste produced in larger restaurants, and by tourists as opposed to local residents. De Clercq et al. (2017) also studied Chinese restaurant waste in the context of biogas production via anaerobic digestion (Chapter 4) and offered six national policy recommendations to accelerate this waste conversion pathway, including improving collection efficiency by incentivizing generators to direct waste from landfills or incinerators to biogas projects. Other recent studies have analyzed food waste resources in targeted Chinese cities, including Beijing (De Clercq et al., 2016; Ek, 2017) and Suzhou City (Wen et al., 2016). Other developing countries with large populations are also focusing their attention on the potential for productively utilizing food waste materials. For example, Salihoglu et al. (2017) recently conducted a comprehensive review of food loss and waste in Turkey and showed that a significant fraction of the national energy demand can be met by utilizing biomass lost in the initial phases of the FSC, combined with dedicated energy crops grown on fallow land.

In many countries of the European Union (EU), mature food waste valorization systems already exist, but research is still being conducted to identify additional opportunities in specific FSC sectors. Taking an EU-wide perspective, Sala et al. (2017) quantified food waste from the macroscale to single stages of the FSC, and used "top-down" and "bottom-up" approaches to provide estimates of waste generation rates. They considered the specific example of tomatoes to compile bottom-up data for losses from agricultural production, manufacturing, distribution, and consumption phases. Eberle and Fels (2016) reported that in Germany, the best opportunities for food waste reduction exist with products of animal origin, and especially in the context of agricultural production and consumption (both in households and out-of-home). Raak et al. (2017) conducted interviews with representatives of 13 German food processing companies and identified a wide range of causes of food waste and loss, resulting in recommendations such as alternate food production pathways for "second choice items" and greater deployment of backup power systems to avoid production waste during loss of grid power. Redlingshöfer et al. (2017) considered losses in the French food supply chain, with a specific focus on the primary production and processing phases. In 2013, up to 12% of fruit and vegetables were lost before the retail phase, and secondary food products and animal feed were identified as two pathways that could play a moderate role in reducing these wastes. In Sweden, Scholz et al. (2015) assessed food waste and the associated carbon footprint (CF) for supermarkets. Their analysis showed that fruits and vegetables contribute 85% of the mass of waste and 46% of the CF, but only three products (tomatoes, peppers, and bananas) accounted for nearly half of this portion of the CF. Conversely, meats contributed 3.5% of the waste mass, but 29% of the total CF.

In the United Kingdom (UK) there has been a strong centralized effort to improve FSC efficiency and minimize food loss, mostly through the efforts of the *Waste and Resources Action Programme* (WRAP), with a goal of a 20% reduction in per capita waste and greenhouse gas emissions associated with the food and beverage industry by 2025.[1] This organization has compiled extensive data to characterize the trends in waste generation, especially at the household level (Quested and Parry, 2017), which accounted for 71% of waste, 7.3 MMT[2] in 2015 (WRAP, 2017). Although there is generally a lack of data available for other stages of the FSC, several studies have focused on opportunities for minimizing food losses in the

1. http://www.wrap.org.uk/food-waste-reduction.
2. MMT = million metric ton. A metric ton is equivalent to 1000 kg and approximately 1.1 short tons.

retail sector and point to the importance of information sharing and management practices on the scale of an individual grocery store (e.g., Mena et al., 2011; Filimonau and Gherbin, 2017).

The overall size of the food industry in the United States is much larger than in the United Kingdom or any single country of the European Union, and thus the amount of food waste resources available are comparatively higher. Several studies have attempted to quantify the waste on a national scale, and published estimates vary somewhat depending on the definitions of food waste and loss, as described previously, and the sources of primary data. For example, in the widely cited report by ReFED (2016a), the total U.S. food waste level was estimated as 56.7 MMT, with agriculture, food processing, consumer-facing businesses (groceries, restaurants, institutional food service, etc.), and households accounting for 9.1, 0.9, 22.7, and 24.5 MMT, respectively. Several other publications have suggested that the ReFED report significantly underestimated the total level of food waste because it quantified only edible material going to landfills and cosmetically imperfect produce left unused on farms and in packing operations that can be repurposed for higher value use. For example, Dou et al. (2016) quantified total U.S. food waste at 96.2 MMT, with industry (handling, processing, manufacturing), retail and consumers accounting for 35.9, 19.5, and 40.8 MMT, respectively. Using primary data from Buzby et al. (2014) (based on data from the U.S. Department of Agriculture's Loss Adjusted Food Availability databases; USDA, 2015) and reports of the Food Waste Reduction Alliance for losses in the food processing, wholesale and retail sectors (BSR, 2014), they estimated food waste across the FSC according to different food groups, as shown in Fig. 2.3. Here the values represent the percentage of food entering each of the supply chain sectors that is wasted by the food group in that sector. For example, in the vegetable food group, 31% of the total mass produced is lost in the industrial stage of the FSC, with retail and consumer stages accounting for 8% and 22%, respectively. In total, industrial, retail, and consumer losses accounted for 15%, 10%, and 21% of the mass of all food produced.

Beyond just quantifying the total mass of food that is wasted, there has also been significant interest in calculating the associated economic value of this waste. For example, Buzby and Hyman (2012) applied retail prices to retail and consumer level losses in the United States from 2008 and computed a staggering total value of $165.6 billion.

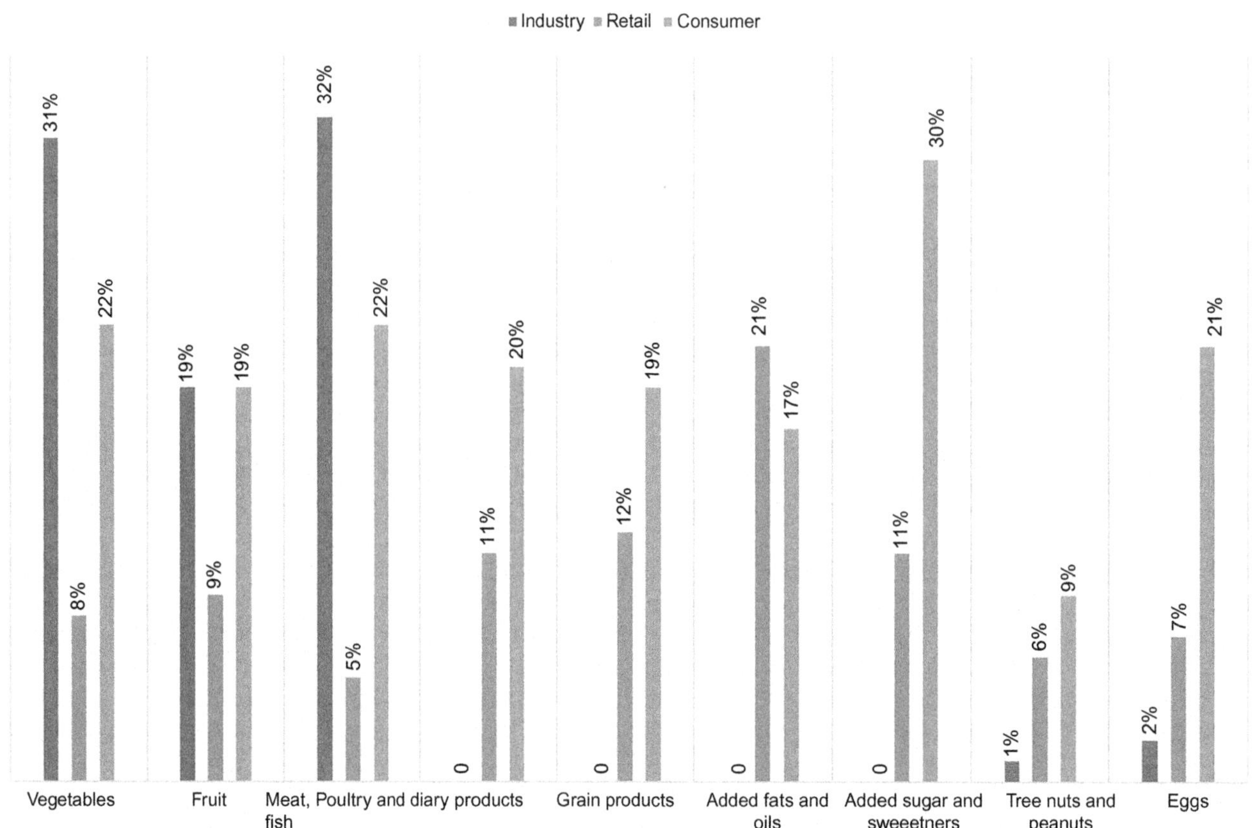

FIG. 2.3 Estimated food losses across the U.S. supply chain, based on percentages of total production in each food group (Dou et al., 2016).

Meat (including poultry and fish), vegetables, and dairy products were the largest contributors, accounting for 41%, 17%, and 14% of the total value, respectively. The ReFED report and associated Technical Appendix (2016a,b) applied a combination of wholesale and retail prices to compute total loss values in agriculture, manufacturing, consumer-facing businesses, and households of $15 billion, $2 billion, $57 billion, and $144 billion, respectively (Fig. 1.8). Others have questioned the methodologies applied in these studies and have concluded that the loss values are highly inflated and thus of limited value to policymakers (e.g., Koester, 2013). In any case, when assigning economic value to food losses in the context of evaluating potential investment in food waste-to-energy systems, it is critical that a consistent and transparent methodology be applied.

An excellent source of information for food supply chain material flows and losses is the United States Department of Agriculture (USDA) Economics, Statistics and Market Information System (ESMIS) that contains nearly 2500 reports and datasets from several agencies within USDA.[3] This resource covers U.S. and international agriculture and provides many time series datasets in spreadsheet format that are updated annually. Using this data, it is possible to develop a detailed picture of how most commodity crops move through the FSC, including where major waste streams are generated in agriculture, processing, retail, and consumption stages. For example, in Fig. 2.4, data for the potato supply chain in the United States are presented as a Sankey diagram, whereby total supply on the left (26.42 Mt) is distributed among fresh consumption, processed consumption, and mass losses, the latter accounting for 41% of total supply. Mass losses during processing were estimated using factors adapted from Hung et al. (2006):

- 15% mass loss in chips/shoestring processing, other frozen and miscellaneous products
- 12% solid mass loss during dehydration,
- 26.5% during frozen French fry production and canning
- 2.5% mass loss in starch/flour production.

Also, mass loss factors of 10% and 6% were used in the retail stage for fresh and processed potatoes, respectively, and for the consumption phase loss factors of 22% and 16% were applied for fresh and processed potatoes (Buzby et al., 2014).

A subset of the data in Fig. 2.4 related to the specific solid, oil based, and aqueous waste streams has been reconfigured in Fig. 2.5 to highlight the waste volumes as well as the potential for waste-to-energy conversion. Potato peel waste quantities were calculated using a factor of 12% (Hung et al., 2006); mass losses from trimming and blanching operations are not

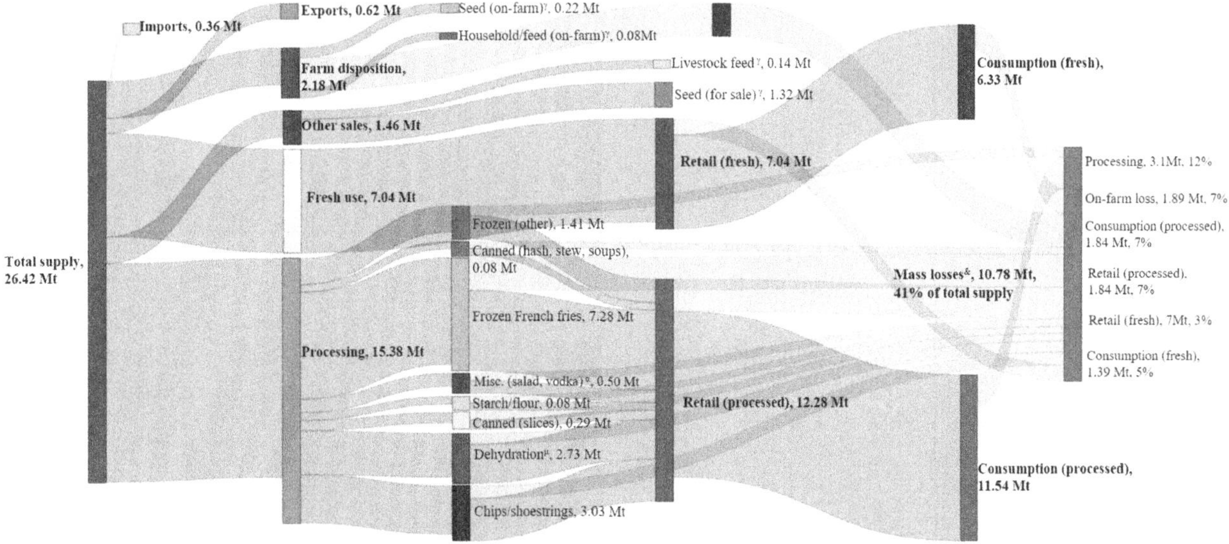

FIG. 2.4 Sankey diagram showing mass losses in U.S. potato supply chain. All mass flows were calculated based on production and utilization data averaged over the 2000–2006 period, as obtained from USDA's food availability database: http://usda.mannlib.cornell.edu/MannUsda/viewDocumentInfo.do?documentID=1235. (γ) Losses not calculated; (&) includes losses due to shrinkage, unavoidable waste during processing, and losses due to cooking, but does not include losses from water evaporation; (μ) approximately 60% of the mass is water evaporated during dehydration (Hung et al., 2006); however, water loss due to evaporation is not included in mass loss calculation as this water does not represent a waste stream. (α) 2015 data (National Potato Council, 2017). Note that "Mt" indicates million short tons, where a short ton is equivalent to 2000 lb., or approximately 0.91 metric ton.

3. The ESMIS is hosted by the Albert R. Mann Library at Cornell University, Ithaca, New York http://usda.mannlib.cornell.edu/MannUsda/homepage.do.

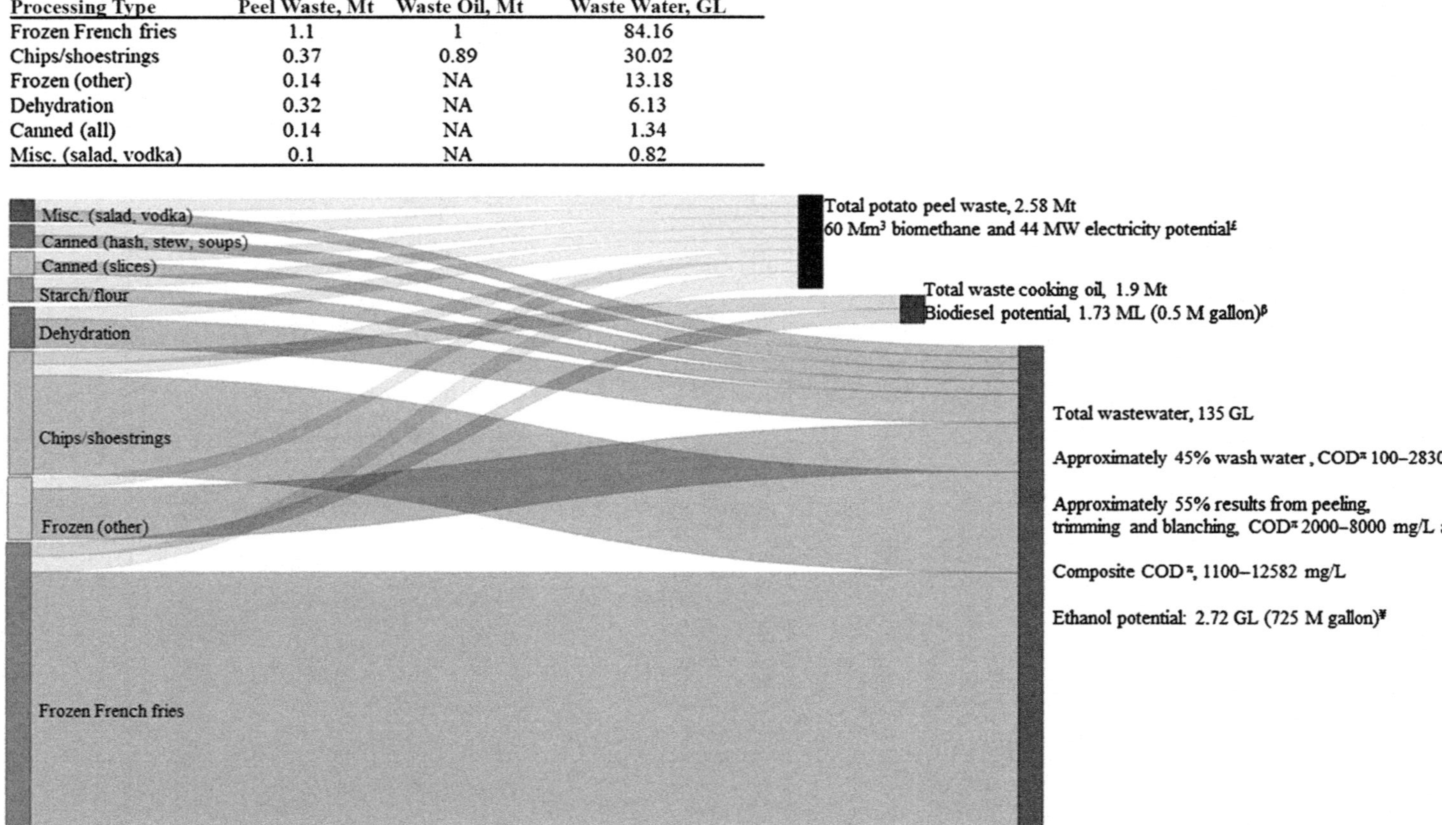

Processing Type	Peel Waste, Mt	Waste Oil, Mt	Waste Water, GL
Frozen French fries	1.1	1	84.16
Chips/shoestrings	0.37	0.89	30.02
Frozen (other)	0.14	NA	13.18
Dehydration	0.32	NA	6.13
Canned (all)	0.14	NA	1.34
Misc. (salad, vodka)	0.1	NA	0.82

FIG. 2.5 Solid (peel) waste, oil based, and aqueous waste streams associated with potato processing operations. (£) Biomethane potential, $23\,m^3$/wet ton potato peels (Hamilton, 2012). Electricity potential was calculated using an energy value of $6.4\,kWh/m^3$ for biomethane and 35% electric conversion efficiency (Deublein and Steinhauser, 2008); (β) estimated using 80% conversion efficiency; biodiesel density is $874.6\,kg/m^3$. (¥) Estimated for wastewater streams (55% of the total) excluding wash water, with the following assumptions/data used: starch content of waste water is 15% (Hung et al., 2006; Beynelmilel Food Corp, 2008), ethanol conversion efficiency of starch (glucose) is 51% (Lee et al., 2015), cellulose density is $1500\,kg/m^3$, and ethanol density is $789\,kg/m^3$. (π) COD values from Hung et al. (2006). Note that "Mt" indicates million short tons, where a short ton is equivalent to 2000 lb., or approximately 0.91 metric ton.

included in solid waste stream, as starch from these operations accumulates in the wastewater stream. The wastewater flows of 5625, 4275, 1668, and 2345 L/raw ton of potatoes were used to calculate wastewater quantities for washing, peeling, trimming, and blanching, respectively (Hung et al., 2006; Beynelmilel Food Corp, 2008). In chips/shoestring production, 530 kg oil are used per ton of potatoes and 63% are wasted; in frozen French fry production, 200 kg oil are used for prefrying one ton of potatoes and 80% of this amount enters the waste stream (Beynelmilel Food Corp, 2008). Potato peel waste quantities were estimated using a factor of 12% mass loss (Hung et al., 2006).

As described in Sections 2.2 and 2.3, many studies have been conducted to characterize waste at the global scale and in certain select countries. However, most of these sources lack the granularity needed to fully understand the nuances of FCS interactions that often lead to food waste and loss. It is also pertinent to note that significant financial resources are required to quantify and characterize waste in a scientifically rigorous manner, and the process of acquiring accurate food waste data may be impractical in many developing countries and less economically advanced regions of developed countries. Moreover, as recently discussed by Reutter et al. (2017) in connection with food waste research in Australia, there are different methods available to quantify food waste at a national scale, and these methods do not necessarily yield consistent results. There is no "one-size-fits-all" solution to food waste valorization, and the most sustainable outcome will depend on the local waste resource availability, existing conversion facilities, demand for energy and fuel, policy and regulatory framework, etc. Therefore, characterizing food waste at a more local scale with the greatest possible level of detail is an important precursor to assessing potential waste-to-energy pathways. Because it is obviously impossible to provide in this document such detailed data for all global regions, we focus our attention in Section 2.4 on New York State as an example of a relatively highly populated region that has a diverse mix of large urban centers with surrounding suburban zones, medium-sized cities, and rural areas comprised of smaller towns and villages. Finally, we proceed to even finer resolution and illustrate the process whereby the available food resources in a smaller, well-defined region can be identified and quantified to support consideration of alternative food waste valorization strategies.

2.4 ASSESSMENT OF STATE AND REGION-SPECIFIC FOOD WASTE RESOURCES

Our group has been actively involved in research focused on food waste valorization in New York State (NYS), an effort largely motivated by the expectation that in the future NYS will join other Northeastern states (Connecticut, Massachusetts, Rhode Island, and Vermont) and California in passing a food waste landfill ban (Jones, 2017; Breunig et al., 2017). This legislation was mentioned in the 2016 State of the State address, in which Governor Andrew M. Cuomo directed that any commercial entity producing more than two tons of food waste per week, such as grocery stores, colleges, hospitals, and restaurants, would be required to donate or recycle these materials and thus divert them from methane-producing landfills (Cuomo, 2016). However, to effectively implement a landfill ban and achieve the desired environmental benefits while avoiding excessive economic burdens on food sector companies and institutions, there needs to be a comprehensive understanding of where waste is generated, the amounts generated (including seasonal variations), and the physical and chemical characteristics of the waste in terms of its phase (solid, liquid, packaged), moisture content, volatile solids content, pH, etc. Acquiring, organizing, analyzing, and disseminating this information requires cooperation and collaboration across the food supply chain, and a willingness among sector stakeholders to work toward the common goal of food system sustainability.

As part of the effort to enhance understanding of state-level food waste challenges, The New York State Pollution Prevention Institute (NYSP2I) at Rochester Institute of Technology has developed the Organic Resource Locator (ORL), a publicly accessible, web-based mapping tool that according to the web site:

> *"... provides information on organic waste resources and utilization pathways in New York State. The goal of the Organic Resource Locator is to enable efficient and increased utilization of organic resources by connecting producers of organics with those who have a use for them, diverting a valuable resource from our landfills. NYSP2I hopes this effort will help reduce environmental impacts, promote economic development and encourage the development of green technologies."*

As shown in Fig. 2.6, this tool enables the user to identify generators in hospitality, restaurant, institutional, and retail sectors, which are clearly concentrated in the Upstate population centers of Buffalo, Rochester, Syracuse, and Albany, as well as in the New York City and Long Island regions.

To further facilitate improved understanding of available food waste resources and potential valorization opportunities (including waste-to-energy), the New York State Energy Research and Development Authority (NYSERDA) recently sponsored a study to quantify the food waste originating from "large" generators, that is, those producing >2 tons/week as earlier defined in the State of the State address (Manson, 2017). This report used data for large food waste generators from an earlier publication by Labuzetta et al. (2016). Because New York City (NYC) already has a landfill ban in place,

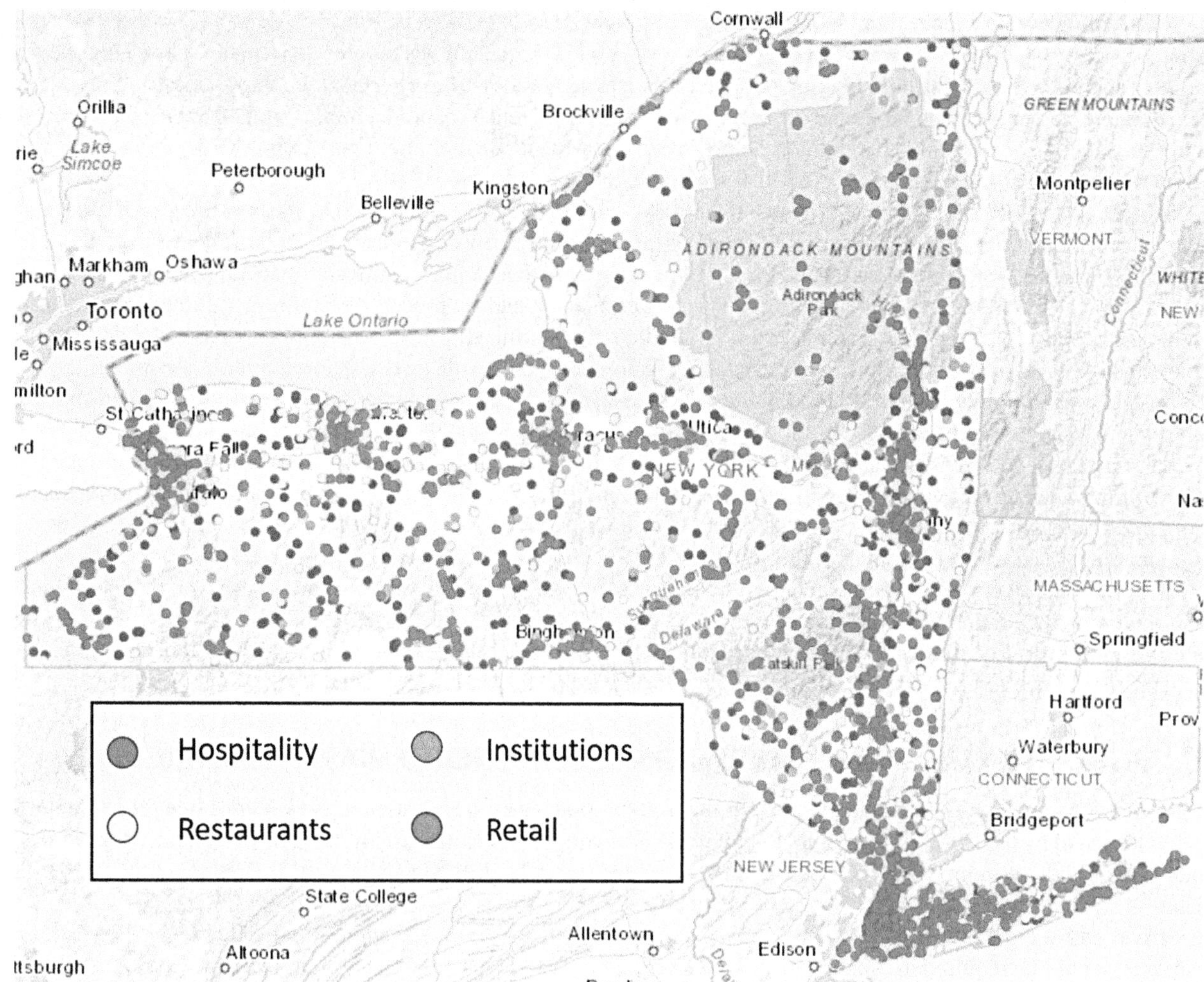

FIG. 2.6 Map of New York State food waste generators by according to FSC sector. https://www.rit.edu/affiliate/nysp2i/OrganicResourceLocator/.

TABLE 2.2 Number of Large Food Waste Generators and Associated Annual Generation Rates in New York State (Outside of New York City)

Sector	# Large Generators	Food Waste Generation (tons/week)	Food Waste Generation (tons/year)
Institutions	333	1420	73,840
Retail	1173	6742	350,584
Service and Hospitality	201	620	32,240
Total	1707	8782	456,664

this report identified the total number of food waste generators outside of NYC that would be affected by a landfill ban, and the associated annual waste generation rate, as summarized in Table 2.2. Because many food processing companies already divert a significant fraction of their waste to animal feed or other valorization pathways, this study focused on three specific sectors:

- *Institutions*, including colleges and universities, hospitals, nursing homes, and correctional facilities.
- *Retail*, including wholesale facilities, big box stores, convenience stores, supermarkets, and supercenters.
- *Service and hospitality*, including hotels/motels and restaurants.

Higher generation rates would be computed for large generators if we included the mass currently being diverted to beneficial use facilities. Labuzetta et al. (2016) estimated that 42% of food waste from wholesale and distribution sectors is diverted to food banks, composting, and anaerobic digestion (AD). If these additional resources were included, the total large generator food waste mass would increase from about 456,000 ton/year to over 588,000 ton/year.

To facilitate meaningful discussion of what technologies may be suitable for valorization of food supply chain resources, data for waste volumes and characteristics need to be available at a scale that is commensurate with the expected size of a deployed conversion facility. For example, to support a centralized anaerobic digestion or fermentation facility, it is likely that all feedstocks would need to be available within about a 50 mile (80 km) radius, otherwise the economic and environmental burdens associated with transport and handling would make the entire enterprise impractical. Also, to achieve broad support among local policymakers and industry stakeholders, it makes sense to align such large investments with existing geographic, political, or commercial boundaries. In New York State, a convenient framework for exploring potential investment opportunities in waste-to-energy or other food valorization projects is the system of 10 Regional Economic Development Council (REDC) zones. The Finger Lakes REDC zone (Fig. 2.7), so-called because it contains finger-shaped lakes formed by the glacial ice sheet that retreated 10,000 years ago, is a 9-County region in the western part of the state, with Rochester as the largest city. The east-west and north-south extents are approximately 90 miles (144 km) and 70 miles (112 km), respectively, so it is conceivable that a centrally located facility could be supplied with feedstocks that originate in any part of this REDC zone.

With our geographical region of interest suitably defined, it was possible to begin the process of identifying potential sources of food waste materials in agriculture, food processing, food distribution, and food service stages of the FSC, as originally outlined by Chan et al. (2013), and later expanded and updated. This analysis also relied in part on a much more extensive and comprehensive analysis conducted by Ebner (2016) that considered all food waste resources throughout New York State. Useful information and methods were also obtained from prior studies published by agencies in different states, including Connecticut (CDEP, 2001), Massachusetts (MDEP, 2002), California (CEPA, 2006), and North Carolina (NCDENR, 2012), but these generally focused on the food waste fraction of municipal solid waste (MSW) and less so on waste generated in other sectors.

2.4.1 Agriculture Sector

- *Crop residues*: Assumed to be insignificant because of generally low volumes and current use as nutrient-rich materials tilled back into the soil for the next growing cycle.
- *Unsold crops*: Also assumed to represent a relatively small volume, and more suitably diverted to donation or animal feed.

2.4.2 Food Processing Sector

This information is generally the most difficult to obtain because it is controlled by private or publicly-traded corporations that may see little financial or competitive value in such disclosures. Also, as mentioned previously in connection with the detailed study by ReFED (2016a), many food processing operations already divert generated waste materials to productive use, often as animal feed. Therefore to estimate the available food processing waste resources, we relied on limited company survey responses, public domain data for wastewater discharges to the publicly owned treatment works (POTW), and empirical relationships reported in earlier publications.

- *Company survey results*: The North American Industry Classification System (NAICS) and *ReferenceUSA* were used to first identify the population of nearly 7000 businesses in the 9-County Finger Lakes region that could potentially be involved in the FSC. This population was then further constrained by eliminating small businesses, defined as those with $<2500\,ft^2$ of operation space or less than $1 million in annual revenue. From this much smaller population, we identified 300 business enterprises in the food processing sector that were determined to be large enough to potentially generate significant waste. Although the survey response rate was rather low (about 10%), the data revealed some interesting trends and also identified several sources of very large wastewater discharges, including nearly 19 million L/year from a dressing and prepared sauce manufacturer, about 7 million L/year from a beverage company, and nearly 4 million L/year from a coffee and tea manufacturing plant. Based on a more extensive surveying effort and using public record resources, Ebner (2016) identified state-wide food manufacturing and processing resources of 777,000 ton/year, but from this data it was not possible to separate out the contribution from the Finger Lakes Region only.

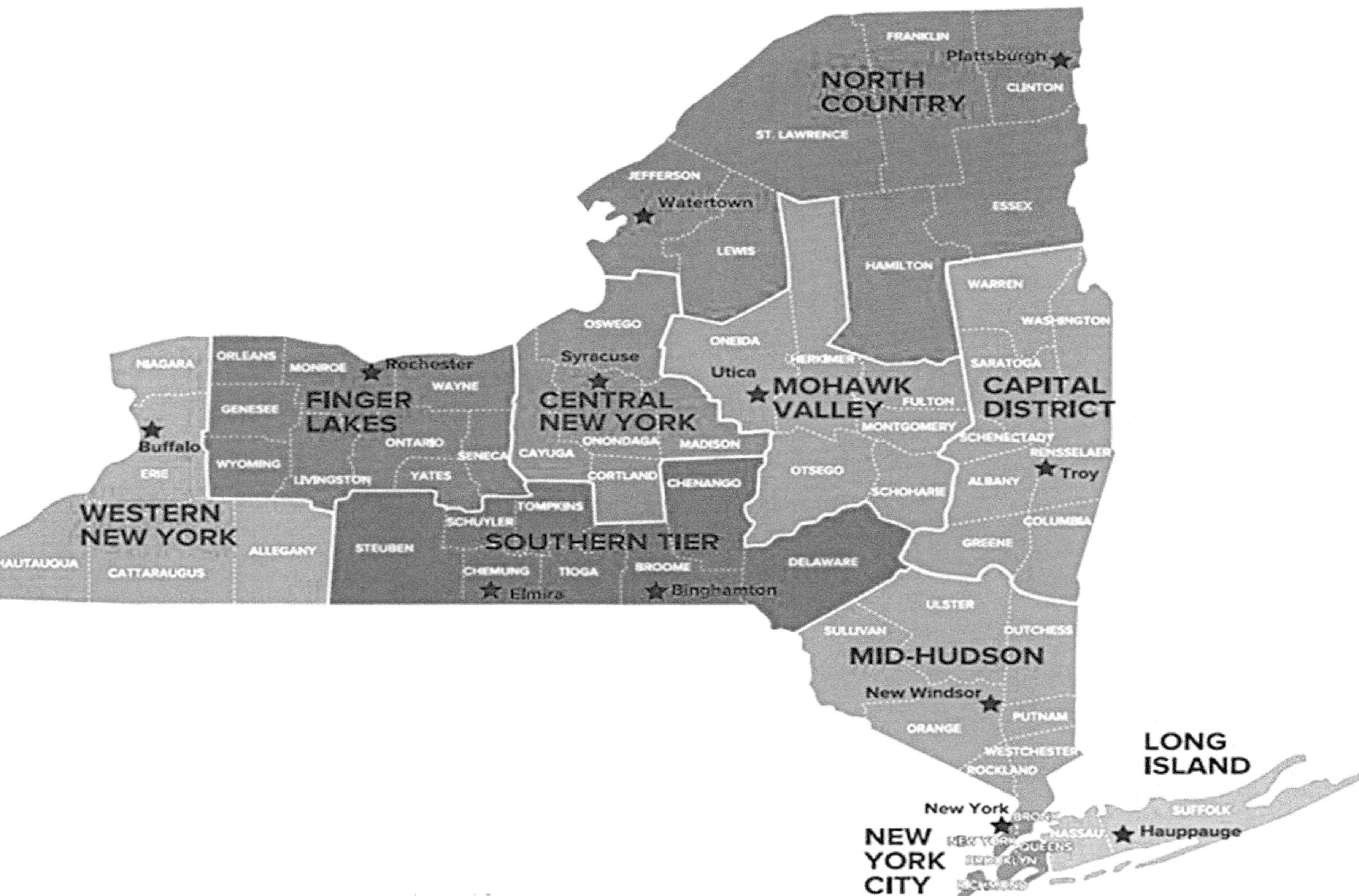

FIG. 2.7 New York State Regional Economic Development Council (REDC) zones, with Finger Lakes the focus of detailed food waste characterization study.

TABLE 2.3 Largest Food Processing Wastewater Surcharges Paid in Monroe County, NY (Trabold et al., 2011)

Food Company Sector	Average BOD[a] (mg/L)	Water Consumption (1000 gal/year)	Wastewater Surcharge[b]
Animal by-products, oils	7003	10,875	$224,637
Milk	1473	25,399	$143,123
Baked goods	1326	19,485	$97,250
Tomato products	426	308,722	$89,468
Baking supplies	2803	11,026	$57,516
Soft drinks	3598	6793	$56,434
Baked goods	9211	1697	$33,387

[a]*BOD, biological oxygen demand.*
[b]*Surcharges based on BOD, suspended solids, and phosphorous content of wastewater, and paid in addition to normal water supply charges.*

- *POTW discharge data*: Freedom of Information Law (FOIL) requests were submitted to County-level environmental services agencies to acquire data for discharges from food processing companies (Trabold et al., 2011; Rankin et al., 2012). Of the 62 counties in New York State, only eight have POTWs that charge wastewater surcharges and thus maintain records for discharges received from specific companies. The most detailed data were available for Monroe County, part of our 9-County study (Table 2.3).
- *Previously published studies*: Because of the difficulty in obtaining data directly from food processing companies, we relied on empirical relationships developed in the above-referenced state reports, as well as the Ph.D. dissertation of Ma (2006) which provided annual waste generation rates based on multiplicative factors applied to the number of employees, as summarized as follows:

Business Sector	Multiplicative Factor (*A*) Waste Generated (kg/year) = # Employees * *A*
Dairy	3541
Bakery	459
Meat processing	9.5
Fruit	2605
Mixed produce	198

Also, CDEP (2001) determined that beverage distributors generate waste at a rate of about 16,500 kg/year per facility.

2.4.3 Food Distribution Sector

This sector is well represented in the state studies mentioned previously and in the dissertation of Ebner (2016), and these sources were used to establish the multiplicative factors summarized later. For wholesale and retail food establishments, the total resources available were considered to be a combination of the materials sent to landfill and diverted to nonenergy beneficial uses.

Business Sector	Multiplicative Factor (*A*) Waste Generated (kg/year) = # Employees * *A*
Wholesalers	478
Retailers (combined supermarkets and convenience stores)	2000

2.4.4 Food Service and Institution Sector

Waste data from this sector is also difficult to obtain, because food materials must be physically separated from the MSW stream during an auditing process that can be quite time consuming and costly. Again, previously published studies were used to develop appropriate multiplicative factors as discussed later and summarized in Table 2.4. These businesses are especially critical to food waste resource quantification and characterization because they are so prevalent and generally have relatively high waste generation rates. More than half of the generators identified in this 9-County study were in this sector.

- *Restaurants*: The study by CEPA (2006) audited 50 restaurants to identify wasted and diverted food materials, the combination of which was used for the multiplicative factor recommended in Table 2.4. It is important to note that a large part of the diverted fraction corresponds to waste cooking oil or other fats, oils and grease (FOG), known to be an effective and productive feedstock for both anaerobic digestion (Chapter 4) and transesterification (Chapter 6).
- *Universities*: In the previously cited study from Connecticut (CDEP, 2001), waste generation rates were proposed based on an assumed generation rate of 0.16 kg/meal and the different number of meals per student at residential and non-residential universities. More recently, Ebner et al. (2014) updated and improved this relationship by conducting bottom-up and top-down assessments of food waste generated at Rochester Institute of Technology, in combination with published studies from 11 other universities and colleges in the United States.
- *Schools*: Block (2000) conducted an assessment of elementary, middle, and high school students in Wichita, Kansas and computed a waste generation rate of 15 kg/year for each enrolled student, assuming 150 meals per year. Clark (2014) reported roughly the same per-student food waste generation rate for a similar study in South Carolina, although an even more recent study of K-12 students in Iowa determined that the waste rate was significantly higher at 34 kg/year (Feeney, 2017). Ebner (2016) recommended a value of 15 kg/student/year based on an average from seven prior state-level studies, and therefore this same value is used in the current analysis.
- *Hospitals and nursing homes*: Ma (2006) developed a waste generation relation based on data provided in CDEP (2001) and MDEP (2002), assuming an average of 5.7 meals/bed/day and 0.27 kg average waste per meal. A similar relation was applied for nursing homes, assuming 3 meals/bed/day and 0.27 kg average waste per meal.
- *Correctional facilities*: As cited by Marion (2000), inmates housed by the New York State Department of Corrections produced food waste at a rate of about 0.45 kg per day, and most of this material originated from scraps generated in food preparation, not consumption.
- *Food pantries*: Based on data obtained directly from our local food bank organization (Foodlink), it was determined that about 6% of the mass of total food donations received is generated as waste or diverted to beneficial uses, and both quantities were considered available for WtE conversion.

From the data summary presented in Table 2.5, it is clear that the majority of food waste originates from a only a few food supply chain subcategories, with retail groceries and restaurants accounting for nearly 86% of the total mass of waste available in the 9-County region. Although waste materials in excess of 100,000 metric tons per year seem like a significant amount, it should be recognized that in many cases there can be significant seasonable variations that must be considered

TABLE 2.4 Computation of Food Waste Generated in the Food Service and Institution Sector

Business or Institution Sector	Relation Used to Compute Waste Generation Rate (kg/year)
Restaurant Full-service Fast food	Number of employees (N_E) × 1538 kg/year Number of employees (N_E) × 1134 kg/year
University	Number of enrolled students (N_S) × 26 kg/year
School	Number of enrolled students (N_S) × 15 kg/year
Hospital	Number of beds (N_B) × 576 kg/year
Nursing home	Number of beds (N_B) × 298 kg/year
Correctional facility	Number of inmates (N_I) × 166 kg/year
Food pantry	Total food bank donations (kg/year) × 0.06

TABLE 2.5 Summary of Food Waste Resources Identified in 9-County Finger Lakes Region

FSC Category	Subcategory	Food Waste Relation[a]	Number in 9-County Region	Waste Quantity (metric ton/year)	% of Total
Food processing	Beverages	$N_F \times$ 16,526 kg/year	90	1490	1.1
	Dairy	$N_E \times$ 3541 kg/year	131	464	0.3
	Bakery	$N_E \times$ 459 kg/year	1032	473	0.4
	Meat	$N_E \times$ 9.5 kg/year	196	1.9	<0.1
	Fruits and vegetables	$N_E \times$ 2605 kg/year	483	1258	0.9
	Mixed produce	$N_E \times$ 198 kg/year	1698	336	0.2
	Wastewater to POTW	Solids content of wastewater discharges reported to Monroe County, obtained via FOIL request (Table 2.3)		472	0.4
Food distribution	Wholesale	$N_E \times$ 478 kg/year	9747	4659	3.5
	Retail	$N_E \times$ 2000 kg/year	19,484	38,968	28.9
Food service and institutions	Restaurants (full-service)	$N_E \times$ 1538 kg/year	43,286	66,574	49.4
	Restaurants (fast food)	$N_E \times$ 1134 kg/year	8822	10,004	7.4
	Universities	$N_S \times$ 26 kg/year	100,979	2625	1.9
	Schools	$N_S \times$ 15 kg/year	104,757	1571	1.2
	Hospitals	$N_B \times$ 576 kg/year	3479	2003	1.5
	Nursing homes	$N_B \times$ 298 kg/year	8457	2520	1.9
	Correctional facilities	$N_I \times$ 166 kg/year	6006	997	0.7
	Food pantries	6% of mass of total donations		424	0.3
Total food waste in 9-County region				134,840	100

[a]N_F, *number of facilities;* N_E, *number of employees;* N_S, *number of students;* N_B, *number of beds;* N_I, *number of inmates.*

when designing conversion systems (including waste-to-energy) that would rely on these feedstocks for continuous operation. Also, these FSC resources should be viewed as only one component of a broader organic biomass ecosystem, where other feedstocks could reasonably be combined to maximize environmental and economic benefits. For example, locally derived lawn and forest residues could serve as cofeeds for thermochemical conversion processes (Chapters 8 and 9), and many existing anaerobic digestion systems (Chapter 4) operate using food waste codigested with livestock manure. In fact, the 9-County region featured in this study has a large milk and yogurt industry that depends on over 100,000 milking cows, producing in excess of 2.5 million ton/year of manure (Chan et al., 2013). It should also be emphasized that some of the FSC resources summarized previously are likely underestimated, especially the food processing sector wastes for which very little primary data is available. Additionally, as discussed by Ebner (2016), wastes generated in households would be a significant contribution to the overall food resource portfolio, but these materials were not included in the earlier analysis.

2.5 CONCLUSIONS

Understanding the quantities and characteristics of waste resources across the food supply chain is an essential first step to exploring the economic and environmental viability of waste-to-energy system deployment. Very large quantities of waste are known to exist on a global scale, but the nature of these materials and their distribution among different FSC sectors varies greatly across global regions. In developing countries, the greatest opportunities for food waste valorization exist in the upstream stages of the supply chain, mostly in connection with agriculture and first-line product handling, whereas consumer-facing businesses and consumption are the stages best positioned for improved efficiency in developed economies. Many published studies have also been conducted with a focus on a specific country or region, with the greatest concentration in Europe and North America, but even finer data granularity is needed to determine where a food waste-to-energy system shall be located to achieve maximum environmental and economic benefits. In the United States, certain states are accelerating the focus on food waste valorization because of existing or impending landfill bans that will necessitate the development of alternative technologies including composting, anaerobic digestion, and fermentation, in addition to the other technologies discussed in Chapters 4–9. These states often have the most developed information and waste data resources, but there is often a need for even more comprehensive data at a localized scale, commensurate with the distance over which feedstocks would be transported to support a centralized waste-to-energy conversion system. Such a study has been conducted in our region of western New York State, and the results indicate that food waste resources on the order of 100,000 ton/year are available, suitable for deployment of one or more centralized systems that could viably serve the region's current needs and support future growth and expansion. In addition to comprehensive data on food waste resources themselves, other data are needed regarding existing facilities and infrastructure, roads and transport/hauling services, distribution of electrical and natural gas service, water pipelines, and locations of natural water bodies (rivers, streams, lakes, ponds) that could potentially be impacted by WtE system outputs, such as the high-strength effluent from anaerobic digesters. Also, whenever considering the potential viability of new technologies, it is essential that the performance and value proposition of existing conversion facilities (e.g., landfills, wastewater treatment, etc.) are fully comprehended. These conventional food waste management technologies are discussed in detail in Chapter 3.

REFERENCES

Barrett, C.B., Dorosh, P.A., 1996. Farmers' welfare and changing food prices: nonparametric evidence from rice in Madagascar. Am. J. Agric. Econ. 78 (3), 656–669.

Bellemare, M.F., Çakir, M., Peterson, H.H., Novak, L., Rudi, J., 2017. On the measurement of food waste. Am. J. Agric. Econ. 99 (5), 1148–1158.

Beynelmilel Food Corp, 2008. Frozen French Fries and Potato Chips Factory—Foodelphi. Retrieved from: https://www.foodelphi.com/frozen-french-fries-and-potato-chips-factory/.

Block, D., 2000. School district supplies organics to commercial composter. BioCycle (August), 57–58.

Breunig, H.M., Jin, L., Robinson, A., Scown, C.D., 2017. Bioenergy potential from food waste in California. Environ. Sci. Technol. 51 (3), 1120–1128.

BSR (Business for Social Responsibility), 2014. Analysis of U.S. Food Waste Among Food Manufacturers, Retailers, and Restaurants. Available at: http://www.foodwastealliance.org/wp-content/uploads/2014/11/FWRA_BSR_Tier3_FINAL.pdf.

Buzby, J.C., Hyman, J., 2012. Total and per capita value of food loss in the United States. Food Policy 37 (5), 561–570.

Buzby, J.C., Wells, H.F., Hyman, J., 2014. The Estimated Amount, Value and Calories of Postharvest Food Losses at the Retail and Consumer Levels in the United States (Economic Information Bulletin Number 121). Economic Research Service/USDA.

CDEP (Connecticut Department of Environmental Protection), 2001. Identifying, Quantifying, and Mapping Food Residuals from Connecticut Businesses and Institutions. Draper/Lennon and Atlantic Geoscience Corp. (September).

CEPA (California Environmental Protection Agency—Integrated Waste Management Board Cascadia Consulting Group), 2006. Targeted Statewide Waste Characterization Study: Waste Disposal and Diversion Findings for Selected Industry Groups. Cascadia Consulting Group (June).

Chan, W.H., Ebner, J., Ramchandra, R., Trabold, T.A., 2013. Food waste resources for sustainable energy production in the Finger Lakes region of New York State.Proceedings of the ASME 7th International Conference on Energy Sustainability & 11th International Fuel Cell Science, Engineering and Technology Conference, Paper ESFuelCell2013-18189, Minneapolis, MN, July 14–19.

Clark, N., 2014. Food scraps diversion goes to school. BioCycle 55 (4), 21.

Corrado, S., Ardente, F., Sala, S., Saouter, E., 2017. Modelling of food loss within life cycle assessment: from current practice towards a systematisation. J. Clean. Prod. 140, 847–859.

Corsten, M.A.M., Worrell, E., Van, D.J.C.M., 2012. The Potential for Waste Management in Brazil to Minimize GHG Emissions and Maximize Re-Use of Materials. Utrecht University, Brazil.

Cuomo, A.M., January 13, 2016. New York State of the State Address—Built to Lead. Albany, NY.

De Clercq, D., Wen, Z., Fan, F., Caicedo, L., 2016. Biomethane production potential from restaurant food waste in megacities and project level-bottlenecks: a case study in Beijing. Renew. Sustain. Energy Rev. 59, 1676–1685.

De Clercq, D., Wen, Z., Fan, F., 2017. Performance evaluation of restaurant food waste and biowaste to biogas pilot projects in China and implications for national policy. J. Environ. Manage. 189, 115–124.

Derqui, B., Fayos, T., Fernandez, V., 2016. Towards a more sustainable food supply chain: opening up invisible waste in food service. Sustainability 8 (7), 693.

Deublein, D., Steinhauser, A., 2008. Biogas from Waste and Renewable Resources. Wiley-VCH, Weinheim.

Dou, Z., Ferguson, J.D., Galligan, D.T., Kelly, A.M., Finn, S.M., Giegengack, R., 2016. Assessing US food wastage and opportunities for reduction. Glob. Food Sec. 8, 19–26.

Eberle, U., Fels, J., 2016. Environmental impacts of German food consumption and food losses. Int. J. Life Cycle Assess. 21 (5), 759–772.

Ebner, J.H., 2016. Sustainable Management of Food Supply-Chain Resources in New York State (Ph.D. Dissertation). Rochester Institute of Technology.

Ebner, J., Win, S.S., Hegde, S., Vadney, S., Williamson, A., Trabold, T.A., 2014. Estimating the biogas potential from colleges and universities. Proceedings of the ASME 8th International Conference on Energy Sustainability & 12th International Fuel Cell Science, Engineering and Technology Conference, Paper ESFuelCell2014-6433, Boston, MA, June 30–July 2.

Ek, A., 2017. Food Waste in Beijing-Life Cycle Assessment Approach to Estimating the Environmental Impact and Resource Utilization of Various Alternatives for Food Waste Treatment in Beijing. Bachelor's thesis, Lund University, Sweden.

FAO (Food and Agriculture Organization of the United Nations), 2013. Food Wastage Footprint: Impacts on Natural Resources: Summary Report. 2013.

Feeney, M., 2017. Tackling food waste in Iowa's K-12 schools. BioCycle 58 (3), 44.

Filimonau, V., Gherbin, A., 2017. An exploratory study of food waste management practices in the UK grocery retail sector. J. Clean. Prod. 167, 1184–1194.

Gandhi, P., Kumar, S., Paritosh, K., Pareek, N., Vivekanand, V., 2017. Hotel generated food waste and its biogas potential: a case study of Jaipur City, India. Waste Biomass Valoriz, 1–10.

Ghosh, P.R., Fawcett, D., Sharma, S.B., Poinern, G.E.J., 2016. Progress towards sustainable utilisation and management of food wastes in the global economy. Int. J. Food Sci.

Gokarn, S., Kuthambalayan, T.S., 2017. Analysis of challenges inhibiting the reduction of waste in food supply chain. J. Clean. Prod. 168, 595–604.

Gustavsson, J., Cederberg, C., Sonesson, U., Van Otterdijk, R., Meybeck, A., 2011. Global Food Losses and Food Waste. FAO, Rome, pp. 1–38.

Hall, M., 2016. Techno-Environmental Analysis of Generating Animal Feed from Wasted Food Products (M.S. Thesis). Rochester Institute of Technology, Rochester, NY.

Hamilton, D.W., 2012. Anaerobic digestion of animal manures: methane production potential of waste materials. Bae 1762 (2), 4–7. Retrieved from: http://pods.dasnr.okstate.edu/docushare/dsweb/Get/Document-8544/BAE-1762web.pdf.

Hung, Y.T., Salman, H., Lo, H.H., Awad, A., 2006. Potato wastewater treatment. In: Handbook of Industrial and Hazardous Waste Treatment. Marcel Dekker, Inc., New York, pp. 193–254.

Jones, C.A., 2017. Food waste infrastructure in disposal ban states. BioCycle (November), 19–23.

Koester, U., 2013. Total and per capita value of food loss in the United States. Food Policy 41, 63–64 (Comments).

Labuzetta, A., Hall, M., Trabold, T., 2016. Initial Roadmap for Food Scrap Recovery and Utilization in New York State. New York State Pollution Prevention Institute, Rochester, New York.

Lee, S., Speight, J.G., Loyalka, S.K., 2015. Handbook of Alternative Fuel Technologies. Physical Chemistry Chemical Physics: PCCP. vol. 9. CRC Press, Boca Raton, Florida.

Lipinski, B., et al., 2013. Reducing Food Loss and Waste (Working Paper, Installment 2 of Creating a Sustainable Food Future). World Resources Institute, Washington, DC. Available online at: http://www.worldresourcesreport.org.

Ma, J., 2006. A Web-Based Spatial Decision Support System for Utilizing Organic Wastes as Renewable Energy Resources in New York State (Ph.D. Dissertation). Cornell University.

Manipadma, J., 2013. India's Food Security Rots in Storage. http://www.ipsnews.net/2013/06/indias-food-security-rots-in-storage.

Manson, C., 2017. Benefit-Cost Analysis of Potential Food Waste Diversion Legislation (NYSERDA Report 17-06). Industrial Economic, Inc. (March).

Marion, J., 2000. Composting 12,000 tons of food residuals a year. BioCycle, 30–35.

MDEP (Massachusetts Department of Environmental Protection), 2002. Identification, Characterization, and Mapping of Food Waste and Food Waste Generators in Massachusetts. Draper/Lennon, Inc. (September 19).

Mena, C., Adenso-Diaz, B., Yurt, O., 2011. The causes of food waste in the supplier–retailer interface: evidences from the UK and Spain. Resour. Conserv. Recycl. 55 (6), 648–658.

Minten, B., Reardon, T., Gupta, S.D., Hu, D., Murshid, K.A.S., 2016. Wastage in food value chains in developing countries: evidence from the potato sector in Asia. In: Schmitz, A., Lynn Kennedy, P., Schmitz, T.G. (Eds.), Food Security in a Food Abundant World (Frontiers of Economics and Globalization). vol. 16. Emerald Group Publishing Limited, Bingley, United Kingdom, pp. 225–238.

National Potato Council, 2017. Potato Utilization, United States 2013–2015. http://www.nationalpotatocouncil.org/potato-facts/.

NCDENR (North Carolina Department of Environment and Natural Resources), 2012. North Carolina 2012 Food Waste Generation Study. (August).

NEA (National Environment Agency), 2013. Food Wastage in Singapore over the Years. 2013.

Ong, K.L., Kaur, G., Pensupa, N., Uisan, K., Lin, C.S.K., 2017. Trends in food waste valorization for the production of chemicals, materials and fuels: case study south and Southeast Asia. Bioresour. Technol. 248, 100–112.

Parfitt, J., Barthel, M., Macnaughton, S., 2010. Food waste within food supply chains: quantification and potential for change to 2050. Philos. Trans. R. Soc. Lond. B: Biol. Sci. 365 (1554), 3065–3081.

Quested, T., Johnson, H., 2009. Household Food and Drink Waste in the UK: Final Report. Wastes & Resources Action Programme (WRAP).

Quested, T., Parry, A., 2017. Household Food Waste in the UK, 2015. Waste and Resources Action Programme (WRAP) (January).

Raak, N., Symmank, C., Zahn, S., Aschemann-Witzel, J., Rohm, H., 2017. Processing-and product-related causes for food waste and implications for the food supply chain. Waste Manag. 61, 461–472.

Ranjith, K.A., 2012. Sustainable Solid Waste Management in India (Master of Science in Earth Resources Engineering). Department of Earth and Environmental Engineering, Columbia University, New York.

Rankin, M.J., Trabold, T.A., Williamson, A.A., Augustine, M., 2012. Analysis of dairy manure and food manufacturing waste as feedstocks for sustainable energy production via anaerobic digestion.Proceedings of the ASME 6th International Conference on Energy Sustainability, Paper ESFuelCell2012-91091, San Diego, CA, July 23–26.

Redlingshöfer, B., Coudurier, B., Georget, M., 2017. Quantifying food loss during primary production and processing in France. J. Clean. Prod. 164, 703–714.

ReFED, Rethink Food Waste through Economics and Data, 2016a. A Roadmap to Reduce U.S. Food Waste by 20 Percent. 2016.

ReFED, Rethink Food Waste through Economics and Data, 2016b. Technical Appendix—A Roadmap to Reduce U.S. Food Waste by 20 Percent. (Latest Revision: March 2016).

Reutter, B., Lant, P.A., Lane, J.L., 2017. The challenge of characterising food waste at a national level—an Australian example. Environ. Sci. Pol. 78, 157–166.

Sala, S., Corrado, S., Caldeira, C., De Laurentis, V., 2017. Food Waste Accounting: Ongoing Research at the Joint Research Centre. Bio-Economy Unit, Joint Research Center (25th September).

Salihoglu, G., Salihoglu, N.K., Ucaroglu, S., Banar, M., 2017. Food loss and waste management in Turkey. Bioresour. Technol. 2017.

Scholz, K., Eriksson, M., Strid, I., 2015. Carbon footprint of supermarket food waste. Resour. Conserv. Recycl. 94, 56–65.

Stenmarck, A., Jensen, C., Quested, T., Moates, G., Buksti, M., Cseh, B., Juul, S., Parry, A., Politano, A., Redlingshofer, B., Scherhaufer, S., 2016. Estimates of European Food Waste Levels. IVL Swedish Environmental Research Institute, Stockholm, Sweden.

Thi, N.B.D., Kumar, G., Lin, C.Y., 2015. An overview of food waste management in developing countries: current status and future perspective. J. Environ. Manage. 157, 220–229.

Trabold, T.A., Ramchandra, R., Haselkorn, M.H., Williamson, A.A., 2011. Analysis of waste-to-energy opportunities in the New York State food processing industry.Proceedings of the ASME 5th International Conference on Energy Sustainability, Paper ESFuelCell2011-54334, Washington, DC, August 7–11.

USDA, 2016. Americans' Budget Shares Devoted to Food Have Flattened in Recent Years. https://www.ers.usda.gov/data-products/chart-gallery/gallery/chart-detail/?chartId=76967. Accessed 22 December 2017.

USDA, U.S. Department of Agriculture Economic Research Service (USDA-ERS), 2015. Food Availability (Per Capita) Data System—Loss-Adjusted Food Availability.

Wang, L.E., Liu, G., Liu, X., Liu, Y., Gao, J., Zhou, B., Gao, S., Cheng, S., 2017. The weight of unfinished plate: a survey based characterization of restaurant food waste in Chinese cities. Waste Manag. 66, 3–12.

Wen, Z., Wang, Y., De Clercq, D., 2016. What is the true value of food waste? A case study of technology integration in urban food waste treatment in Suzhou City, China. J. Clean. Prod. 118, 88–96.

WRAP, 2017. Estimates of Food Surplus and Waste Arisings in the UK. WRAP, Banbury. Available from: http://www.wrap.org.uk/sites/files/wrap/Estimates_%20in_the_UK_Jan17.pdf.

FURTHER READING

Ebner, J.H., Williamson, A.A., Trabold, T.A., 2015. Quantifying the greenhouse gas impacts of pathways for treatment of secondary resources generated in the food supply chain.Proceedings of the ASME Power and Energy Conversion Conference, Paper PowerEnergy2015-49559, San Diego, CA, June 28–July 2.

Quested, T.E., Parry, A.D., Easteal, S., Swannell, R., 2011. Food and drink waste from households in the UK. Nutr. Bull. 36 (4), 460–467.

Chapter 3

Conventional Food Waste Management Methods

Thomas A. Trabold and Vineet Nair
Golisano Institute for Sustainability, Rochester Institute of Technology, Rochester, NY, United States

3.1 INTRODUCTION

As described in Chapter 2, food loss and waste occur at every stage of the farm-to-fork spectrum, and the quantities and characteristics of these materials can vary significantly, both geographically and temporally. It has been recognized for quite some time that most current practices of handling food system wastes are environmentally and economically detrimental, and more sustainable solutions are needed. However, before considering options for waste-to-energy technologies or other seemingly preferable food waste valorization strategies that are beyond the scope of this book, it is imperative that conventional waste management methods are understood, as well as the barriers to moving away from some of these methods on a large scale.

In the past several years, there has been a surge of activity in communicating the scope and impact of the food waste problem, and many publications have reported that from 30% to 40% of food resources produced are never consumed by humans (Buzby et al., 2014; Gunders, 2012; ReFED, 2016). Moreover, with a particular focus on the consumption end of the farm-to-fork spectrum, it has been reported that as much of 97% of food waste is sent to landfills where the potential for fugitive methane emissions makes this end-of-life strategy a great source of concern (ReFED, 2016). But such broad, generalized statistics hide the nuances of the food waste management ecosystem and overlook other pathways that are currently utilized and, depending on the particular system input/output details, may offer more sustainable alternatives to landfilling or other "disposal" methods. Fig. 3.1 presents a modified food waste hierarchy that considers the available waste management options after source reduction (the top alternative in the US Environmental Protection Agency hierarchy, Fig. 1.9) has been exploited to the greatest possible extent. At the top of the hierarchy is food donation, followed by animal feed production, and both of these options are considered favorable to the waste-to-energy (WtE) technologies to be covered in Chapters 4–9. Below WtE are less desirable alternatives for food waste management, including composting, wastewater treatment, incineration, and lastly landfill. Wastewater treatment and incineration may have some redeeming qualities because of the potential for useful coproducts such as thermal energy or fertilizer, and the greenhouse gas emissions that do occur are usually in the form of CO_2 instead of methane, the latter having a much greater global warming potential. An important consideration in determining which of these various pathways are suitable for a particular waste stream is the material *phase* of the waste. While solid (S) and packaged (P) food wastes have a variety of potential outlets, for liquid waste streams (L) the only conventional option is wastewater treatment, which may involve onsite aerobic conversion in a lagoon or similar structure, or transport via pipeline or truck to a centralized processing facility, often referred to as the publicly owned treatment works (POTW).

To help understand the current ecosystem of food waste management pathways, it is instructive to consider as a representative case study our region of Upstate New York (Fig. 3.2). Rochester is the major city in the region, near the center of Monroe County, with a total county-wide population of approximately 750,000. The County is bordered on the north by Lake Ontario, the smallest and most eastward of the five Great Lakes, and is a mix of urban, suburban, and rural communities. Beyond the significant amount of food waste generated at the consumption phase (as stated in Chapter 2, very geographically dispersed and heterogeneous), more concentrated sources of waste are available at hospitality, food processing, restaurant, institutional, and retail facilities. These facilities are densely concentrated near the population centers, but also extend along the main highways into suburban and rural sections. Once waste is generated, or edible food is no longer desired by its intended consumer, there are three primary outlets available within Monroe County: donation, wastewater treatment, and landfilling. The primary donation organization is FoodLink, a centralized food bank facility located on the

Sustainable Food Waste-to-Energy Systems. https://doi.org/10.1016/B978-0-12-811157-4.00003-6

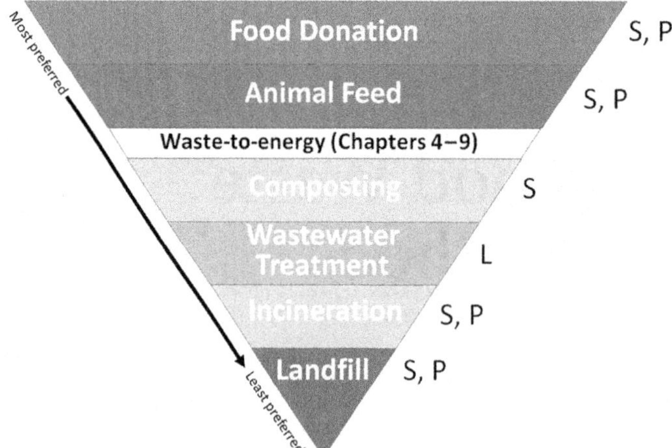

FIG. 3.1 Modified food waste hierarchy showing conventional management options after source reduction has been maximized. According to the well-known EPA hierarchy (https://www.epa.gov/sustainable-management-food/food-recovery-hierarchy), "industrial uses" including waste-to-energy (Chapters 4–9) fall between animal feed and composting. For each option, the applicable food waste phases are indicated: solid (S), liquid (L), and packaged (P).

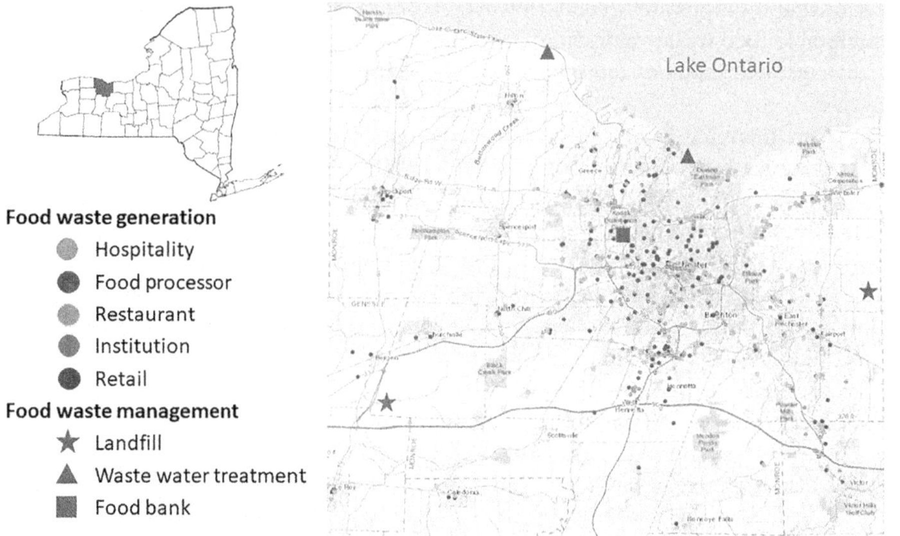

FIG. 3.2 Locations of large nonresidential food waste generation and waste management facilities in Monroe County, New York.

west side of Rochester that provides mostly donated (and some procured) food to various food cupboards, outreach centers, and community and religious organizations. Inedible or low-value waste has four main destinations; liquid-phase wastes are sent to one of two wastewater treatment plants located near the Lake Ontario shoreline, while solid and packaged waste materials go to landfills located near the eastern and western County borders. The geographic distribution among the waste generation sites and potential treatment facilities is indicative of the significant energy and fuel costs (and associated environmental impacts) of the conventional waste management ecosystem. For example, a restaurant located in the southwestern quadrant of Rochester would likely have its solid and packaged food waste transported over 16 miles (26 km) by truck to Millseat Landfill near the western County border, and its wash water and liquid food waste discharged down the drain to be pumped to Van Lare Sewage Treatment plant, about 8 miles (13 km) away. If the business owner desired to pursue options that may be more sustainable, such alternatives exist but relative cost and transport distance may present significant barriers. In this case, the nearest commercial composting facility is about 85 miles (136 km) to the east near Syracuse, NY, and the closest anaerobic digester (Chapter 4) is over 30 miles (48 km) to the southwest. If more conventional waste management facilities are closer and offer competitive tipping fees, it will be difficult to justify other pathways unless policy initiatives are applied to incentivize the more sustainable options or disincentivize the incumbent technology (Chapter 12).

Before considering alternative food waste-to-energy strategies in Chapters 4–9, it is essential that the current system for food waste management is understood on a national scale. To support this objective, data and analyses for the United States were obtained from two key reference sources:

- ReFED, Rethink Food Waste through Economics and Data, 2016. *A Roadmap to Reduce U.S. Food Waste by 20 Percent*, including Technical Appendix dated March 2016 (ReFED, 2016).
- Report prepared by the Food Waste Reduction Alliance, *Analysis of U.S. Food Waste Among Food Manufactuers, Retailers and Restaurants, A joint project by the Food Marketing Institute, the Grocery Manufacturers Association & the National Restaurant Association* (FWRA, 2016).

Using these references, coupled with data on food waste volumes and physical/chemical characteristics presented in Chapter 2, it was possible to obtain a reasonably accurate understanding of the main end-of-life strategies employed in the major sectors of the food supply chain (Table 3.1). The *Agriculture* and *Household* sectors are fairly straightforward in that a single end-of-life pathway accounts for nearly all of the food waste management infrastructure. There is almost no landfilled food waste originating in the *Agriculture* sector because most nonconsumed material is tilled back into the farm soil. Conversely, the heterogeneous nature of household waste limits the range of potential beneficial uses available, and therefore almost all of this material is mixed with the rest of the municipal solid waste (MSW) sent to landfill. However, the situation with *Manufacturing* businesses and *Consumer-Facing Organizations* is more complicated because a number of competing food waste management strategies are in play, as illustrated in Figs. 3.3 and 3.4. The *Manufacturing* sector presently extracts value from food wastes or unavoidable losses resulting from, for example, peeling fruits and vegetables prior to packaging. Most of these materials are directed to animal feed, often in a synergistic relationship with farmers from whom they procure commodity food products. Donation and composting also play a role, as does land application of liquid-phase waste, although the latter data from the Food Waste Reduction Alliance report (FWRA, 2016) were skewed because almost all of the land application volume originated from a single food-processing company out of the nine survey respondents. In the *Wholesale/Retail* sector, >70% of food waste is diverted from landfills and distributed fairly evenly among donation, animal feed, composting and incineration pathways. For restaurants, the scenario is almost the exact opposite of the food waste management strategy in the manufacturing sector, because the vast majority of waste (>90%) is landfilled.

Although it is reasonable to assume that most of the decisions regarding conventional food waste management strategies are based on economic considerations, it is pertinent to note that such decisions may have significant environmental and social implications. For example, food donations are at the top of the food waste hierarchy, but companies typically do not extract secondary revenue from this pathway. Also, as discussed in detail in Chapter 10 (Environmental aspects of food waste-to-energy conversion), there are many published studies on life cycle assessment (LCA) of food waste management methods reporting wide ranges of environmental impacts. While most options are considered environmentally favorable to

TABLE 3.1 Summary of Waste and Landfill Diversion Pathways Across the Food Supply Chain

Food Supply Chain Sector	Landfilled Waste (million tons/year) (ReFED, 2016)		Landfill Diversion Pathways
Agriculture	0		10.1 million tons of waste (often resulting from cosmetic imperfection of fruits and vegetables) mostly tilled back into farm soil
Manufacturing	1.1		20 million tons presently diverted; 85% to animal feed (ReFED, 2016)
Consumer-facing organizations	24.8		
Wholesale/retail		8.0	Although this sector accounts for about 40% of landfilled waste, in some subsectors rates of diversion to donation, animal feed, composting, etc. are relatively high
Restaurants		11.4	
Institutions		4.9	
Government		0.5	
Households	26.5		30% of food waste generated is currently disposed of down the drain (ReFED, 2016). Few other diversion methods are utilized

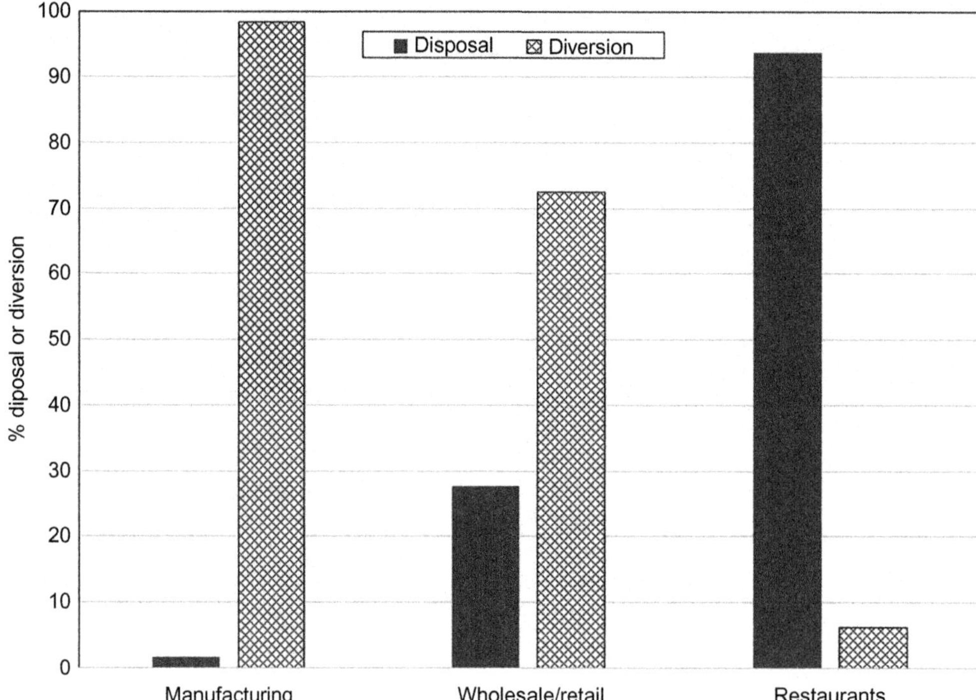

FIG. 3.3 Landfill disposal and diversion in manufacturing, wholesale/retail, and restaurant sectors (FWRA, 2016).

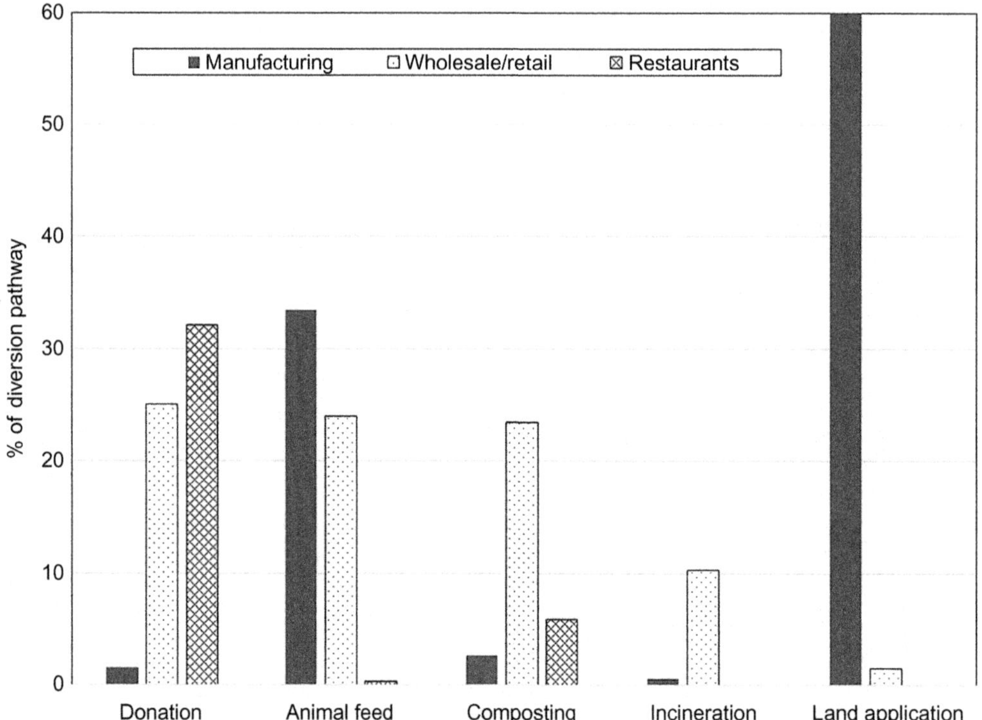

FIG. 3.4 Primary landfill diversion pathways in manufacturing, wholesale/retail, and restaurant sectors (FWRA, 2016).

landfilling, as discussed by Bernstad and la Cour Jensen (2012) many comparative LCAs suffer from inconsistencies in definition of system boundary, assumptions related to the characteristics of treated food waste, impacts of pretreatment, emissions associated with different subsystems and process steps, etc. Therefore, it is not possible based on the existing literature to state with confidence which conventional food waste management pathways are "best." The remaining sections of this chapter provide details on the relative advantages and disadvantages of these pathways, organized based on the ranking from most to least preferred, inferred from the modified EPA waste hierarchy (Fig. 3.1): donation, animal feed, composting, wastewater treatment, incineration, and landfilling.

3.2 FOOD DONATION

The Economic Research Service of the United States Department of Agriculture (USDA) predicts that the number of food-insecure people in 76 low- and medium-income countries will grow from 13.9% of the population in 2014 to 14.6% by 2024 (Rosen et al., 2016). Even in the developed economies of the world, food insecurity is a major social problem despite the fact that 30%–40% of food produced is never consumed by humans, largely the result of postconsumer waste. For example, the rate of food insecurity in the United States was 13.4% in 2015 (Feeding America, 2017), while nearly 44 million people in Europe experienced food insecurity in 2014, 8.6% of the total European Union population at the time (Sert et al., 2014). The paradox of food overabundance and undersupply coexisting in close proximity certainly points to the potential for food donation as one important approach to solving the food waste problem.

After efforts to reduce food waste at its source, food donation is considered the top priority in the waste hierarchy, because the food material is still reaching the consumer even if the market potential (and producer profit) is not achieved as intended. Additionally, while contributing to solving the food waste problem, this practice simultaneously addresses the major social problem of food insecurity described earlier. Vitiello et al. (2015) defined food banks as "nonprofit organizations designated by state governments to distribute federal and state food aid to food cupboards (also known as pantries and sometimes confusingly also called 'food banks'), soup kitchens, homeless shelters, and other emergency food and feeding organizations." A Citizen's Guide to Food Recovery (USDA, 1996), classified four major types of food recovery for the purpose of feeding human beings:

- Field gleaning—collection of crops from farmers' fields that have already been harvested or on fields where it is not economically profitable to harvest.
- Perishable food rescue or salvage—collection of perishable produce from wholesale and retail sources such as supermarkets.
- Food rescue—collection of prepared foods from the foodservice industry, including restaurants and hotels.
- Nonperishable food collection—collection of processed foods with longer shelf lives.

In general terms, gleaning refers to the collection of goods that their producers or resellers cannot or choose not to sell. It most commonly refers to the harvesting of crops from farmers' fields, but also from gardens and other sites. Gleaning is becoming more widely practiced as a means of recovering fresh, nutritious food directly from agricultural operations and gardens (Lee et al., 2017; Sönmez et al., 2015).

Although much of the literature on food donations has focused on various aspects of the social benefits that are achievable, recent studies have considered the potential economic benefit that food waste producers can derive by including donations as part of a broader food waste reduction strategy. For example, Giuseppe et al. (2014) and Muriana (2015) have developed an economic model to optimize profit to producers based upon balancing tax incentives associated with food donations and direct sale of these materials for livestock feed, and selecting the best time to remove nearly expired products from store shelves. Also, Bilska et al. (2016) have identified opportunities where donations can be increased and indeed encouraged across the food supply chain.

Food donations have clearly had a large and measurable impact on providing nutritional benefits to people in need, but there is an upper limit to how much food waste can be effectively managed using this method. Although laws in a number of countries have protected food waste donators from legal liability (e.g., Bill Emerson Good Samaritan Food Donation Act passed in the United States in 1996[1]), many retailers still have concerns with potential negative publicity or damage to corporate reputation if donated food is improperly handled. Beyond these legal and social perception issues, there is a physical limit to how much excess food can be consumed by the existing population of people in need. For example, Scherhaufer and Schneider (2011) have concluded that there is presently a significant oversupply of donated bread products

1. Public Law 104–210—OCT. 1, 1996.

that far exceeds that needed by people experiencing food insecurity. In addition, in some cases, large food donators combine relatively high nutritional quality products with lower value foods (e.g., soft drinks, candy, etc.) that food banks often have policies against distributing. In these cases, the food bank is serving to offset the disposal fees that would have otherwise been borne by the donating organization itself. Despite these barriers, it is expected that there are opportunities for expanding food donations in the future. As outlined in ReFED (2016), based on stakeholder interviews, it is possible to increase food donations by factors of 15, 10, and 2 in the agriculture, restaurant/food service, and grocery retail sectors, respectively, which could potentially result in an increase in total donated food in the United States from 1.1 million tons (2016) to 5.5 million tons. Even at this higher projected level, food donations would still represent an outlet for $<10\%$ of the currently generated US food waste that ends up in landfills.

3.3 ANIMAL FEED PRODUCTION

It is widely assumed that feeding food waste to livestock is common in the United States and other countries, especially on farms raising pigs. However, the practice of feeding food scraps to animals has declined significantly in the past three decades, after several disease outbreaks were linked to animal byproducts in livestock feed, including foot-and-mouth disease and bovine spongiform encephalopathy (BSE), the latter commonly referred to as "mad cow disease." Most states now impose varying levels of regulatory control to minimize the spread of animal feed-borne diseases originating from food waste (Leib et al., 2016):

- Full prohibition on both animal and vegetable waste (10 states including Alabama, Illinois, and Wisconsin).
- Prohibition of animal waste only (six states including Delaware, Nebraska, and Vermont).
- Heat-treating requirement for both animal and vegetable waste (12 states including Arkansas, Illinois, and Wisconsin).
- Heat-treating requirement for animal waste only (Puerto Rico and 20 states including Arizona, New York and Washington).
- The remaining states either require only licensure to feed food waste to swine (Georgia), or have no regulations at all (Alaska and Utah).

As recently reported by Salemdeeb et al. (2017), in many other parts of the world animal feed is not a common food waste management option because of similar disease concerns, and the need to heat treat before being suitable for livestock consumption. For example, of the 89 to 100 million tons of food waste produced per year in the European Union, only around 3 million tons are recycled as animal feed (zu Ermgassen et al., 2016). The percentage of food waste converted to animal feed in this study is roughly equivalent to the percentage of hog farmers who fed food scraps to their livestock in the United States in 2007 (Federal Register, 2009). Additionally, as described by Kawashima (2004), global economics can also play a significant role when the relatively low cost of imported animal feed products makes efficient domestic waste-to-animal feed conversion impractical.

Despite the increased global restrictions on converting food waste into animal feed, the practice has not disappeared completely, but rather it has changed in scale and become more industrialized (Leib et al., 2016). In fact, in many regions, the types of food waste materials being used for animal feed have changed from minimally controlled scraps and "garbage" to large-scale food processing and packaged products that are well controlled in terms of both pathogen control and nutritional content. As reported by Hall (2016), as much as 26% of animal feed used in the US originated as byproducts of other industrial processes, including distillers grains from corn ethanol production (Wisner, 2010), bakery and bread wastes, and sugar beet processing waste (Harris and Staples, 1991). In the European Union, there is better centralized organization of the processes and facilities that produce animal feed from so-called former foodstuffs, including bread and bakery wastes, breakfast cereals, snack foods (crisps), chocolate and dairy products (Bouxin, 2012). In 2011, processing facilities in a number of European countries produced up to 3.5 million tons of animals feed from these materials (Table 3.2).

The greatest success in developing animal feed products from food waste has been achieved in Japan and Korea. Unlike the approaches generally taken in the United States and Europe that restrict the types of waste materials that can enter the animal feed supply chain, these Asian countries rely more heavily on pretreatment and rigorous testing before animal feed products are made available for sale. In Japan, the government has fostered the growth of the food waste-to-animal feed industry by two key policy initiatives: dictating specific food waste reduction targets for all industry sectors and entities of all sizes, and promoting (with associated economic incentives) a certified animal feed (Ecofeed) that is required to contain at least 20% food waste (Sugiura et al., 2009; zu Ermgassen et al., 2016). As of 2007, there were 171 Ecofeed producers in Japan manufacturing over 150,000 tons of product, with the 2015 goal of utilizing over 5 million metric tons of food waste (45% of Japan's total generation) in animal feed (Sugiura et al., 2009). There is a wide diversity of food waste materials currently converted in this manner, including

TABLE 3.2 Former Foodstuffs Used for Animal Feed in the European Union (Bouxin, 2012)

Country	Number of Processors	Volume Processed (1000 tons)	% of Total
United Kingdom	20	500	14.3
Germany	11	800	22.9
The Netherlands	10	300	8.6
Italy	10	200	5.7
Spain	9	300	8.6
France	4	200	5.7
Denmark	3	95	2.7
Ireland	2	20	0.6
Portugal	2	30	0.9
Belgium	1	125	3.6
Other	20–30	450–950	12.9–27.1
Total	90–100	3000–3500	100

- Food-processing industry byproducts: rice bran, rice noodle debris, soystarch and beer residue, wheatbran, sesame oil residue, corn gluten meal, juice pulp, bean curd, tofu residue, molasses, etc.
- Surplus food not used during processing: rice, bread, noodles, bean curd, milk, ice cream, delicatessen items, juice, boxed lunches, etc.
- Cooking waste and leftovers produced by catering businesses and households.

In Korea, government policy initiatives were also largely responsible for first mandating broad reductions in waste across the food supply chain, and promoting conversion of an increasing fraction of this material into animal feed. As described by Hall (2016), The Food Waste Reduction Master Plan of 1996 required commercial food waste generators to recycle their food wastes, and food waste generated at households was required to be separated and recovered from municipal solid waste (MSW). This initiative was followed in 2005 by a national food waste landfilling ban, and in 2010 a volume-based Food Waste Fee Systems was introduced, similar to the policy for general MSW for households and small businesses that had been in place since the mid-1990s. The resulting impact on food waste management was significant and rapid. As shown in Fig. 3.5, over a 5-year period (2001–06), landfilling and incineration were nearly completely eliminated, while composting and animal feed accounted for 42% and 43%, respectively, of food waste management by mass as the total waste increased from 11,237 to 13,372 tons/day.

As the technical and policy aspects of food waste-to-animal feed have been widely researched and documented, there is also growing interest in understanding and quantifying the environmental benefits of this food waste management option relative to the other more conventional practices outlined in this chapter. Several recent studies have reported life cycle assessments of food waste-to-animal feed processes that generally show the potential for deep reductions in greenhouse gas emissions, especially when compared to landfilling and incineration (e.g., Ebner, 2016; Hall, 2016; Salemdeeb et al., 2017; San Martin et al., 2016). Hall (2016) conducted an extensive literature review of earlier LCA studies and concluded that on average both dry and wet feed processes are better than landfilling in regards to total global warming potential, but there is no clear distinction among animal feed, composting, and anaerobic digestion. This work also included a case study of an animal feed plant in Upstate New York that uses wood chips and packaging from food waste to dry the incoming feedstock. In this case, negative global warming potentials were computed for food waste moisture contents of 20% to 80%, with avoidance of conventional animal feed being the largest contribution. However, as illustrated by Ebner (2016), the relative benefit of diverting food waste from landfill to animal feed is strongly dependent on the physical and chemical characteristics of the feedstock material itself. Fairly small benefits were realized for higher moisture content materials like salad, whey, and apple pomace, but found to be very large for dry goods, cereals, and baked goods.

As the environmental and economic benefits of food waste-to-animal feed production become more widely known, it is reasonable to expect that this waste management strategy will play a bigger role in the future, as has clearly been the case in

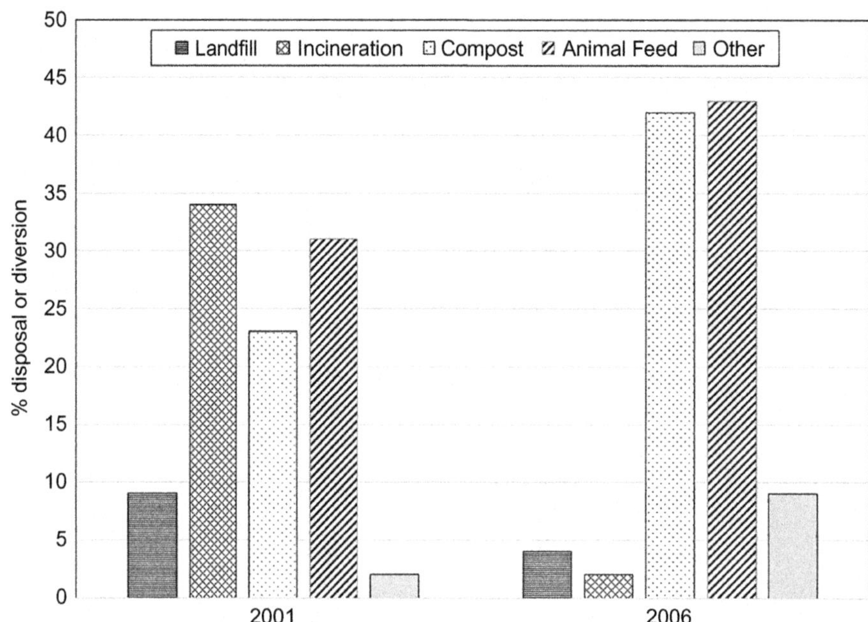

FIG. 3.5 Evolution of Korean food waste management, 2001–06 (Kim and Kim, 2010; Kim et al., 2011).

Japan and Korea. Also, it is pertinent to note that there are efforts underway to explore other food sector waste materials for their suitability as animal feed products. For example, significant sources of food waste can be refined to produce protein-rich compounds and other additives for animal feed, as Xavier et al. (2017) have reported in manufacturing cattle silage by treating squid processing waste with formic acid. However, any new food-waste-derived animal feed products must be subjected to rigorous testing to confirm its safety and digestibility for the targeted livestock population (Russick and Urriola, 2017).

3.4 COMPOSTING

When one thinks of conventional food waste management strategies, composting is often considered to be the most simple and sustainable. Of all the methods described in this chapter, composting is also the only one viable over a wide range of scales, from the simple backyard version (Fig. 3.6) to commercial windrow systems processing thousands or millions of tons per year. As defined by Platt et al. (2014), composting is "the controlled aerobic, or oxygen-requiring, decomposition of organic materials by microorganisms, under controlled conditions."

BioCycle magazine publishes an annual report entitled "The State of Organics Recycling" that provides detailed data for composting and anaerobic digestion facilities in the United States. From the most recent version (Goldstein, 2017), it was found that of the 4713 operating facilities, only about 18% (869 facilities) accepted either a combination of yard trimming and food waste, or "multiple organics" comprising these materials plus manure, wood shavings, and industrial organics. Moreover, the vast majority of these facilities are concentrated in a few regions, with nearly 75% located in only 10 states (Massachusetts, California, Ohio, Washington, New York, Louisiana, Kansas, Oregon, North Carolina, and Wisconsin; Fig. 3.7). Alternately, analyzing the data in terms of the amount of food waste processed per year (Fig. 3.8), it was found that less than half of states process at least 1000 tons/year, with only four (California, Massachusetts, Washington, and Florida) processing >100,000 tons/year. The estimated 1.8 million tons of total food waste managed via composting is a small fraction of the 52 million currently landfilled (Table 3.1), although by tilling food waste back into the soil, the agricultural sector is essentially composting the bulk of its waste (10.1 million tons). It is also pertinent to note that although Fig. 3.8 includes data for both composting and anaerobic digestion (discussed in Chapter 4), the latter is associated with only 112 facilities processing either food waste only, or codigesting food waste with livestock manure. Most of these existing AD facilities are concentrated in California, New York, Pennsylvania, and Wisconsin.

One major challenge with composting is that its sustainability potential is largely dependent on how effectively nutrients are recycled to the agricultural operations instead of simply being used for landscaping. Anecdotal information obtained from our local food and agriculture industry partners suggests that it will be difficult to use compost to

FIG. 3.6 Typical household-scale food waste composting bin fabricated from wooden pallets.

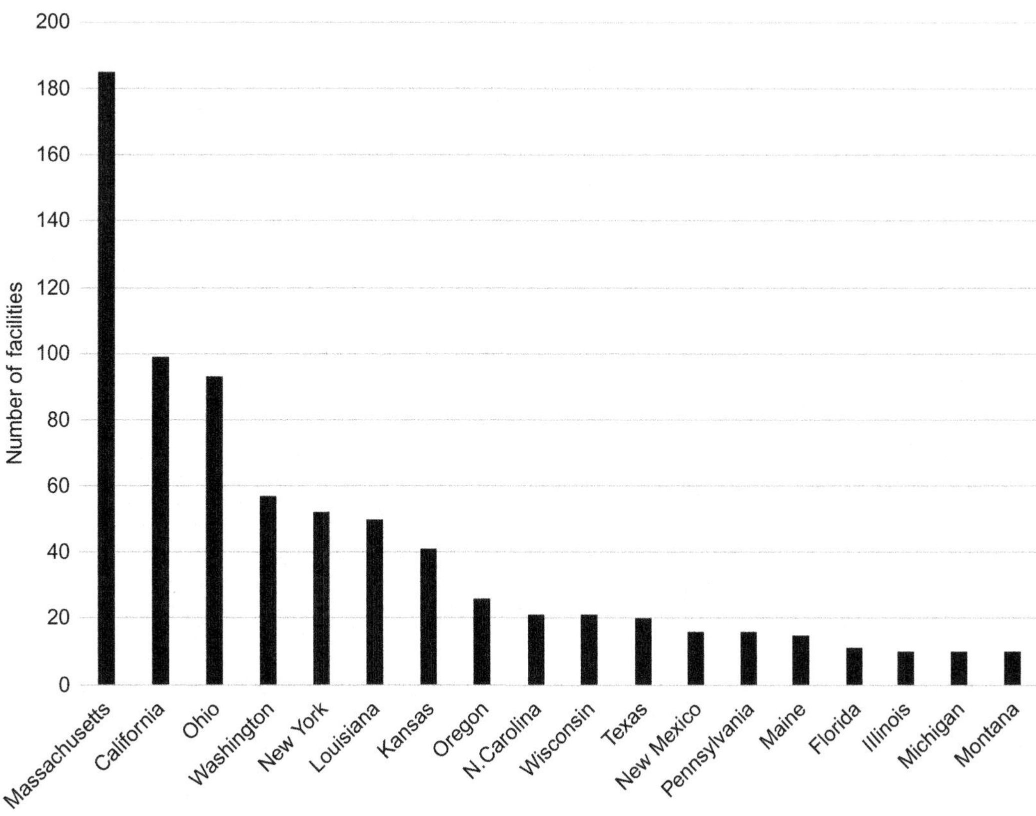

FIG. 3.7 States with 10 or more facilities composting combined yard and food wastes, or multiple organics (Goldstein, 2017).

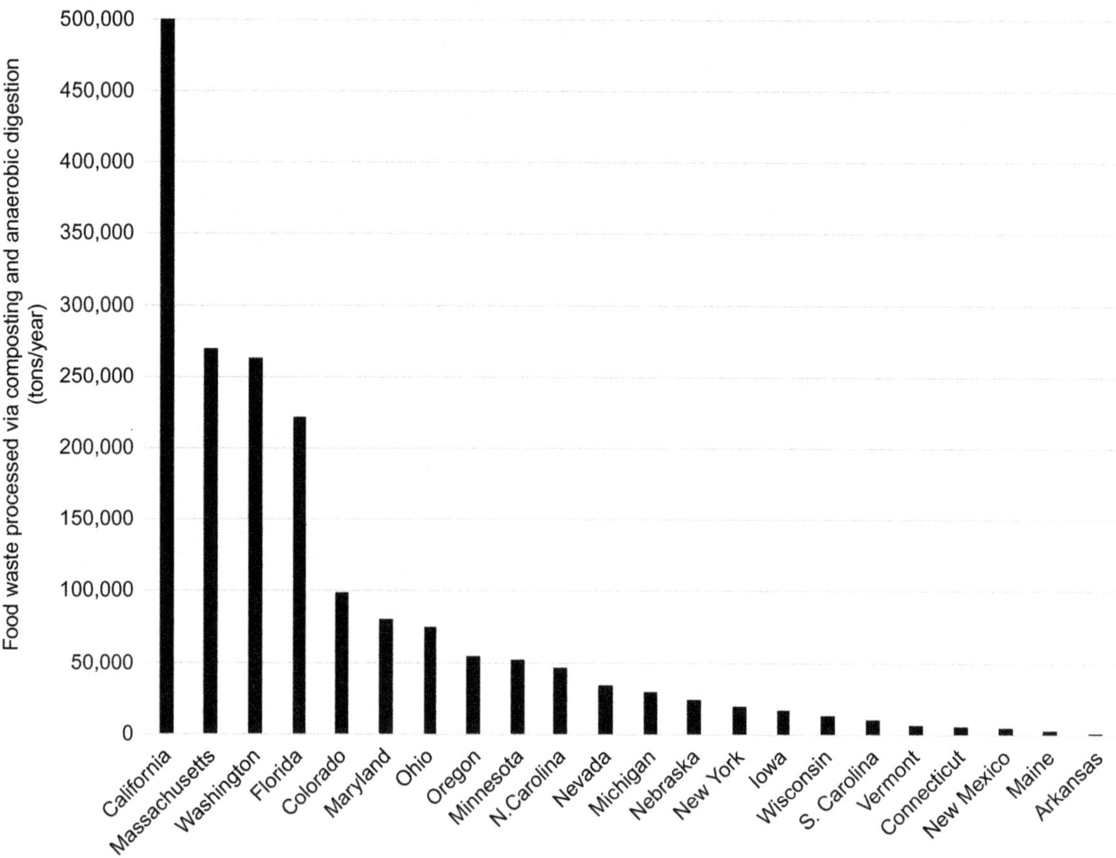

FIG. 3.8 States processing >1000 tons/year of food waste via composting and anaerobic digestion (Goldstein, 2017).

significantly displace synthetic fertilizer in conventional farming, in part because the challenge in precisely matching macronutrient profiles of nitrogen, phosphorus, and potassium (NPK). Also, unless there is a compelling cost benefit to be realized, there may be little incentive to deviate from existing practices that the farmer has relied on for many growing seasons. However, if barriers to broad acceptance in the marketplace can be overcome, and specific technical challenges associated with food waste composting can be addressed (Cerda et al., 2017; Li et al., 2013), the potential environmental benefits are significant. For example, if food-derived compost is used to displace soil amendment material like peat, Saer et al. (2013) demonstrated large benefits from eliminating the impacts associated with peat mining, processing, and transport. Mu et al. (2017) also showed large environmental and economic benefits associated with in-vessel composting relative to landfilling, but in this case the constrained system (a university campus) both generated the food waste and composted it for use in an onsite vegetable garden.

3.5 WASTEWATER TREATMENT

Food waste has inherently variable composition, high moisture content, and low calorific value, which constitute an impediment to the development of robust, large-scale, and efficient industrial processes. Traditionally, liquid-phase food waste (also referred to as food waste recycled wastewater, FRW) has often been discharged into nearby water bodies, but this adversely affects the environment, so an alternative method of treating this material is required (Shin et al., 2010). Domestic wastewater is typically treated using aerobic biological methods to ensure adequate oxygen content for the growth of microorganisms (Metcalf & Eddy, Inc, 2003). After removal of the relatively clean supernatant, the high-solid sludge remaining at the bottom of the treatment units may be collected and sent for anaerobic digestion (Chapter 4), where options include onsite anaerobic conversion, or transport to a centralized processing facility. This method also has the additional benefit of generating combustible biogas, which can be used to produce energy. Previous reports have demonstrated that food wastes are desirable substrates for anaerobic digesters (Wang et al., 2002; Zhang et al., 2007), but raw wastewater is rarely treated anaerobically, mainly due to its low strength and solids content, leading to inadequate gas generation and low process efficiency (Ghanimeh et al., 2017). Despite the challenges in using conventional wastewater treatment for managing liquid food wastes, the environmental benefits of diverting food waste from conventional disposal methods such as landfilling and incineration are significant. It is well known that many landfills produce significant methane leaks, even with well-designed landfill capture systems, and the high water content of food waste (typically 80%–90% of the total weight) often renders incineration an energy-intensive and costly option. The cost implications of wastewater treatment can, however, also be severe, especially for many food processing companies in urban or suburban regions that have few available options for handling liquid-phase waste streams. Because such wastes are often "high strength," their treatment at a centralized wastewater treatment plant (WWTP) requires additional energy and flow capacity. Therefore, in many regions of the U.S., the local municipal or county-level water authority charges an extra surcharge (i.e., beyond the cost of basic water supply) to companies that discharge relatively large volumes of food processing wastewater. For example, in Monroe County, New York, the multiplicative surcharge factor (SF) applied to the standard water supply rate depends on the level of biological oxygen demand (BOD), suspended solids, and phosphorous in excess of 300, 300, and 10 mg/L, respectively (Monroe County, 2016). In 2010, seven of the ten largest wastewater surcharges (in the range of $33,000 to $245,000) were applied to small- and medium-sized food processing companies, and presented a significant financial burden and barrier to growth (Trabold et al., 2011).

The United States has over 2000 sites producing biogas: 239 anaerobic digesters on farms, 1241 wastewater treatment plants using an anaerobic digester (~860 currently use the biogas they produce), and 636 landfill gas projects. There is potential for 2440 wastewater treatment plants (including ~381 that are making biogas but not using it), which could support a digester and 450 untapped landfill gas projects. In comparison, Europe has over 10,000 operating digesters and some communities are essentially fossil fuel free because of them (USDA/EPA/DOE, 2014). Water Resource Recovery Facilities (WRRFs) have been using anaerobic digestion to produce biogas for heat and power since the 1920s. It has become a common practice for WRRFs to enhance biogas production by adding fats, oil, and grease (FOG) to digesters. The codigestion of other types of food waste, including municipally collected food waste and byproducts from food processing facilities, is also becoming increasingly common at WRRFs. Codigesting organic waste materials allows facilities to save and make money while reducing greenhouse gas emissions, providing a renewable energy source, and diverting valuable resources from landfills and sewer pipes (EPA, 2016). In September 2014, EPA published a report, "Food Waste to Energy: How Six Water Resource Recovery Facilities are Boosting Biogas Production and the Bottom Line," examining the codigestion practices, performance, and experiences of six WRRFs (three in California, two in Wisconsin, and one in Indiana). All six of the WRRFs interviewed for the report stated that although codigestion presents challenges, the benefits outweigh the difficulties and they would recommend that other facilities pursue codigestion (EPA, 2014).

Wastewater treatment is vital for environmental protection, but it is also an energy-intensive activity. Municipal water systems account for about 4% of global and US electricity consumption and 5% of global greenhouse gas emission (Copeland and Carter, 2017; IEA, 2016). Thus, a viable and pragmatic approach is to codigest municipal organic wastes in combination with municipal wastewater sludge using the spare digestion capacity at waste water treatment plants (WWTPs) to increase biogas production. This approach not only allows WWTPs to significantly improve their energy efficiency, but also reduce the cost of municipal organic waste management while facilitating nutrient recycling (Nghiem et al., 2017). Diggelman and Ham (2003) also demonstrated that in such an arrangement, food waste and the associated carrier water is a small fraction of the total flows and solids passing through the WWTP over its 30-year design life, and the resulting processing cost is on the order of $0.50 per 100kg of food waste. A number of recent studies have addressed the technical challenges and environmental/economic impacts of combining food waste with urban wastewater, or anaerobic co-digestion of food waste and biosolids downstream of the aerobic wastewater treatment process (e.g., Pretel et al., 2016; Becker et al., 2017; Edwards et al., 2017; Moñino et al., 2017; Zamorano-López et al., 2018).

Based on the renewable energy potential of combining wastewater treatment sludge and food waste, it is reasonable to consider separation of food waste from municipal solid waste (MSW) and combining this material in WWTPs with existing excess capacity. Food waste can be sorted and processed either onsite, or delivered in a pretreated liquid form. In addition to sorting and processing equipment, onsite processing also requires several auxiliary facilities including a weighing bridge and possibly airlock passages to the receiving bay. Thus, onsite processing of waste materials is mainly suitable for large plants. For a small plant, onsite waste processing can be inefficient in terms of equipment and manpower utilization, which has been cited as a major reason for small plants to change its operation from onsite processing to only accepting pretreated liquid food waste (Nghiem et al., 2017).

Beyond conventional wastewater treatment at publicly owned treatment works (POTW), as described earlier, it is pertinent to note that there are other methods of managing liquid-phase food waste resources. In some cases, particularly in rural areas where access to a POTW is not readily available, many food-processing facilities discharge the waste to onsite lagoons, which provide on a smaller scale similar aerobic conversion conditions to those found at centralized facilities.

However, these plants often have high electricity demand to run large compressors required for waste aeration and fluid mixing. Also, if a food processor is located in close proximity to one or more farms, field spreading is another option for liquid-phase food waste (Chandrasekaran, 2012).

3.6 INCINERATION

Incineration systems provide controlled combustion of waste at high temperature, and were originally intended to reduce the volume of municipal solid waste (MSW). However, incineration is generally regarded as unfavorable from an environmental standpoint, as it has been reported that 1 kg of solid waste can produce 0.51 kg of CO_2 equivalent emissions when incinerated (Wang and Geng, 2015). In 2011, about 12% of the total MSW generated (17.9% of the MSW disposed) was converted through combustion with energy recovery. In 2011, 34.9 million tons of food waste were disposed of through landfilling or incineration (EPA, 2013).

The successful outcome of a waste incineration project depends on fairly accurate data on the projected waste quantities and its characteristics. Food waste is indeed an untapped resource with great potential for energy production, but utilization of food waste for energy conversion currently represents a challenge for several reasons. For example, to be suitable for incineration, waste must meet certain basic requirements. In particular, the energy content of the waste (lower calorific value, LCV) must be above a minimum level. The specific composition of the waste is also important, including its variable composition, high moisture content, and lower calorific value, which constitute an impediment to the development of robust, large-scale, and efficient industrial processes. Beyond the technical requirements for the success of waste incineration projects, there are other issues also that need to be addressed, as outlined below (World Bank, 1999).

Energy market: The main benefit of food waste incineration is the possibility of reusing the waste as fuel for energy production. Waste incineration may reduce methane gas emissions at landfills and displace fossil fuel, reducing the emission of greenhouse gases overall. Incineration plants consume and generate large amounts of energy and are therefore important players in the local energy market, especially in relatively small communities. The design and layout of incineration plants are based on continuous operation at 100% load. The waste energy can therefore be regarded as a supplement to other fossil-fuel-based energy sources that are operated at a load corresponding to the actual energy demand. There are several issues to be addressed here based on political and socioeconomic considerations. Depending on the price pattern, the price of the waste-generated energy will reflect this base load status.

Economics: Waste incineration involves high investment and operating and maintenance (O&M) costs. Hence, the resulting net treatment cost per ton of waste incinerated is rather high compared to the alternative (usually landfilling). The higher net treatment cost is a critical issue when considering implementing a waste incineration plant. Financing can be structured in terms of tipping fees, a general levy, public subsidies, and combinations thereof. At the same time, when applying waste incineration, the economic risk in case of project failure is high because of special requirements in terms of quantity and composition (for example, minimum net calorific value) and the need for a comprehensive waste management system and institutional setup in general.

Pollution: Incineration generates large volumes of flue gases. The pollutants and their concentration depend on the composition of the waste incinerated and the combustion conditions. However, these gases always carry ash, heavy metals, and a variety of organic and inorganic compounds. While the combustion of solid wastes is an established technique, its use as a viable waste management strategy is still not fully accepted, especially in the European Union. The reluctance of some countries to rely on waste incineration is related to toxic air emissions containing dioxins and heavy metals generated from the earlier equipment and technologies (Katami et al., 2004).

Deployment location: Incineration plants are generally located close to densely populated areas for economic reasons (e.g., sale of energy, minimizing waste transport distance, etc.). Thus, any negative environmental effects of such plants can influence a great number of people. The flue-gas cleaning residues are much more soluble than the original wastes and have to be disposed of in an appropriately designed and operated landfill to avoid negative impacts on ground or surface water. Special precautions need to be taken to minimize such risks. Occupational health risk and protective measures for the rest of the plant are similar to those required at large-scale energy facilities such as coal-fired power plants.

Few studies focusing on direct energy recovery from food waste by incineration are available in the literature. This may be because food waste alone seems ill-suited for incineration, due to its high moisture content and noncombustible components. A Korean study conducted by Lee et al. (2007) determined that the moisture content is around 23% for separated food waste and combustible content reduces from 32.9% to 15.7%. This lowers the lower heating value (LHV) from 3525 to 1748 kJ/kg, which in turn reduces the effectiveness of incineration plants. For this reason, in countries with limited or nonexistent landfill capacity, food waste is typically discarded into the general flow of MSW and converted into heat and energy by incineration. Kim et al. (2013) studied food waste disposal options in Korea with respect to global warming and energy recovery. The results of incineration analysis revealed that 37.7 kJ of heat was produced from drying and incineration of 1 g of food waste. The environmental credit was found to be 315 kg of CO_2-eq for the dryer-incineration system from 1 ton of food waste by electricity, thermal energy generated, and primary materials avoided. It was also concluded that the dryer-incineration system seemed to be the best option for food waste recycling in a metropolitan area in Korea. However, these positive results were derived from incineration of MSW combined with dried food wastes, and energy recovery through incineration of solely food wastes is not always feasible, generally due to the energy loss to evaporate the large water content in these organic wastes. In another study, Caton et al. (2010) investigated energy recovery from postconsumer food waste by direct combustion. They utilized the thermal losses from the combustion process, through the housing of the system or expelled through the exhaust, to dehydrate the influent food waste. The results showed that energy recovery from wasted food could facilitate cost savings by offsetting traditional fuel use and by reducing disposal costs.

In some situations, incineration may be an efficient way to reduce waste volume and demand for landfill space in regions where very limited land is available, such as Singapore (Khoo et al., 2010). In comparison with other treatments, incineration is a capital-intensive method with high maintenance costs. It also requires highly technical operations and costly instruments for controlling gas emission residues (World Bank Technical Report, 1999). A situation in which food waste is collected and incinerated separately is also not thought to be realistic, and it is more natural to assume that it is incinerated together with other solid wastes. A particularly challenging problem involves packaged food waste management. Whereas both the organic food fraction and plastic, paper or metallic packaging materials may have upcycling options if the "phases" of the system can be efficiently separated, such de-packaging operations are often either too expensive or simply not practical. For example, in the case of small plastic containers of sauces or syrups available at many fast-food restaurants, food-packaging separation is essentially impossible. Anecdotal information from food processers and retailers in our region suggests that in such cases, paying high transport costs to dispose of waste at centralized incineration facilities is the only available option.

3.7 LANDFILLING

The practice of landfilling food waste is generally considered to the least environmentally sustainable option, and thus occupies the bottom rung of the waste hierarchy (Fig. 3.1). Several published metadata studies (e.g., Bernstad and la Cour Jansen, 2012; Morris et al., 2011; Schott et al., 2016) have compiled the results of a large number of prior research efforts, and the consensus is that landfilling food waste has among the greatest relative impacts in terms of net greenhouse gas (GHG) emissions. However, it is important to note that landfilling also presents other risks that must be comprehended in any comparative analysis among possible food waste management strategies. For example, leaching of rainwater through the landfill structure creates a liquid-phase waste (leachate) that can be particularly hazardous, especially in aging landfills (Brennan et al., 2016; Clarke et al., 2015).

As the global population increases and becomes more urbanized, there are serious concerns with the expanding role of open waste dumps and landfills in handling concentrated food waste resources (Adhikari et al., 2006). Because of the widely recognized adverse impacts of food waste landfilling and the sheer magnitude of waste currently handled using this method in the United States (Table 3.1) and elsewhere, the practice has been declining in many developed economies for a number of years. This is especially true in densely populated regions of Asia such as Japan, Singapore, and Korea (Fig. 3.5) where open, unoccupied land is either very expensive or simply not available. In the rapidly expanding economies of China and India, the situation is quite different. For example, China's volume of municipal solid waste (MSW), including a large fraction of food scraps, has been expanding by 8%–10% annually since the 1980s, and the vast majority is disposed of in landfills (Cheng and Hu, 2010). Many of these landfills were not designed to minimize environmental impacts, including capturing and utilizing methane-rich landfill gases, which is common practice in the United States and Europe (Zhang et al., 2010). For India, Annepu (2012) reported a 50% increase in MSW from 2001 to 2011, and projected that "business as usual" will result in a nearly fourfold additional increase by 2041. This study also emphasized the role of landfill gas management in determining the overall environmental impact of a landfill facility. While so-called unsanitary landfills and open burning produce the highest levels of GHG emissions (accounting for over 90% of MSW management in Indian cities and towns; Sharholy et al., 2008), impacts can be significantly minimized by employing "modern" landfill designs that recover and either flare the gas or use it for electricity generation, heating, etc.

In Europe and many regions of North America, there is a strong trend toward reducing the amount of waste sent to landfills, but many food materials are still handled using this option. For example, in the 27 countries of the European Union, there was a 60% overall reduction from 1995 to 2015 in the per capita landfill rate, but over 40% of biowaste (defined as garden and park waste, food and kitchen waste from households, restaurants, caterers, retail businesses, and

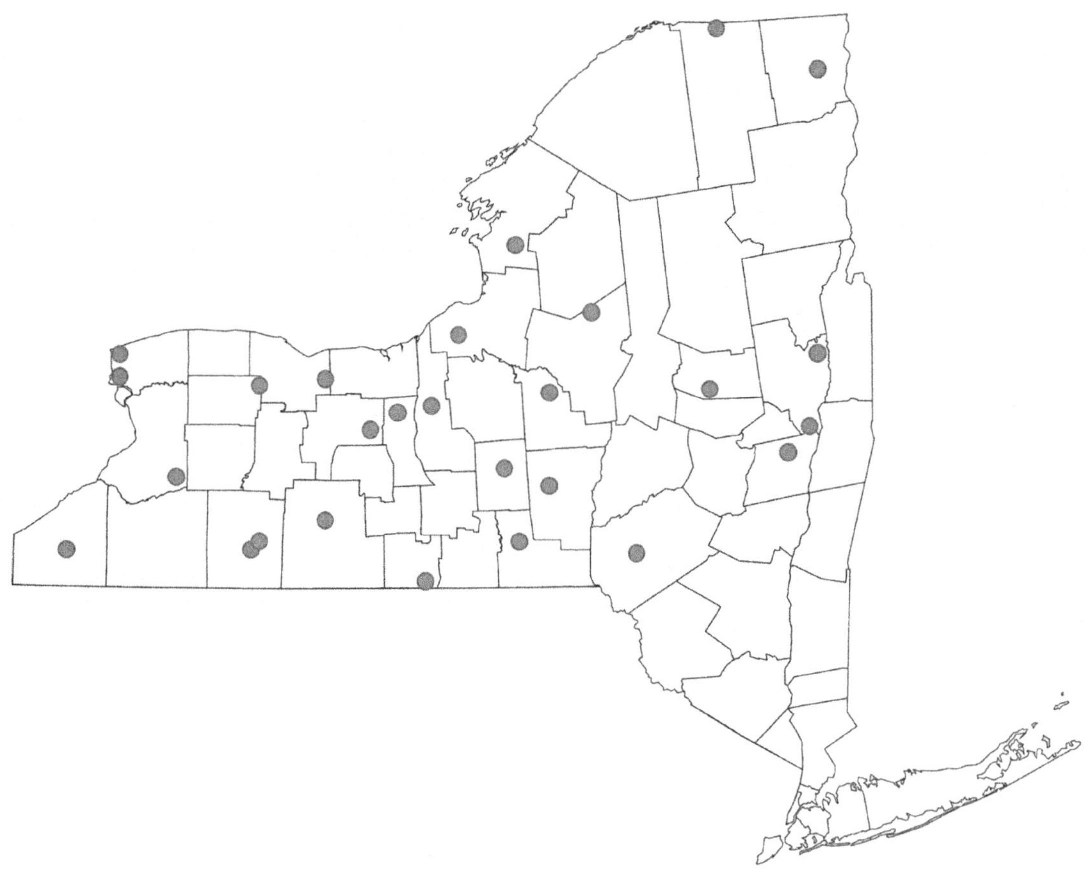

FIG. 3.9 Locations of 27 active landfills in New York State as of January 2016 (New York State Department of Environmental Conservation, http://www.dec.ny.gov/chemical/32501.html).

food-processing plants) was still sent to landfills (Righi et al., 2013). In North America., there is significant regional variation in the fraction of food waste sent to landfills (Levis et al., 2010). Even though the national average food waste landfilling fraction in the United States exceeds 90%, many regions are beginning to impose city- or state-level landfilling bans, especially on large operations such as food processors, consumer-facing businesses and institutions that exceed certain generation limits. Four New England states (Connecticut, Massachusetts, Rhode Island, Vermont) and California currently have some form of food waste disposal ban (Jones, 2017), and similar legislation is being considered in the state of New York after a ban went into effect in 2017 in New York City (NYC). As shown in Fig. 3.9, the ban in NYC was motivated in part by economics, because the lack of active landfill sites in close proximity to the population centers in the southern downstate counties and on Long Island. The need to transport large volumes of food waste over long distances became an economic burden and motivated serious consideration of other methods such as composting and anaerobic digestion (Chapter 4).

Despite the growing interest in Europe and North America to divert food waste to more sustainability management options, landfills in these two regions are generally well designed and managed, and in many cases productively utilize landfill gas for electricity generation, direct heating, natural gas pipeline injection (after appropriate clean-up), or synfuel production (Themelis and Ulloa, 2007). Therefore, as technologies such as waste-to-energy are considered as potential alternatives to current practice, it is critically important that all sources of environmental impact are comprehended and quantified to ensure that long-term environmental benefits are truly achieved as intended.

3.8 CONCLUSIONS

This chapter provided an overview of conventional food waste management strategies that are currently in use on a commercial scale, and must be fully understood before considering alternative methods that may offer sustainability benefits. After first reducing the total amount of food waste generated, the most favorable conventional method is feeding human beings that face food insecurity, a serious problem even in developed economies of the world. The stringent requirements on food quality and safety limit how much food waste material can be managed in this manner, although opportunities for increasing donations exist, especially among agricultural and restaurant/food service operations. Animal feed is also an environmentally favorable food waste pathway, and a significant proportion of food processing waste is already being managed in this manner, especially in Korea and Japan. Composting offers another potentially beneficial pathway for food waste management, but its application varies greatly among different regions, and in the United States most of the composting activity is concentrated in a few states. Also, to fully realize the sustainability benefits, composting facilities must be well designed and maintained to ensure conversion occurs only under aerobic condition (to generate CO_2 instead of CH_4), and markets must be developed to encourage use of compost in farming, thereby creating a closed-loop cycle of macronutrients.

The remaining conventional food waste management methods (wastewater treatment, incineration, and landfilling) are generally considered less favorable from an environmental standpoint, especially when gaseous, liquid, and solid effluents are not properly handled, as is the case with open waste dumps in many parts of the developing world. However, if modern environmental controls are used and the various system outputs (including methane-rich biogas, syngas, waste heat, etc.) are managed and utilized, the adverse environmental impacts of these methods can be minimized. In this context, it is critically important when assessing the relative merits of competing technologies that system boundaries and material/energy balances are compared on a common basis. It should not be automatically assumed that seemingly sustainable technologies are superior to the conventional methods outlined in this chapter. For the food waste-to-energy technologies discussed in Chapters 4–9, all inputs must be controlled and monitored, and all coproducts must be valorized to the greatest extent possible to displace products derived from fossil resources, including fuels for transportation and electricity generation, as well as synthetic fertilizer. This also implies that deployment of sustainable technology is not sufficient in its own right, there must also be a priori effort made to establish reliable and steady sources of food waste feedstock and markets for all coproducts.

REFERENCES

Adhikari, B.K., Barrington, S., Martinez, J., 2006. Predicted growth of world urban food waste and methane production. Waste Manag. Res. 24 (5), 421–433.

Annepu, R.K., 2012. Sustainable Solid Waste Management in India. (MS thesis), Columbia University, New York.

Becker Jr, A.M., Yu, K., Stadler, L.B., Smith, A.L., 2017. Co-management of domestic wastewater and food waste: a life cycle comparison of alternative food waste diversion strategies. Bioresour. Technol. 223, 131–140.

Bernstad, A., la Cour Jansen, J., 2012. Review of comparative LCAs of food waste management systems-current status and potential improvements. Waste Manag. 32 (12), 2439–2455.

Bilska, B., Wrzosek, M., Kołożyn-Krajewska, D., Krajewski, K., 2016. Risk of food losses and potential of food recovery for social purposes. Waste Manag. 52, 269–277.

Bouxin, A., 2012. Feed use of former foodstuffs converting food into food.Advisory Group for the Food Chain and Animal and Plant Health, Working Group on Food Losses and Food Waste.

Brennan, R.B., Healy, M.G., Morrison, L., Hynes, S., Norton, D., Clifford, E., 2016. Management of landfill leachate: the legacy of European Union directives. Waste Manag. 55, 355–363.

Buzby, J.C., Wells, H.F., Hyman, J., 2014. The estimated amount, value, and calories of postharvest food losses at the retail and consumer levels in the United States. EIB-121, U.S. Department of Agriculture, Economic Research Service, February.

Caton, P.A., Carr, M.A., Kim, S.S., Beautyman, M.J., 2010. Energy recovery from waste food by combustion or gasification with the potential for regenerative dehydration: a case study. Energy Convers. Manag. 51, 1157–1169. https://doi.org/10.1016/j.enconman.2009.12.025.

Cerda, A., Artola, A., Font, X., Barrena, R., Gea, T., Sánchez, A., 2017. Composting of food wastes: status and challenges. Bioresour. Technol. 248, 57–67.

Chandrasekaran, M. (Ed.), 2012. Valorization of Food Processing by-Products. CRC Press.

Cheng, H., Hu, Y., 2010. Municipal solid waste (MSW) as a renewable source of energy: current and future practices in China. Bioresour. Technol. 101 (11), 3816–3824.

Clarke, B.O., Anumol, T., Barlaz, M., Snyder, S.A., 2015. Investigating landfill leachate as a source of trace organic pollutants. Chemosphere 127, 269–275.

Copeland, C., Carter, N.T., 2017. Energy-Water Nexus: The Water Sector's Energy Use. (Congressional Research Service).

Diggelman, C., Ham, R.K., 2003. Household food waste to wastewater or to solid waste? That is the question. Waste Manag. Res. 21 (6), 501–514.

Ebner, J.H., 2016. Sustainable Management of Food Supply-Chain Resources in New York State. (Ph.D. dissertation), Rochester Instutute of Technology, Rochester, NY.

Edwards, J., Othman, M., Crossin, E., Burn, S., 2017. Anaerobic co-digestion of municipal food waste and sewage sludge: a comparative life cycle assessment in the context of a waste service provision. Bioresour. Technol. 223, 237–249.

EPA, U.S. Environmental Protection Agency, 2016. Waste to Biogas Mapping Tool. https://epamap21.epa.gov/biogas/index.html.

EPA, U.S. Environmental Protection Agency, 2014. Food Waste to Energy: How Six Water Resource Recovery Facilities are Boosting Biogas Production and the Bottom Line. http://www.casaweb.org/documents/food_waste_to_energy.pdf.

EPA, U.S. Environmental Protection Agency, 2013. Municipal Solid Waste: In the United States: 2011 Facts and Figures. NEPIS. https://nepis.epa.gov/Exe/ZyPURL.cgi?Dockey=P100GMT6.txt.

Federal Register, 2009. Swine Health Protection; Feeding of Processed Product to Swine. 6 C.F.R. § 166. https://www.aphis.usda.gov/animal_health/animal_dis_spec/swine/downloads/interim_rule_pro-products.pdf.

Feeding America, 2017. Map the Meal Gap 2017. Highlights of Findings for Overall and Child Food Insecurity. www.feedingamerica.org.

FWRA, Food Waste Reduction Alliance, 2016. Analysis of U.S. Food Waste Among Food Manufactures. A joint project by the Food Marketing Institute, the Grocery Manufacturers Association & the National Restaurant Association.

Ghanimeh, S., Khalil, C.A., Mosleh, C.B., Habchi, C., 2017. Optimized anaerobic-aerobic sequential system for the treatment of food waste and wastewater. Waste Manag. 71, 767–774.

Giuseppe, A., Mario, E., Cinzia, M., 2014. Economic benefits from food recovery at the retail stage: an application to Italian food chains. Waste Manag. 34 (7), 1306–1316.

Goldstein, N., 2017. The state of organics recycling in the U.S. BioCycle (October), 22–26.

Gunders, D., 2012. Wasted: how America is losing up to 40 percent of its food from farm to fork to landfill. NRDC Issue Paper. August 2012 IP:12-06-B.

Hall, M., 2016. Techno-Environmental Analysis of Generating Animal Feed From Wasted Food Products. (M.S. thesis)Rochester Institute of Technology, Rochester, NY.

Harris, B., Staples, C.R., 1991. Energy and milling by-product feedstuffs for dairy cattle. Rep.University of Florida.

IEA, 2016. International Energy Agency, Water Energy Nexus, Excerpt from the World Energy Outlook. International Energy Agency, Paris.

Jones, C.A., 2017. Organics disposal bans and processing infrastructure. BioCycle 58 (8), 54.

Katami, T., Yasuhara, A., Shibamoto, T., 2004. Formation of dioxins from incineration of foods found in domestic garbage. Environ. Sci. Technol. 38, 1062–1065. https://doi.org/10.1021/es030606y.

Kawashima, T., 2004. The use of food waste as a protein source for animal feed—current status and technological development in Japan.Protein Sources for the Animal Food Industry, Expert Consultation and Workshop.

Khoo, H.H., Lim, T.Z., Tan, R.B.H., 2010. Food waste conversion options in Singapore: environmental impacts based on an LCA perspective. Sci. Total Environ. 408 (6), 1367–1373.

Kim, M.H., Song, Y.E., Song, H.B., Kim, J.W., Hwang, S.J., 2011. Evaluation of food waste disposal options by LCC analysis from the perspective of global warming: Jungnang case, South Korea. Waste Manag. 31 (9), 2112–2120.

Kim, M.H., Kim, J.W., 2010. Comparison through a LCA evaluation analysis of food waste disposal options from the perspective of global warming and resource recovery. Sci. Total Environ. 408 (19), 3998–4006.

Kim, M.H., Song, H.B., Song, Y., Jeong, I.T., Kim, J.W., 2013. Evaluation of food waste disposal options in terms of global warming and energy recovery: Korea. Int. J. Energy Environ. Eng. 4 (2013), 1–12.

Lee, D., Sönmez, E., Gómez, M.I., Fan, X., 2017. Combining two wrongs to make two rights: mitigating food insecurity and food waste through gleaning operations. Food Policy 68, 40–52.

Lee, S.-H., Choi, K.-I., Osako, M., Dong, J.-I., 2007. Evaluation of environmental burdens caused by changes of food waste management systems in Seoul, Korea. Sci. Total Environ. 387, 42–53.

Leib, E.B., et al., 2016. Leftovers for Livestock: A Legal Guide for Using Excess Food as Animal Feed. Harvard Food Law and Policy Clinic and University of Arkansas School of Law.

Levis, J.W., Barlaz, M.A., Themelis, N.J., Ulloa, P., 2010. Assessment of the state of food waste treatment in the United States and Canada. Waste Manag. 30 (8), 1486–1494.

Li, Z., Lu, H., Ren, L., He, L., 2013. Experimental and modeling approaches for food waste composting: a review. Chemosphere 93 (7), 1247–1257.

Metcalf & Eddy, Inc, 2003. Wastewater Engineering: Treatment and Reuse. McGraw-Hill, Boston.

Moñino, P., Aguado, D., Barat, R., Jiménez, E., Giménez, J.B., Seco, A., Ferrer, J., 2017. A new strategy to maximize organic matter valorization in municipalities: combination of urban wastewater with kitchen food waste and its treatment with AnMBR technology. Waste Manage. 62, 274–289.

Monroe County, 2016. Scale of Charges. Monroe County, NY.

Morris, J., Matthews, S., Morawski, C., 2011. Review of LCAs on Organics Management Methods and Development of an Environmental Hierarchy. Sound Resource Management Group 1–68.

Muriana, C., 2015. Effectiveness of the food recovery at the retailing stage under shelf life uncertainty: an application to Italian food chains. Waste Manag. 41, 159–168.

Mu, D., Horowitz, N., Casey, M., Jones, K., 2017. Environmental and economic analysis of an in-vessel food waste composting system at Kean University in the US. Waste Manag. 59, 476–486.

Nghiem, L.D., Koch, K., Bolzonella, D., Drewes, J.E., 2017. Full scale co-digestion of wastewater sludge and food waste: bottlenecks and possibilities. Renew. Sust. Energ. Rev. 72, 354–362.

Platt, B., Goldstein, N., Coker, C., Brown, S., 2014. State of Composting in the U.S. What, Why, Where & How. Institute for Local Self-Reliance, Minneapolis, MN.

Pretel, R., Moñino, P., Robles, A., Ruano, M.V., Seco, A., Ferrer, J., 2016. Economic and environmental sustainability of an AnMBR treating urban wastewater and organic fraction of municipal solid waste. J. Environ. Manage. 179, 83–92.

ReFED, Rethink Food Waste through Economics and Data, 2016. A roadmap to reduce U.S. Food Waste by 20 percent. includes Technical Appendix, dated March.

Righi, S., Oliviero, L., Pedrini, M., Buscaroli, A., Della Casa, C., 2013. Life cycle assessment of management systems for sewage sludge and food waste: centralized and decentralized approaches. J. Clean. Prod. 44, 8–17.

Rosen, S., Meade, B., Fuglie, K., Rada, N., 2016. International food security assessment, 2014-2024. Econ. Res. 2014, 2024.

Russick, D., Urriola, P.E., 2017. Digestibility of food waste-derived animal feed. BioCycle (October), 30–32.

Saer, A., Lansing, S., Davitt, N.H., Graves, R.E., 2013. Life cycle assessment of a food waste composting system: environmental impact hotspots. J. Clean. Prod. 52, 234–244.

Salemdeeb, R., zu Ermgassen, E.K., Kim, M.H., Balmford, A., Al-Tabbaa, A., 2017. Environmental and health impacts of using food waste as animal feed: a comparative analysis of food waste management options. J. Clean. Prod. 140, 871–880.

San Martin, D., Ramos, S., Zufía, J., 2016. Valorisation of food waste to produce new raw materials for animal feed. Food Chem. 198, 68–74.

Scherhaufer, S., Schneider, F., 2011. Prevention, recycling and disposal of waste bread in Austria. In: Cossu, R. et al. (Eds.). Sardinia 2011—Thirteenth International Waste Management and Landfill Symposium (October 3–7, 2011, S. Margherita di Pula, Sardinia, Italy). Executive Summaries CISA Environmental Sanitary Engineering Centre. ISBN 978-88-6265-000-7.

Schott, A.B.S., Wenzel, H., la Cour Jansen, J., 2016. Identification of decisive factors for greenhouse gas emissions in comparative life cycle assessments of food waste management—an analytical review. J. Clean. Prod. 119, 13–24.

Sert, S., Garrone, P., Melacini, M., 2014. Keeping food alive: surplus food management. Eur. J. Sustain. Dev. 3 (4), 339.

Sharholy, M., Ahmad, K., Mahmood, G., Trivedi, R.C., 2008. Municipal solid waste management in Indian cities—a review. Waste Manag. 28 (2), 459–467.

Shin, S.G., Han, G., Lim, J., Lee, C., Hwang, S., 2010. A comprehensive microbial insight into two-stage anaerobic digestion of food waste-recycling wastewater. Water Res. 44.4838–49.

Sönmez, E., Lee, D., Gómez, M.I., Fan, X., 2015. Improving food bank gleaning operations: an application in New York state. Am. J. Agric. Econ. 98 (2), 549–563.

Sugiura, K., Yamatani, S., Watahara, M., Onodera, T., 2009. Ecofeed, animal feed produced from recycled food waste. Vet. Ital. 45 (3), 397–404.

Themelis, N.J., Ulloa, P.A., 2007. Methane generation in landfills. Renew. Energy 32 (7), 1243–1257.

Trabold, T.A., Ramchandra, R., Haselkorn, M.H., Williamson, A.A., 2011. Analysis of waste-to-energy opportunities in the New York State food processing industry. In: Proceedings of the ASME 5th International Conference on Energy Sustainability, August 7–11, Washington, D.C Paper ESFuelCell 2011-54334.

U.S. Department of Agriculture, U.S. Environmental Protection Agency, U.S. Department of Energy, 2014. Biogas Opportunities Roadmap Voluntary Actions to Reduce Methane Emissions and Increase Energy Independence. Washington, D.C.

USDA, U.S. Department of Agriculture, 1996. Food recovery and gleaning initiative. In: A Citizen's Guide to Food Recovery. Washington, D.C.

Vitiello, D., Grisso, J.A., Whiteside, K.L., Fischman, R., 2015. From commodity surplus to food justice: food banks and local agriculture in the United States. Agric. Hum. Values 32 (3), 419–430.

Wang, Z., Geng, L., 2015. Carbon emissions calculation from municipal solid waste and the influencing factors analysis in China. J. Clean. Prod. 104, 177–184.

Wang, J.Y., Xu, H.L., Tay, J.H., 2002. A hybrid two-phase system for anaerobic digestion of food waste. Water Sci. Technol. 45 (12), 159–165.

Wisner, R., 2010. Distillers grain price relationships and export developments. AgMRC Renewable Energy and Climate Change Newsletter. December 2010.

World Bank, 1999. Municipal Solid Waste Incineration: Technical Guidance Report. The International Bank for Reconstruction and Development. https://goo.gl/kABenL.

Xavier, K.M., Geethalekshmi, V., Senapati, S.R., Mathew, P.T., Joseph, A.C., Nair, K.R., 2017. Valorization of squid processing waste as animal feed ingredient by acid ensilaging process. Waste Biomass Valoriz. 8 (6), 2009–2015.

Zamorano-López, N., Moñino, P., Borrás, L., Aguado, D., Barat, R., Ferrer, J., Seco, A., 2018. Influence of food waste addition over microbial communities in an Anaerobic Membrane Bioreactor plant treating urban wastewater. J. Environ. Manage. 217, 788–796.

Zhang, D.Q., Tan, S.K., Gersberg, R.M., 2010. Municipal solid waste management in China: status, problems and challenges. J. Environ. Manag. 91 (8), 1623–1633.

Zhang, R., El-Mashad, H.M., Hartman, K., Wang, F., Liu, G., Choate, C., Gamble, P., 2007. Characterization of food waste as feedstock for anaerobic digestion. Bioresour. Technol. 98 (4), 929–935.

zu Ermgassen, E.K., Phalan, B., Green, R.E., Balmford, A., 2016. Reducing the land use of EU pork production: where there's swill, there's a way. Food Policy 58, 35–48.

FURTHER READING

Bazerghi, C., McKay, F.H., Dunn, M., 2016. The role of food banks in addressing food insecurity: a systematic review. J. Community Health 41 (4), 732–740.

Mousa, T.Y., Freeland-Graves, J.H., 2017. Organizations of food redistribution and rescue. Public Health 152, 117–122.

Chapter 4

Sustainable Waste-to-Energy Technologies: Anaerobic Digestion

Rodrigo A. Labatut*,† and Jennifer L. Pronto†
Department of Hydraulic and Environmental Engineering, Pontifical Catholic University of Chile, Santiago, Chile, †Bioprocess Analytics LLC, Gansevoort, NY, United States

4.1 INTRODUCTION

Anaerobic digestion (AD) is a biochemical treatment process that allows stabilizing a myriad of organic wastes, from complex lignocellulosic materials to easily degradable food wastes, while simultaneously producing renewable energy, recovering fibers and nutrients for soil amendment, and offsetting GHG emissions. As such, AD may be the most adequate and environmentally sustainable technology to tackle the challenge of treating the 1.3 billion tons of food waste produced worldwide every year (FAO, 2013). According to the Food and Agriculture Organization (FAO), this amount corresponds to one-third of all food produced for human consumption (FAO, 2013). FAO also estimates that in Europe and North America the per capita food waste at the consumer level is 95–115 kg/year (Gustavsson et al., 2011). For homes, businesses, and institutions, food waste has been historically disposed of along with other (nonorganic) municipal solid waste (MSW) in landfills or incinerators. In the United States, over 40% of the food produced ends up in a landfill without ever reaching a table (Gunders, 2012), whereas in China, nearly 56% (96 million tons) of the MSW is comprised of household kitchen food waste, which is entirely disposed of via landfilling or incineration (De Clercq et al., 2017). These practices have far-reaching implications regarding food security, efficient use of resources, air pollution, and climate change. Wholesome foods that can be used to feed the hungry, and key resources such as water and nutrients, are lost and cannot be recovered. In addition, food waste, being the primary and most biodegradable component of MSW, is the main source of methane, a greenhouse gas (GHG) with a global warming potential (GWP) 28–36 times greater than CO_2 on a 100-year time scale (USEPA, 2017). This makes landfills the third largest contributor of atmospheric methane in the United States, accounting for 18% of the total emissions of this gas in the country. These figures can be even higher in large developing countries, such as China, India, Turkey, or Sudan, where the fraction of food waste in MSW is over 50% (Iacovidou et al., 2012). On the other hand, the high moisture content of food waste, which averages between 80% and 90% of the total weight, renders incineration an energy intensive, highly pollutant, and costly option (Nghiem et al., 2017). Fortunately, over the past several decades, the negative impacts of landfilling and incineration have come to light, and globally, legislation has been enacted to divert organic waste streams from these historically conventional methods. The Landfill Directive is a piece of European legislation that requires member countries to reduce the volume of biodegradable municipal waste to 35% by July 2016 (Commission, 2005). Notably, Germany has banned all biodegradable waste from entering landfills since 2005, motivated by this legislation. The EU Landfill Directive is arguably the most significant driver for anaerobic digestion of food waste via codigestion (Nghiem et al., 2017). Following the European trend, several U.S. states have passed legislation to reduce the amount of organic waste deposited in landfills. Currently, there are five states that have instituted bans on landfilling organic waste, and each ban has different regulations on what waste types and sources are covered under the legislation (Leib et al., 2016).

Diverting food waste from landfilling not only benefits the environment, but it also provides an opportunity to implement sustainable waste management solutions that better fit a circular economy.[1] Here, anaerobic digestion is a central example. Through AD, food waste and any other type of organic matter can be converted into biogas, a gas mainly consisting of methane, carbon dioxide, and water vapor. Methane is the combustible portion of biogas, which is

1. *Circular economy*: materials that are at the end of their service life, or become residues/wastes, are turned into resources for other purposes, thus closing loops in industrial ecosystems while minimizing waste (Stahel, 2016).

Sustainable Food Waste-to-Energy Systems. https://doi.org/10.1016/B978-0-12-811157-4.00004-8

subsequently converted into a usable form of energy (e.g., electrical, thermal). Since food waste is mainly comprised of easily degradable lipids, proteins, and carbohydrates, it constitutes an ideal substrate for anaerobic digestion, as it can be quickly stabilized and converted into biogas (Jia et al., 2017). Over the last 20 years, AD has become the most accepted and widely used technology for the treatment of the organic fraction of MSW (De Baere and Mattheeuws, 2012). This is primarily because AD can effectively treat a variety of organic wastes, while simultaneously producing energy and recovering valuable products. In doing so, AD can also recover fibers and nutrients for soil amendment, offset GHG emissions, and displace fossil fuel-derived energy. This represents an environmentally and economically competitive advantage over landfilling, incineration, and other waste-to-energy (WTE) technologies. Germany, Spain, England, and Korea are all equipped with full-scale AD plants with a capacity of 2500 tons per year or larger (Li et al., 2017). As of 2014, there were 244 anaerobic digestion facilities for the treatment of the organic fraction of MSW in Europe, accounting for nearly 8 million tons of organics treated per year via anaerobic digestion (De Baere and Mattheeuws, 2012). In the United States, anaerobic digestion has been widely applied for the conversion of food waste and other readily biodegradable organic waste streams into methane, which is then used as a source for heat and electricity generation in combined heat and power (CHP) systems (Labatut et al., 2011). There are mainly three types of AD systems that include food waste in the United States: (1) stand-alone digestion, processing food wastes only; (2) farm-based codigestion, processing animal manure with food wastes; and (3) wastewater treatment plant (WWTP)-based codigestion systems, processing waste activated sludge (WAS) with food wastes. According to data obtained from the U.S. EPA website, as of July 2016 there were 41 stand-alone systems and 24 WWTP-based systems (USEPA, 2016). According to the USDA's AgStar database, as of 2016, of the 242 farm-based anaerobic codigestion (AcoD) systems operating in the United States, 67 included at least one type of food waste in their operation (AgSTAR, 2016). From these operations, 95% used the biomethane produced to generate some type of energy, with the balance flaring the gas to convert CH_4 to CO_2 before discharge to the environment (Fig. 4.1). These on-farm systems treat livestock manure with a variety of easily degradable food residues, including cheese and yogurt whey, and wastes from ice cream, potatoes, and onion processors (Gooch and Pronto, 2009).

4.1.1 Sources of Food Waste in AD Systems

Food waste generators include businesses and households, institutions, supermarkets, farms, and food processors. Types of food waste include fresh vegetables and fruits, meats, baked goods, and dairy products. Commercial food waste is mainly composed of retail food waste and food service waste. Food processing waste originates from food factories/producers, including (among many others) the beverage, dairy, meat, and canning industries. Supermarkets are a large source of retail food waste consisting of rotting produce, damaged packaged goods, or otherwise unmarketable products. Food service waste consists of scraps generated during food preparation as well as postconsumer plate waste and unserved food (Ebner et al., 2016). Retail food waste consists of anything from stale baked goods, damaged canned goods, or rotting bagged lettuce (Ebner et al., 2016).

The desirability of different food waste streams for use in both mono- (stand-alone) and codigestion systems depends on their consistency and characteristics. For example, postconsumer food waste is more challenging to manage since this

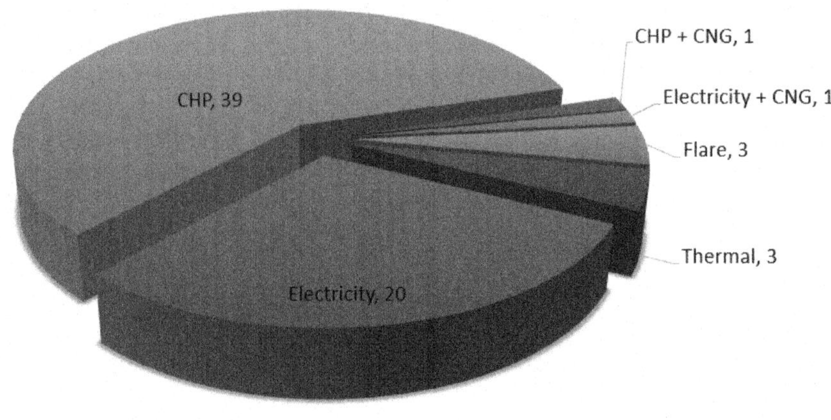

FIG. 4.1 Type of energy produced in the 67 AcoD systems receiving food waste currently operating in the United States. *CHP*, combined heat and power, *CNG*, compressed natural gas, *thermal*, hot water.

material often includes contamination and/or unwanted content (i.e., plastic utensils or aluminum foil wrapping). Retail food waste and preconsumer food preparation waste are more consistent and far less likely to contain unwanted or undigestable materials.

Synergy Biogas, LLC located in New York State, United States, is an anaerobic digester which codigests animal manure, specifically dairy cow manure, with up to 14 different food waste substrates simultaneously. Table 4.1 includes data from six operational anaerobic codigestion systems in New York State, and the associated food wastes that were actively used in each anaerobic codigestion system at the time of reporting.

Food waste in developing countries is similar in quantity to that of industrialized nations; however, the proportion of how it is distributed varies greatly. In developing countries, >40% of food losses occur postharvest and during processing, while in industrialized countries, >40% of food losses occur at the retail and consumer level (Fusions, 2016). In Europe, the

TABLE 4.1 Six Anaerobic Codigestion Systems and Associated Food Waste Sources Utilized

Anaerobic Digester	Food Waste Utilized	Source Industry	Composition
Synergy Biogas, LLC[a]	Acid whey	Dairy	Sugars, proteins
	Grease trap cleanings	Prepared foods	Diluted FOG
	Rendering solids	Prepared foods	Diluted FOG
	Processed foods	Prepared foods	Dense FOG
	Wine pulp	Alcohol	Sugars, alcohol
	Hog processing waste	Animal processing	Proteins, diluted FOG
	Cream waste and wash water from line cleaning	Prepared foods	Dense FOG
	Tomato paste	Prepared foods	Sugars, starches
	Grease trap cleanings	Prepared foods	Diluted FOG
	Yogurt waste/whey and wash water from line cleaning	Dairy	Sugars, proteins
	Dairy waste and wash water from line cleaning	Dairy	Sugars, proteins
	Alcohol-mix drink waste	Alcohol	Sugars, alcohol
	Vegetable waste	Restaurants/grocery stores	Fibers, sugars, starches
	Grease trap cleanings, soup waste	Restaurants/grocery stores	Diluted FOG
Ridgeline[b]	Ice cream wastewater	Restaurant/grocery stores	Sugars, proteins, diluted FOG
	Salad dressing waste	Prepared foods	Sugars, diluted FOG
	Hog processing waste	Food/animal processing	Proteins, diluted FOG
Noblehurst[b]	Hog processing waste	Food/animal processing	Proteins, diluted FOG
Patterson[b]	Milk slop, whey, permeate, and concentrate	By-products from cheese processing plant	Sugars, proteins
	Corn syrup waste	Food production	Sugars
	Onion waste	Agricultural, food production	Fibers, sugars, starches
Sunny Knoll[b]	Grease trap cleanings	Restaurant	Diluted FOG
	Corn syrup waste	Food production	Sugars
Emerling[b]	Used cooking oil	Restaurant	Dense FOG

[a]*Labatut and Gooch (2014).*
[b]*Gooch et al. (2011).*

main sources of the food waste come from households (43%) and food processing industries (39%) (Koch et al., 2016). In the United States, 60 million tons, or 31% of the food waste, originates at the retail and consumer levels, whereas >35 million tons of food waste are contributed by food processing industries (Gunders, 2012).

4.1.2 Characteristics of Food Waste Typically Used in AD Systems

Anaerobic digestion is a robust and flexible technology that allows stabilizing a wide variety of food wastes, from diluted, soluble dairy waste, to insoluble, concentrated fat, oil, and grease (FOG). Some of the physical and chemical characteristics of the most representative food wastes subjected to anaerobic digestion are shown in Table 4.2. The table shows the characteristics of individual food wastes as well as mixed food wastes obtained from retail and commercial locations

TABLE 4.2 Characterization of Representative Food Wastes Subjected to Anaerobic Digestion

	BOD (g/kg)	COD (g/kg)	TS (g/kg)	VS (g/kg)	BOD/ COD	VS/ TS	VS/ COD	Reference[a]
Individual food wastes								
Beet molasses		80.5	109.0	79.0				3
Cabbage, whole		90.9	78.6	72.0		0.9	0.8	1
Cheese whey	64.9	128.3	71.4	59.8	0.5	0.8	0.5	1
Cheese whey		68.8	1.3	0.9				4
Corn silage			217.3	200.7		1.0	0.9	1
Ice cream waste		266.8	113.8	109.1		1.0	0.4	1
Onion, peels		652.8	960.0	518.4				5
Onion, pulp		73.5	70.0	65.1				5
Pasta, meat	205.8	562.8	381.8	340.6	0.4	0.9	0.6	1
Pasta, plain	188.7	934.3	422.6	407.7	0.2	1.0	0.4	1
Potatoes, whole	53.5	261.8	177.4	163.5	0.2	0.9	0.6	1
Cola soda		121.5	93.6	88.7		1.0	0.7	1
Tomato paste		7.0						6
Vegetable oil, used		2880.0	991.0	988.8		1.0	0.3	1
Mixed commercial and plant-processing food waste								
Assorted canned goods (e.g., fish, soups, beans, fruits)		120.0	105.5	95.6		0.9	0.8	2
Assorted grains (e.g., rice, oat meal, bread crumbs)		1092.6	923.9	904.0		1.0	0.8	2
Assorted kitchen waste (e.g., melon rinds and seeds, assorted fresh fruits and vegetables, olives, coffee grounds)		183.6	142.9	142.9		1.0	0.8	2
Assorted postconsumer waste (e.g., pizza, French fries, cheese)		711.9	465.9	452.3		1.0	0.6	2
Cooked vegetable waste		498.0	415.0					7
Distillery wastewater		48.7						8
Fresh vegetable waste		108.0	108.0				1.0	7
Fruit and vegetable waste		87.3	76.6	71.5		0.9	0.8	2
Fruit and vegetable waste		126.0	90.4					9

TABLE 4.2 Characterization of Representative Food Wastes Subjected to Anaerobic Digestion—cont'd

	BOD (g/kg)	COD (g/kg)	TS (g/kg)	VS (g/kg)	BOD/COD	VS/TS	VS/COD	Reference
Fats, oil, and greases (FOG), suspended	30,7	600,1	267.2	229.7	0,1	0.9	2.6	1
Fats, oil, and greases (FOG), settled	28,8	290,0	128.4	112.6	0,1	0.9	2.6	1
Hog-processing waste (i.e., slaughterhouse waste)		2.6	2.5					10
Ice cream processing wastewater	2.5	5.2	3.9					11
Restaurant grease dewatered		439.0	972.0	972.0		1.0		12
Rotten and expired salad mixes		47.3	40.6	36.5		0.9	0.8	2
Spent coffee grounds and filter paper		365.1	293.0	290.9		1.0	0.8	2
Stale bagels, muffins and donuts		1096.7	915.6	888.9		1.0	0.8	2
Sweetened breakfast cereals		1085.7	926.7	880.0		0.9	0.8	2
Winery waste		15.4						13
Yogurt and frozen dairy deserts		396.2	309.1	302.4		1.0	0.8	2

[a](1) Labatut et al. (2011); (2) Ebner et al. (2016); (3) Jiménez et al. (2003); (4) Malaspina et al. (1995); (5) Lubberding et al. (1988); (6) Gohil and Nakhla (2006); (7) Carucci et al. (2005); (8) Krzywonos et al. (2009); (9) Bouallagui et al. (2005); (10) Massé and Masse (2000); (11) Carvalho et al. (2013); (12) Long et al. (2012); (13) Shepherd et al. (2001).

(source-separated) and from food processing industries. Also, FOG is produced from grease traps located in restaurants, or as a result of separation processes, such as dissolved air flotation (DAF) systems. It is apparent that the variation in the characteristics among different food wastes is significant, and even for the same type of food waste there can be significant variation. This is the case of cheese whey, which depends on the source and type of cheese produced, and will result in a product with a higher or lower solid and chemical oxygen demand (COD) content (Table 4.2). This also occurs for the products from DAF systems, which produce a settled, and more diluted FOG, and a suspended, and denser FOG.

The physicochemical characteristics of the influent waste stream will directly influence the performance of the anaerobic digester. Assuming steady-state operating conditions (e.g., solids retention time, temperature), the methane production of an anaerobic digester is primarily determined by the characteristics of the substrate, and particularly, the chemical strength and biodegradability of the material. The chemical strength can be defined in terms of the chemical oxygen demand (COD), a parameter that can be determined via standard analytical methods or calculated from the chemical composition of the substrate. The COD defines the energy density of the substrate, and thus, its theoretical methane yield. However, only a fraction of the influent substrate is comprised of solid material. Most substrates, especially food wastes, have a high moisture content and consists primarily of water. This is evidenced by the total solids (TS) content (Table 4.2). Likewise, not all the substrate TS is organic, a portion consists of ashes and it is inorganic. This is evidenced by the volatile solids (VS) content (Table 4.2). Finally, because of the limitations in enzymatic hydrolysis, only a fraction of the substrate VS is ultimately stabilized, and therefore converted to biomethane.

4.2 ANAEROBIC DIGESTION PROCESS

Anaerobic digestion is a biologically mediated process that relies on a consortium of microorganisms, namely, the microbiota, to stabilize the organic matter under oxygen-free conditions. During anaerobic digestion, organic substrates undergo a series of biochemical transformations, where methane and carbon dioxide are the main gas products (Fig. 4.2).

In the first step, biopolymers are hydrolyzed to soluble products via extracellular enzymes that allow their passage across the cell membrane. Once in the cell, sugars, amino acids, and long-chain fatty acids are fermented (or oxidized) to volatile fatty acids, alcohols, carbon dioxide, molecular hydrogen, as well as nitrogenous and sulfurous compounds. With the exception of acetate, volatile fatty acids are converted to additional acetate, hydrogen, and carbon dioxide. In the final step, that is, methanogenesis, molecular hydrogen (H_2), CO_2, and acetate (CH_3COO^-), are converted to CH_4 (Fig. 4.2).

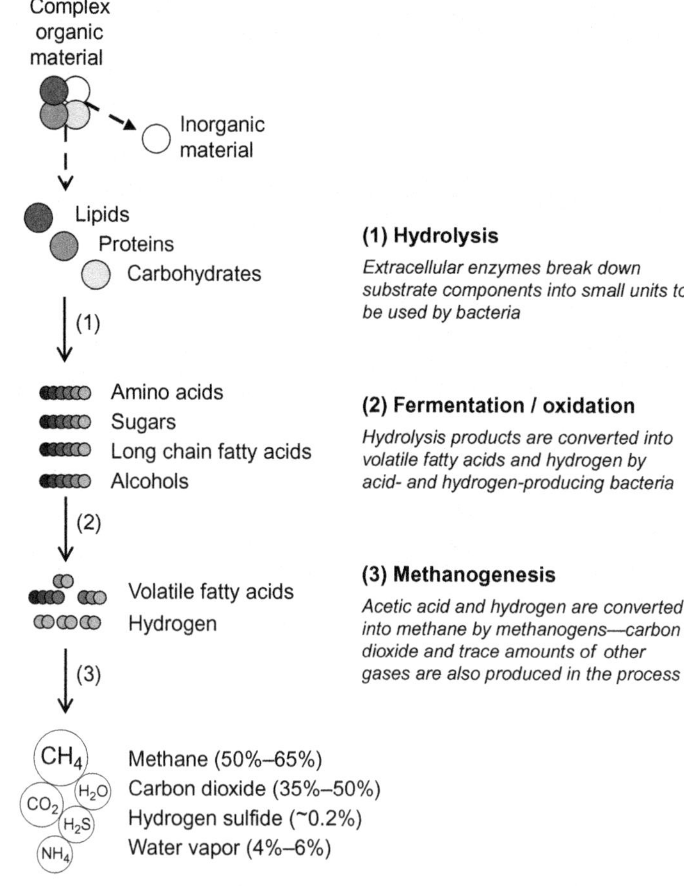

FIG. 4.2 Major bioconversion pathways involved in the anaerobic digestion of complex substrates. *LCFA*, long-chain fatty acids, *Ac*, acetate, *Pr*, propionate, *Bu*, butyrate, *Va*, valerate. *(Adapted from McCarty, P.L., Smith, D.P., 1986. Anaerobic waste-water treatment. Environ. Sci. Technol. 20(12), 1200–1206; Pavlostathis, S.G., Giraldo-Gomez, E., 1991. Kinetics of anaerobic treatment—A critical review. Crit. Rev. Environ. Control 21(5–6), 411–490; and Batstone, D.J., Keller, J., Angelidaki, I., Kalyuzhnyi, S.V., Pavlostathis, S.G., Rozzi, A., Sanders, W.T.M., Siegrist, H., Vavilin, V.A., 2002. The IWA anaerobic digestion model No. 1 (ADM1). Water Sci. Technol. 45(10), 65–73.)*

Because of their central role in methanogenesis, propionate, acetate, and hydrogen are probably the most important intermediate products of anaerobic digestion, and therefore key indicators to monitor for process stability of the system. About 64% of the methane produced during anaerobic digestion comes from acetate, while the remaining 36% comes from hydrogen (Batstone et al., 2002). Propionate is an important precursor of acetate and hydrogen; approximately 30% of the electron flow directly related to methane production goes through propionate (Jeris and McCarty, 1965; McCarty and Smith, 1986).

4.2.1 Process Parameters and Operating Conditions

The anaerobic digestion of complex substrates, such as food wastes, is a highly dynamic, multistep process, where physicochemical and biochemical reactions take place in a sequential and parallel way resulting from continuous interactions between substrates, the reactor's microbiome, and the environment. Therefore to achieve a stable and highly performing anaerobic digestion process, key operating and process parameters must be closely controlled and monitored.

The following section discusses the effects of selected parameters on the stability and performance of an anaerobic digester. Also, characteristic parameter values, as reported in the literature for a variety of AD systems including food waste, are presented in Table 4.3.

4.2.1.1 OLR/HRT

The organic loading rate (OLR) is a measure of the quantity of influent substrate entering the digester per unit of time (g/L-d). For a particular influent substrate concentration and digester hydraulic retention time (HRT), the OLR can be calculated from the following equation:

TABLE 4.3 Key Design and Operating Parameters of Bench-, Pilot-, and Commercial-Scale Anaerobic Digesters Including Food Waste

System Type	Feedstock(s)	Reactor Type	Reactor Scale—Volume	Pretreatment	OLR	HRT	T	VFA	TAN	LCFA	Methane in Biogas	Methane Production (@STP)	Reference
Farm-based AcoD	Dining hall food waste codigested with cow manure at ratio 9:91 (VS/VS)	CSTR	Bench—4.5 L	Ground in blender	2.2 g VS/L-d	25 d	37°C	160 mg/L as HAc	1310 mg/L as N	74 mg/L	59%	231 L/kg VS	1
Farm-based AcoD	Dining hall food waste codigested with cow manure at ratio 17:83 (VS/VS)	CSTR	Bench—4.5 L	Ground in blender	2.42 g VS/L-d	25 d	37°C	170 mg/L as HAc	1480 mg/L as N	78 mg/L	58%	245 L/kg VS	1
Farm-based AcoD	Dining hall food waste codigested with cow manure at ratio 25:75 (VS/VS)	CSTR	Bench—4.5 L	Ground in blender	2.66 g VS/L-d	25 d	37°C	170 mg/L as HAc	1580 mg/L as N	77 mg/L	56%	255 L/kg VS	1
Farm-based AcoD	Dining hall food waste codigested with cow manure at ratio 32:68 (VS/VS)	CSTR	Bench—4.5 L	Ground in blender	2.93 g VS/L-d	25 d	37°C	170 mg/L as HAc	1610 mg/L as N	132 mg/L	57%	259 L/kg VS	1
Farm-based AcoD	Dining hall food waste codigested with cow manure at ratio 38:62 (VS/VS)	CSTR	Bench—4.5 L	Ground in blender	3.22 g VS/L-d	25 d	37°C	190 mg/L as HAc	1620 mg/L as N	174 mg/L	58%	282 L/kg VS	1
Farm-based AcoD	Dining hall food waste codigested with cow manure at ratio 44:56 (VS/VS)	CSTR	Bench—4.5 L	Ground in blender	3.54 g VS/L-d	25 d	37°C	200 mg/L as HAc	1820 mg/L as N	427 mg/L	58%	289 L/kg VS	1
Farm-based AcoD	Dining hall food waste codigested with cow manure at ratio 49:51 (VS/VS)	CSTR	Bench—4.5 L	Ground in blender	3.9 g VS/L-d	25 d	37°C	250 mg/L as HAc	1930 mg/L as N	700 mg/L	58%	297 L/kg VS	1
Farm-based AcoD	Codigestion of manure, slaughterhouse waste, and carcasses	CSTR	Commercial—3700 m^3	None	2.5 g VS/L-d	25 d	NA	NA	NA	NA	NA	NA	2
Farm-based AcoD	Food processing, restaurant, retail, slaughterhouse waste (details on Table 4.1)	CSTR	Commercial—8330 m^3	Grounding and 60-min pasteurization	2.1 g VS/L-d	29 ± 7 d	41 ± 2°C	878 mg/L as HAc	1550 mg/L as N	NA	60%	19.7 L/L reactor-d	3
Stand-alone AD	Dining hall food waste	Batch Single stage	Bench—1.6 L	Short-term hydrothermal pretreatment	NA	159 h	35°C	NA	NA	NA	NA	1760 mL	4
Stand-alone AD	Dining hall food waste	Batch Single stage	Bench—1.6 L	None	NA	209 h	35°C	NA	NA	NA	NA	1658 mL	4

Continued

TABLE 4.3 Key Design and Operating Parameters of Bench-, Pilot-, and Commercial-Scale Anaerobic Digesters Including Food Waste—cont'd

System Type	Feedstock(s)	Reactor Type	Reactor Scale—Volume	Pretreatment	OLR	HRT	T	VFA	TAN	LCFA	Methane in Biogas	Methane Production (@STP)	Reference
Stand-alone AD	Dining hall food waste	Batch Two-stage	Bench—1.6 L	Short-term hydrothermal pretreatment	NA	187 h	35°C	NA	NA	NA	77%	3519 mL	4
Stand-alone AD	Dining hall food waste	Batch Two-stage	Bench—1.6 L	None	NA	208 h	35°C	NA	NA	NA	NA	2638 mL	4
Stand-alone AD	Source-separated food waste collected from restaurants	CSTR	Bench—10 L	Ground to 2 mm	2.16 g VS/L-d	100 d	38°C	160 mg/L as HAc	NA	NA	41%	70 L/kg VS	5
Stand-alone AD	Source-separated food waste collected from restaurants	CSTR	Bench—10 L	Ground to 2 mm	3.58 g VS/L-d	60 d	38°C	207 mg/L as HAc	NA	NA	61%	180 L/kg VS	5
Stand-alone AD	Source-separated food waste collected from restaurants	CSTR	Bench—10 L	Ground to 2 mm	7.18 g VS/L-d	30 d	38°C	1849 mg/L as HAc	NA	NA	58%	290 L/kg VS	5
Stand-alone AD	Source-separated food waste collected from restaurants	CSTR	Bench—10 L	Ground to 2 mm	8.62 g VS/L-d	25 d	38°C	5447 mg/L as HAc	NA	NA	67%	430 L/kg VS	5
Stand-alone AD	Source-separated food waste collected from restaurants	CSTR	Bench—10 L	Ground to 2 mm	2.16 g VS/L-d	100 d	55°C	148 mg/L as HAc	NA	NA	42%	70 L/kg VS	5
Stand-alone AD	Source-separated food waste collected from restaurants	CSTR	Bench—10 L	Ground to 2 mm	3.58 g VS/L-d	60 d	55°C	247 mg/L as HAc	NA	NA	55%	140 L/kg VS	5
Stand-alone AD	Source-separated food waste collected from restaurants	CSTR	Bench—10 L	Ground to 2 mm	7.18 g VS/L-d	30 d	55°C	1775 mg/L as HAc	NA	NA	53%	260 L/kg VS	5
Stand-alone AD	Source-separated food waste collected from restaurants	CSTR	Bench—10 L	Ground to 2 mm	8.62 g VS/L-d	25 d	55°C	6288 mg/L as HAc	NA	NA	62%	470 L/kg VS	5
Stand-alone AD	Solid slaughterhouse waste	CSTR	Bench—2 L	None	0.8 g VS/L-d	50 d	35°C	NA	NA	NA	NA	520–550 L/kg VS	2
Stand-alone AD	Organic fraction MSW: fresh vegetable waste, cooked food waste, egg shells, and coffee	Batch	Bench—2 L	None	NA	15 d	35°C	Initial: 581 mg/L as HAc Final: 192 mg/L as HAc	Initial: 3348 mg/L as N Final: 5134 mg/L as N	NA	76.5%	165 L/kg VS	6
Stand-alone AD	Organic fraction MSW: fresh vegetable waste, cooked food waste, egg shells, and coffee	Batch	Bench—2 L	None	NA	18 d	35°C	Initial: 640 mg/L as HAc Final: 224 mg/L as HAc	Initial: 3287 mg/L as N Final: 4480 mg/L as N	NA	80.8%	351 L/kg VS	6

Stand-alone AD	Organic fraction MSW: fresh vegetable waste, cooked food waste, egg shells, and coffee	Batch	Bench—2 L	None	NA	17 d	35°C	Initial: 269 mg/L as HAc Final: 184 mg/L as HAc	Initial: 3527 mg/L as N Final: 4244 mg/L as N	NA	79.7%	397 L/kg VS	6
Stand-alone AD	Organic fraction MSW: fresh vegetable waste, cooked food waste, egg shells, and coffee	Batch	Bench—2 L	None	NA	12 d	35°C	Initial: 280 mg/L as HAc Final: 168 mg/L as HAc	Initial: 3226 mg/L as N Final: 4619 mg/L as N	NA	78.7%	276 L/kg VS	6
Stand-alone AD	Dining hall food waste: grains, vegetables and meats	UASB	Bench—41 L	Hydrolysis acidification	15.8 g COD/L-d	7.9 h	37°C	180 mg/L as COD	NA	NA	77–81%.	226 L/d (maximum)	7
Stand-alone AD	Sheep tallow	CSTR-batch	Bench—CSTR: 12 L/ batch: 12 L	None	5–20 g/L tallow (2.84 g COD/g)	NA	35–55°C	NA	NA	NA	NA	NA	2
Stand-alone AD	Cattle blood and rumen paunch contents	Two-stage HFR—CSTR	Bench—HFR: 4 L/CSTR: 1 L	None	2.6 g TS/L-d	Solid 2–30 d; liquid 2–10 d	35°C	NA	NA	NA	NA	270 L/kg TS	2
Stand-alone AD	Poultry mortalities	Two-stage LB—UASB	Bench—LB: 10 L/UASB: 3 L	None	(1) 2279 g (2) <2 g COD/L-d	NA	35–55°C	NA	NA	NA	NA	201 L/kg wet weight	2
Stand-alone AD	Poultry mortalities	Two-stage LB—UASB	Bench—NA	None	NA	NA	NA	NA	NA	NA	NA	254 L/kg wet weight	2
Stand-alone AD	Cattle and lamb paunch contents, blood, and process wastewaters	CSTR	Commercial—105 m^3	None	0.4 g COD/m^3-d	43 d	NA	NA	NA	NA	NA	180 L/kg COD	2
Stand-alone AD	Domestic food waste mixed with small amounts of commercial food waste and municipal green waste	CSTR	Pilot—900 m^3	None	2.5 kg VS/m3 day	80 d	42°C	NA	NA	NA	63%	402 L/kg VS	8
WWTP-based AcoD	Hog stomach contents codigested with WAS	CSTR	Pilot—2 m^3	None	2.9 g TS/L-d	17 d	37°C	NA	NA	NA	NA	230 L/kg TS	2
WWTP-based AcoD	Floating tailings codigested with WAS	CSTR	Pilot—2 m^3	None	1.5 g TS/L-d	15 d	37°C	NA	NA	NA	NA	320 L/kg TS	2

OLR, organic loading rate; *HRT*, hydraulic residence time; *T*, temperature; *VFA*, total volatile fatty acids; *TAN*, total ammonia nitrogen; *LCFA*, total long-chain fatty acids.
References: (1) Usack and Angenent (2015); (2) Salminen and Rintala (2002); (3) Labatut and Gooch (2014); (4) Jia et al. (2017); (5) Nguyen et al. (2017); (6) Pavi et al. (2017); (7) Shin et al. (2001); (8) Banks et al. (2011).

$$\mathrm{OLR} = \frac{S_0}{\mathrm{HRT}} = S_0 \cdot \frac{Q}{\forall} \tag{4.1}$$

where OLR = organic loading rate, VS or COD basis (g/L-d); S_0 = influent substrate concentration, VS or COD basis (g/L); Q = flow rate (L/d); HRT = hydraulic retention time (d); $\forall$ = anaerobic digester volume (L).

As shown, the OLR can be increased not only by increasing the influent flow rate (Q), but also by increasing S_0. Provided there is no inhibition due to substrate overloading, as the OLR increases, the biogas production increases. In general, the optimal OLR to be used in a digester will depend primarily on the chemical characteristics of the food waste; therefore long-term, bench-scale studies are usually needed to determine the optimal OLR for a particular condition. For example, a bench-scale, long-term anaerobic codigestion (AcoD) study evaluating the codigestion of cow manure and dining hall food waste in a continuous stirred tank reactor (CSTR) showed no indication of inhibition or instability when digested up to a total OLR of 3.9 g VS/L-d (Usack and Angenent, 2015). Also, a commercial-scale, CSTR anaerobic digester codigesting 96% household waste mixed with commercial food waste at an average OLR of 2.5 g VS/L-d (80-d HRT) showed no inhibition and stable biogas production during a 426-d performance monitoring period (Banks et al., 2011) (Table 4.3). Likewise, a bench-scale study investigating dry anaerobic digestion of source-separated food waste from restaurants at mesophilic and thermophilic temperatures showed the maximum performance at an OLR of 8.62 g VS/L-d (25-d HRT), provided that a (long) period of adaptation for the microbiota was allowed (Table 4.3) (Nguyen et al., 2017).

4.2.1.2 Nutrients and C/N Ratio

The performance of AD is significantly affected by the carbon-to-nitrogen (C/N) ratio of the influent substrate (Zhang et al., 2014). The C/N ratio determines the proportion between substrate and nutrients; the latter, required for microbial synthesis and to provide alkalinity, via ammonia metabolism. In general, it is thought that a C/N ratio between 20 and 30 is optimal for the AD process (Puyuelo et al., 2011). However, performance and stability of the AD process is rather dependent on the characteristics of the food wastes being digested and the type and amount of microbial biomass (inoculum) in relation to the influent substrate. For example, in the study of Pavi et al. (2017), who performed codigestion of the organic fraction of MSW with fruit and vegetable waste (FVW), the highest biogas production was observed at a C/N of 35. Also, as discussed later, trace elements such as Ni and Co are essential for methanogens and therefore are considered vital for adequate performance of the AD process (Agler et al., 2008). If these elements are not supplied by the substrate or cosubstrates (in the case of codigestion), they need to be supplemented to the anaerobic digester to avoid eventual failure of the digestion process.

4.2.1.3 Volatile Fatty Acids

Volatile fatty acids (VFAs) are the intermediary products of the anaerobic digestion process and are mainly produced by acidogenic and acetogenic bacteria. The accumulation of VFAs in anaerobic digester systems can stem from many different causes, such as changes in temperature, high OLRs, low inoculum-to-substrate ratio, or the accumulation of toxic compounds (Nguyen et al., 2017). The accumulation of VFAs in both mesophilic and thermophilic anaerobic digester systems can result in an imbalance in the fermentation process, which could ultimately lead to digester failure (Labatut et al., 2014). Thus as a process stability indicator, the concentration of VFAs is one of the most sensitive parameters to monitor. VFAs encompass a group of six acids: acetic, propionic, butyric, valeric, caproic, and enanthic, of which acetic acid is predominant. It is well known that within the digestion process, the fermentation of hydrolyzed organics to VFAs may result in accumulation and an associated drop in pH, if the acids are not metabolized by methanogens (Schievano et al., 2010). Among all six VFAs, propionate is known to have the lowest tolerance level for anaerobic bacteria and tends to linger in AD systems of complex substrates (Labatut et al., 2014; Speece et al., 2006). When an AD system is overloaded, propionate tends to accumulate and persist in the reactor, which makes its removal difficult during system recovery (Shin et al., 2001).

In a correctly designed and well operated digester, the concentration of total VFAs is typically below 500 mg/L as acetic acid. However, if the reactor is undersized for the actual organic loading rate, this concentration can be higher. At VFA concentrations over 1500–2000 mg/L, biogas production may be limited by inhibition. Yet, rather than a specific concentration, it is a sudden and steady increase of VFAs in the effluent that can be a sign of digester instability (Labatut and Gooch, 2012). Thus for high organic load systems (including many of those associated with codigestion of high strength substrates) it is essential to monitor VFAs periodically in order to detect problems in a timely manner, and to make the necessary operational changes prior to potential digester failure.

4.2.1.4 Alkalinity

The buffering capacity of an anaerobic digester is determined by the amount of alkalinity present in the system. The bicarbonate ion (HCO_3^-) is the main source of buffering capacity to maintain the system's pH in the range of 6.5–7.6. The concentration of HCO_3^- in solution is related to the percent of carbon dioxide in the gas phase. In a digester with an optimal pH of 7.4 and CO_2 concentration of 35%, the bicarbonate alkalinity is about 5500 mg/L as $CaCO_3$. Such alkalinity usually provides enough buffering capacity to withstand moderate shock loads of VFAs. Here is where anaerobic codigestion can be an essential strategy to efficiently convert high-strength, easily degradable food wastes into energy. Substrates with high alkalinity (e.g., cow manure) are included in the mixture to increase the buffering capacity and pH of the influent mixture, thereby maximizing performance and process stability (see Section 4.3).

4.2.1.5 Long-Chain Fatty Acids

Long-chain fatty acids (LCFAs) are fatty acids with aliphatic tails consisting of 13–21 carbon atoms, resulting from the hydrolysis of neutral lipids. Three of the most abundant LCFAs in anaerobic digesters are palmitate, stereate, and oleate. Lipid-based, or fats, oil, and grease (FOG) wastes, are typically added as cosubstrates in full-scale manure-based AcoD systems; mainly sourced from olive oil production, and food- and fish-processing industries. Lipid-rich wastes can be highly inhibitory, particularly of the β-oxidation and methanogenesis steps (Hanaki et al., 1981; Neves et al., 2009), due to the accumulation of LCFAs. As with ammonia, it has also been shown that the inhibitory effect of LCFAs is more pronounced at thermophilic temperatures (Hwu and Lettinga, 1997). LCFA accumulation occurs when molecular hydrogen (and/or acetate), a major product of β-oxidation, accumulates to thermodynamically limiting levels that prevent LCFAs (and propionate) to be oxidized. LCFAs can be further converted to hydrogen and acetate by acetogenic bacteria through a β-oxidation process, and finally to methane by methanogenic archaea (Zhang et al., 2014). Theoretically, the methane potential of lipids is 1014 L/kg-VS, a value clearly higher than that of carbohydrates (e.g., 370 L/kg VS for glucose) (Chen et al., 2014).

4.2.1.6 Ammonia

Ammonia is released upon hydrolysis of proteinaceous substrates. Food wastes with a high protein content are likely to produce problems in anaerobic digesters due to their nitrogen content. With high nitrogen concentrations, elevated ammonia concentrations are likely to be produced in a digester. The two principal forms of inorganic ammonia-nitrogen in an aqueous solution are the ammonium ion (NH_4^+) and free ammonia (NH_3). NH_3 has been suggested to be the main cause of inhibition since it is freely membrane permeable (Chen et al., 2008). The distribution of the two species and their relative toxicity is pH dependent, with the more toxic form dominating at higher pH. Among the anaerobic microorganisms, the methanogens are likely the most affected by ammonia inhibition (Chen et al., 2008). However, there is inconsistency regarding the concentration at which ammonia becomes inhibitory to methanogenesis, and this is reflected in the various threshold values that can be found in the literature (Banks et al., 2011). Studies on the AD of the organic fraction of MSW revealed that inhibition occurs at total ammonia-nitrogen (TAN) concentrations of 1200 mg/L and above (Kayhanian, 1999), while Angelidaki et al. (2005) in a study of 18 full-scale biogas plants in Denmark codigesting manure and organic waste, reported a decrease in digester efficiency when TAN concentrations were higher than 4000 mg NH_3-N/L. Others have reported that ammonia inhibition occurs at free ammonia concentrations higher than 650 mg/L NH_3-N (Hartmann and Ahring, 2005).

4.2.1.7 pH

Maintenance of the system pH in the proper range is required for efficient anaerobic digestion. The generally accepted values are in the neutral range, between 6.5 and 7.6. The anaerobic digestion of complex organic substrates, such as food wastes, requires the joint work of several groups of microorganisms. The methanogens are the most sensitive to pH variations. Changes in digester operating conditions or introduction of toxic substances may result in process imbalance and accumulation of VFAs. Unless the system contains enough buffer capacity (i.e., alkalinity), the pH will drop below optimal levels and digester performance will decline. Depending on the pH magnitude and the duration of the drop, biogas production will decrease to a point where it may completely cease. On the contrary, in a well operated system digesting (protein-rich) food waste, a slight increase of the digester's effluent pH is expected, because organisms produce alkalinity as they consume nitrogen-rich organic matter.

4.3 PERFORMANCE OF ANAEROBIC DIGESTION SYSTEMS

The performance of an anaerobic digestion system is primarily evaluated on the basis of the efficiency of substrate stabilization (or treatment), and the methane production (or energy output). The substrate stabilization efficiency of an anaerobic digester operating at steady-state conditions can be expressed as:

$$E = \frac{S_0 - S}{S_0} \times 100 \tag{4.2}$$

where E = substrate stabilization (treatment) efficiency (%) and S = effluent substrate concentration, VS or COD basis (g/L).

The substrate stabilization efficiency is simply the extent to which the treatment of organic matter has been accomplished. Such extent is directly related to the methane production of an anaerobic digester, and consequently to the energy output of the system as a whole. In other words, the higher the methane production, the more usable energy that can be subsequently generated. In a well-designed and operated anaerobic digester, both biogas flow rate and methane concentration should be fairly stable over time. An anaerobic digester containing food waste should have at least 60% (v/v) methane concentration, with the remaining volume consisting primarily of carbon dioxide and water vapor. Thus the energy content of biogas prior to utilization should be ca. 20 MJ/m^3 after water vapor has been removed.[2]

At steady-state conditions, the methane production is primarily dependent on the chemical strength and biodegradability of the food waste being digested (see Section 4.1.2); however, the actual methane produced will be ultimately determined by the extent of organic matter stabilized (i.e., destroyed) during the anaerobic digestion process, as shown by Eq. (4.2). Theoretically, 350 L of methane are to be produced per kg of COD stabilized at standard conditions, that is, 0°C, 1 atm (McCarty, 1972). This considers that all the COD stabilized through AD is allocated to the production of energy (as opposed to the synthesis of new cells).

4.3.1 Methane Production Potential

In order to anticipate the methane production of specific substrates, and thus the energy output in commercial anaerobic digesters, lab-scale anaerobic digestion tests can be conducted. The most widely used and accepted test is the biochemical methane potential (BMP) assay, first described by Owen et al. (1979). Since then, numerous BMP studies on a variety of food wastes have been reported in the literature. Table 4.4 presents the BMP-obtained methane production and corresponding equivalent energy output for selected food waste mixtures, as listed on Table 4.2. BMP results are traditionally reported as volume of methane produced per mass of VS subjected to digestion (i.e., L CH_4/kg VS) to contrast results on the basis of what it is potentially digestible (organic matter), that is, not including water and ashes. However, Table 4.4 presents the results of the BMP on both, VS and *as is* basis, specifically to highlight the influence of the substrate's TS and VS on the methane production, as discussed in Section 4.1.2. When expressed on a VS basis, the methane production of most food wastes is relatively high (ca. 350–500 L CH_4/kg VS food waste), suggesting a good potential for energy recovery (ca. 12–17 MJ/kg VS food waste), particularly in relation to other organic substrates, such as cow manure (ca. 8 MJ/kg VS). However, when expressed on an *as is* basis, the methane production not only is significantly lower among the different types of food wastes, but it also varies on a considerably wider range. *As is* normalized values are indeed more representative of the actual methane production that would be obtained from food wastes as they are fed to a digester. Food wastes with high moisture content (i.e., low VS, TS), such as salads or whey, may have methane yields of up to 10 times lower than those observed for high-solid content food wastes, such as grains or cereals. The VS and TS concentration as well as methane potential of the intended food wastes to subject to anaerobic digestion are key aspects to consider when sizing the reactors and especially to determine optimal hydraulic retention times (HRTs) and organic loading rates (OLRs).

4.3.2 Performance of Bench-, Pilot-, and Commercial-Scale Anaerobic Digesters

In a UASB reactor digesting dining hall food waste at 15.8 g COD/L-d, a maximum daily value of 279 L-d of biogas production was realized, with an average methane concentration between 71% and 81% (Shin et al., 2001) (Table 4.3). Using CSTR reactors, Usack and Angenent (2015) evaluated the AcoD of cow manure and dining hall food waste by increasing the OLR in a stepwise manner during 564 days of study period. As they increased the OLR, the authors observed a

2. Based on a lower heating value (LHV) for pure methane of 33.39 MJ/m^3 (Avallone et al., 2006).

TABLE 4.4 Methane Production Potential (L @STP) and Energy Output (kJ) on a Volatile Solid (VS) and *as Is* Basis for Selected Food Waste Mixtures, as Listed in Table 4.2

	L CH_4/kg VS Food Waste	kJ/kg VS Food Waste	L CH_4/kg Food Waste	kJ/kg Food Waste
Rotten and expired salad mixes	375 (21)	12,534 (695)	14 (1)	458 (25)
Cheese whey	424 (58)	14,144 (1950)	25 (3)	846 (117)
Fruit and vegetable waste	422 (63)	14,088 (2113)	30 (5)	1007 (151)
Cola soda	373 (126)	12,457 (4197)	33 (11)	1105 (372)
Assorted kitchen waste	251 (40)	8395 (1324)	36 (6)	1199 (189)
Assorted canned goods	436 (10)	14,546 (338)	42 (1)	1391 (32)
Grocery store	371 (26)	12,378 (871)	46 (3)	1524 (107)
Fats, oil, and greases (FOG)—settled	413 (104)	13,804 (3482)	47 (12)	1555 (392)
Ice cream waste	502 (9)	16,772 (311)	55 (1)	1830 (34)
Restaurant food waste	419 (7)	13,994 (217)	57 (1)	1897 (29)
Corn silage	296 (5)	9886 (160)	59 (1)	1984 (32)
Fats, oil, and greases (FOG)—suspended	402 (120)	13,432 (3999)	92 (28)	3086 (919)
University dining hall	383 (28)	12,785 (928)	101 (7)	3388 (246)
Spent coffee grounds and filter paper	365 (10)	12,189 (318)	106 (3)	3546 (92)
Hotel food waste	488 (10)	16,291 (341)	126 (3)	4196 (88)
Yogurt and frozen dairy deserts	454 (6)	15,166 (185)	137 (2)	4587 (56)
Assorted postconsumer waste	546 (14)	18,236 (457)	247 (6)	8248 (207)
Sweetened breakfast cereals	362 (36)	12,081 (1197)	318 (32)	10,631 (1054)
Assorted grains	377 (89)	12,579 (2974)	341 (81)	11,372 (2688)
Stale bagels, muffins, and donuts	465 (26)	15,537 (854)	414 (23)	13,810 (759)
Used vegetable oil	649 (83)	21,654 (2789)	641 (83)	21,411 (2758)

Values in parenthesis represent the standard deviation.
Sources: Ebner, J.H., Labatut, R.A., Lodge, J.S., Williamson, A.A., Trabold, T.A., 2016. Anaerobic co-digestion of commercial food waste and dairy manure: characterizing biochemical parameters and synergistic effects. Waste Manag. 52, 286–294; Labatut, R.A., Angenent, L.T., Scott, N.R., 2011. Biochemical methane potential and biodegradability of complex organic substrates. Bioresour. Technol. 102(3), 2255–2264.

proportional increase in the methane production, reaching 297 L/kg VS (2 L biogas/L digester-d) at an OLR of 3.9 g VS/L-d, which was equivalent to a codigestion ratio of 51:49 manure:food waste (VS basis) (Usack and Angenent, 2015) (Table 4.3). A 900-m^3 CSTR digesting primarily household food wastes achieved an average methane production of 642 L/kg VS with a methane concentration of 62% when using an OLR of 2.5 g VS/L-d (Banks et al., 2011) (Table 4.3). In a study using a bench-scale, dry mesophilic and thermophilic semicontinuous anaerobic digestion, the conversion of food waste to methane was evaluated over varying OLRs, HRTs, and influent solids concentrations (Nguyen et al., 2017). Results showed that the highest average methane production was observed at an OLR of 8.62 g VS/L-d (HRT = 25 d), where the mesophilic and thermophilic digesters achieved 430 and 470 L/kg VS, respectively (Nguyen et al., 2017) (Table 4.3). As these results suggest, biogas, methane yields, and the organic matter removal efficiency are significantly affected by the OLR. In general, studies show that as the OLR is increased (or the HRT is decreased), the biogas/biomethane production is increased until inhibition ensues due to substrate overloading.

4.4 PROCESS STABILITY

Anaerobic digestion is a mature technology; however, poor process stability leading to low system performance is still frequently observed in full-scale plants worldwide. Many of these problems occur as a result of inadequate operational management or lack of process control (Labatut and Gooch, 2012). However, for food waste-based AD this is not always the case. Stand-alone food waste anaerobic digesters are in particular more susceptible to instability than, for example, manure- or WAS-based anaerobic digesters. Food wastes are usually characterized by high hydrolysis rates and low alkalinity (Murto et al., 2004). Thus if no alkalinity is externally provided to stand-alone food waste digesters, there may not be enough buffering capacity to counteract the production of VFA, resulting in acidification, decreasing digester pH, and eventually failure of the digestion process. Also, if anaerobic digestion of food wastes is to be performed without the use of a primary codigestate, such as cow manure, addition of essential nutrients as well as alkalinity may be required. Inhibition is usually the main cause of performance instability. It normally arises when the accumulation of intermediate products during digestion disrupts the homeostatic equilibrium of the microbiota and/or create thermodynamic limitations for syntrophic biochemical reactions. There are several substrates capable of producing intermediate products with the potential to cause process inhibition and instability (Chen et al., 2008). The most important ones were discussed in Section 4.2.1. In particular, urea- and protein-containing food wastes can create high levels of ammonia, which is particularly inhibitory to methanogens in its unionized form, and can cause instability (Angelidaki and Ahring, 1993). Also, lipid-rich food wastes have the potential to create inhibition of both acetogenesis and methanogenesis steps via LCFA accumulation (Palatsi et al., 2010). Inhibition can also occur when inhibitory or toxic chemical compounds are present in the influent substrate, having similar consequences as previously described.

Process monitoring and control of key parameters are widely accepted methods for improving AD stability and efficiency. Some of the indicators of a process imbalance were described in Section 4.2.1, for example, accumulation of VFAs, particularly acetate and propionate, accumulation of LCFAs, increase of ammonia concentrations, decrease in pH, among others. In addition, a steady drop of the methane production below the digester's average daily value is usually an indicator of digester upset. This condition is characteristic of increasing OLRs in anaerobic digesters. For example, when performing CSTR-based AD studies on source-separated food waste (Table 4.2), Nguyen et al. (2016, 2017) reported that the daily production of biogas/methane and the VS reduction by the digester often experienced a period of instability following each increase in the OLR. A similar situation may also occur when (high-strength) organic wastes are fed intermittently, which may produce a sudden decrease in both methane production and concentration at the times coinciding with digester loading. As an alternative to the conventional physicochemical parameters monitoring to assess process stability as described previously, a novel and more powerful technique has been recently explored based on the characterization of the digester's microbial communities. This comes from the realization that the myriad of biochemical reactions that occur during anaerobic digestion strongly rely on a consortium of microorganisms continuously interacting through syntrophic reactions. It is now understood that the dynamics of microbial communities in digesters are crucial for optimizing AD (Li et al., 2017). Certain communities can be used as process indicators that relate environmental factors and AD performance (De Vrieze et al., 2016, 2015).

4.5 ANAEROBIC CODIGESTION

In the United States, most anaerobic digesters treat food waste in conjunction with other substrates, mainly in farm-based operations. In this practice, commonly known as anaerobic codigestion (AcoD), food wastes are codigested with animal manures, as a primary substrate in terms of proportional influent mass. This allows increasing organic loading rates and improving performance relative to mono-digestion by diluting toxic or inhibitory compounds and/or providing macro- or micronutrients that may not be present in the food waste alone (Labatut et al., 2011). In addition, codigesting recalcitrant manure with easily degradable food wastes in farm-based digesters can improve the economic viability of the farm operation; first, by increasing biomethane yields, which can then be converted to electricity and sold to the utility company, and second, by receiving additional income in the form of tipping fees for the imported food waste. Also, animal manures, although fairly undegradable, are particularly suitable for codigestion with food waste. Cow manure, for example, can provide the necessary nutrients (e.g., N, Ni, Co) and environmental conditions (e.g., alkalinity, pH) to improve process stability, maximize waste stabilization, and ultimately increase biomethane production. This is especially true for sugar- or lipid-rich food wastes, where nitrogen is not present and needs to be externally supplemented to provide the buffering capacity (in the form of bicarbonate alkalinity) required for the process pH to be maintained within the optimal range. Also, trace element deficiency can occur during long-term anaerobic digestion of certain food wastes lacking an essential element for a specific metabolic process during AD. This was reported by Agler et al. (2008) while digesting corn-derived waste in high-rate anaerobic reactors, where an increase in the volatile fatty acids concentration and a decrease in biogas production were observed after interrupting external cobalt supplementation (Agler et al., 2008). Additionally, the formation of LCFAs during the digestion of lipid-rich food wastes, such as olive mill waste, can lead to severe inhibition of the anaerobic process, particularly at thermophilic temperatures (Angelidaki et al., 1997). Labatut et al. (2014) reported that increased hydrolysis rates in thermophilic digesters can produce significant accumulation of LCFAs. While codigesting a synthetic, lipid-rich food waste with cow manure in a thermophilic digester, the authors observed highly stable conditions when high ratios of manure-to-food waste (25:75 VS basis) were used; contrarily, digester failure quickly occurred when a lower proportion of manure was used in the influent mixture. Rather than dilution, the authors attributed the positive effects of manure to an increase in surface area provided by the manure fibers, which allowed the LCFAs to adsorb and decrease their concentration in solution, thereby providing stability to the process. In summary, there seems to be a consensus among researchers that codigesting food waste with an appropriate primary substrate may be necessary to enhance process stability and performance of organic matter biodegradation, leading to an increase of the energy output of the plant (Zhang et al., 2014).

Synergy Biogas, LLC is a current example of an anaerobic codigestion facility that successfully codigests up to 14 different waste streams at a time together with dairy manure, the primary substrate. This anaerobic codigestion facility utilizes an anaerobic digester tank with a treatment volume of $8330\,m^3$ and produces enough biogas to power a 1.4-MW internal combustion engine-generator set (Labatut and Gooch, 2014). The anaerobic digestion tank is shown in Fig. 4.3, with the three pasteurization units shown in the foreground. All influent streams, including the dairy manure, are commingled and subsequently pasteurized prior to inclusion to the digester.

4.6 BIOGAS UTILIZATION

Biogas is primarily composed of methane and carbon dioxide, but it also contains water vapor and traces of nitrogen, hydrogen sulfide, and other gases. The exact composition depends on the biomass source and the technology used to produce and/or extract the biogas. The energy in biogas comes from methane—its combustion will produce heat and release uncombusted and flue gases (e.g., H_2, CO). Biogas has between 30% and 40% less methane (v/v) than fossil fuel-derived natural gas, and therefore a proportionally lower heating value. However, methane concentration in the biogas produced by some operations can be as high as 70%. The highest methane concentration and production of biogas are normally obtained via AD systems (as opposed to landfill gas recovery systems), because operating parameters are tightly controlled and high-strength, high-energy yielding feedstocks are usually introduced.

The most widely used energy generation technologies that are suited for pairing with an anaerobic codigestion facility to convert the methane gas produced to heat and/or electricity currently include: combined heat and power (CHP), reciprocating engines, gas turbines, microturbines, and fuel cells, although fuel cells are still considered an emerging technology. Most anaerobic codigestion facilities that pursue energy generation fall under the classification of a distributed energy

FIG. 4.3 Anaerobic codigestion system at Synergy Biogas, LLC. *(Photo credit: R. Labatut.)*

TABLE 4.5 Comparison Between Existing Combined Heat and Power (CHP) Technologies for Biogas

	Gas Turbines	Microturbines	Reciprocating Engines	Fuel Cells
Capital cost ($/kW)	700–2000	1100–2000	800–1500	1000–5000
O&M costs ($/kW-h)	0.006–0.011	0.008–0.02	0.008–0.025	0.03–0.04
Electrical efficiency (%)	22–36	25–35	22–45	40–60
Emissions (NOx, SOx, CO_2, PM, etc.)	Significant	Moderate	Significant	Low
Fuel flexibility	Fair	Fair	Fair	High

Sources: EERE, 2017. Fuel Cells. Office of Energy Efficiency & Renewable Energy; energy.ca.gov., 2017. Vol. 2017, California Energy Commission; Fuelcells.org., 2017. Fuel Cells 2000, Vol. 2017. http://hfcarchive.org/fuelcells/; USEPA, 2007. Biomass Combined Heat and Power Catalog of Technologies. U.S. Environmental Protection Agency.

generation facility, as opposed to the traditionally employed centralized power plant model. Distributed renewable energy generation technologies are emerging as the ideal approach to incorporate small-scale renewable energy sources to the main power grid.

Multiple technologies are compared in Table 4.5. As shown, reciprocating engines have good operational efficiency and are cost competitive, but have low fuel flexibility, high O&M costs, and high emissions. The cost of the fuel cells that have been operated with biogas, that is, phosphoric acid and molten carbonate, is still very high. There has recently been increased interest in the option of combining AD systems with solid oxide fuel cells (SOFCs) due to their relatively high operating temperature (nominally 800°C) which makes them more robust to fuel impurities (Cozzolino et al., 2017; Rayner et al., 2017; Rillo et al., 2017).

In a study performed in the United Kingdom, domestic food wastes were collected and incorporated into a full-scale food waste digester (900 m^3 complete mixed tank reactor), from which biogas was harvested and used to produce electricity using a 195-kW combined heat and power (CHP) unit with an assumed electrical conversion efficiency of 32% when run at full capacity, and the potential for 53% recovery of heat. Electricity produced by the CHP unit and imports and exports to the grid were all metered. The power requirements of the plant were calculated as follows: CHP generator meter + grid import meter − grid export meter. A portion of the heat produced by the CHP unit was reincorporated back into the process (Banks et al., 2011).

4.7 FUTURE PERSPECTIVE AND RESEARCH NEEDS

The case for employing AD to treat food waste has been made worldwide. There is a massive surplus of organic waste that is currently landfilled around the world, taking up precious volume and adding to the damaging release of one of the more potent GHGs, methane. Additionally, the ever growing need to supply the main power grid with renewable energy is pressing, as citizens, businesses, and countries push ambitious goals to increase the portion of energy supplied by renewable sources, increasing resiliency and energy independence in an age of uncertainty.

Table 4.6 lists the annual food waste generation potential for 25 countries (10 developed and 15 developing) (Thi et al., 2015). Based on the total mass of food waste projected and specific common characteristics for the food waste generated, the associated total annual energy potential of this waste was calculated, as was the annual electrical and thermal energy generation potentials. This is an important exercise in order to promote the future employment of AD technology for food waste management applications. In particular, large, developing countries, such as China, show enormous annual energy generation potential, with a total of over a million terajoules, or a theoretical power capacity of 12,000 MW. This is more than three times greater than the total annual energy generation potential of the United States.

Despite the clear potential for future implementation of new stand-alone food waste AD plants worldwide, and the clear advantages of doing so, there are currently barriers to widespread adoption. Two main reasons are primarily, lack of policy directives, and secondarily, lack of funding resources which are unlikely to happen without there being a clear policy on the issue. In examining current policies worldwide, there is a disparity as to the maturity of each countries' waste management initiatives. For example, in Costa Rica, the 2010 status was that a law was passed to mandate "…the necessary legislation to regulate and organize a comprehensive national plan for the management of solid waste" (Thi et al., 2016), clearly showing that no initiatives or organized plan exists, since there is no national policy on the matter. However, in Thailand, the "3Rs Strategy" aims to increase organic waste utilization by 50% before 2026 (Thi et al., 2016), which shows that the country has

TABLE 4.6 Annual Food Waste Generation Potential for 25 Countries, and the Associated Annual Total, Electrical and Thermal Energy Potentials, and the Theoretical Power Capacities

Country[a]	Total FW[b] (ton/year)	Annual Total Energy Potential[c] (GJ/year)	Annual CHP Electrical Energy (35%) Produced (GWh/year)	Annual CHP Thermal Energy (45%) Produced (GJ/year)	Theoretical Power Capacity (MW)
Developed countries					
Australia	2,261,061	12,749,708	1240	5,737,369	142
Denmark	790,502	4,457,640	434	2,006,139	50
Sweden	1,915,460	10,800,927	1050	4,860,417	120
Singapore	796,000	4,488,644	437	2,020,092	50
United States	60,849,145	343,117,161	33,359	154,402,723	3808
The Netherlands	8,841,307	49,854,508	4847	22,434,528	553
Germany	12,257,998	69,120,601	6720	31,104,270	767
United Kingdom	14,257,000	80,392,606	7816	36,176,673	892
South Korea	6,241,500	35,194,673	3422	15,837,603	391
Taiwan	2,318,169	13,071,730	1271	5,882,278	145
Developing countries					
Brazil	33,489,000	188,838,325	18,359	84,977,246	2096
Turkey	12,375,000	69,780,354	6784	31,401,159	774
Malaysia	5,477,263	30,885,281	3003	13,898,376	343
Mexico	19,916,000	112,302,669	10,918	50,536,201	1246
Costa Rica	903,375	5,094,132	495	2,292,589	57
Romania	3,573,481	20,150,204	1959	9,067,592	224
South Africa	9,040,000	50,974,901	4956	22,938,705	566
Belarus	903,690	5,095,908	496	2,293,388	57
China	195,000,000	1,099,569,212	106,903	494,806,146	12,203
Thailand	9,312,788	52,513,102	5105	23,630,896	583
Jamaica	433,333	2,443,564	238	1,099,714	27
Ukraine	4,440,000	25,036,345	2434	11,266,355	278
Nigeria	25,000,000	140,970,412	13,705	63,436,685	1565
India	71,952,838	405,728,848	39,446	182,577,982	4503
Vietnam	5,743,056	32,384,039	3148	14,572,817	359

[a]*The sampling of countries and associated FW generation quantities from Thi et al. (2015).*
[b]*A sampling of 24 different FW streams from various sources was used to calculate an average methane potential, as determined via BMP assays, to be 408.1 L CH_4/kg VS). Similarly, average VS concentrations for each waste were averaged to yield an overall average VS concentration of 413.8 g/kg.*
[c]*Assumed LHV CH_4 = 33.39 kJ/L CH_4.*

drafted a plan and has steps and actionable goals to meet the initiative. South Korea has a very mature, organized directive to integrate AD technology; "the government target is to increase the mandatory supply quantity of biogas to 10% of total power generation in 2022" (De Clercq et al., 2017). The country has pursued multiple initiatives in support of meeting this goal, and recognize the importance of different types of support, including monetary support as well as research support.

A few of the more notable economic initiatives include: no tariffs or subsidies on biogas, government funding support for 60%–80% of agricultural biogas plant capital needs, enforcement of renewable portfolio standards, and creation of a feed-in tariff (De Clercq et al., 2017). Various government agencies also show support through initiatives such as the creation of an "Organic Waste to Energy" research center and the construction of a pilot AD plant specifically for food waste application (De Clercq et al., 2017). All of these initiatives are great examples of actions that can follow the implementation of a national goal or strategy to deal with organic waste management.

Despite the maturity of AD technology, applications to food waste continue to have some major considerations and challenges moving forward to expand the use of the technology. Process instability is a main cause of failure in operational systems, which is prevented by robust process monitoring and control, and by careful microbial management. When a survey was performed of 400 full-scale AD waste water treatment plants, the extent to which the systems were monitored was minimal beyond basic measures such as pH, temperature, water flow, biogas flow, level, and pressure (Li et al., 2017). Some of the plants were even found not to have the capability to provide in situ or real-time measurements of the process. Even when a robust monitoring system is in place, subsequent ability to adjust control of the system to remedy or prevent an impending failure is a necessary feature. Knowing what to do and equally important, having the ability to follow through with necessary adjustments or repairs to a reactor's process, are imperative to restore normal function (Labatut and Gooch, 2012; Li et al., 2017). Li et al. (2017) cite three requirements for properly managing microbial populations within an AD: (1) understanding the microbial community makeup, (2) establishing early warning microbial indicators, and (3) developing methods to enhance the biological function of the system.

In terms of future research, selection and proliferation of target microorganisms is the basis for long-term biotechnological intensification. Future research should focus on microbial growth kinetics, ecology, and competition to help control operating conditions and create a competitive advantage for target microorganisms (Li et al., 2017). Codigestion of food waste streams with other substrates, namely, agricultural wastes and residues, is likely to be the leading trend of future systems, as opposed to stand-alone food waste AD. Optimization of the mixture and ratio of food waste codigestion with other substrates will continue to be a leading area of future efforts, to further reduce impacts of toxicity effects and to enhance micronutrient profiles needed in the process.

As countries continue to put more emphasis on sustainable practices, including the reduction of landfilled organic waste and the reduction of GHG emissions, and begin to focus more on the opportunities provided by reusing food waste and other organic substrates as resources, opportunities will continue to emerge. Economic systems worldwide are slowly adapting to provide incentives and benefits for waste-to-energy systems, and revenue opportunities are beginning to outweigh the initial capital costs for nontraditional energy generation systems such as anaerobic digestion.

ACKNOWLEDGMENTS

This study was supported by the Chilean Fund for Science and Technology (FONDECYT) – project # 1161697. Special thanks to Andrés Holtheuer (PUC) for the through review and compilation of information which was critical for the realization of this work.

REFERENCES

Agler, M.T., Garcia, M.L., Lee, E.S., Schlicher, M., Angenent, L.T., 2008. Thermophilic anaerobic digestion to increase the net energy balance of corn grain ethanol. Environ. Sci. Technol. 42 (17), 6723–6729.

AgSTAR, 2016. Livestock Anaerobic Digester Database. vol. 2017. United States Environmental Protection Agency. https://www.epa.gov/agstar/livestock-anaerobic-digester-database.

Angelidaki, I., Ahring, B.K., 1993. Thermophilic anaerobic-digestion of livestock waste—the effect of ammonia. Appl. Microbiol. Biotechnol. 38 (4), 560–564.

Angelidaki, I., Ellegaard, L., Ahring, B.K., 1997. Modelling anaerobic codigestion of manure with olive oil mill effluent. Water Sci. Technol. 36 (6–7), 263–270.

Angelidaki, I., Boe, K., Ellegaard, L., 2005. Effect of operating conditions and reactor configuration on efficiency of full-scale biogas plants. Water Sci. Technol. 52 (1–2), 189–194.

Avallone, E., Baumeister, I., Sadegh, A., 2006. Marks' Standard Handbook for Mechanical Engineers, tenth ed. McGraw-Hill, New York.

Banks, C.J., Chesshire, M., Heaven, S., Arnold, R., 2011. Anaerobic digestion of source-segregated domestic food waste: performance assessment by mass and energy balance. Bioresour. Technol. 102 (2), 612–620.

Batstone, D.J., Keller, J., Angelidaki, I., Kalyuzhnyi, S.V., Pavlostathis, S.G., Rozzi, A., Sanders, W.T.M., Siegrist, H., Vavilin, V.A., 2002. The IWA anaerobic digestion model No. 1 (ADM1). Water Sci. Technol. 45 (10), 65–73.

Bouallagui, H., Touhami, Y., Cheikh, R.B., Hamdi, M., 2005. Bioreactor performance in anaerobic digestion of fruit and vegetable wastes. Process Biochem. 40 (3), 989–995.

Carucci, G., Carrasco, F., Trifoni, K., Majone, M., Beccari, M., 2005. Anaerobic digestion of food industry wastes: effect of codigestion on methane yield. J. Environ. Eng. 131 (7), 1037–1045.

Carvalho, F., Prazeres, A.R., Rivas, J., 2013. Cheese whey wastewater: characterization and treatment. Sci. Total Environ. 445, 385–396.

Chen, Y., Cheng, J.J., Creamer, K.S., 2008. Inhibition of anaerobic digestion process: a review. Bioresour. Technol. 99 (10), 4044–4064.

Chen, W.-T., Zhang, Y., Zhang, J., Yu, G., Schideman, L.C., Zhang, P., Minarick, M., 2014. Hydrothermal liquefaction of mixed-culture algal biomass from wastewater treatment system into bio-crude oil. Bioresour. Technol. 152, 130–139.

Commission, E., 2005. On the National Strategies for the Reduction of Biodegradable Waste Going to Landfills Pursuant to Article 5(1) of Directive 1999/31/EC on The Landfill of Waste.

Cozzolino, R., Lombardi, L., Tribioli, L., 2017. Use of biogas from biowaste in a solid oxide fuel cell stack: application to an off-grid power plant. Renew. Energy 111, 781–791.

De Baere, L., Mattheeuws, B., 2012. Anaerobic digestion of the organic fraction of municipal solid waste in Europe—status, experience and prospects. In: Thomé-Kozmiensky Karl, J., Thiel, S. (Eds.), Recycling and Recovery. vol. 3. pp. 517–526.

De Clercq, D., Wen, Z., Gottfried, O., Schmidt, F., Fei, F., 2017. A review of global strategies promoting the conversion of food waste to bioenergy via anaerobic digestion. Renew. Sust. Energ. Rev. 79, 204–221.

De Vrieze, J., Saunders, A.M., He, Y., Fang, J., Nielsen, P.H., Verstraete, W., Boon, N., 2015. Ammonia and temperature determine potential clustering in the anaerobic digestion microbiome. Water Res. 75, 312–323.

De Vrieze, J., Raport, L., Roume, H., Vilchez-Vargas, R., Jáuregui, R., Pieper, D.H., Boon, N., 2016. The full-scale anaerobic digestion microbiome is represented by specific marker populations. Water Res. 104, 101–110.

Ebner, J.H., Labatut, R.A., Lodge, J.S., Williamson, A.A., Trabold, T.A., 2016. Anaerobic co-digestion of commercial food waste and dairy manure: characterizing biochemical parameters and synergistic effects. Waste Manag. 52, 286–294.

FAO, 2013. Food Wastage Footprint: Impacts on Natural Resources: Summary Report. FAO.

Fusions, 2016. Estimates of European Food Waste Levels. European Commission.

Gohil, A., Nakhla, G., 2006. Treatment of tomato processing wastewater by an upflow anaerobic sludge blanket–anoxic–aerobic system. Bioresour. Technol. 97 (16), 2141–2152.

Gooch, C., Pronto, J., 2009. Anaerobic Digestion at Sunny Side Dairy, Inc.: Case Study. Cornell University.

Gooch, C., Pronto, J., Labatut, R., 2011. Evaluation of Seven On-Farm Anaerobic Digestion Systems Based on the ASERTTI Monitoring Protocol: Consolidated Report and Findings Cornell University.

Gunders, D., 2012. Wasted: How America Is Losing up to 40 Percent of its Food from Farm to Fork to Landfill.

Gustavsson, J., Cederberg, C., Sonesson, U., Van Otterdijk, R., Meybeck, A., 2011. Global Food Losses and Food Waste. FAO, Rome.

Hanaki, K., Nagase, M., Matsuo, T., 1981. Mechanism of inhibition caused by long-chain fatty-acids in anaerobic-digestion process. Biotechnol. Bioeng. 23 (7), 1591–1610.

Hartmann, H., Ahring, B.K., 2005. A novel process configuration for anaerobic digestion of source-sorted household waste using hyper-thermophilic post-treatment. Biotechnol. Bioeng. 90 (7), 830–837.

Hwu, C.S., Lettinga, G., 1997. Acute toxicity of oleate to acetate-utilizing methanogens in mesophilic and thermophilic anaerobic sludges. Enzym. Microb. Technol. 21 (4), 297–301.

Iacovidou, E., Ohandja, D.-G., Gronow, J., Voulvoulis, N., 2012. The household use of food waste disposal units as a waste management option: a review. Crit. Rev. Environ. Sci. Technol. 42 (14), 1485–1508.

Jeris, J.S., McCarty, P.L., 1965. The biochemistry of methane fermentation using C^{14} tracers. J. Water Pollut. Control Fed. 37 (2), 178–192.

Jia, X., Xi, B., Li, M., Xia, T., Hao, Y., Liu, D., Hou, J., 2017. Evaluation of biogasification and energy consumption from food waste using short-term hydrothermal pretreatment coupled with different anaerobic digestion processes. J. Clean. Prod. 152, 364–368.

Jiménez, A.M., Borja, R., Martı´n, A., 2003. Aerobic–anaerobic biodegradation of beet molasses alcoholic fermentation wastewater. Process Biochem. 38 (9), 1275–1284.

Kayhanian, M., 1999. Ammonia inhibition in high-solids biogasification: an overview and practical solutions. Environ. Technol. 20 (4), 355–365.

Koch, K., Plabst, M., Schmidt, A., Helmreich, B., Drewes, J.E., 2016. Co-digestion of food waste in a municipal wastewater treatment plant: comparison of batch tests and full-scale experiences. Waste Manag. 47, 28–33.

Krzywonos, M., Cibis, E., Miskiewicz, T., Ryznar-Luty, A., 2009. Utilization and biodegradation of starch stillage (distillery wastewater). Electron. J. Biotechnol. 12 (1), 6–7.

Labatut, R., Gooch, C., 2012. In: Monitoring of anaerobic digestion process to optimize performance and prevent system failure.Proceedings of Got Manure? Enhancing Environmental and Economic Sustainability, pp. 209–225.

Labatut, R., Gooch, C., 2014. Evaluation of the Continuously-Mixed Anaerobic Digester System at Synergy Biogas Following the Protocol for Quantifying and Reporting the Performance of Anaerobic Digestion Systems for Livestock Manures. Cornell University, p. 65961.

Labatut, R.A., Angenent, L.T., Scott, N.R., 2011. Biochemical methane potential and biodegradability of complex organic substrates. Bioresour. Technol. 102 (3), 2255–2264.

Labatut, R.A., Angenent, L.T., Scott, N.R., 2014. Conventional mesophilic vs. thermophilic anaerobic digestion: a trade-off between performance and stability? Water Res. 53, 249–258.

Leib, E.B., Rice, C., Mahoney, J., 2016. Fresh look at organics bans and waste recycling laws. Biocycle 57, 16.

Li, L., Peng, X., Wang, X., Wu, D., 2017. Anaerobic digestion of food waste: a review focusing on process stability. Bioresour. Technol.

Long, J.H., Aziz, T.N., Francis, L., Ducoste, J.J., 2012. Anaerobic co-digestion of fat, oil, and grease (FOG): a review of gas production and process limitations. Process Saf. Environ. Prot. 90 (3), 231–245.

Lubberding, H.J., Gijzen, H.J., Heck, M., Vogels, G.D., 1988. Anaerobic digestion of onion waste by means of rumen microorganisms. Biol. Wastes 25 (1), 61–67.

Malaspina, F., Stante, L., Cellamare, C.M., Tilche, A., 1995. Cheese whey and cheese factory wastewater treatment with a biological anaerobic–aerobic process. Water Sci. Technol. 32 (12), 59–72.

Massé, D., Masse, L., 2000. Characterization of raw wastewater from hog slaughterhouses in Eastern Canada and evaluation of their in-plant wastewater treatment system. Can. Agric. Eng. 42, 139–146.

McCarty, P.L., 1972. Energetics of organic matter degradation. In: Mitchell, R. (Ed.), Water Pollution Microbiology. In: vol. 5. Wiley-Interscience, New York, pp. 91–118.

McCarty, P.L., Smith, D.P., 1986. Anaerobic waste-water treatment. Environ. Sci. Technol. 20 (12), 1200–1206.

Murto, M., Björnsson, L., Mattiasson, B., 2004. Impact of food industrial waste on anaerobic co-digestion of sewage sludge and pig manure. J. Environ. Manag. 70 (2), 101–107.

Neves, L., Oliveira, R., Alves, M.M., 2009. Fate of LCFA in the co-digestion of cow manure, food waste and discontinuous addition of oil. Water Res. 43 (20), 5142–5150.

Nghiem, L.D., Koch, K., Bolzonella, D., Drewes, J.E., 2017. Full scale co-digestion of wastewater sludge and food waste: bottlenecks and possibilities. Renew. Sust. Energ. Rev. 72, 354–362.

Nguyen, D.D., Chang, S.W., Jeong, S.Y., Jeung, J., Kim, S., Guo, W., Ngo, H.H., 2016. Dry thermophilic semi-continuous anaerobic digestion of food waste: performance evaluation, modified Gompertz model analysis, and energy balance. Energy Convers. Manag. 128, 203–210.

Nguyen, D.D., Chang, S.W., Cha, J.H., Jeong, S.Y., Yoon, Y.S., Lee, S.J., Tran, M.C., Ngo, H.H., 2017. Dry semi-continuous anaerobic digestion of food waste in the mesophilic and thermophilic modes: new aspects of sustainable management and energy recovery in South Korea. Energy Convers. Manag. 135, 445–452.

Owen, W.F., Stuckey, D.C., Healy, J.B., Young, L.Y., Mccarty, P.L., 1979. Bioassay for monitoring biochemical methane potential and anaerobic toxicity. Water Res. 13 (6), 485–492.

Palatsi, J., Illa, J., Prenafeta-Boldu, F.X., Laureni, M., Fernandez, B., Angelidaki, I., Flotats, X., 2010. Long-chain fatty acids inhibition and adaptation process in anaerobic thermophilic digestion: batch tests, microbial community structure and mathematical modelling. Bioresour. Technol. 101 (7), 2243–2251.

Pavi, S., Kramer, L.E., Gomes, L.P., Miranda, L.A.S., 2017. Biogas production from co-digestion of organic fraction of municipal solid waste and fruit and vegetable waste. Bioresour. Technol. 228, 362–367.

Puyuelo, B., Ponsá, S., Gea, T., Sánchez, A., 2011. Determining C/N ratios for typical organic wastes using biodegradable fractions. Chemosphere 85 (4), 653–659.

Rayner, A.J., Briggs, J., Tremback, R., Clemmer, R.M.C., 2017. Design of an organic waste power plant coupling anaerobic digestion and solid oxide fuel cell technologies. Renew. Sust. Energ. Rev. 71, 563–571.

Rillo, E., Gandiglio, M., Lanzini, A., Bobba, S., Santarelli, M., Blengini, G., 2017. Life cycle assessment (LCA) of biogas-fed solid oxide fuel cell (SOFC) plant. Energy 126, 585–602.

Salminen, E., Rintala, J., 2002. Anaerobic digestion of organic solid poultry slaughterhouse waste—a review. Bioresour. Technol. 83 (1), 13–26.

Schievano, A., D'Imporzano, G., Malagutti, L., Fragali, E., Ruboni, G., Adani, F., 2010. Evaluating inhibition conditions in high-solids anaerobic digestion of organic fraction of municipal solid waste. Bioresour. Technol. 101 (14), 5728–5732.

Shepherd, H.L., Grismer, M.E., Tchobanoglous, G., 2001. Treatment of high-strength winery wastewater using a subsurface-flow constructed wetland. Water Environ. Res. 73 (4), 394–403.

Shin, H., Han, S., Song, Y., Lee, C., 2001. Performance of UASB reactor treating leachate from acidogenic fermenter in the two-phase anaerobic digestion of food waste. Water Res. 35 (14), 3441–3447.

Speece, R.E., Boonyakitsombut, S., Kim, M., Azbar, N., Ursillo, P., 2006. Overview of anaerobic treatment: thermophilic and propionate implications. Water Environ. Res. 78 (5), 460–473.

Stahel, W.R., 2016. Circular economy: a new relationship with our goods and materials would save resources and energy and create local jobs. Nature 531 (7595), 435–439.

Thi, N.B.D., Kumar, G., Lin, C.-Y., 2015. An overview of food waste management in developing countries: current status and future perspective. J. Environ. Manag. 157, 220–229.

Thi, N.B.D., Lin, C.-Y., Kumar, G., 2016. Electricity generation comparison of food waste-based bioenergy with wind and solar powers: a mini review. Sustain. Environ. Res. 26 (5), 197–202.

Usack, J.G., Angenent, L.T., 2015. Comparing the inhibitory thresholds of dairy manure co-digesters after prolonged acclimation periods: part 1. Performance and operating limits. Water Res. 87, 446–457.

USEPA, 2016. Anaerobic digestion facilities processing food waste. In: Anaerobic Digestion Facilities Processing Food Waste Tracking Project. vol. 2017. United States Environmental Protection Agency.

USEPA, 2017. Understanding global warming potentials. In: Greenhouse Gas Emissions. vol. 2017. U.S. Environmental Protection Agency. https://www.epa.gov/ghgemissions/understanding-global-warming-potentials.

Zhang, C., Su, H., Baeyens, J., Tan, T., 2014. Reviewing the anaerobic digestion of food waste for biogas production. Renew. Sust. Energ. Rev. 38, 383–392.

FURTHER READING

EERE, 2017. Fuel Cells. Office of Energy Efficiency & Renewable Energy.

energy.ca.gov., 2017. Vol. 2017, California Energy Commission.

Fuelcells.org, 2017. Fuel Cells 2000. Vol. 2017. http://hfcarchive.org/fuelcells/.

Pavlostathis, S.G., Giraldo-Gomez, E., 1991. Kinetics of anaerobic treatment—a critical review. Crit. Rev. Environ. Control 21 (5–6), 411–490.

USEPA, 2007. Biomass Combined Heat and Power Catalog of Technologies. U.S. Environmental Protection Agency.

Chapter 5

Sustainable Waste-to-Energy Technologies: Fermentation

Swati Hegde and Thomas A. Trabold
Golisano Institute for Sustainability, Rochester Institute of Technology, Rochester, NY, United States

5.1 INTRODUCTION

The depletion of petroleum sources and rising prices of raw materials have increased the demand for fuels derived from renewable energy sources (Raganati et al., 2014). Ethanol is the most commonly used fuel alcohol, now widely applied as an oxygenating blending component for gasoline, produced using corn as the primary feedstock in the United States, and sugarcane molasses and biomass in Brazil (Wang et al., 2012). Advanced biofuels like propanol, 1-butanol, and 3-methyl 1-butanol are also receiving attention from alcohol producers, and production of biofuels, particularly fuel alcohols using lignocellulosic biomass, has gained increased interest over the last decade. Significant research and development efforts are underway for the large-scale production of butanol from lignocellulosic materials, though butanol has not yet been regulated as a fuel blend in the United States. There are limited efforts on using propanol as a fuel, possibly due to its relatively high viscosity. Methanol is used mainly as an industrial commodity instead of a fuel and is typically produced by chemical synthesis. Though alcohol production from food waste has been explored since the 1980s (e.g., Schoutens et al., 1985; Voget et al., 1985; Hang et al., 1981), a continued research effort is required to reduce the production cost of higher alcohols to achieve commercial viability.

The Renewable Fuel Standard (RFS) in the United States requires an increase in the amount of renewable fuels blended with gasoline and diesel, and sets an increased target for biofuels every year. While the target production volume for conventional renewable fuel ethanol from corn starch has remained constant since 2015, the targets for cellulosic and advanced biofuels are increasing every year, as shown in Fig. 5.1. Advanced biofuels use renewable raw materials as feedstocks and are expected to result in an emission reduction of 50% or greater (RFA, 2017a). Food waste is a potential raw material for the production of advanced biofuels as it is generated in large quantities and diverting food wastes from landfill can help achieve a greater environmental benefit.

While fermentative alcohol production is well understood and developed, hydrogen production from fermentation is under research. Hydrogen is an alternative fuel that can be produced from a variety of feedstocks. Although hydrogen is relatively new to the market as a transportation fuel, governments and industry are working toward widespread distribution of hydrogen for fuel cell electric vehicles. The increasing demand for electric vehicles increases the need to develop safer, cleaner, and more economical production of hydrogen. Currently the majority of hydrogen in the United States is produced using steam reforming of methane in natural gas. Hydrogen can also be produced from electrolysis of water but this method is very energy intensive and not presently economically viable at scale. Fermentative production of hydrogen is an alternative option to derive energy value from a heterogeneous mix of organic waste materials like food waste which would otherwise be landfilled.

Due to increased interest in advanced biofuels, there is a need to evaluate cheaper and renewable raw materials to produce higher fuel alcohols and hydrogen. One such raw material is food waste that, in many cases, provides a rich nutrient source for microbes and can be converted into useful products through biodegradation pathways. Producing fuels out of wastes can not only contribute to meeting current and future fuel demand but also helps in reducing the wastes produced by the food sector. Commercializing biofuels using food waste raw material can be very economical since food waste can be obtained at lower cost relative to other potential feedstocks. Fuel alcohol production is a potential option to treat liquid and semisolid food wastes rich in carbohydrates. Food waste is cheaper compared to all other feedstocks currently being used, for example, corn, lignocellulosic biomass, and molasses. However, optimizing the production process plays a major role in accommodating spatial and temporal variations in food waste composition while still achieving acceptable yield.

Sustainable Food Waste-to-Energy Systems. https://doi.org/10.1016/B978-0-12-811157-4.00005-X

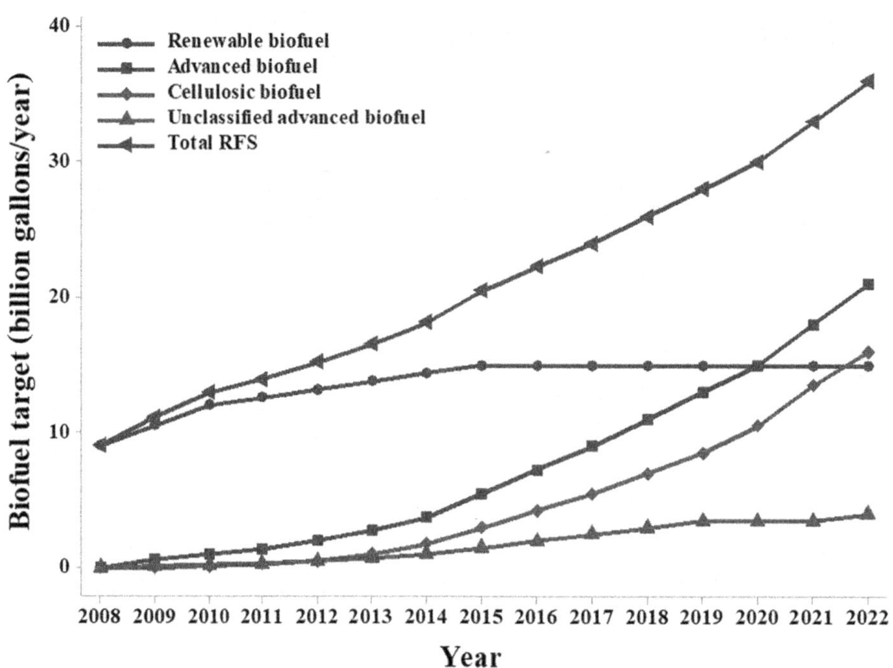

FIG. 5.1 RFS target for biofuel production (biodiesel not shown). *Source: RFA, 2017. Renewable Fuel Standard. Accessed June 2017. http://www.ethanolrfa.org/policy/regulations/renewable-fuel-standard/.*

This chapter discusses the potential of using food wastes in biofuel production, the generalized approach to producing ethanol from food waste, by-products of ethanol production, other biofuels from fermentation (butanol and hydrogen), challenges in large-scale production, and techno-economic considerations.

5.2 BIOETHANOL FROM FOOD WASTE

Ethanol that is produced from renewable raw materials using fermentation processes is called bioethanol, and its production has increased significantly over the last decade, indicating its growing demand as an oxygenating agent and flex fuel for transportation applications. There are 209 operating ethanol plants in the United States located as shown in Fig. 5.2, and as of September 2017 the total ethanol production from all the operating plants was 15.6 billion gallons per year (59 billion liters per year); NREL (2017). The United States is the largest producer of bioethanol in the world, followed by Brazil, the Europe Union, China, and Canada. In 2015, the United States produced 56.7% of the total ethanol produced globally, and Brazil's ethanol production accounted for 27.6% (EIA, 2017). Together the United States and Brazil accounted for nearly 85% of the global ethanol production with corn as the primary feedstock in the United States and sugarcane molasses in Brazil.

5.2.1 Food Waste as Feedstock for Bioethanol Production

The US ethanol plants primarily use corn as the feedstock, accounting for 95% of the total production while only 1% of the total bioethanol produced comes from food and beverage waste as shown in Fig. 5.3. In 2016, 38% of the total corn grown in the United States was utilized for ethanol production which is threefold higher compared to 12.4% in 2006 (USDA ERS, 2017). The National Renewable Energy Laboratory (NREL) Biofuel Atlas and Ethanol Producer Magazine databases (EPM, 2017) list only eight ethanol plants that use food and beverage waste as the feedstock (NREL, 2017). These eight plants produce 78 million gallons ethanol per year that accounts for $<1\%$ of the total production. It is important to note that most of these plants are not independent biorefineries, but small-scale plants colocated at a food processing facility to treat food waste generated on-site. Table 5.1 lists the ethanol plants using food waste as the raw material. Despite this very low total production capacity, there is significant research effort worldwide to use food waste as the feedstock for the production of ethanol and 1-butanol. However, before using food wastes for any production process, it is of utmost importance to characterize the amount and types of macronutrients available. Understanding the composition enables careful design of an optimal process and enhances economic viability.

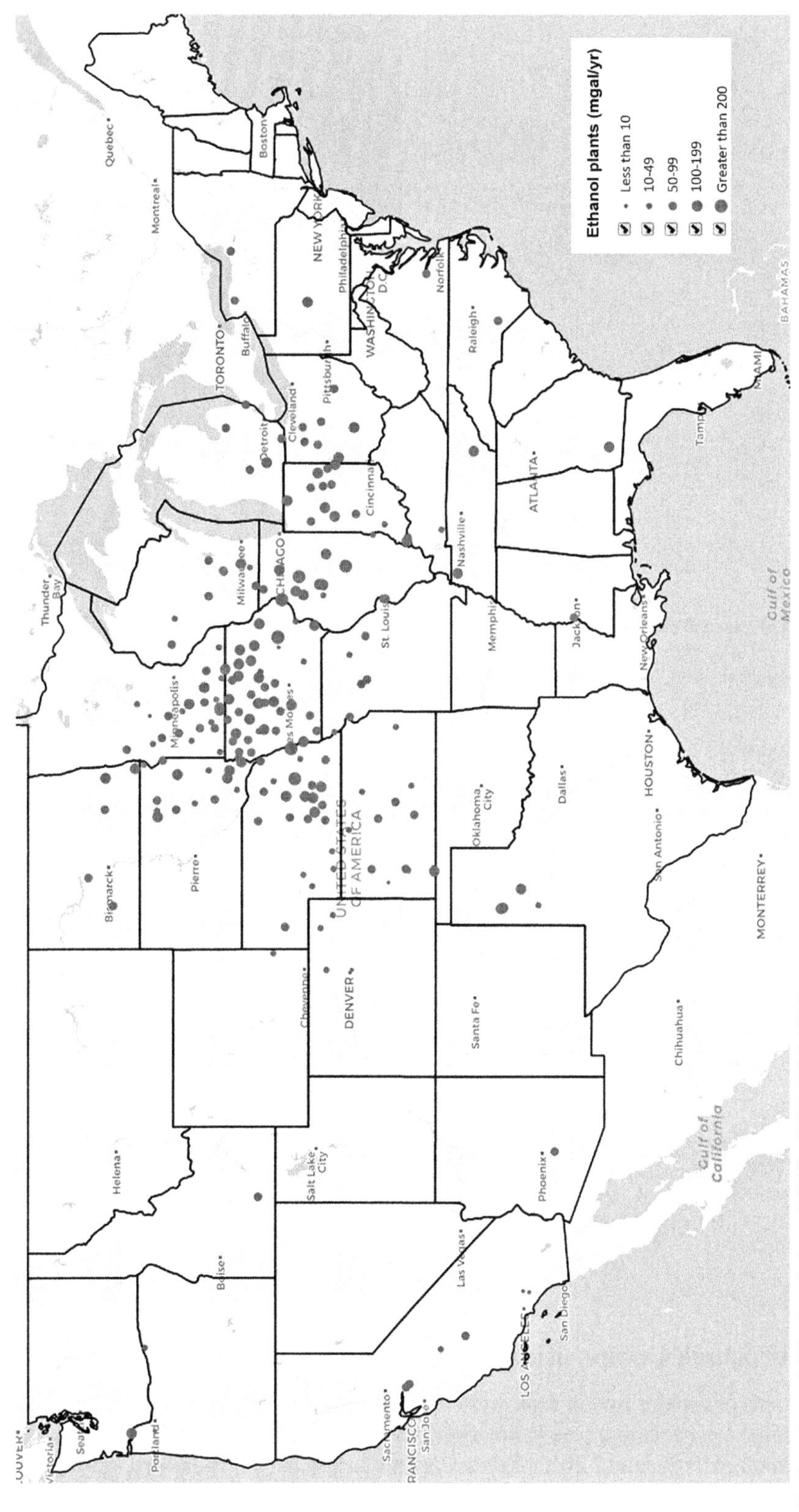

FIG. 5.2 Locations of US ethanol plants (NREL, 2017; EPM, 2017).

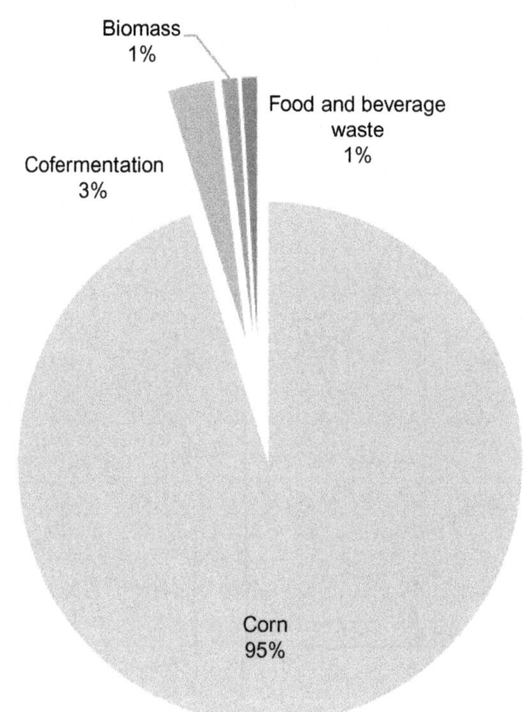

FIG. 5.3 Percent share of feedstocks in total US ethanol production. *Source: NREL, 2017. The Biofuels Atlas. Accessed September 2017. https://maps.nrel.gov/biofuels-atlas/#/?aL=yilN7K%255Bv%255D%3Dt&bL=groad&cE=0&lR=0&mC=40.21244%2C91.625976&zL=4; EPM, 2017. Ethanol Producer Magazine: US ethanol plants. http://www.ethanolproducer.com/plants/listplants/US/Operational/All.*

TABLE 5.1 Commercial Ethanol Production Plants Utilizing Food Waste as Feedstock

Company Name	Feedstock	Plant Capacity (Million Gallons/Year)	Location
Golden Cheese company	Cheese whey	19.0	Corona, CA
Land O' Lakes	Cheese whey	9.9	Melrose, MN
Merrick and Company	Waste beer	11.4	Aurora, CO
Parallel Products	Waste beverage	9.1	Louisville, KY
Parallel Products	Waste beverage	11.4	Rancho Cucamonga, CA
Wind Gap Farms	Brewery waste	1.5	Baconton, GA
Summit Natural Energy Inc	Fruit syrup and fruit processing waste	1.0	Cornelius, OR
Veolia ES Technical Solutions	Organic waste stream (not specified)	6.0	Medina, OH

5.2.2 Effects of Feedstock Composition

The organic composition plays a major role in food waste treatment and valorization. A higher reducing sugar content is preferred to produce alcohols, especially ethanol; however, very high sugar levels inhibit the production by negatively affecting microbial metabolism (Lin et al., 2012; Pérez-Carrillo et al., 2011). Therefore, during fermentation, the production medium should have a free sugar concentration of 15%–20% (Tahir et al., 2010). If food waste substrates do not contain the required amount of reducing sugars, a pretreatment like acid hydrolysis, heat treatment, or enzymatic hydrolysis is required to convert complex sugars into simpler mono and oligosaccharides. It is also important to balance the ratio of carbohydrates to proteins, as proteins are a rich source of nitrogen. The substrate should always provide a balanced carbon-to-nitrogen ratio (C/N) in a single substrate or cofermentation mix. For example, when corn starch was cofermented with a protein-rich soy skim milk, ethanol production rate increased from 18% to 25% (Yao et al., 2012). In some cases, the interactions between nutrients affect both pretreatment and fermentation processes. Certain proteins have been observed to interfere with enzymatic hydrolysis of maize starch (Zhao et al., 2009). Such effects reduce the efficiency of pretreatment, making a smaller fraction of the sugars from the feedstock available for fermentation. Therefore a thorough understanding of carbon and nitrogen sources is essential to avoid antagonistic effects. Certain lipid molecules are known to produce an altered stimulatory effect on yeast growth. Lipids rich in free fatty acids have been found to be controlling factors in ethanol tolerance of yeast. The ethanol tolerance was increased from 13% to 14% when lecithin, palmitic acid, and cholesterol were added to the production medium. While low concentrations of lipids had stimulatory effect, high concentrations were inhibitory (Ghareib et al., 1988). Lipids affect fermentation as antifoaming agents, improving the ethanol tolerance of yeast and as growth stimulants. It should be noted that food waste organic composition can vary with season, place, and time of collection and storage conditions. Thus continual monitoring of the physical and chemical properties of feedstock is an essential part of any well-designed food waste fermentation process.

Various metal ions contribute to the inorganic content of food waste, and they play a major role in the microbial metabolism, as they take part in a variety of biochemical reactions, and act as cofactors for enzymes responsible for growth and product formation. Metal ions are important in maintaining the osmotic pressure of the cells in production medium. However, both mineral deficiency and overload can cause cell damage (Xu et al., 2016) and hence limit alcohol production. Therefore mineral composition of feedstock can significantly affect the yield of alcohol. The presence of Mg^{2+}, Mn^{2+}, Fe^{2+}, and K^{+} was shown to be essential for stabilizing the yeast strain and improving ethanol production using mango residues (Somda et al., 2011). Udeh et al. (2014) observed a positive effect of barium and magnesium on ethanol fermentation, but no effect from zinc. Mineral salts have also been shown to have an effect on glucose and xylose consumption by ethanol-producing yeast (Casey et al., 2013). Metal absorption is hypothesized as a controlling mechanism in alcohol tolerance levels in microbes (Chin et al., 2013). This mechanism can be used to genetically modify microbes to absorb minerals better, and thus acquire improved tolerance to alcohols, especially butanol. These minerals can be supplemented in the form of micronutrient solution to achieve higher alcohol tolerance, and hence improve yield. Some food waste materials like acid whey and tofu processing waste water naturally contain high concentrations of minerals and could act as supplements to more conventional substrates (Hegde et al., 2018).

5.2.3 Microbiology and Biochemistry of Ethanol Production

A eukaryotic microbe, *Saccharomyces cerevisiae* of the yeast family, is used to commercially produce ethanol when the fermentation is starch, glucose, or sucrose based. Food waste is rich in carbohydrates of various types that *S. cerevisiae* alone is not capable of utilizing, for example, xylose and lactose. Therefore other microorganisms have been investigated for ethanol production from different food waste materials. Other yeasts belonging to *Kluyveromyces* species is known to utilize lactose as a carbon source for growth. Species like *K. marxianus* and *K. lactis* (Hadiyanto et al., 2014; Zafar and Owais, 2006) are capable of metabolizing lactose, which *S. cerevisiae* is not. Food processing wastes like whey from milk and cheese manufacturing are rich in lactose that can be metabolized by *Kluyveromyces* strains to produce ethanol. Scientists have developed a few strains of the versatile bacteria *Escherichia coli* capable of tolerating high concentrations of ethanol and utilizing a broad range of sugars (Förster and Gescher, 2014; Ingram and Clark, 1991). Another bacteria that is capable of producing ethanol is *Zymomonas mobilis* (Yang et al., 2016; Letti et al., 2012; Panesar et al., 2006; Baratti and Bu'lock, 1986). Since *Z. mobilis* is a prokaryote, like *E. coli*, it offers advantages of easy genetic manipulation. *S. cerevisiae*, however, is still the most commonly used organism due to its nonspecific growth requirements and high ethanol tolerance. Other organisms play a major role in food waste fermentation since *S. cerevisiae* cannot easily metabolize all the nutrients present in food waste. When using mixed food waste as the feedstock, a cofermentation approach may be necessary where a mixed culture of microbes is applied to utilize different nutrients present in food waste.

Irrespective of the initial feedstock complexity, substrates are converted into hexoses after pretreatment (physical, chemical, mechanical, or enzymatic) which are fermented by relevant microorganisms into ethanol (Sarris and Papanikolaou, 2016). First, polymeric materials like starch or cellulose from the feedstock convert into oligomers and dextrin. The oligomers produced are repeating units of glucose but less branched than starch and cellulose. The oligomers are then broken down into simple sugars like hexoses (e.g., glucose and fructose), pentoses (xylose), disaccharides (sucrose), and glycerol. *S. cerevisiae* and bacterium *Z. mobilis* ferment glucose into ethanol. Certain yeasts like *Pachysolen tannophilus*, *Candida shehatae*, and *Pichia stipis* are capable of converting xylose into ethanol. Alternatively, xylose can be isomerized into xylulose with the enzyme glucose or xylose isomerase, and subsequent fermentation with *S. cerevisiae*

(Hahn-HäGerdal et al., 1991). Finally, there have been some efforts to convert glycerol produced as a by-product of waste cooking oil to biodiesel (via transesterification, Chapter 6) into ethanol using yeast and bacteria (Azreen et al., 2014; Ito et al., 2005).

5.3 ETHANOL PRODUCTION PROCESS DESCRIPTION

Ethanol is produced commercially using corn as the primary feedstock. Corn contains starches and sugars that are relatively easily converted into ethanol. Ethanol production using lignocellulosic biomass is in its early stages, and there are only a few pilot plants around the world. Food waste is viable as a feedstock for ethanol production because it is generally a rich mixture of starch, sugars, and cellulosic materials. However, because of its complexity, food waste often needs several pretreatments steps before it can be converted into ethanol. A few food processing industries in the United States have built ethanol plants that operate at a small scale, as outlined in Table 5.1. It is important to note that all these existing ethanol plants are an integral part of a food processing facility and not operating as independent biorefineries. The commercial production of ethanol involves several steps: feedstock preparation, pretreatment, inoculum development, fermentation, and downstream processing. The following sections describe the unit processes involved in ethanol production, with a general process flow diagram presented in Fig. 5.4.

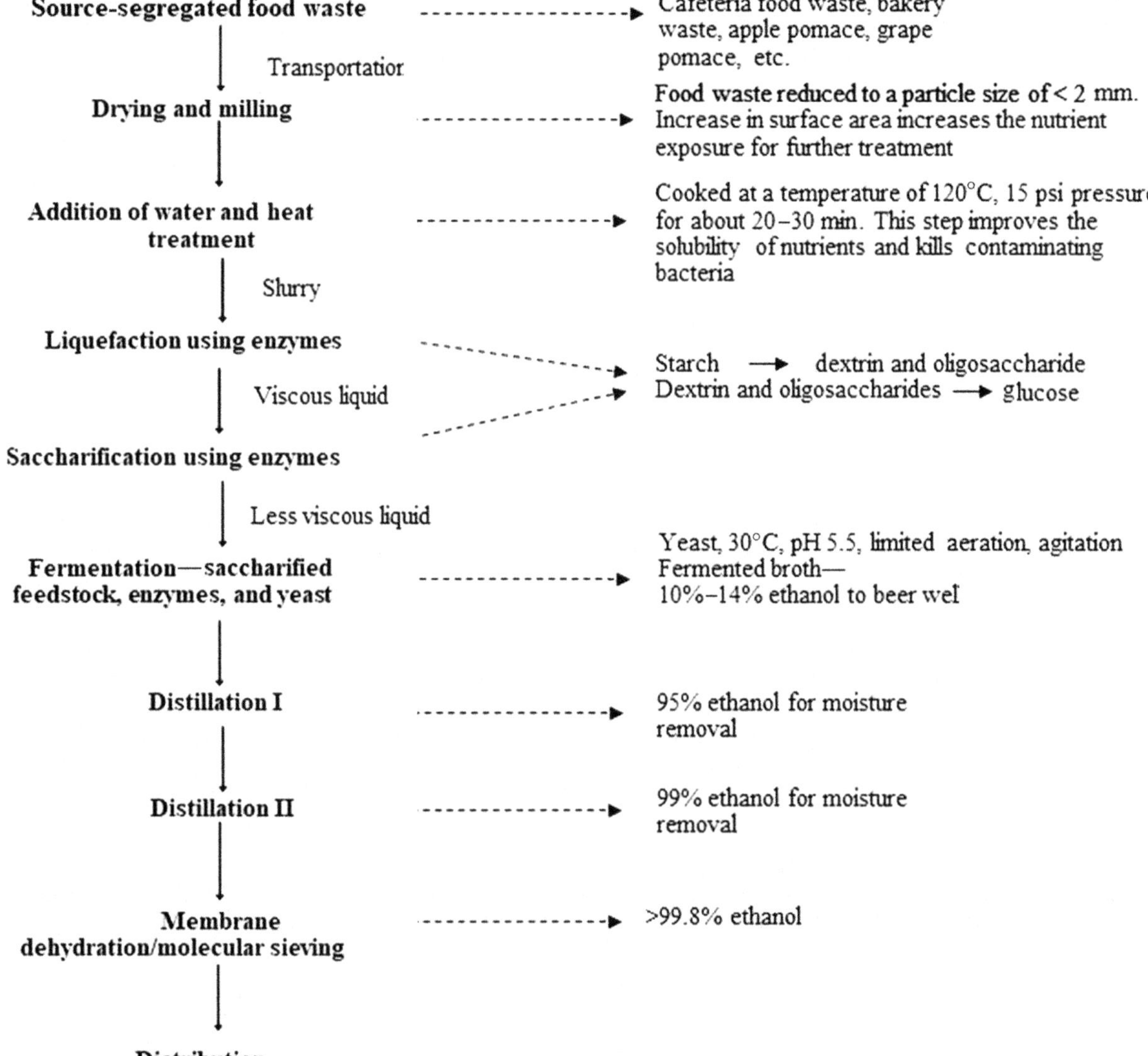

FIG. 5.4 General process flow for ethanol production with food waste as primary feedstock (Mosier and Ileleji, 2002; RFA, 2017a, b; Chopade et al., 2015; Sivasakthivelan et al., 2014).

5.3.1 Feedstock Preparation

As stated earlier, in commercial ethanol production corn is used as the major feedstock in the United States. Starch is the major carbon source in corn and must be made available to further hydrolysis by grinding the grain. A hammer mill is most commonly used industrially to grind the corn grains to a preset particle size. The ground feedstock is passed through a screen to select the particle size, with any material that did not pass through the screen subjected to grinding again. The average particle size of the feedstock is important in determining the efficiency of the process. A particle size of 0.5 mm corn yielded an ethanol concentration of 12.6% (v/v), while a particle size of 5 mm yielded only 1.62%(v/v) (Naidu et al., 2007). The smaller particle size increases soluble solids concentration, easing the mass transfer of nutrients from medium to the microbial cells. Several studies involving food waste as the feedstock have reported a range of particle sizes varying between 355 μm and 3 mm (Meenakshi and Kumaresan, 2017; Matsakas et al., 2014; Moukamnerd et al., 2013). A particle size >3 mm is not advisable for food waste since it contains complex nutrients which would otherwise remain unexposed to microbial activity, leading to incomplete conversion.

5.3.2 Pretreatment

Pretreatment begins with cooking the corn in water at a temperature of up to 120°C. This process improves the solubility of starch (gelatinization). Without heat treatment, the enzymes cannot break down the microcrystalline structure in starch. Therefore the feedstock is heated at high pressures before enzymatic treatment. The heat treatment also sterilizes the feedstock, avoiding any contamination by stray bacteria present. The glycosidic bonds called $\alpha(1\rightarrow6)$ and $\alpha(1\rightarrow4)$ link the repeating units of glucose in starch. Enzymatic hydrolysis using glucoamylase breaks down the $\alpha(1\rightarrow6)$ bonds and α-amylase breaks the $\alpha(1\rightarrow4)$ bonds (Onuki, 2017). The action of α-amylase is called liquefaction where starch is transformed into shorter chain glucose units called dextrin. Glucoamylase is used in the next step to hydrolyze the starch and dextrins into glucose further. The saccharification step reduces the viscosity of pretreated materials which helps improve the mass transfer during fermentation. The glucose produced during this step is directly fermented into ethanol. Often, the saccharification and fermentation steps occur simultaneously in the same reactor. Since food waste contains both starch and cellulose, further hydrolysis is necessary to break down the cellulose. Cellulose can be converted to glucose units either by acid hydrolysis (Dussan et al., 2014; Binder and Raines, 2010) or cellulase enzyme (Yamada et al., 2013; Sánchez and Cardona, 2008). Because food waste contains numerous microbial species, heat treatment must be done to kill the contaminating microbes. Moon et al. (2009) reported a high glucose yield of 46% from food waste with a combination of the enzymes amyloglucosidase and carbohydrase. The amount of carbohydrase determined the degree of saccharification. Acid and enzymatic hydrolysis are also well-adapted pretreatment methods in commercial-scale ethanol production. Other reported pretreatment methods are dilute acid hydrolysis, alkaline pretreatment, steam explosion, ammonia and carbon dioxide pretreatment, organosolv pretreatment, use of ionic liquids, sulfite and inorganic salt pretreatment, irradiation and microwave-assisted chemical conversion (Tutt et al., 2012; Ghasemzadeh et al., 2014; Bensah and Mensah, 2013).

5.3.3 Fermentation

Yeast is a facultative anaerobe, and under aerobic conditions, it produces carbon dioxide and water, while under anaerobic conditions, it converts sugars into ethanol and carbon dioxide. Therefore it is important to exclude oxygen in the fermenter during industrial ethanol production. A propagation step is employed before fermentation to reduce the lag time. In the propagation tank, the feedstock slurry, enzymes, water, and yeast are mixed to activate the yeast. In the absence of this step, a significant lag phase will occur in the fermenter. The fermenter conditions must be maintained at the optimum temperature, pH, reducing sugar levels, and osmotic stress. The yeast fermentation requires a pH of 5–5.5, and bacterial fermentation requires a pH of 6–6.5 (Sivasakthivelan et al., 2014; Narendranath and Power, 2005). Very low or very high pH of the fermenter makes the yeast expend extra energy in balancing the internal pH, hence reducing the productivity. High temperatures cause the yeast cells to die, and low temperatures make the enzymes deactivate which result in incomplete degradation of sugars into alcohol. Not maintaining appropriate conditions leads to contamination by other unwanted species of bacteria causing a batch failure. Fermentation is an exothermic reaction causing the process temperature to rise over time. Therefore the fermenters are equipped with cooling jackets in commercial facilities. After the fermentation time, the fermented broth is emptied into a "beer well." The beer well stores fermented broth from different batches to provide a continuous stream for ethanol recovery operations including distillation (Mosier and Ileleji, 2006). Ethanol production from food wastes with solid content as high as 35% w/w has been studied using mashed potatoes, sweet corn, and white bread collected from cafeterias, households, and retail stores (Huang et al., 2015a, b; Matsakas et al., 2014).

Ebner et al. (2014) have reported an ethanol yield of 295 L/dry ton of retail food waste cofermented with conventional feedstock. They also indicated that this process resulted in negative greenhouse gas emissions impact and approximately 500% improvement in environmental benefits compared to the conventional corn ethanol process. Though higher ethanol yield can potentially be achieved using food wastes, pretreatment is an essential step that accounts for a significant fraction of the production cost. If pretreatment methods are optimized to reduce the cost, these wastes can be potential substrates for production at commercial scale. Whey is typically used for production in the form of permeate where the sugars present are in concentrated form (Koushki et al., 2012). It is possible to reach the maximum theoretical yield of ethanol using whey as the feedstock, although the final production volume may be lower. Various researchers were able to achieve ethanol concentration of 8– 46 g/L using crude and pretreated whey (Hadiyanto et al., 2014; Koushki et al., 2012). Whey could act as a potential substrate for a small-scale ethanol production unit distributing the product locally. Apple pomace was explored as a substrate for alcohol production decades ago (Voget et al., 1985; Hang et al., 1981), but further efforts in research and development have been limited. One study has shown that apple pomace can be fermented in solid state, with or without pretreatment and supplements, to obtain a higher yield of ethanol (Hang et al., 2006). The maximum reported ethanol concentration from apple pomace is 190 g/kg using enzyme pretreatment and a vacuum extraction recovery method (Parmar and Rupasinghe, 2013). Therefore apple pomace appears to be a promising substrate; however, seasonal availability of apple pomace could pose a potential challenge in its viability as a substrate. The possibility of using tomato pomace has also been investigated (Lenucci et al., 2013); however, the yield achieved by conversion of untreated tomato pomace is not economically viable. Tomato pomace has not been directly used to produce ethanol, though a serum obtained from supercritical extraction of tomato pomace yielded an ethanol concentration of 46.8 g/L (Lenucci et al., 2013). Table 5.2 contains a summary of prior studies of ethanol production from food wastes.

5.3.4 Downstream Processing

Downstream processing involves all unit operations after fermentation that improve the purity of the final ethanol product. Ethanol is typically purified using combinations of distillation and molecular sieving. The fermented broth from the beer well is sent to a traditional distillation column to obtain 95% ethanol. Since this ethanol contains 5% moisture, it is dehydrated using azeotropic distillation. The presence of water increases the molecular polarity of ethanol, making it separate when mixed with gasoline (Onuki, 2017). In azeotropic distillation, a third chemical called an entrainer (e.g., benzene or cyclohexane) is added. Since azeotropic distillation is complicated and expensive, molecular sieving methods have been implemented recently in large-scale ethanol purification. The molecular sieves let the smaller water molecules (0.28 nm) pass through while retaining dehydrated ethanol (0.44 nm) (Chopade et al., 2015). It is possible to regenerate molecular sieves by heating or applying vacuum. As indicated in Table 5.2, many food waste feedstocks have ethanol yields lower than 5% w/v, illustrating the need to optimize more economical downstream processing technologies.

5.3.5 By-Products of Ethanol Production

Irrespective of the feedstock, carbon dioxide is continuously produced as a result of microbial metabolism during ethanol fermentation. Carbon dioxide is considered a secondary revenue source by the ethanol producers and in 2012, nearly one-third of the North American CO_2 supply was fulfilled by ethanol industries (Lane, 2012). The exhaust gases from fermentation processes contain up to 99% CO_2, which is therefore regarded as high grade after some minor processing to remove trace impurities. A typical ethanol plant with a capacity to produce 50 million gallons of ethanol a year will also produce about 150,000 metric ton of CO_2. According to current practice, ethanol producers sell CO_2 to a third-party refinery that purifies it and then sells for beverage, medical, or dry ice applications (Mosier and Ileleji, 2006). However, the revenue from refined CO_2 typically does not accrue to the ethanol producer, rather to refined CO_2 supplier. Therefore ethanol producers often cannot justify the large capital investment involved in CO_2 refining and marketing (Lane, 2012).

Solutions4CO_2 Inc., (S4CO2, now called BlueOcean NutraSciences) has developed an integrated biogas refinery process that takes stillage left over after fermentation to produce biogas. The company has also developed a process to infuse carbon dioxide into the stillage and use this for the cultivation of microalgae. Microalgae are marketed as high-value nutraceutical products. This low capital technology can be located at the ethanol plant to derive further economic value which accrues to the ethanol producer. The energy produced from biogas plants could also partially or completely fulfill the energy requirement of the ethanol plant and approach net-zero energy operation. Integrated ethanol fermentation and anaerobic digestion (Chapter 4) have also gained increased interest because fermentation results in a high chemical oxygen demand (COD) stillage suitable for anaerobic digestion, which can convert more than 50% of the inherent COD from stillage to energy-rich biogas (Cesaro and Belgiorno, 2015; Wilkie et al., 2000). Stillage properties from corn and

TABLE 5.2 Reported Studies of Bioethanol Production From Food Wastes

Food Waste Type	Organism	Pretreatment Method	Process Conditions	Fermentation Time (h)	Ethanol Output: Concentration (g/L)	Ethanol Output: g/g Wet FW	References
Kitchen waste	*S. cerevisiae*	Enzymatic hydrolysis	Batch fermentation 100 mL, 30°C, 150 rpm	58.8	32.2	0.58	Uncu and Cekmecelioglu (2011)
Kitchen garbage	*Zymomonas mobilis*	Enzymatic hydrolysis using protease and amylase	Batch fermentation, 150 mL, 30°C	40	53.4		Ma et al. (2008)
Household food waste	Baker's yeast	Enzymatic and microwave heat treatment	Batch fermentation, 100 mL, 30°C, 100 rpm	15	42.8	0.14	Matsakas et al. (2014)
Cafeteria food waste (Korea)	*S. cerevisiae*	Grinding and enzymatic hydrolysis	Batch fermentation in 5 L fermenter with 3 L working volume, 35°C, pH 4.5	24	28.32	0.16	Kim et al. (2011a, b)
Retail store waste (mainly containing mashed potatoes, sweet corn, and bread)	NR	Enzymatic hydrolysis with stargen (starch hydrolyzing enzyme) and protease	3 L fermenter, 32°C, 30 rpm	72	144	0.33	Huang et al. (2015a, b)
Dining hall food waste (China)	*S. cerevisiae*	Enzymatic	1000 L fermenter with 700 L working volume	60	94	0.10	Yan et al. (2013)
Bread waste	Yeast (sp not mentioned)	Amylase and protease	Batch fermentation, 300 mL, 35°C, 150 rpm	72	279.6[a]	0.35	Pietrzak and Kawa-Rygielska (2014)
Bread crust	Commercial yeast "Super Camellia"	Enzymatic hydrolysis with α-amylase and glucoamylase	Fed-batch fermentation rotating drum reactor with a humidifier and condenser, 32°C, 5 rpm	30	NR	0.27	Moukamnerd et al. (2013)
Bakery wastes (bread, biscuits, buns, cakes, donuts, potato chips, and flour) and cheese whey	*S. cerevisiae* (distiller's yeast)	Enzymatic hydrolysis using termamyl/AMG at 70°C and pH 5.0	Batch fermentation in 14 L Microferm benchtop fermenter, 30°C, pH 5.0	14	255.9[a]	0.25	Kumar et al. (1998)

Continued

TABLE 5.2 Reported Studies of Bioethanol Production From Food Wastes—cont'd

Food Waste Type	Organism	Pretreatment Method	Process Conditions	Fermentation Time (h)	Ethanol Output		References
					Concentration (g/L)	g/g Wet FW	
Potato chips	Commercial yeast "Super Camellia"	Enzymatic hydrolysis with α-amylase and glucoamylase	Continuous fermentation in a rotating drum reactor, 32°C, 5 rpm	NR	32	0.24	Moukamnerd et al. (2013)
Waste potato mash	*S. cerevisiae*	Liquefaction and saccharification using amylase and amyloglucosidase	Batch fermentation in 2.5 L fermenter at 30°C, pH 5.5 and 400 rpm	18	31	0.21	Izmirlioglu and Demirci (2012)
Potato peel waste	*S. cerevisiae* var. *bayanus*	Enzymatic hydrolysis with a combination of three enzymes	Batch fermentation in 250 mL Erlenmeyer flask at 30°C, 100 rpm and pH 5.0	48	7.6	<0.1	Izmirlioglu and Demirci (2012)
Sweet potato waste	*S. cerevisiae*	Enzymatic hydrolysis using cellulase and pectinase	Batch fermentation, 30°C	72	79	0.23	Wang et al. (2016)
Apple pomace	*S. cerevisiae*	Dilute acid hydrolysis and laccase enzyme	Batch fermentation with 5 L working volume fermentation bucket, 25°C	168	149.9[a]	0.4	Parmar and Rupasinghe (2013)
Apple pomace	*S. cerevisiae* Montrachet strain 522 w	None	Solid-state fermentation in 3.8 L mason jars fitted with a CO_2 exhaust tube at 30°C		43		Hang et al. (1981)
Apple pomace amended with molasses	*S. cerevisiae*	Enzymatic hydrolysis of apple pomace with α-amylase and then amended with molasses	Batch fermentation, 2 L, 30°C and 40 rpm	62.1	72		Kanwar et al. (2012)
Grape pomace	*S. cerevisiae*	None	Solid-state fermentation (SSF), 28°C	48	39.9[a]	0.05	Rodríguez et al. (2010)
Grape pomace	*Pichia rhodanensis*	Enzymatic hydrolysis using cellulase and pectinase	Batch fermentation at 30°C	48	18.1		Korkiel and Janse (2002)

Carrot pomace	Thermo-tolerant *Kluyveromyces marxianus*	Enzymatic hydrolysis with Accellerase 1000 and pectinase	Solid-state fermentation in 1 L custom made jar fermenters at 42°C, 680 rpm and initial pH of 5.0	37	42	0.18	Yu et al. (2013)
Cola-based sweet beverage	*Saccharomyces cerevisiae* var. *Windsor*	The sweet beverage medium was degassed before use	Batch fermentation, 500 mL, 30°C	8	55	0.51	Isla et al. (2013)
Tomato serum left after sauce production	Immobilized *S. cerevisiae*	Enzymatic hydrolysis using Driselase	Batch fermentation	10	23.7		Lenucci et al. (2013)
Whey permeate	*K. marxianus*	Whey was deproteinated by boiling and filtration and concentrated using vacuum evaporation	Batch fermentation at 30°C, pH 4.6	60	46		Koushki et al. (2012)
Cheese whey	*K. marxianus*	Cheese whey supplemented with yeast extract	Fed-batch fermentation	4	8	0.21	Hadiyanto et al. (2014)
Lipid extracted coffee grounds	*S. cerevisiae*	Dilute sulfuric acid hydrolysis and autoclaving	Batch fermentation, 250 mL, 30°C	10	17.2		Rocha et al. (2014)
Pineapple and banana peels	Coculture of *Aspergillus niger* and *S. cerevisiae*	Oven dried and ground before fermentation. *A. niger*-assisted hydrolysis	Batch fermentation, 28°C, pH 5.5	96	32–49.34		Shilpa et al. (2013)
Banana peels from a processing plant	*S. cerevisiae*	Drying, grinding, steam treatment, and acid hydrolysis	Batch fermentation, 250 mL, 30°C, 200 rpm, pH 5–5.5	NR	NR	0.45	Gebregergs and Sahu (2016)

NA, not applicable; *NR*, not reported.
[a]*Concentration converted from g/kg to g/L using density of ethanol.*

lignocellulosic biomass have been widely studied (e.g., Wilkie et al., 2000), but the properties of food waste residue from fermentation are not well known. Western Plains Energy, based in Oakley, Kansas started producing biogas from manure, stillage, and food waste to power its own 50 million gallon per year ethanol plant in 2013 (Kryzanowsky, 2013). Ethanol plants need approximately 20–30 MWh energy to operate (Solda, 2010) and colocating an anaerobic digester to generate heat or electricity can offer significant economic benefits.

5.4 BIOBUTANOL FROM FOOD WASTE

Butanol, a four-carbon alcohol, is a priority bulk material and an industrial solvent for the synthesis of a variety of chemical products. Butanol is also a high-octane biofuel with properties theoretically superior to those of ethanol (Dürre, 2011). These properties include higher specific energy content, noncorrosiveness, ability to be used without engine modifications, nonhygroscopic nature, and ability to be used in existing pipelines (Stombaugh et al., 2012; Nigam and Singh, 2011; Jin et al., 2011). Butanol, like ethanol, helps to reduce carbon monoxide levels in the engine exhaust since it is a good oxygenator (Stombaugh et al., 2012) and therefore is a potential substitute for ethanol. Its nonhygroscopic nature makes fuels with butanol have a longer shelf life. Butanol is less corrosive than ethanol, which means there would be more possibility for distributing butanol through pipelines instead of batch movement by truck or rail, as is typically done for ethanol. The barriers to large-scale fermentative production of butanol are the lower yield and low butanol tolerance of microbes. If these issues could be addressed effectively, fermentative production is in principle a more economical option than chemical synthesis methods. Currently, butanol is synthesized chemically for industrial uses, and much of the research is based on fermentation of lignocellulosic biomass. Several researchers, however, have reported the use of food waste feedstocks, including apple pomace (Kwak and Kang, 2006) and bakery waste (Ujor et al., 2014a), and dairy waste (Ujor et al., 2014b) in 1-butanol production. Whey from cheese processing has been reported by several researchers as a potential feedstock for butanol production. In addition to 1-butanol, there are limited research efforts on fermentative production of 3-methyl 1-butanol, also known as isobutanol. This branched chain alcohol has several properties superior to those of 1-butanol: higher heating value and carbon content, and greater stability as it has more side chains than 1-butanol. However, only a limited number of species of microorganisms, particularly ethanol-producing yeasts and recombinant bacteria, possess the capability to produce isobutanol (Branduardi et al., 2013). Although some food wastes have shown potential to produce isobutanol, the observed yields are impractically low to be economical; however, there could be a compelling research opportunity to genetically modify microbes to produce high amounts of isobutanol. One such effort explored production of isobutanol using genetically modified *E. coli* to ferment protein-rich wastes from food processing operations. These researchers were able to obtain an isobutanol concentration of 4 g/L through this research (Huo et al., 2011). Their research is promising in exploring the use of protein-rich food waste materials like meat processing waste in alcohol production that is otherwise unsuitable for anaerobic digestion or composting. Although it is known that butanol may be a superior fuel to ethanol, the primary barrier to its large-scale production is that the yield is often considerably lower due to poor tolerance of the microbes producing it. The highest concentration of butanol that can be tolerated by microbes is 2% v/v (López-Contreras et al., 2000). The production rate of butanol starts to drop after 1% v/v butanol is produced in the fermentation broth. The lower yields increase the energy requirement of distillation, thus increasing the overall cost of production. Reported data on 1-butanol and isobutanol production from various food wastes are listed in Table 5.3.

5.5 BIOHYDROGEN FROM FOOD WASTE FERMENTATION

Hydrogen is one of the cleanest fuels since it only produces water when combusted. Hydrogen production is a possible platform for efficient utilization of food waste due to the possibility of using different food waste materials. Three different mechanisms have been proposed in the literature for the production of biohydrogen: photolysis of water by algae, dark fermentation during the acidogenic phase of anaerobic digestion (hybrid biogas and biohydrogen), and two-stage dark/photo fermentation (Kapdan and Kargi, 2006). At present, most hydrogen is commercially produced by steam reforming of methane in natural gas, often extracted by hydro-fracking operations. Biohydrogen production is an attractive alternative to demand for fossil natural gas, since many species of bacteria have the capability to produce hydrogen via fermentation of a broad range of organic feedstocks. The hydrogen production from a few representative food wastes is outlined in Table 5.4, and a comprehensive review of hydrogen production from food wastes and food processing wastes is available in the literature (Yasin et al., 2013). The methods used for hydrogen production vary widely compared to alcohol production and have not yet been standardized. The lack of a standard method for production and uncertainty in yields makes it difficult to commercialize hydrogen production for energy generation at this time. There are a limited number of research efforts to directly couple hydrogen fermentation with fuel cell technologies. Rahman et al. (2015) provided an overview of

TABLE 5.3 Reported Data on Production of Biobutanol From Food Wastes

Food Waste	Product	Yield (g/L)	References
Retail waste	ABE	20.9	Huang et al. (2015b)
Apple and pear peels	ABE	20	Raganati et al. (2016)
Household organic waste	ABE	75[a]	López-Contreras et al. (2000)
Milk dust powder	1-Butanol	7.3	Ujor et al. (2014a)
Bakery waste (inedible dough, batter liquid, and waste breading)	1-Butanol	14.4–15.1	Ujor et al. (2014b)
Apple droppings and waste corn	1-Butanol	9.8	Jesse et al. (2002)
Waste potatoes	1-Butanol	4.7	Patáková et al. (2009)
Beverage waste	1-Butanol	12.8–14	Raganati et al. (2014)
Waste potato starch	1-Butanol	15.3	Kheyrandish et al. (2015)
Wastewater from palm oil processing	1-Butanol	10.4	Hipolito et al. (2008)
Cheese whey	1-Butanol	5	Raganati et al. (2013)
Apple pomace	1-Butanol	17.6[b]	Voget et al. (1985)
Acid whey	1-Butanol	12.2	Stoeberl et al. (2011)
Apple pomace	3-Methyl 1-butanol	0.12	Hang et al. (1981)
Whey permeate	3-Methyl 1-butanol	0.02	Parrondo et al. (2000)
Protein waste from food processing	3-Methyl 1-butanol	4	Huo et al. (2011)

ABE, acetone-butanol-ethanol.
[a]*Yield in g/kg.*
[b]*Units converted from g/kg to g/L using density of 1-butanol.*

various technologies available for biohydrogen production and their integration into fuel cell systems. It has been reported that up to 5 kW power generation is possible using a proton exchange membrane fuel cell integrated with a dark fermentation bioreactor of 1500 L volume (Rahman et al., 2015). An integrated biohydrogen refinery approach suggested a combined fermentation and hydrothermal process for biohydrogen production from food waste, coupled with power generation using a fuel cell. In this study, an experimental energy balance on the integrated biorefinery showed that the energy output from the fuel cell would be 157 kWh as electricity and 52 kWh as waste heat when processing 1 ton of food waste (Redwood et al., 2012). An integrated system was also studied for fermentative hydrogen production of other organic wastes like sludge and continuous electricity generation using a proton exchange membrane fuel cell (Guo et al., 2012). Campo et al. (2012) used a high-temperature proton exchange membrane fuel cell to utilize biohydrogen from fruit juice industry wastewater and reported 35% system efficiency.

5.6 FUTURE PERSPECTIVE AND RESEARCH NEEDS

Although food wastes are rich in organic content and have great potential for waste-to-energy conversion, storage and transportation pose significant operational challenges. It is always difficult and inefficient to transport liquid food wastes, unless there are dedicated pipelines available to pump material from food processing operations to a centralized conversion facility. Solid food wastes must be dried before they are stored and transported to avoid faster decay of organic materials, but drying incurs additional handling and energy costs. The production capacity of corn ethanol outnumbers all other technologies including lignocellulose, municipal and food processing wastes. The potential barriers of each new raw material in the production of alcohols would need careful consideration and systematic planning. Fig. 5.5 lists the barriers to commercial production of biofuels by fermentation using food waste as a feedstock.

TABLE 5.4 Representative Studies of Hydrogen Production From Food Wastes

Food Waste	Fermentation Type	Hydrogen Yield	References
Canteen waste	Batch fermentation with simultaneous enzymatic hydrolysis	74 mL H_2/g FW[a]	Han et al. (2016a, b)
Cafeteria food waste	Dark fermentation using *Clostridium beijerinckii*	55 mL H_2/g COD of FW	Kim et al. (2008)
Canteen waste	Combined solid-state and dark fermentation	39 mL H_2/g FW	Han et al. (2015)
Tofu processing wastewater	Anaerobic fermentation using mixed microbial culture	37 mL/ g FW[b]	Kim et al. (2011a, b)
Tofu processing wastewater	Continuous fermentation with sewage sludge inoculum	107.5 mL H_2/g COD of FW	Lay et al. (2013)
Cheese whey	Continuous dark fermentation under thermophilic conditions	20 mL H_2/L cheese whey[c]	Azbar et al. (2009)
Apple waste	Batch fermentation using photosynthetic bacteria	111.9 mL H_2/g total solids	Lu et al. (2016)
Apple pomace	Anaerobic fermentation with river sludge	101 mL H_2/g total solids	Feng et al. (2010)
Waste bread	Continuous anaerobic fermentation using sludge	109 mL/g FW	Han et al. (2016b)
Pineapple waste	Anaerobic mixed culture fermentation	77.3 mL H_2/g volatile solids	Reungsang (2013)
Mixed fruit peel waste	Continuous fermentation using up-flow anaerobic contact filter	42.8 mL H_2/g FW[d]	Vijayaraghavan et al. (2007)
Banana peels	Two-phase batch anaerobic fermentation for coproduction of methane and hydrogen	209.9 mL H_2/g volatile solids	Nathoaa et al. (2014)

[a]*Converted from mL H_2/g glucose using 0.3 g glucose/g FW as reported by the authors.*
[b]*Converted from mL H_2/gVS using the reported VS data.*
[c]*Converted from m mol/g COD using COD data reported in the article for cheese whey and hydrogen density.*
[d]*Converted using reported biogas yield of 0.73 m^3/kg with 63% hydrogen and 9.3% VS in the waste.*

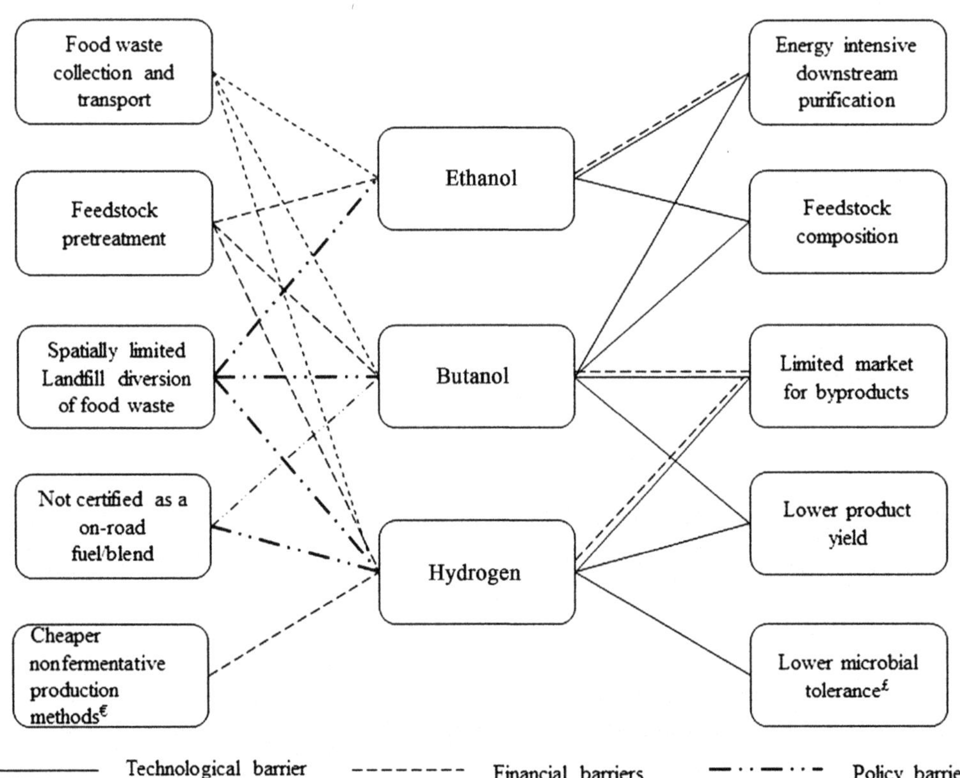

FIG. 5.5 Technological, economic and policy barriers to fermentative production of biofuels £butanol and hydrogen inhibit microbial metabolism after they reach a certain concentration in fermenter; €steam reforming or other fossil fuel-based methods.

Food waste is not currently mandated under the RFS as a feedstock for biofuel production which makes the use of food waste even more challenging. The environmental impact of producing various biofuels from food wastes has not been thoroughly analyzed, even though it is apparent that using food waste offers greater benefits compared to purpose grown crops. Certain policy implementations are necessary for encouraging biofuel production using waste feedstocks. For example, the city of Chicago acquired 1860 light duty E85 vehicles in 1995 as a part of emission reduction initiative and this has led to an annual reduction of 1.2 million gallons of gasoline usage (AFDC, 2014). Such initiatives would help build the local economy by encouraging biofuel production using food waste generated within a locality. Butanol is classified as a fuel under RFS, but currently large refineries cannot sell butanol as a blend for on-road applications since it is not registered under 40 CFR (Code of Federal regulations) (AFDC, 2017). However, smaller manufacturers that are not regulated under 40 CFR have sold bio-butanol as a jet fuel, marine fuel, and bulk material. Owing to the current regulations, biobutanol production is feasible only as a small-scale decentralized option. If the technological and regulatory barriers for large-scale production of butanol are overcome, the existing ethanol plants can be retrofitted to produce butanol.

The US Department of Energy (USDOE) has set a research and development goal to achieve an untaxed cost of hydrogen of $4 per gasoline gallon equivalent (DOE-EERE, 2017) and 1 kilogram of hydrogen is effectively equivalent to 1 gallon of gasoline on an energy content basis. Currently, cost is the overall challenge to hydrogen production owing to feedstock pretreatment, inefficiency of biochemical conversion processes, and requirement for downstream purification. The technical requirements specifically set by USDOE for microbial production of hydrogen by 2020 target a twofold increase in the yield of hydrogen relative to the maximum reported yield, and reducing feedstock cost by 20% compared to the 2015 target (DOE, 2015). Based on the current status of biohydrogen production, an integrated approach could help fulfill the parasitic load on process energy requirement, however, may not be able to compete with renewable electricity infrastructure. Small-scale application offers a decentralized solution for food processing operations to manage their waste and generating energy to fulfill or partially fulfill a plant's electricity or steam requirements.

Food waste is available as a distributed resource, as opposed to conventional feedstocks that makes its transportation and storage challenging, but using food waste as raw material poses other logistical constraints. There is need to establish the transportation routes from food waste generators to centralized fermentation plants for efficient utilization of this resource. In addition, several technological barriers make the large-scale production of advance biofuels from food waste challenging. While ethanol production from food waste can be integrated with existing corn ethanol plants, production of advanced alcohols needs to overcome several technological and economic barriers. There is a need for genetically modifying microbes to tolerate high concentrations of butanol. Catalytic conversion of bioethanol to higher alcohols like butanol could offer benefits in terms of energy value, transport and storage, as well as minimizing requirements for modification of conventional gasoline engines (Bernstein, 2013). Catalytic conversion could also enable integration of butanol production into the existing ethanol infrastructure. As demonstrated by several researchers, the yield of higher alcohols from food waste is low unless a product recovery technique is employed, but such techniques are often energy intensive and increase the cost of downstream processing (Patáková et al., 2009; Jesse et al., 2002). There is a significant need for developing an economic product recovery method to make food waste-based alcohol production feasible, as well as further research to achieve efficient use of by-products of the process to improve economics. The efficient technology integration to upgrade hydrogen to beneficial use needs further research, as does assessment of the relative merits of centralized or independent small- and medium-scale biofuel production facilities.

5.7 CONCLUSIONS

Waste composition varies widely across the various sectors of the food supply chain, which makes standardizing a food waste fermentation production process difficult. Food waste can be utilized in ethanol production to a significant extent, but producing other higher alcohols needs extensive research. Though there are a number of published research articles in this area, only a few studies have shown promising results in terms of alcohol yields. Food waste generation will continue to increase in the future and must be effectively utilized. Comprehensive understanding of waste composition and generation quantities would help in standardizing a production process, and an industrial ecology approach should be applied to map the food waste resource flow in a particular region, and thereby provide an estimate of the share of food wastes available for the production of various alcohols. Before considering use of food wastes for large-scale production of biofuels, it is important to evaluate the life cycle energy requirements and environmental impact of the production process. A thorough economic evaluation of the technology, particularly for upstream processing, is important as using food waste could significantly reduce the cost of raw material and production.

REFERENCES

AFDC. 2014. City of Chicago Program Encourages Petroleum Displacement and Collaboration Between Departments. Alternative Fuel Data Center. May 16, 2014. https://www.afdc.energy.gov/case/1844.

AFDC. 2017. Alternative Fuels Data Center: Ethanol Production. Accessed May 2017. https://www.afdc.energy.gov/fuels/emerging_biobutanol.html.

Azbar, N., Dokgöz, T., Keskin, R., Eltem, K., Korkmaz, S., Gezgin, Y., Akbal, Z., 2009. Comparative evaluation of bio-hydrogen production from cheese whey wastewater under thermophilic and mesophilic anaerobic conditions. Int. J. Green Energy 6 (2), 192–200. https://doi.org/10.1080/15435070902785027.

Azreen, N.A., Norliana, A.S., Abd-Aziz, S.S., Hassan, M.A., 2014. Optimization of bioethanol production from glycerol by Escherichia coli SS1. Renew. Energy 66, 625–633. https://doi.org/10.1016/j.renene.2013.12.032.

Baratti, J.C., Bu'lock, J.D., 1986. Zymomonas mobilis: a bacterium for ethanol production. Biotechnol. Adv. 4 (1), 95–115. https://doi.org/10.1016/0734-9750(86)90006-6.

Bensah, E., Mensah, M., 2013. Chemical pretreatment methods for the production of cellulosic ethanol: technologies and innovations. Int. J. Chem. Eng. 2013, 1–21. https://doi.org/10.1155/2013/719607.

Bernstein, M., 2013. Cost-saving measure to upgrade ethanol to butanol—a better alternative to gasoline. American Chemical Society News. April 11, 2013, https://www.acs.org/content/acs/en/pressroom/newsreleases/2013/april/cost-saving-measure-to-upgrade-ethanol-to-butanol-a-better-alternative-to-gasoline.html.

Binder, J.B., Raines, R.T., 2010. Fermentable sugars by chemical hydrolysis of biomass. Proc. Natl Acad. Sci. U. S. A. 107 (10), 4516–4521. National Academy of Sciences, https://doi.org/10.1073/pnas.0912073107.

Branduardi, P., Longo, V., Berterame, N.M., Rossi, G., Porro, D., 2013. A novel pathway to produce butanol and isobutanol in Saccharomyces cerevisiae. Biotechnol. Biofuels 6, 68. https://doi.org/10.1186/1754-6834-6-68.

Campo, A.G., Cañizares, P., Lobato, J., Rodrigo, M.A., Fernadez, F.J., 2012. Electricity production by integration of acidogenic fermentation of fruit juice wastewater and fuel cells. Int. J. Hydrogen Energy 37 (11), 9028–9037. https://doi.org/10.1016/j.ijhydene.2012.03.007.

Casey, E., Mosier, N., Adamec, J., Stockdale, Z., Ho, N., Sedlak, M., 2013. Effect of salts on the co-fermentation of glucose and xylose by a genetically engineered strain of Saccharomyces cerevisiae. Biotechnol. Biofuels 6, 83. https://doi.org/10.1186/1754-6834-6-83.

Cesaro, A., Belgiorno, V., 2015. Combined biogas and bioethanol production: opportunities and challenges for industrial application. Energies 8121–8144. https://doi.org/10.3390/en8088121.

Chin, W., Lin, K., Chang, J., Huang, C., 2013. Improvement of N-butanol tolerance in Escherichia coli by membrane-targeted tilapia metallothionein. Biotechnol. Biofuels 6, 130. https://doi.org/10.1186/1754-6834-6-130.

Chopade, V.J., Khandetod, Y.P., Mohod, A.G., 2015. Dehydration of ethanol-water mixture using 3a zeolite adsorbent. Int. J. Emerg. Technol. Adv. Eng. 5 (11), 152–155. http://www.ijetae.com/files/Volume5Issue11/IJETAE_1115_27.pdf.

DOE. 2015. Hydrogen production. https://energy.gov/sites/prod/files/2015/06/f23/fcto_myrdd_production.pdf.

DOE-EERE. Office of Energy Efficiency and Renewable Energy. 2017 Hydrogen production. Accessed September 2017. https://energy.gov/eere/fuelcells/hydrogen-production.

Dürre, P., 2011. Fermentative production of butanol—the academic perspective. Curr. Opin. Biotechnol. 22 (3), 331–336. https://doi.org/10.1016/j.copbio.2011.04.010.

Dussan, K.J., Silva, D., Moraes, E.J., Arruda, P.V., Filipe, M., 2014. Dilute acid hydrolysis of cellulose to glucose from sugarcane bagasse. Chem. Eng. Trans. 38, 433–438. https://doi.org/10.3303/CET1438073.

Ebner, J., Babbitt, C., Winer, M., Hilton, B., Williamson, A., 2014. Life cycle greenhouse gas (GHG) impacts of a novel process for converting food waste to ethanol and co-products. Appl. Energy 130, 86–93. https://doi.org/10.1016/j.apenergy.2014.04.099.

EIA. 2017. Monthly Energy Review. US Energy Information Administration. https://www.eia.gov/totalenergy/data/monthly/#renewable.

EPM, 2017. Ethanol Producer Magazine: US ethanol plants. http://www.ethanolproducer.com/plants/listplants/US/Operational/All.

Feng, X., Wang, H., Wang, Y., Wang, X., Huang, J., 2010. Biohydrogen production from apple pomace by anaerobic fermentation with river sludge. Int. J. Hydrogen Energy 35 (7), 3058–3064. https://doi.org/10.1016/J.IJHYDENE.2009.07.015.

Förster, A.H., Gescher, J., 2014. Metabolic engineering of Escherichia coli for production of mixed-acid fermentation end products. Front. Bioeng. Biotechnol. 2, 16. https://doi.org/10.3389/fbioe.2014.00016.

Gebregergs, A., Sahu, O., 2016. Industrial ethanol from banana peels for developing countries: response surface methodology. Pac. Sci. Rev. A: Nat. Sci. Eng. 18 (1), 22–29. https://doi.org/10.1016/J.PSRA.2016.06.002.

Ghareib, M., Youssef, K.A., Khalil, A.A., 1988. Ethanol tolerance of Saccharomyces cerevisiae and its relationship to lipid content and composition. Folia Microbiol. 33 (6), 447–452. https://doi.org/10.1007/BF02925769.

Ghasemzadeh, R., Kargar, A., Lotf, M., 2014. Comparison of pretreatment methods for biofuel production. International Conference on Agriculture, Food and Environmental Engineering, Jan 15–16, 2014, Kuala Lumpur. http://psrcentre.org/images/extraimages/3114517.pdf (Accessed: 10 May 2017).

Guo, K., Hassett, D.J., Gu, T., 2012. Microbial fuel cells: electricity generation from organic wastes by microbes. In: Arora, R. (Ed.), Microbial Biotechnology: Energy and Environment. CAB International, Oxon, UK, pp. 162–189. ISBN 978-1845939564.

Hadiyanto, D., Ariyanti, D., Aini, A.P., Pinundi, D.S., 2014. Optimization of ethanol production from whey through fed-batch fermentation using Kluyveromyces marxianus. Energy Procedia 47, 108–112. https://doi.org/10.1016/j.egypro.2014.01.203.

Hahn-HäGerdal, B., Lindén, T., Senac, T., Skoog, K., 1991. Ethanolic fermentation of pentoses in lignocellulose hydrolysates. Appl. Biochem. Biotechnol. 28–29 (1), 131–144. https://doi.org/10.1007/BF02922595.

Han, W., Ye, M., Zhu, A., Zhao, H.T., Li, Y.F., 2015. Batch dark fermentation from enzymatic hydrolyzed food waste for hydrogen production. Bioresour. Technol. 191, 24–29. https://doi.org/10.1016/j.biortech.2015.04.120.

Han, W., Huang, J., Zhao, H., Li, Y., 2016a. Continuous biohydrogen production from waste bread by anaerobic sludge. Bioresour. Technol. 212, 1–5. https://doi.org/10.1016/j.biortech.2016.04.007.

Han, W., Yan, Y., Shi, Y., Gu, J., Tang, J., Zhao, H., 2016b. Biohydrogen production from enzymatic hydrolysis of food waste in batch and continuous systems. Sci. Rep. 6 (1), AN 38395, https://doi.org/10.1038/srep38395.

Hang, Y.D., Lee, C.Y., Woodams, E.E., Cooley, H.J., 1981. Production of alcohol from apple pomace. Appl. Environ. Microbiol. 42 (6), 1128–1129. http://www.ncbi.nlm.nih.gov/pubmed/16345905.

Hang, Y.D., Lee, C.Y., Woodams, E.E., 2006. A solid state fermentation system for production of ethanol from Apple Pomace. Food Sci. 1851–1852. https://doi.org/10.1016/j.rser.2017.07.012.

Hegde, S., Lodge, J.S., Trabold, T.A., 2018. Characteristics of food processing wastes and their use in sustainable alcohol production. Renew. Sust. Energy Rev. 81, 510–523. https://doi.org/10.1016/j.rser.2017.07.012.

Hipolito, C.N., Carmen, E.C., Badillo, M., Zarrabal, O.C., Mora, M., Flores, P.G., Cortazar, M.H., Ishizaki, A., 2008. Bioconversion of industrial wastewater from palm oil processing to butanol by Clostridium saccharoperbutylacetonicum N1-4 (ATCC 13564). J. Clean. Prod. 16 (5), 632–638. https://doi.org/10.1016/J.JCLEPRO.2007.02.005. https://www.ers.usda.gov/data-products/us-bioenergy-statistics/.

Huang, H., Qureshi, N., Chen, M., Liu, W., Singh, V., 2015a. Ethanol production from food waste at high solids content with vacuum recovery technology. J. Agric. Food Chem. 63 (10), 2760–2766. https://doi.org/10.1021/jf5054029.

Huang, H., Singh, V., Qureshi, N., 2015b. Butanol production from food waste: a novel process for producing sustainable energy and reducing environmental pollution. Biotechnol. Biofuels 8. https://doi.org/10.1186/s13068-015-0332-x.

Huo, Y., Cho, K.M., Rivera, J., Monte, E., Shen, C., Yan, Y., Liao, J., 2011. Conversion of proteins into biofuels by engineering nitrogen flux. Nat. Biotechnol. 29 (4), 346–351. https://doi.org/10.1038/nbt.1789.

Ingram, L., Clark, D. 1991. Ethanol production using engineered mutant *E. coli*. US 5028539. Issued 1991. https://www.google.com/patents/US5028539.

Isla, M., Comelli, R.N., Seluy, L.G., 2013. Wastewater from the soft drinks industry as a source for bioethanol production. Bioresour. Technol. 136, 140–147. https://doi.org/10.1016/j.biortech.2013.02.089.

Ito, T., Nakshimada, Y., Senba, K., Matsui, T., Nishio, N., 2005. Hydrogen and ethanol production from glycerol-containing wastes discharged after biodiesel manufacturing process. J. Biosci. Bioeng. 100 (3), 260–265. https://doi.org/10.1263/jbb.100.260.

Izmirlioglu, G., Demirci, A., 2012. Ethanol production from waste potato mash by using Saccharomyces cerevisiae. Appl. Sci. 2 (4), 738–753. https://doi.org/10.3390/app2040738.

Jesse, T.W., Ezeji, T.C., Qureshi, N., Blaschek, H.P., 2002. Production of butanol from starch-based waste packing peanuts and agricultural waste. J. Ind. Microbiol. Biotechnol. 29 (3), 117–123. https://doi.org/10.1038/sj.jim.7000285.

Jin, C., Yao, M., Liu, H., Lee, C.F., Ji, J., 2011. Progress in the production and application of N-butanol as a biofuel. Renew. Sust. Energy Rev. 15 (8), 4080–4106. https://doi.org/10.1016/j.rser.2011.06.001.

Kanwar, S., Kumar, G., Sahgal, M., Singh, A., 2012. Ethanol production through Saccharomyces based fermentation using apple pomace amended with molasses. Sugar Tech 14 (3), 304–311. https://doi.org/10.1007/s12355-012-0163-z.

Kapdan, I.K., Kargi, F., 2006. Bio-hydrogen production from waste materials. Enzyme Microb. Technol. 38, 569–582. https://doi.org/10.1016/j.enzmictec.2005.09.015.

Kheyrandish, M., Asadollahi, M., Jeihanipour, A., Doostmohammadi, M., Rismani-Yazdi, M., Karimi, K., 2015. Direct production of acetone–butanol–ethanol from waste starch by free and immobilized Clostridium acetobutylicum. Fuel 142, 129–133. https://doi.org/10.1016/j.fuel.2014.11.017.

Kim, S., Han, S., Shin, H., 2008. Optimization of continuous hydrogen fermentation of food waste as a function of solids retention time independent of hydraulic retention time. Process Biochem. 43 (2), 213–218. https://doi.org/10.1016/J.PROCBIO.2007.11.007.

Kim, J.H., Lee, J.C., Pak, D., 2011a. Feasibility of producing ethanol from food waste. Waste Manag. 31 (9), 2121–2125. https://doi.org/10.1016/j.wasman.2011.04.011.

Kim, M.S., Lee, D.Y., Kim, D.H., 2011b. Continuous hydrogen production from tofu processing waste using anaerobic mixed microflora under thermophilic conditions. Int. J. Hydrogen Energy 36 (14), 8712–8718. https://doi.org/10.1016/j.ijhydene.2010.06.040.

Korkiel, L.J., Janse, B.J.H., 2002. Utilising grape pomace for ethanol production. S. Afr. J. Enol. Vitic. 23 (1), 31–36. http://www.sawislibrary.co.za/dbtextimages/17546.pdf.

Koushki, M., Jafari, M., Azizi, M., 2012. Comparison of ethanol production from cheese whey permeate by two yeast strains. J. Food Sci. Technol. 49 (5), 614–619. https://doi.org/10.1007/s13197-011-0309-0.

Kryzanowsky, T., 2013. Himark BioGas closes loop between Kansas feedlot and ethanol plant. Full Circle. March 11, 2013, https://www.manuremanager.com/beef/full-circle-13074.

Kumar, J.V., Mathew, R., Shahbazi, A., 1998. Bioconversion of solid food wastes to ethanol. Analyst 123 (3), 497–502. https://doi.org/10.1039/a706088b.

Kwak, W.S., Kang, J.S., 2006. Effect of feeding food waste-broiler litter and bakery by-product mixture to pigs. Bioresour. Technol. 97 (2), 243–249. https://doi.org/10.1016/j.biortech.2005.02.008.

Lane, J., 2012. By-products: the key to ethanol's struggles: biofuels digest. *Biofuel Digest*, December 12, 2012, http://www.biofuelsdigest.com/bdigest/2012/12/12/by-products-the-key-to-ethanols-struggles/.

Lay, C.H., Sen, B., Huang, S.C., Chen, C.C., Lin, C.Y., 2013. Sustainable bioenergy production from tofu-processing wastewater by anaerobic hydrogen fermentation for onsite energy recovery. Renew. Energy 58, 60–67. https://doi.org/10.1016/j.renene.2013.03.011.

Lenucci, M.S., Durante, M., Anna, M., Dalessandro, G., Piro, G., 2013. Possible use of the carbohydrates present in tomato pomace and in byproducts of the supercritical carbon dioxide lycopene extraction process as biomass for bioethanol production. J. Agric. Food Chem. 61 (15), 3683–3692. https://doi.org/10.1021/jf4005059.

Letti, L.A., Karp, S.G., Woiciechowski, A., Soccol, C., 2012. Ethanol production from soybean molasses by Zymomonas mobilis. Biomass Bioenergy 44, 80–86. https://doi.org/10.1016/j.biombioe.2012.04.023.

Lin, Y., Zhang, W., Li, C., Sakakibara, K., Tanaka, S., Kong, H., 2012. Factors affecting ethanol fermentation using Saccharomyces cerevisiae BY4742. Biomass Bioenergy 47, 395–401. https://doi.org/10.1016/j.biombioe.2012.09.019.

López-Contreras, A.M., Claassen, P., Mooibroek, H., De Vos, W., 2000. Utilisation of saccharides in extruded domestic organic waste by Clostridium acetobutylicum ATCC 824 for production of acetone, butanol and ethanol. Appl. Microbiol. Biotechnol. 54, 162–167. https://doi.org/10.1007/s002530000374.

Lu, C., Zhang, Z., Ge, X., Wang, Y., Zhou, X., You, X., Liu, H., Zhang, Q., 2016. Bio-hydrogen production from apple waste by photosynthetic bacteria HAU-M1. Int. J. Hydrogen Energy 41 (31), 13399–13407. https://doi.org/10.1016/j.ijhydene.2016.06.101.

Ma, J., Wambeke, M.V., Carballa, M., Verstraete, M., 2008. Improvement of the anaerobic treatment of potato processing wastewater in a UASB reactor by co-digestion with glycerol. Biotechnol. Lett. 30 (5), 861–867. https://doi.org/10.1007/s10529-007-9617-x.

Matsakas, L., Kekos, D., Loizidou, M., Christakopoulos, P., 2014. Utilization of household food waste for the production of ethanol at high dry material content. Biotechnol. Biofuels 7 (1), 4. https://doi.org/10.1186/1754-6834-7-4.

Meenakshi, A., Kumaresan, A., 2017. Ethanol production from corn, potato peel waste and its process development. Int. J. ChemTech Res. 6 (5), 974–4290. http://sphinxsai.com/2014/vol6pt5/4/(2843-2853)S-2014.pdf.

Moon, H., Song, S., Kim, J., Shirai, Y., Lee, D., Kim, J., Chung, S., Kim, K., Oh, K., Cho, Y., 2009. Enzymatic hydrolysis of food waste and ethanol fermentation. Int. J. Energy Res. 33 (2), 164–172. https://doi.org/10.1002/er.1432.

Mosier, N.S., Ileleji, K., 2002. Purdue extension. Nutrition, pp. 10–12. Available at: https://www.extension.purdue.edu/extmedia/id/id-328.pdf. (Accessed: 17 September 2017).

Mosier, N., Ileleji, K. 2006. How fuel ethanol is made from corn. Bioenergy Purdue Extension. https://www.extension.purdue.edu/extmedia/id/id-328.pdf.

Moukamnerd, C., Kawahara, H., Katakura, Y., 2013. Feasibility study of ethanol production from food wastes by consolidated continuous solid-state fermentation. J. Sustain. Bioenergy Syst. 3, 143–148. https://doi.org/10.4236/jsbs.2013.32020.

Naidu, K., Singh, V., Johnston, D., Rausch, K., Tumbleson, M.E., 2007. Effects of ground corn particle size on ethanol yield and thin stillage soluble solids. Cereal Chem. J. 84 (1), 6–9. https://doi.org/10.1094/CCHEM-84-1-0006.

Narendranath, N.V., Power, R., 2005. Relationship between pH and medium dissolved solids in terms of growth and metabolism of Lactobacilli and Saccharomyces cerevisiae during ethanol production. Appl. Environ. Microbiol. 71 (5), 2239–2243. https://doi.org/10.1128/AEM.71.5.2239-2243.2005.

Nathoaa, C., Sirisukpocab, U., Pisutpaisala, N., 2014. Production of hydrogen and methane from banana peel by two phase anaerobic fermentation. Energy Procedia 50, 702–710. https://doi.org/10.1016/j.egypro.2014.06.086.

Nigam, P.S., Singh, A., 2011. Production of liquid biofuels from renewable resources. Prog. Energy Combust. Sci. 37 (1), 52–68. https://doi.org/10.1016/j.pecs.2010.01.003.

NREL (2017) The Biofuels Atlas. Accessed September 2017. https://maps.nrel.gov/biofuels-atlas/#/?aL=yilN7K%255Bv%255D%3Dt&bL=groad&cE=0&lR=0&mC=40.21244%2C-91.625976&zL=4.

Onuki, S. 2017. Bioethanol: industrial production process and recent studies. Accessed May 10, 2017. http://www.public.iastate.edu/~tge/courses/ce521/sonuki.pdf.

Panesar, P.S., Marwaha, S., Kennedy, J., 2006. Zymomonas mobilis: an alternative ethanol producer. J. Chem. Technol. Biotechnol. 81 (4), 623–635. John Wiley & Sons, Ltd. https://doi.org/10.1002/jctb.1448.

Parmar, I., Rupasinghe, V., 2013. Bio-conversion of apple pomace into ethanol and acetic acid: enzymatic hydrolysis and fermentation. Bioresour. Technol. 130, 613–620. https://doi.org/10.1016/j.biortech.2012.12.084.

Parrondo, J., Garcia, L., Diaz, M., 2000. Production of an alchoholic beverage by fermentation of whey permeate with Kluyveromyces fragilis I: primary metabolism. J. Inst. Brew. 106 (6), 367–375. https://doi.org/10.1002/j.2050-0416.2000.tb00527.x.

Patáková, P., Lipovský, J., Čížková, H., Fořtová, J., Rychtera, M., Melzoch, K., 2009. Exploitation of food feedstock and waste for production of biobutanol. Czech J. Food Sci. 27 (4), 276–283.

Pérez-Carrillo, E., Cortés-Callejas, M.L., Sabillón-Galeas, L.E., Montalvo-Villarreal, J.L., Canizo, J., Moreno-Zepeda, M.G., Serna-Saldivar, S.O., 2011. Detrimental effect of increasing sugar concentrations on ethanol production from maize or decorticated sorghum mashes fermented with Saccharomyces cerevisiae or Zymomonas mobilis: biofuels and environmental biotechnology. Biotechnol. Lett. 33 (2), 301–307. https://doi.org/10.1007/s10529-010-0448-9.

Pietrzak, W., Kawa-Rygielska, J., 2014. Ethanol fermentation of waste bread using granular starch hydrolyzing enzyme: effect of raw material pretreatment. Fuel 134, 250–256. https://doi.org/10.1016/j.fuel.2014.05.081.

Raganati, F., Olivieri, G., Procentese, A., Russo, M.E., Salatino, P., Marzocchella, A., 2013. Butanol production by bioconversion of cheese whey in a continuous packed bed reactor. Bioresour. Technol. 138, 259–265. https://doi.org/10.1016/j.biortech.2013.03.180.

Raganati, F., Procentese, A., Montagnaro, F., Olivieri, G., Marzocchella, A., 2014. Butanol production from leftover beverages and sport drinks. Bioenergy Res. 8 (1), 369–379. https://doi.org/10.1007/s12155-014-9531-8.

Raganati, F., Procentese, A., Olivieri, G., Russo, M., Marzocchella, A., 2016. Butanol production by fermentation of fruit residues. Chem. Eng. Trans. 49, 229–234. https://doi.org/10.3303/CET1649039.

Rahman, S.N.A., Masdar, M.S., Rosli, M.I., Majlan, E.H., Husaini, T., 2015. Overview of biohydrogen production technologies and application in fuel cell. Am. J. Chem. 5 (3A), 13–23. https://doi.org/10.5923/c.chemistry.201501.03.

Redwood, M.D., Orozco, R.L., Majewski, A.J., Macaskie, L.E., 2012. An integrated biohydrogen refinery: synergy of photofermentation, extractive fermentation and hydrothermal hydrolysis of food wastes. Bioresour. Technol. 119, 384–392. https://doi.org/10.1016/j.biortech.2012.05.040.

Reungsang, A., 2013. Bio-hydrogen production from pineapple waste extract by anaerobic mixed cultures. Energies 6, 2175–2190. https://doi.org/10.3390/en6042175.

RFA, 2017a. Renewable Fuel Standard. http://www.ethanolrfa.org/policy/regulations/renewable-fuel-standard/. Accessed June 2017.

RFA, 2017b. Renewable Fuel Association. How ethanol is made. http://www.ethanolrfa.org/how-ethanol-is-made/. Accessed June 2017.

Rocha, M., Matos, L., Lima, L., Figueiredo, P., Lucena, I., Fernandes, F., Gonçalves, L., 2014. Ultrasound-assisted production of biodiesel and ethanol from spent coffee grounds. Bioresour. Technol. 167, 343–348. https://doi.org/10.1016/j.biortech.2014.06.032.

Rodríguez, L.A., Toro, M.E., Vazquez, F., Correa-Daneri, M.L., Gouiric, S.C., Vallejo, M.D., 2010. Bioethanol production from grape and sugar beet pomaces by solid-state fermentation. Int. J. Hydrogen Energy 35 (11), 5914–5917. https://doi.org/10.1016/j.ijhydene.2009.12.112.

Sánchez, Ó.J., Cardona, C.A., 2008. Trends in biotechnological production of fuel ethanol from different feedstocks. Bioresour. Technol. 99 (13), 5270–5295. https://doi.org/10.1016/J.BIORTECH.2007.11.013.

Sarris, D., Papanikolaou, S., 2016. Biotechnological production of ethanol: biochemistry, processes and technologies. Eng. Life Sci. 16 (4), 307–329. https://doi.org/10.1002/elsc.201400199.

Schoutens, G.H., Nieuwenhuizen, M.C.H., Kossen, N.W.F., 1985. Continuous butanol production from whey permeate with immobilized Clostridium beyerinckii LMD 27.6. Appl. Microbiol. Biotechnol. 21 (5), 282–286. https://doi.org/10.1007/BF00252705.

Shilpa, C., Malhotra, G., Chanchal, 2013. Alcohol production from fruit and vegetable waste. Int. J. Appl. Eng. Res. 8, 1749–1756. http://www.ripublication.com/ijaer.html.

Sivasakthivelan, P., Saranraj, P., Sivasakthi, S., 2014. Production of ethanol by Zymomonas mobilis and Saccharomyces cerevisiae using sunflower head wastes—a comparative study. Int. J. Microbiol. Res. 5 (3), 208–216. https://doi.org/10.5829/idosi.ijmr.2014.5.3.8476.

Solda M. 2010. Ethanol from corn: analysis of using corn stover to supply heat and power to ethanol plants. http://tesi.cab.unipd.it/28973/1/Ethanol_from_corn_-.

Somda, M.K., Savadogo, A., Barro, N., Thinart, P., Traore, T.A., 2011. Effect of minerals salts in fermentation process using mango residues as carbon source for bioethanol production. Asian J. Ind. Eng. 3 (1), 29–38.

Stoeberl, M., Werkmeister, R., Faulstich, M., Russ, W., 2011. Biobutanol from food wastes—fermentative production, use as biofuel an the influence on the emissions. Procedia Food Sci. 1, 1867–1874. https://doi.org/10.1016/j.profoo.2011.09.274.

Stombaugh, T., Montross, M., Nokes, S., Gray, K. 2012. Butanol: the new biofuel. Coop Extension Service. U of Kentucky College of Agr. http://www2.ca.uky.edu/agcomm/pubs/aen/aen111/aen111.pdf.

Tahir, A., Aftab, M., Farasat, T., 2010. Effect of cultural conditions on ethanol production by locally isolated Saccharomyces cerevisiae BIO-07. J. Appl. Pharm. 3 (2), 72–78.

Tutt, M., Kikas, T., Olt, J., 2012. Engineering for rural development comparison of different pretreatment methods on degradation of rye straw. Engineering for Rural Development, pp. 412–416. http://www.tf.llu.lv/conference/proceedings2012/Papers/072_Tutt_M.pdf.

Udeh, H O., Kgatla, T E., Jideani, AIO. 2014. Effect of mineral ion addition on yeast performance during very high gravity wort fermentation. World Acad. Sci. Eng. Technol. 8 (11): 1208–16. doi:scholar. waset.org/1999.1/9999716.

Ujor, V., Bharathidasan, A.K., Cornish, K., Ezeji, T.C., 2014a. Evaluation of industrial dairy waste (milk dust powder) for acetone-butanol-ethanol production by solventogenic Clostridium species. SpringerPlus 3 (1), 387. https://doi.org/10.1186/2193-1801-3-387.

Ujor, V., Bharathidasan, A.K., Cornish, K., Ezeji, T.C., 2014b. Feasibility of producing butanol from industrial starchy food wastes. Appl. Energy 136, 590–598. https://doi.org/10.1016/j.apenergy.2014.09.040.

Uncu, O.N., Cekmecelioglu, D., 2011. Cost-effective approach to ethanol production and optimization by response surface methodology. Waste Manag. 31, 636–643. https://doi.org/10.1016/j.wasman.2010.12.007.

USDA ERS 2017 U.S. Bioenergy Statistics. Accessed September 2017. https://www.ers.usda.gov/data-products/us-bioenergy-statistics/us-bioenergy-statistics/#Supply and Disappearance.

Vijayaraghavan, K., Ahmad, D., Soning, C., 2007. Bio-hydrogen generation from mixed fruit peel waste using anaerobic contact filter. Int. J. Hydrog. Energy 32 (18), 4754–4760. https://doi.org/10.1016/j.ijhydene.2007.07.001.

Voget, C.E., Mignone, C.F., Ertola, R.J., 1985. Butanol production from apple pomace. Biotechnol. Lett. 7 (1), 43–46. Kluwer Academic Publishers, https://doi.org/10.1007/BF01032418.

Wang, M., Han, J., Dunn, J.B., Cai, H., Elgowainy, A., 2012. Well-to-wheels energy use and greenhouse gas emissions of ethanol from corn, sugarcane and cellulosic biomass for US use. Environ. Res. Lett. 7 (4), 45905. IOP Publishing, https://doi.org/10.1088/1748-9326/7/4/045905.

Wang, F., Jiang, Y., Guo, W., Niu, K., Zhang, R., Hou, S., Wang, M., 2016. An environmentally friendly and productive process for bioethanol production from potato waste. Biotechnol. Biofuels 9. https://doi.org/10.1186/s13068-016-0464-7.

Wilkie, A.C., Riedesel, K.J., Owens, J.M., 2000. Stillage characterization and anaerobic treatment of ethanol stillage from conventional and cellulosic feedstocks. Biomass Bioenergy 19 (2), 63–102. https://doi.org/10.1016/S0961-9534(00)00017-9.

Xu, Q., Wu, M., Hu, J., Gao, M.T., 2016. Effects of nitrogen sources and metal ions on ethanol fermentation with cadmium-containing medium. J. Basic Microbiol. 56 (1), 26–35. https://doi.org/10.1002/jobm.201500470.

Yamada, R., Nakatani, Y., Ogino, C., Kondo, A., 2013. Efficient direct ethanol production from cellulose by cellulase- and cellodextrin transporter-co-expressing *Saccharomyces cerevisiae*. AMB Express 3, 34. Springer, https://doi.org/10.1186/2191-0855-3-34.

Yan, S., Chen, X., Wu, J., Wang, P., 2013. Pilot-scale production of fuel ethanol from concentrated food waste hydrolysates using Saccharomyces cerevisiae H058. Bioprocess Biosyst. Eng. 36 (7), 937–946. https://doi.org/10.1007/s00449-012-0827-9.

Yang, S., Fei, Q., Zhang, Y., Lopez-Contreras, M., Utturkar, S.M., Brown, S.D., Himmel, M.E., Zhang, M., 2016. Zymomonas mobilis as a model system for production of biofuels and biochemicals. Microb. Biotechnol. 9 (6), 699–717. https://doi.org/10.1111/1751-7915.12408.

Yao, L., Lee, S.L., Wang, T., de Moura, J., Johnson, L.A., 2012. Effects of fermentation substrate conditions on corn-soy co-fermentation for fuel ethanol production. Bioresour. Technol. 120, 140–148. https://doi.org/10.1016/j.biortech.2012.04.071.

Yasin, N.H.M., Mumtaz, T., Hassan, M., Rahman, N., 2013. Food waste and food processing waste for biohydrogen production: a review. J. Environ. Manag. 130, 375–385. https://doi.org/10.1016/j.jenvman.2013.09.009.

Yu, C.Y., Jiang, B.H., Duan, K.J., 2013. Production of bioethanol from carrot pomace using the thermotolerant yeast Kluyveromyces marxianus. Energies 6 (3), 1794–1801. https://doi.org/10.3390/en6031794.

Zafar, S., Owais, M., 2006. Ethanol production from crude whey by Kluyveromyces marxianus. Biochem. Eng. J. 27 (3), 295–298. https://doi.org/10.1016/j.bej.2005.05.009.

Zhao, R., Bean, S.R., Wang, D., Park, S.H., Schober, T.J., Wilson, J.D., 2009. Small-scale mashing procedure for predicting ethanol yield of Sorghum grain. J. Cereal Sci. 49, 230–238. https://doi.org/10.1016/j.jcs.2008.10.006.

FURTHER READING

Dimitris, S., Papanikolaou, S., 2016. Biotechnological production of ethanol: biochemistry, processes and technologies. Eng. Life Sci. 16 (4), 307–329. https://doi.org/10.1002/elsc.201400199.

Dwidar, M., Lee, S., Mitchell, R., 2012. The production of biofuels from carbonated beverages. Appl. Energy 100, 47–51. https://doi.org/10.1016/j.apenergy.2012.02.054.

Gebregergs, A., Gebresemati, M., Sahu, O., 2016. Industrial ethanol from banana peels for developing countries: response surface methodology. Pacific Sci. Rev. A Nat. Sci. Eng. 18, 22–29.

Krishnan, V., Ahmed, D., Soning, C., 2007. Bio-hydrogen generation from mixed fruit peel waste using anaerobic contact filter. Int. J. Hydrog. Energy 32 (18), 4754–4760. https://doi.org/10.1016/J.IJHYDENE.2007.07.001.

Li, X., Mu, H., Chen, Y., Zheng, X., Luo, J., Zhao, S., 2013. Production of propionic acid-enriched volatile fatty acids from co-fermentation liquid of sewage sludge and food waste using Propionibacterium acidipropionici. Water Sci. Technol. 68 (9), 2061–2066. https://doi.org/10.2166/wst.2013.463.

Walther, T., François, J.M., 2016. Microbial production of propanol. Biotechnol. Adv. 34, 984–996. https://doi.org/10.1016/j.biotechadv.2016.05.011.

Chapter 6

Sustainable Waste-to-Energy Technologies: Transesterification

Shwe Sin Win and Thomas A. Trabold
Golisano Institute for Sustainability, Rochester Institute of Technology, Rochester, NY, United States

6.1 INTRODUCTION

As a result of rising demand for energy from various industries and broad global modernization, the price of crude oil is steadily increasing and driving volatility in cost of finished fuel products such as gasoline and diesel. Prices of crude oil are influenced by short-term impacts of world supply and demand for petroleum and other liquids, and production decisions by OPEC (DoE/EIA, 2014). Alternative fuels could provide a means to decrease reliance on fossil fuels, while minimizing net greenhouse gas (GHG) emissions from fuel combustion, thus improving air and water quality. Due to the growing demand from the renewable energy sector and high crude oil prices, global biodiesel production has increased by 23% per year on average between 2005 and 2015 (Naylor and Higgins, 2017). Biodiesel can be produced from vegetable and plants oils, or animal fats, and is an excellent alternative renewable energy source to nonrenewable petroleum-derived diesel fuel.

In August 2017, the Energy Information Administration (EIA) estimated that the United States produces 149 million gallons (564 million liters) of biodiesel produced from 1140 million pounds (517 million kg) of feedstock (EIA, 2017). The U.S. Department of Energy Clean Cities Alternative Fuel Price Report (2014) stated that the average B20 (i.e., 20% biodiesel, 80% petroleum diesel) price at market nationwide was \$3.81/gal, while the market price of B100 was \$4.21/gal, and diesel \$3.77/gal. Over the past 5 years, the cost of pure biodiesel (B100) has been approximately 1.2–1.5 times higher than petroleum diesel, depending on feedstock cost, plant size, and the value of by-product glycerol. Early adopters and environmentally minded consumers may be willing to pay a premium price for a renewable energy resource, but biodiesel production cost is high due to the generally low production volume of biodiesel relative to petroleum diesel. The National Renewable Energy Laboratory (NREL) reported that in 2018 there were 138 biodiesel plants across the United States, producing approximately 1 billion gallons of biodiesel per year (3.8 billion liters/year) from soybean, canola oil, corn oil, animal fats, yellow waste, and waste vegetable oils (NREL, 2018). This production volume is <2/3 of the total production capacity of nearly 2.5 billion gallons/year (9.5 billion liters/year).[1] About 40% of production plants use multiple feedstocks and diversifying feedstock portfolio will reduce price volatility and dependency on a single product (Lim and Teong, 2010; Chai et al., 2014). Soybean oil is still the largest biodiesel feedstock, with 608 million pounds (276 million kg) consumed,[2] and many biodiesel plants are located in close proximity to the centers of plant and animal feedstocks production in the Midwestern states (Fig. 6.1). However, waste cooking oil (WCO) and fats, oils, and grease (FOG) are common waste streams generated by many industrial, commercial and institutional food sector operations, and with appropriate pretreatment can also be suitable feedstocks for biodiesel production. These materials should be considered integral to the portfolio of resources available for sustainable food waste-to-energy conversion.

As an increase in global biodiesel production capacity has been observed, production of crude glycerol, a by-product of the transesterification reaction, is also expected to increase. However, the demand for crude glycerol has not kept pace, and because of its limited use, an oversupply of crude glycerol has had significant downward pressure on the market price. In 2008 crude glycerol from the rapidly growing biodiesel industry became the main player in the global glycerol market, which has driven some traditional glycerol production plants out of business (Gholami et al., 2014). As the value of crude glycerol has fallen to near zero, biodiesel producers are in many cases required to pay tipping fees to remove the material from their plants for disposal or incineration (Quispe et al., 2013). Efficient utilization of crude glycerol would potentially

1. http://www.biodieselmagazine.com/plants/listplants/USA/
2. https://www.eia.gov/biofuels/biodiesel/production/

Sustainable Food Waste-to-Energy Systems. https://doi.org/10.1016/B978-0-12-811157-4.00006-1

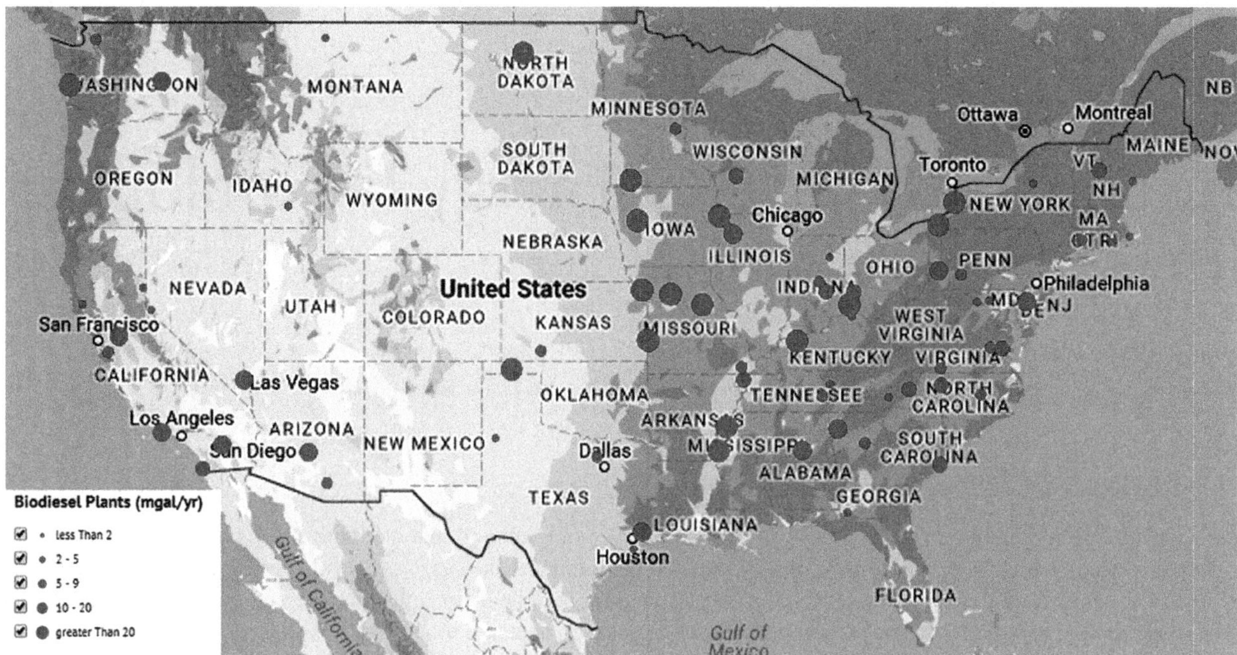

FIG. 6.1 Map of U.S. biodiesel plants. Size of dot indicates the volume of biodiesel produced (NREL, 2018).

help to lower the overall production cost of biodiesel, which is already high compared to diesel. But as the price of crude glycerol has fallen, the prices of alcohols and catalysts needed for purification and upcycling are generally unchanged (Mohammadshirazi et al., 2014). Crude glycerol from biodiesel production often contains a high level of contaminants, which further increases the purification cost. New applications for crude glycerol have been considered, and utilizing low-value crude glycerol in a community-based biodiesel production system is presented later in this chapter.

6.2 POTENTIAL FEEDSTOCKS FOR BIODIESEL PRODUCTION

It has been reported that >350 crops are available as feedstocks for biodiesel production (Atabani et al., 2013). Globally, 30% of biodiesel is produced from soybean oil, 25% from rapeseed oil, 18% from palm oil, 10% from recycled oils, 6% from animal fats, and 11% from other sources (Naylor and Higgins, 2017). In Europe, rapeseed oil is the primary source, while in the United States soybean oil accounts for 51% of total production on average, fluctuating between 40% and 60% of total feedstock used. Palm and coconut oils are widely used in Indonesia and Malaysia (Lim and Teong, 2010).

In many parts of the world, edible oils are considered essential to the daily diet as primary food resources. The demand for edible oils for biodiesel feedstock has been rising and that may create upward pressure on the price of the food commodity. The cost of raw material (vegetable oil or animal fat) is the primary factor driving the high biodiesel production cost, and biodiesel produced from virgin vegetable oil is often sold at a higher price than conventional petroleum diesel oil. Currently, the feedstock used for biodiesel production is primarily edible oils which account for 70%–95% of total production (Math et al., 2010).

The prospective feedstocks for biodiesel production can be categorized into three groups: edible, nonedible, and other sources, as summarized in Table 6.1. The cultivation of edible crops requires arable land which competes with food crops in agricultural land usage, and introduces associated environmental issues such as soil erosion, and pesticide leakage into the surface and groundwater (EEA, 2017). Nonedible biodiesel feedstocks such as jatropha, castor bean, and rubber seed do not directly compete for land needed for food production. Third-generation biofuels derived from alternative feedstocks such as algae seek to address the concerns associated with both edible and nonedible feedstocks. Algae have been considered as a viable feedstock to potentially replace soybean, rapeseed, and palm oils (Lim and Teong, 2010). However, despite high per acre productivity, fast growth rates, and ability to be grown on nonarable land areas, cultivation, harvesting, and valorization of algae biomass require high water use and high capital and operation cost, which make it economically challenging

TABLE 6.1 Summary of Feedstocks for Biodiesel Production

Edible feedstock (1st generation)	Canola, coconut, cottonseed, corn, soybean, sunflower, peanut, rapeseed, palm, mustard
Nonedible feedstock (2nd generation)	*Calophyllum inophyllum* L, castor bean, *Croton megalocarpus* L, *Jatropha curcas* L, *Moringa oleifera* L, karanja (pongamia oil), *Pongamia pinnata,* patchouli, *Sterculia foetida* L, rubber seed, sea mango, jojoba
Other sources (3rd generation)	Microalgae, spent coffee grounds (SCG), fats, oils and grease (FOG; including animal tallow, white grease, brown grease, yellow grease, fish oil), waste cooking oil (WCO)

Adapted and revised from Gui et al. (2008), Atabani et al. (2013), and Verma and Sharma (2016).

(Mussgnug et al., 2010). Nonetheless, utilization of algae for large-scale biodiesel production is under active research and development in many regions.

Approximately 8 million tons of coffee are produced per year and global coffee consumption is predicted to continue to increase (Vardon et al., 2013). A substantial quantity of spent coffee grounds (SCG) enters the global waste stream and in many regions is recognized as a major disposal challenge. Oil derived from spent coffee grounds (SCG) has been identified as a potential waste-based feedstock for biodiesel production due to its favorable oil content, which can be between 15 and 28 wt% depending on the coffee species, roasting and brewing processes, and extraction methods. One of the drawbacks to SCG as a biodiesel feedstock is that the sources can be widely dispersed, and transporting small quantities of SCG from multiple locations to a central biodiesel plant for further oil extraction is still a questionable economic proposition. Like other edible and nonedible feedstocks, oil from SCG can be extracted using n-hexane as a solvent which is only economically viable at relatively large production scale (Tuntiwiwattanapun et al., 2017).

The problem of high raw material cost can be addressed by using lower cost (or even "free") feedstocks such as animal fat and waste vegetable oils. The production cost of biodiesel can be effectively reduced by 60%–70% by using waste cooking oil (WCO) instead of refined virgin oils (Math et al., 2010). Even when including the cost of pretreatment needed for refining relatively impure waste feedstocks, the overall production cost can still be reduced by 45% (de Araújo et al., 2013). The feedstock used for biodiesel production varies across countries depending on their geographical locations and agricultural practices (Lim and Teong, 2010), and the selection of raw material plays a significant role in effective biodiesel production depending on oil content, process chemistry, and economy of the process (Karmakar et al., 2010). For example, the composition of fatty acids in the raw feedstock influences the physiochemical properties of the final biodiesel product. Fig. 6.2 shows the diverse sources of feedstocks used for biodiesel production in the United States. Free fatty acid (FFA) content, moisture content, and other impurities from oils can inhibit the transesterification reaction (described in Section 6.3) and affect the purity and quality of both biodiesel and crude glycerol coproducts. Table 6.2 summarizes typical oil contents and estimated prices of select feedstocks for biodiesel production.

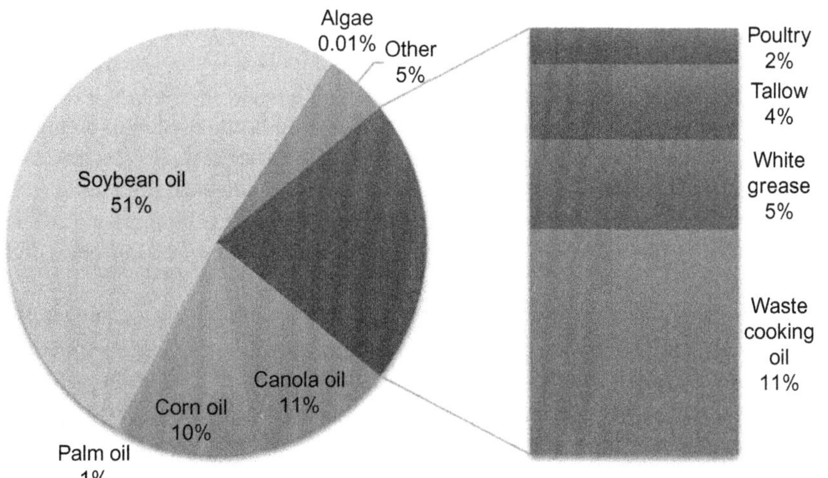

FIG. 6.2 Production of biodiesel from different feedstocks in the United States (Brorsen, 2015).

TABLE 6.2 Estimated Oil Content and Average International Crude Vegetable Oil Price of Different Feedstocks and the Corresponding Biodiesel Price

Feedstock	Oil Concentration (%w/w)	Price of Crude Feedstock (US dollar/ton)	Price of B100 Biodiesel (US dollar/ton)	References
Edible oils				
Soybean	15–20	735	800–805	[1,2]
Rapeseed	38–46	815–829	940–965	[1,2]
Palm oil	45–55	610	720–750	[1,2]
Nonedible oil				
Jatropha carcus	Seed 35–40 Kernel 50–60		400–500	[2,3]
Other sources				
Microalgae (*Botryococcus* sp.)	26.3	–	–	[4]
Microalgae (*Chlorella* sp.)	10–48	–	–	[5]
Spent coffee ground	10–15	–	–	[6]
Animal tallow		245	500	[2]
Yellow grease	–	412	–	[7]
Waste cooking oil	–	224–360	600	[2,7]

[1] Baskar and Aiswarya (2016); [2] Lim and Teong (2010); [3] Gui et al. (2008); [4] Ashokkumar et al. (2014); [5] Islam et al. (2017); [6] Somnuk et al. (2017); [7] Demirbas (2009).

6.3 TRANSESTERIFICATION OF WASTE COOKING OIL (WCO)

Oils and Fats International has predicted that global total vegetable oil production in 2017/2018 will be 195 million tons (Lim, 2017). It is reasonable to expect that such large oil production levels will ultimately translate into availability of large volumes of WCO, also referred to as yellow grease, typically derived from edible oil that has been used for a certain period in a deep-frying process. Brown grease is often collected at grease traps in sink drains to prevent discharges of fats, oils, and grease (FOG) from entering and clogging sewer pipes. NREL has estimated that between 3 and 21 lbs (1.4 and 9.5 kg/person/year) of yellow grease from fast-food restaurants are produced in the metropolitan areas of the United States (Wiltsee, 1999). The average amount of grease discharged from each restaurant to sewage treatment plants ranged from 800 to 17,000 lbs/year (363 to 7718 kg/year). Similarly, Wiltsee (1999) reported total per capita production of 4.1 kg/year of yellow grease and 5.9 kg/year of brown grease (grease trap waste), and based on total US population this translates to an estimated 1.23 billion kg (358 million gallons) of waste oil and 1.81 billion kg (525 million gallons) of grease trap waste per year. It is therefore likely that a significant volume of used grease and oil resources is available for conversion to biodiesel fuel. A fraction of the material originating in the nonresidential sector is already repurposed for beneficial uses, but most of the rest is sent to wastewater treatment facilities and landfills. Some of the yellow grease resources are used for animal feed supplements, biodiesel production, and feedstock for anaerobic digestion system, but the bulk of WCO generated in the residential sector is discharged into the sewer systems where it can accumulate and cause blockages when left untreated (Wallace et al., 2017).

Waste cooking oil as a triglyceride supply from domestic and food service industries is the most economically viable feedstock because it is essentially "free" and potentially lowers the biodiesel production cost while also solving a serious waste disposal issue. However, the main disadvantages of waste cooking oil are the high free fatty acid (FFA), water and food residue content, depending on the cooking process and the storage type. Yellow grease (waste cooking oil) typically contains $<15\%$ FFA, while brown grease; fats, oils, and grease (FOGs) from grease traps or sewer systems often have $>15\%$ FFA. The content of FFA in the oil significantly increases during heating and frying processes over a long period. High FFA and water content and other impurities can present challenges to producing high-quality biodiesel (Wallace et al., 2017).

6.3.1 Basic Transesterification Process

Biodiesel production from vegetable oils and fats has been investigated for many years, and different production methods have been proposed to utilize triacylglycerol-containing materials (Knothe and Razon, 2017). Direct use and blending with petroleum-derived diesel fuel (petrodiesel), microemulsification, thermal cracking, and transesterification are the four basic production technologies. Among them, transesterification is the most economical technology, because of its low temperature and pressure conditions and relatively high yield. Biodiesel is chemically known as the mono-alkyl ester of fatty acids, or fatty acid (m)ethyl ester derived from triacylglycerol of vegetable/plant oils, and animal fats produced by transesterification (Knothe and Razon, 2017). During the transesterification process, also known as alcoholysis, triglycerides of fatty acids are reacted with alcohols in the presence of a catalyst to form monomethyl esters (biodiesel) and glycerol as coproducts, as shown in Fig. 6.3. The most commonly used low molecular weight, short-chain alcohols are methanol and ethanol, with methanol generally being less expensive with better reactivity.

Fundamentally, there are five main parameters that influence the biodiesel production process: free fatty acid (FFA) content of the feedstock, the type and amount of alcohol needed to support the reaction, type and amount of catalyst, molar ratio (alcohol:oil), and reaction time and temperature (Verma and Sharma, 2016). The stoichiometry of the reaction is 3:1 M ratio of alcohol to triglyceride in the oil; however, in practice the most recommended molar ratio is 6:1, in the temperature range of 50–60°C, with reaction time of 1 h with 1% w/w sodium hydroxide or potassium hydroxide since the reaction is reversible and excess methanol is required. <0.5 wt% of free fatty acid content is recommended with zero or low moisture content (Knothe and Razon, 2017). Type of feedstock, choice of catalyst, alcohol-to-oil ratio, reaction temperature, and time influence the transesterification and side reactions (Keera et al., 2011).

Transesterification processes can be categorized into catalytic and noncatalytic reactions. As expected, the presence of a catalyst should noticeably accelerate the reaction (Keera et al., 2011). Catalytic transesterification can be further divided into homogeneous acid and base catalytic reactions, heterogeneous acid and catalysts, and enzyme catalyst (organic in nature) to a lesser extent. The most common and simple noncatalyzed process uses supercritical methanol (MeOH), but it is still a very expensive process. The choice of catalyst for the transesterification reaction is mainly based on the free fatty acid (FFA) composition of the feedstock, and the type of catalyst influences the composition of the biodiesel product (Lam et al., 2010). All these methods have the ability to produce biodiesel from any triglyceride; however, they each have advantages and disadvantages, as summarized in Table 6.3.

6.3.2 Conversion of Institutional WCO

As described previously, heating conditions, FFA composition, and water content can significantly influence conversion of waste cooking oil to biodiesel. In general, waste cooking oils from restaurants and institutional food service operations such as cafeterias often contain large amount of free fatty acids and water. The free fatty acids can react with the alkaline catalyst to produce undesirable reaction by-products such as soap, and therefore the FFA value is expected to be <1 wt% prior to initiating the conversion process. A triglyceride contains a chain of three fatty acid molecules and one glyceride molecule. During the transesterification reaction, triglycerides (TG) are converted to diglycerides (DG) and then DG to monoglycerides (MG), and finally to glycerol. This three-stepwise reaction is reversible, and excess methanol is required to shift the equilibrium to the desired fatty product (Fig. 6.4).

When the raw WCO material contains a high percentage of free fatty acids and water, the water can hydrolyze with the triglyceride to produce diglycerides and form more long-chain free fatty acids while homogenous alkali catalyt (KOH) will react with the FFA to form soaps and water that reduces the ester yield. Strong acid catalysts like sulfuric acid (H_2SO_4) are often first used for the esterification reaction to reduce the FFA content to less than approximately 1% (Fig. 6.5), followed

$$\begin{array}{lcccl}
CH_2\text{-}O\text{-}\overset{\overset{O}{\|}}{C}\text{-}R_1 & & & CH_3\text{-}O\text{-}\overset{\overset{O}{\|}}{C}\text{-}R_1 & \\
| & & & & CH_2\text{-}OH \\
CH\text{-}O\text{-}\overset{\overset{O}{\|}}{C}\text{-}R_2 & + \; 3\,CH_3OH & \underset{\text{(Catalyst)}}{\longrightarrow} & CH_3\text{-}O\text{-}\overset{\overset{O}{\|}}{C}\text{-}R_2 & + \; CH\text{-}OH \\
| & & & & CH_2\text{-}OH \\
CH_2\text{-}O\text{-}\overset{\overset{O}{\|}}{C}\text{-}R_3 & & & CH_3\text{-}O\text{-}\overset{\overset{O}{\|}}{C}\text{-}R_3 & \\
\text{Triglyceride} & \text{Methanol} & & \text{Mixture of fatty esters} & \text{Glycerol}
\end{array}$$

FIG. 6.3 General chemical equation for the overall transesterification reaction, where R_1, R_2, R_3 are long-chain hydrocarbons (fatty acid chains) (Van Gerpen et al., 2004).

TABLE 6.3 Various Transesterification Methods for Biodiesel Production From Waste Oils and Fats

Chemical Process	Catalyst Materials	Advantages	Disadvantages
Homogeneous acid catalyst	• H_2SO_4 • sulfonic acid • hydrochloric acid (HCL)	• high conversion yield • relatively less sensitivity to FFA and water than base catalyst • low cost	• slow reaction rate • high reaction temperature and pressure • difficult to separate and recover glycerol • H_2SO_4 is corrosive and can lead to corrosion of equipment
Homogeneous base catalyst	• NaOH • KOH • Sodium methoxide • Potassium methoxide	• high conversion rate • high catalytic efficiency • faster mass transfer rate • short retention time • low reaction temperature and pressure • low cost	• high sensitivity to FFA and water • highly energy intensive • difficult to recover glycerol
Heterogeneous acid catalyst	• Zirconium oxide (ZrO_2) • Titanium oxide (TiO_2) • Tin oxide (SnO_2) • Zeolites • Sulfonic ion-exchange resin • Sulfonated carbon-based catalyst • Heteropolyacids (HPAs)	• high conversion rate • insensitivity to FFA and water • catalyze esterification and transesterification simultaneously • easy to separate and reuse catalyst	• slow reaction time • high operation cost • energy intensive • high alcohol-to-oil molar ratio
Heterogeneous base catalyst	• Basic zeolites • Alkaline earth metal oxides • Hydrotalcites (Mg/Al) • Calcium oxide (CaO)	• less sensitivity to FFA and water content • high glycerol purity • easy to separate and reuse catalyst from the product	• slow reaction time and high temperature • slow mass transfer rate • high alcohol-to-oil molar ratio
Enzyme (biocatalyst)	• *Mucor miehei* (Lipozym IM60) • *Pseudomonas cepacia* (PS 30) • *Rhizopus oryzae* • *Penicillium expansum*	• high conversion yield • insensitivity to FFA and water • low operating temperature (lower than homogenous base catalyst) • absence of by-product • easier purification	• long reaction time (slower than acid-catalyzed transesterification) and slower reaction rate • sensitivity to alcohol (methanol deactivates the enzyme) • high cost of lipase • technology is still under development
Noncatalyzed (supercritical MeOH)		• high conversion yield • faster reaction rate and very short reaction time • absence of catalyst • eliminates FFA and water content issues • easy to separate by-products	• high reaction temperature and pressure • requires large amount of MeOH • high cost of operation • high alcohol-to-oil molar ratio • technology is still under development

Adapted and revised from Lam et al. (2010), Leung et al. (2010), Gui et al. (2008), Tsoutsos et al. (2016), Gnanaprakasam et al. (2013), and Yaakob et al. (2013).

Triglyceride + ROH ⇄ Monoglyceride + R'CPOOR

Diglyceride + ROH ⇄ Monoglyceride + R'CPOOR

Monoglyceride + ROH ⇄ Glycerol + R'CPOOR

FIG. 6.4 Three stepwise transesterification reactions (Enweremadu and Mbarawa, 2009).

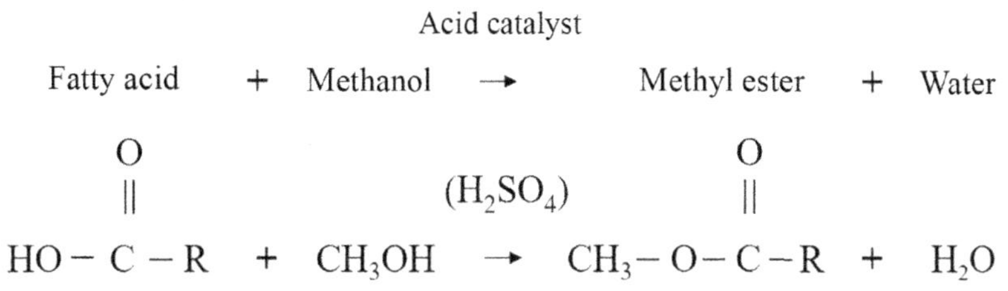

Fatty acid + Methanol $\xrightarrow{\text{Acid catalyst}}$ Methyl ester + Water

$$HO-\overset{\overset{\displaystyle O}{\|}}{C}-R + CH_3OH \xrightarrow{(H_2SO_4)} CH_3-O-\overset{\overset{\displaystyle O}{\|}}{C}-R + H_2O$$

FIG. 6.5 Acid-catalyzed esterification reaction.

by base-catalyzed transesterification (Lam et al., 2010). NREL has recommended a recipe of reactants for the FFA pretreatment which is widely adopted by most biodiesel manufacturers: 2.25 g methanol and 0.05 g sulfuric acid for every gram of FFA in the WCO, equivalent to 19.8:1 methanol-to-FFA molar ratio and 5% acid-to-FFA weight percentage (Chai et al., 2014).

6.3.2.1 Determination of Acid Value/Fatty Acid Content

Free fatty acids content of waste cooking oil from institutional cafeterias has been determined to be lower than many other establishments (i.e., fast-food restaurants) as the cooking oil replacement rate should be faster. However, there are no specific guidelines or regulations on how often cooking oil should be replaced. Sanli et al. (2011) investigated the FFA content from different facilities; however, there are very limited data on free fatty acids values of WCO from institutions such as universities. Frank (2014) and Bruton (2014) measured FFA content of WCO from a university cafeteria, and measured 4.42 and 3.30 wt%, respectively. The acid value is defined as the amount of free fatty acid present in the oil requiring a computed mass of potassium hydroxide to neutralize it. This value is used to determine the amount of base catalyst to neutralize the acidity of a gram of raw material (Luque and Melero, 2012). The acid value (*AV*) is determined by applying Eq. (6.1) (Banani et al., 2015). A titration method is typically used to measure the FFA concentration and determine the fatty acid conversion during esterification. This method provides the amount of excess base catalyst required to neutralize the FFAs based on the initial concentration in the WCO. The extent of the acid esterification reaction can be determined by the acid value and the FFA conversion rate (Chai et al., 2014).

$$AV\,(\text{mg KOH/g}) = \frac{56.1 \times C_{\text{KOH}} \times V_{\text{KOH}}}{m} \tag{6.1}$$

where

56.1 = molecular weight of the solution employed for titration (g/mol)
C_{KOH} = concentration of the titration KOH solution (g/mol)
V_{KOH} = volume of solution employed for titration (mL).
m = mass of the fatty acid sample (g)

The free fatty acid (FFA) conversion rate is calculated by Eq. (6.2):

$$\text{FFA conversion} = \frac{(\text{Initial FFA} - \text{Final FFA})}{\text{Initial FFA}} \times 100\% \tag{6.2}$$

where

Initial FFA = initial free fatty acid value (mg KOH/g)
Final FFA = final free fatty acid value (mg KOH/g)

6.3.2.2 Two-Step Esterification/Transesterification

The conventional method of producing biodiesel from edible virgin oils today is homogeneous base-catalyzed transesterification on a larger production scale. In such systems, common base catalysts are sodium hydroxide (NaOH), sodium methoxide (CH_3NaO), potassium hydroxide (KOH), and potassium methoxide (CH_3KO). These catalysts are widely used because they provide fast reaction rate for a short process time (30–60 min), ability to run the process at lower temperature and atmospheric pressure, higher product yield and purity (Math et al., 2010). However, base catalysts are more effective with high purity reactants having <1 wt% free fatty acids and have generally performed better with virgin vegetable oils (Demirbas, 2009). When the waste cooking oil FFA content exceeds 1 wt%, a two-step catalyzed process is recommended. Acid catalysts are a better option if the initial feedstock oil has high FFA and/or water content. The most widely used acid catalysts are sulfuric acid, sulfonic acid, and hydrochloric acid. Unlike base catalysts, they are generally insensitive to FFA content and both oil and FFA can lead directly to the methyl ester (Chai et al., 2014). Longer reaction times, the requirement of a large amount of alcohol and lower yield makes acid-catalyzed reactions unattractive for large-scale biodiesel production. Fig. 6.6 shows the schematic flow diagram of two steps (esterification and transesterification process) biodiesel production from institutional waste cooking oil.

When the feedstock contains a significant amount of free fatty acids, the acid pretreatment step (acid esterification) is recommended to neutralize the free fatty acid content before proceeding to the "standard" transesterification reaction. As stated previously, many nonedible oils and low-quality feedstocks such as grease trap waste, animal fats, and waste cooking oils contain high free fatty acid content. Acceptable levels of FFA concentration in WCO should be lower than <5 wt% to achieve acceptable biodiesel yields (Lam et al., 2010). In the pretreatment process, free fatty acids in the oil are esterified to methyl ester with acid catalyst (H_2SO_4) in the presence of methanol while reducing the acid value (Fig. 6.5). Without this pretreatment step, excess free fatty acid in the raw oil would likely react with the alkali catalyst (KOH) through the saponification reaction to form alkali salts of fatty acids (soap) and water; Fig. 6.7. During this reaction, the catalyst is utilized by FFA and therefore water also inhibits the reaction which reduces the catalyst activity and leads to more difficulty in the separation process. Additionally, when there is water in the raw oil or it is generated during the saponification reaction, it will hydrolyze the triglycerides to form additional free fatty acid (Fig. 6.8).

As discussed in Section 6.2, feedstock characteristics have an important influence in determining the effectiveness of the biodiesel production process. Direct use of vegetable oils and animal fats as combustible fuel is generally not possible due to their high kinematic viscosity and low volatility compared to No. 2 diesel (Math et al., 2010). The reduction of viscosity can be achieved by transesterification of oil to produce biodiesel which has properties similar to diesel fuel. Furthermore, some feedstocks like jatropha, WCO, and animal fat have higher acid/FFA content which requires additional pretreatment process for biodiesel production. Table 6.4 shows typical physical and chemical properties of some common feedstocks. Generally, the reported properties of WCO, such as water, insolubles, unsaponifiables, phosphorous, and sulfur levels are all within acceptable ranges for biodiesel production and similar to the virgin oils. As shown in Table 6.5, the

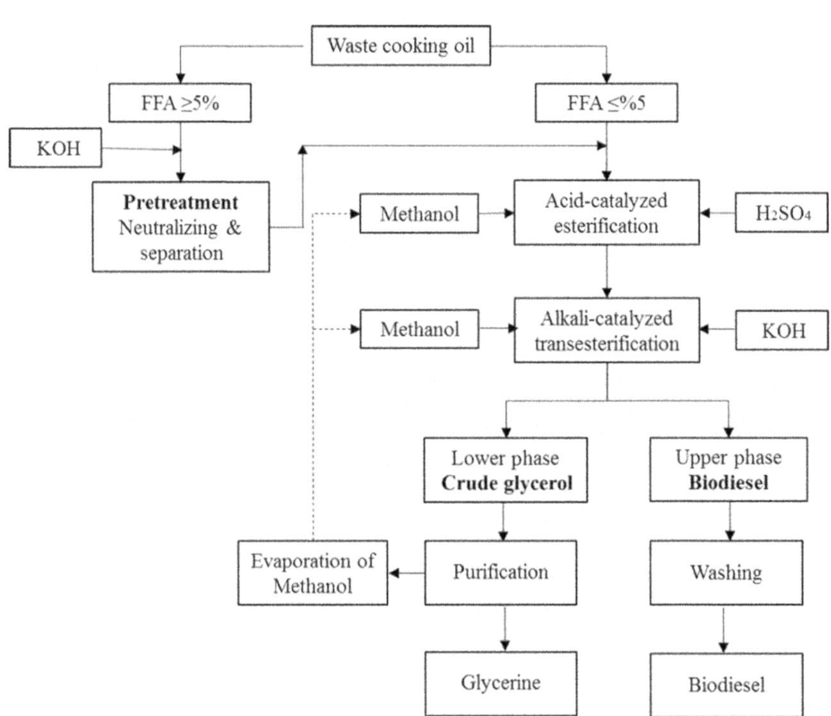

FIG. 6.6 Overall schematic process flow for biodiesel production using institutional WCO.

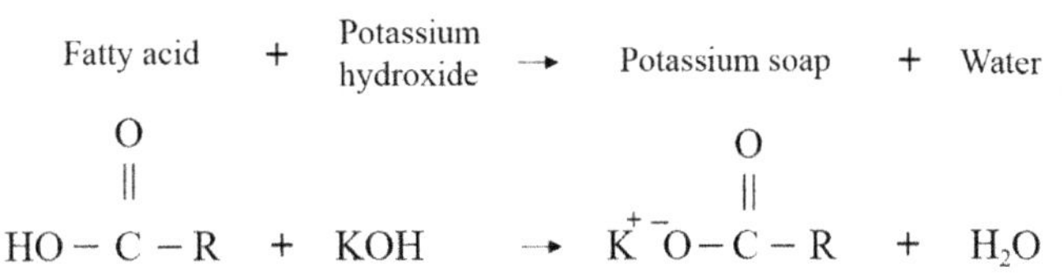

$$\text{Fatty acid} + \text{Potassium hydroxide} \rightarrow \text{Potassium soap} + \text{Water}$$

$$HO-\overset{\overset{\displaystyle O}{\|}}{C}-R + KOH \rightarrow \overset{+}{K}\,\overset{-}{O}-\overset{\overset{\displaystyle O}{\|}}{C}-R + H_2O$$

FIG. 6.7 Saponification reaction (Van Gerpen, 2005).

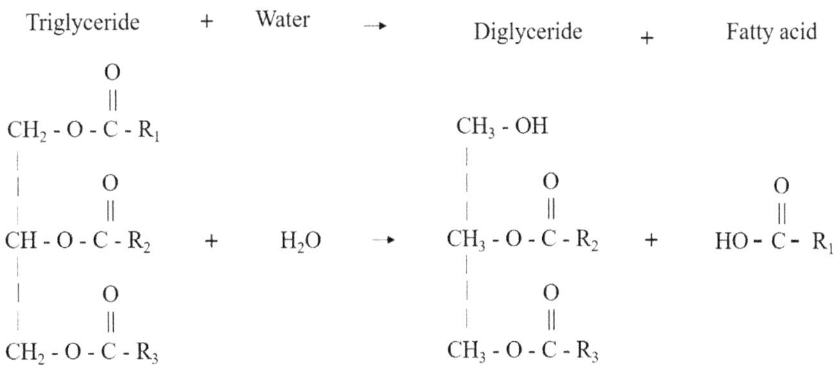

$$\text{Triglyceride} + \text{Water} \rightarrow \text{Diglyceride} + \text{Fatty acid}$$

$$\begin{array}{l} CH_2-O-\overset{\overset{O}{\|}}{C}-R_1 \\ | \\ CH-O-\overset{\overset{O}{\|}}{C}-R_2 \\ | \\ CH_2-O-\overset{\overset{O}{\|}}{C}-R_3 \end{array} + H_2O \rightarrow \begin{array}{l} CH_3-OH \\ | \\ CH_3-O-\overset{\overset{O}{\|}}{C}-R_2 \\ | \\ CH_3-O-\overset{\overset{O}{\|}}{C}-R_3 \end{array} + HO-\overset{\overset{O}{\|}}{C}-R_1$$

FIG. 6.8 Hydrolysis of triglycerides (Van Gerpen et al., 2004).

TABLE 6.4 Comparison of Physiochemical Properties of Different Feedstocks for Biodiesel Production

Parameters	Units	Soybean	Palm	Jatropha	Microalgae (*Spirulina platensis*)	Spent Coffee Oil	Waste Cooking Oil
Heating value	MJ/kg	39.6	39.3	37.01–38.73	41.36	23.10–38.22	41.40
Density (40 °C)	kg/m^3	890–913	881–919	919.5–932	860	925*	924**
Kinematic Viscosity (40 °C)	mm^2/s	28.08–32.6	39.4	35.62–51	5.66	42.65–49.64	36.4
Pour point	°C	−12.2	–	2	−18	7	11
Flash point	°C	254	252–267	242–274	130	>200	212
Cloud point	°C	−3.9	31.0	–	–	11–12.2	–
Cetane Number		37.9	42.0	–	–	–	49
Acid value	mg KOH/g oil	0.709	0.1	42.78	0.45	9.85–11.27	43.46
Free fatty acid	%	0.3545	7.5	21.5	–	0.412	1.32–5
References		[1]	[2]	[3]	[1]	[4]	[5]

[1] Verma and Sharma (2016); Kayode and Hart (2017); [2] Chai et al. (2014); Verma and Sharma (2016); Kayode and Hart (2017); [3] Verma and Sharma (2016); Chai et al. (2014); [4] Haile (2014); Somnuk et al. (2017) (*Density at 30 °C); Vardon et al. (2013); [5] Demirbas (2009) (**Density at 15°C); Chai et al. (2014); Verma and Sharma (2016).

TABLE 6.5 Selected Literature on Operating Parameters of Biodiesel Production From Different Feedstocks

Feedstock	Method	Catalyst		Alcohol	Acid Catalyst (wt%)	Base Catalyst (wt%)	*T* (°C)	Reaction Time	Molar Ratio (Alcohol to Oil Ratio)	Yield (%)
Soybean oil [1]	Esterification	Homogeneous acid	H_2SO_4	Methanol	1	–	65	69 h	–	>90
	Transesterification	Heterogeneous base	PbO	–	–	–	215		7:1	89
		Heterogeneous acid	$Al_2O_3/ZrO_2/WO_3$	Methanol			250		40:1	90
		Biocatalyst (enzyme)	Lipozyme RMIM	Methanol	7	–	50	4 h	3:1	60
Rapeseed oil [2]	Transesterification	Homogeneous base	KOH	Methanol		0.7	30	2 h	9:1	97–98
		Heterogeneous acid	ZnO	Methanol	–	–	225–230	6 h	20:1	94.3
Palm oil [3]	Esterification	Ion-exchange resin catalyst	Amberlyst-15	Methanol	4	–	65	90 min	15:1	
	Transesterification	Homogeneous base	KOH	Methanol		1	60	60 min	6:1	88
		Heterogeneous base	KNO_3/KL zeolite				200	4 h	–	74
Jatropha [4]	Transesterification	Homogeneous base	KOH	Methanol	–	9	60–80	8 h	10:1	
		Heterogeneous base	Hydrotalcite particles with Mg/Al	Methanol		1	45	1.5 h	4:1	95.2
S. platensis algae [5]	Transesterification	–	H_2SO_4	Methanol	60	–	55	90 min	4:1	75
Microalgae oil	Supercritical transesterification	–		Methanol			270–350		10:1–42:1	90.8

Spent coffee grounds [6]	In situ transesterification	–	H_2SO_4	Methanol	20	–	70	12 h	28.87 ml methanol/g oil	98.61
	In situ transesterification	–	NaOH	Methanol		3:1 catalyst to oil mole ratio	60	90 min	400:1	96.0
	Transesterification	1. Homogeneous acid 2. Homogeneous base	1. HCL 2. KOH	Methanol	10	1	54	90 min	20:1 9:1	82
WCO [7]	Two-step: 1. Acid esterification 2. Alkaline transesterification	1. Homogeneous acid 2. Homogeneous base	1. H_2SO_4 2. KOH	Methanol	5	1.47	80	90 min	5:1	96
	Transesterification (used canola)	Homogenous base	NaOH	Methanol	–	0.08	60	20 min	28 mL per 100 mL of oil	94.5
	Transesterification (used sunflower)	Homogenous base	KOH	Methanol	–	1	25	30 min	6:1	Maximum
	Supercritical methanol transesterification			Methanol			286	2 h	6:1–41:1	99.6
	Transesterification	Heterogeneous base	CaO	Methanol		0.85%	60–65	1 h	12:1	66
	Transesterification	Heterogeneous acid	Zeolite Y (Y756)	Methanol	–	–	460	0.37 min	6:1	26.6
	Transesterification	Biocatalyst (enzyme)	Novozyme 435	–	4% based on oil mass	–	40	12 h	4:1	88
	Microwave-assisted transesterification	Heterogeneous alkali on the carbonized coconut shell	KOH/CS	Methanol	–	5	80	40 min	12:1	91.3

[1] Baskar and Aiswarya (2016); Kayode and Hart (2017); [2] Verma and Sharma (2016); Kayode and Hart (2017); [3] Gan et al. (2012); Verma and Sharma (2016); Kayode and Hart (2017); [4] Deng et al. (2011); Verma and Sharma (2016); [5] Verma and Sharma (2016); [6] Haile (2014); Liu et al. (2017); Najdanovic-Visak et al. (2017); [7] Demirbas (2009); Lam et al. (2010); Math et al. (2010); Bruton (2014); Kayode and Hart (2017); Babel et al. (2018).

TABLE 6.6 Physiochemical Properties of Biodiesel Produced From Different Feedstocks

		Edible Oils			Nonedible Oils		Other Sources (Waste Based)	
Parameters	**Units**	**Soybean**	**Rapeseed**	**Palm**	**Jatropha**	**Microalgae *Chlorella* sp.**	**Spent Coffee Oil**	**WCO**
HHV	MJ/kg	39.8–41.3	–	–	–	39.5	35.4	42.65
LHV	MJ/kg	37	32.8–36.7	32.4–33.5	39.23		18.8	38.05
Density (15°C)	kg/m^3	865–880	882*	873–877	875	883	880	897
Kinematic viscosity (40°C)	mm^2/s	4–5.2	4.83	4.71	4.4	4.73	5.4	5.3
Pour point	°C	−4	−10.8	12	2	–	2.0	−11.15
Flash point	°C, min	168–185	155	135	163	179	200	196
Cloud point	°C	−0.5	−4	16	4	–	13	–
Cetane number		45–60.9	52.9	57.3	57.1	–	505–51.9	54
Acid value	mg KOH/g	–	–	0.08	0.4	0.37	0.3–0.7	0.10
Reference		[1]	[2]	[3]	[4]	[5]	[6]	[7]

[1] Lin et al. (2011); Bruton (2014); [2] Fukuda et al. (2001); Lin et al. (2011); (*Density at 21°C); [3] Fukuda et al. (2001); Gui et al. (2008); Lin et al. (2011); [4] Gui et al. (2008); Lin et al. (2011); [5] Islam et al. (2017); [6] Kondamudi et al. (2008); Vardon et al. (2013); Haile (2014); Kang et al. (2017); Liu et al. (2017); Somnuk et al. (2017); [7] Demirbas (2009); Bruton (2014).

varying feedstock properties dictate the use of different transesterification processes and operating parameters, which in turn result in final biodiesel products with different ranges of critical properties such as density, viscosity, pour point, flash point, and cetane number (Table 6.6). Comparison between the data in Tables 6.4 and 6.6 indicates that the relatively high kinematic viscosity of vegetable oil feedstocks is reduced after transesterification.

6.4 USES OF BIODIESEL

Global demand for transportation fuels has been growing and is expected to increase from 124 billion liters in 2015 to 202 billion liters by 2024 (Verma and Sharma, 2016). The rise in crude oil demand, and the attendant GHG emissions, provide strong motivations to find alternative fuels. The U.S. Renewable Fuel Standard (RFS) mandates that renewable energy fuels in the U.S. transportation sector should provide a minimum use of 136 billion liters (36 billion gallons) by 2022. Biodiesel has been recognized as the best alternative to ultra-low sulfur diesel (ULSD) and provides many advantages over diesel fuel which include lower sulfur content, no aromatic content, biodegradability, and miscibility with petroleum-based diesel in all blend ratios. Also, the lack of significant modifications required for diesel engines or gen-sets makes biodiesel the most attractive renewable fuels over other available options, although there is an approximately 10% lower energy content compared to petroleum diesel fuel (Math et al., 2010). Global biodiesel production has continued to rise even as crude oil prices have dropped since mid-2014. As of 2015, the United States produced 4.8 billion liters (1.3 billion gallons) of biodiesel and contributed 15% of global production (Naylor and Higgins, 2017).

Biodiesel has largely developed as an alternative transportation fuel, although it has numerous potential uses which include space heating, and a small quantity of biodiesel is used in lubricants, plasticizers, high boiling absorbents for cleaning of gaseous industrial emissions, and other solvent applications (Knothe and Razon, 2017). Current utilization pathways for biodiesel are in automotive diesel engines for transportation and diesel generators for off-grid or stand-alone

TABLE 6.7 Characterization of Waste Cooking Oil-Based Biodiesel From an Institution and Standard Specification for Biodiesel (Demirbas, 2009; Bruton, 2014)

Test Name	Units	Test Methods	ASTM D6751	WCO	Biodiesel From WCO	No. 2 Diesel Oil
Free glycerin	(mass %)	D6584	MAX 0.020	–	0.000	–
Monoglycerides	(mass %)	D6584	N/A	–	0.078	–
Diglycerides	(mass %)	D6584	N/A	–	0.013	–
Triglycerides	(mass %)	D6584	N/A	–	0.007	–
Total glycerin	(mass %)	D6584	MAX 0.240	–	0.098	–
Density (15°C)	kg/m^3	D1298	–	924	890	75–840
Viscosity (40°C)	mm^2/s	D445	1.9–6.0	36.4	5.3	1.9–4.1
Pour point	(°C)	D93	MIN 93	11	-11.2	−13 to −19
Flash point	(°C)	D93	MIN 93	212	196	67–85
Cloud point	(°C)	D2500	N/A	–	1	–
Cetane number		D976	MIN 47	49	54	40–46
Sulfur content	%	D5453	MAX 15	0.09	0.001	0.35–0.55
Carbon residue	(mass %)	D4530	MAX 0.050	0.46	0.05	0.35–0.40
Water content	(volume %)	D2709	MAX 0.050	0.42	0	0.02–0.05
Acid value/FFA	(mg KOH/g)	D664	MAX 0.50	1.32	0.10	–
HHV	(MJ/g)	–	–	41.40	38	45.62–46.48

electricity generation, especially in developing countries. The most considerable constraint encountered in converting waste cooking oil to biodiesel for heating fuel is blending regulations to produce diesel-based heating oil (commonly referred to as No. 2 heating oil). In the United States, a mixture blend constraint exists for heating oil, because only 2%–10% v/v biodiesel blend is required (EIA, 2017).

European standard EN 14214 and American standard ASTM D6751 specify quality metrics for biodiesel and have served as reference standards worldwide. These standards consider the variation of fuel quality influenced by pretreatment processes for feedstock oils, the transesterification reaction itself, and any posttreatment processes (Bart et al., 2010). For instance, density, kinematic viscosity, cetane number (CN), cold flow, and oxidative stability are all comprehended in the standards and are directly related to the composition of the input feedstock. Biodiesel (B20) can viably be used without the limitation of cloud point and pour point and may also be directly used in engines without further modifications. The disadvantages are cold flow issues at low temperatures and higher concentrations of oxides of nitrogen (NO_x) during combustion. As shown in Table 6.7, reported properties of crude WCO, biodiesel from WCO, and commercial diesel fuel were in accordance with the ASTM standards.

6.5 UTILIZATION OF BY-PRODUCT GLYCEROL

Glycerol (1,2,3-propanetriol) is the simplest trihydric alcohol containing two primary and one secondary hydroxyl groups, and is the main component of triglycerides, usually found in vegetable oils and animal fats. It can be obtained from soap manufacturing, fatty acid production, fatty ester production, microbial fermentation and can be synthesized from propylene oxide (Crocker, 2010). Similar to petroleum fuel production, processing of biodiesel generates by-products, with glycerol being one of the main by-products generated from the homogeneous base-catalyzed transesterification reaction. Fundamentally, for every 3 mol of methyl esters, 1 mol of glycerol is produced, which is approximately 10–20 wt% of the total product (Gholami et al., 2014; Quispe et al., 2013). Every 100 lbs (45 kg) of biodiesel produced generates approximately 10 lbs (4.5 kg) of glycerol. Crude glycerol typically contains 65%–85% w/w glycerol content, with other constituents including methanol, salt, and various impurities. One of the primary drawbacks of homogenous base-catalyzed transesterification is that glycerol separation is not efficient and requires significant secondary processing to achieve low levels of contamination.

6.5.1 Types of Glycerol and Their Uses

Glycerol generated as a by-product of the transesterification reaction can be categorized into three main groups: crude glycerol, technical glycerol, and purified glycerol (Fig. 6.9). In the current market, crude glycerol generated from biodiesel production is sold with methanol concentration of 0.3% (max) and glycerol concentration of 80%–88% purity (min) for lower grade applications. It has very little economic value because of various impurities, but can be further purified to achieve a higher market level. Technical grade glycerol is sold with methanol concentration of 0.1% (max) and glycerol concentration of 95% purity (min) for industrial applications. United States Pharmacopeia (USP) and Food Chemicals Codex (FCC) grade refined glycerol with 99.7% purity (min) is used in cosmetics, pharmaceuticals, and food (Gholami et al., 2014; Quispe et al., 2013).

Glycerol is used in many applications from energy bars to cough syrups to sealants for boat coatings (Gholami et al., 2014). However, the low-grade crude glycerol available directly from biodiesel production cannot be used for food, pharmaceutical, and cosmetic products because of its associated impurities. For large-scale biodiesel producers, crude glycerol can be purified by conventional methods in their manufacturing sites or sent to large refineries to produce a higher grade coproduct that can then be distributed in markets for other industries. Due to the current surplus of crude glycerol on the worldwide market, the capacity of refineries could well reach their production limits. The process of converting and refining crude glycerol into a pure form is usually cost prohibitive for small- and medium-sized scale biodiesel producers. Hence, the development of effective utilization strategies for crude glycerol might be a possible solution for producers in improving the economics of biodiesel manufacturing.

6.5.2 Process of Refining Crude Glycerol

Various kinds of glycerol purification methods and technologies are available in the market today, such as neutralization, acidification, ion exchange resins, vacuum distillation/evaporation, and membrane separation. The combination of more than one method can enhance the process and increase the purity level to as high as 99.2% (Wan Isahak et al., 2014). Distillation is the most common and mature technology, but the drawback is that it is highly energy intensive and consumes a large amount of water during condensation. Purification processes of crude glycerol generally require significant energy input, are high in chemical consumption, and have high production costs. The development of cheap and efficient purification processes that can be deployed at scale is essential to the biodiesel industry because they offer potential for additional economic benefits in generating secondary value-added products (Saifuddin et al., 2014).

Purification of crude glycerol generated by an institutional-scale biodiesel process developed in our laboratory was achieved by chemical treatment combined with vacuum distillation, as described by Hunsom et al. (2013) with additional modifications extracted from the literature (Tiangfen et al., 2013; Manosak et al., 2011; Kongjao et al., 2010; Marbun et al., 2014; Hajek and Skopal, 2010; Yang et al., 2013), following these steps:

(1) Acidification
(2) Neutralization
(3) Vacuum distillation, and
(4) Activated carbon adsorption.

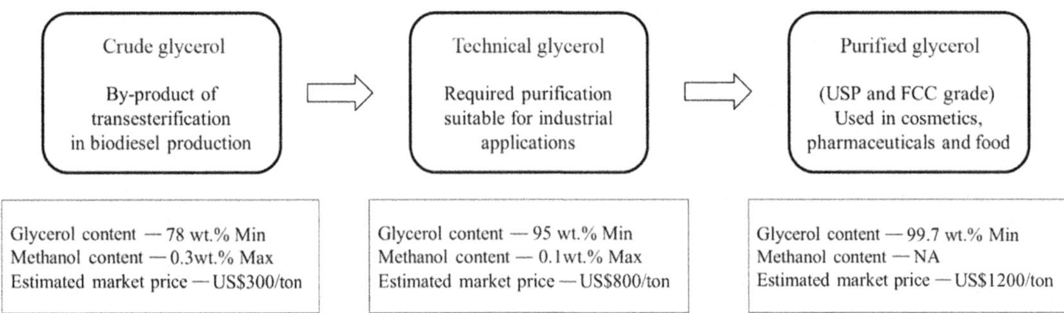

FIG. 6.9 Types of glycerol with their end applications.

FIG. 6.10 (1) Waste cooking oil (WCO); (2) biodiesel; (3) crude glycerol; (4) separated layers: (a) FFA; (b) glycerol; (c) salt; (5) refined glycerol; (6) liquid glycerin soap.

Figs. 6.10 and 6.11 illustrate the overall process of converting crude glycerol to purified glycerol in an institutional-scale process. Crude glycerol from biodiesel production was acidified with H_3PO_4 and left for 12 h before phase separation. After complete separation, the upper phase consisted of free fatty acid, methanol, water, and acids, while the middle phase was a glycerol-rich layer with some methanol, water, and salt, and it had a dark brown color. The lower phase also had a small fraction of an inorganic salt-rich layer with a high concentration of potassium phosphate. The final purified glycerol by-product was obtained from the middle layer after adsorption with activated carbon. Each of these process steps is described in more detail as follows.

Step 1: Acidification.
A phosphorus acid solution (85% H_3PO_4, Sigma-Aldrich) was added to improve the acidity of crude glycerol. This produced three distinct layers: free fatty acid top layer (FFA), glycerol-rich middle layer, and inorganic salt-rich bottom layer. A filter funnel with pore size between 70 and 100 μm was used to remove the precipitated salt from the mixture and to separate the upper layer from the glycerol-rich middle layer. These two layers were then poured into the separation funnel to remove the glycerol-rich layer for further processing.
Step 2: Neutralization.
The glycerol layer was then neutralized by the addition of 12.5 M NaOH to achieve the required pH and remove the precipitated salts.
Step 3: Vacuum distillation.
The methanol and water from the glycerol were removed using two methods. In the first method, the glycerol solution was placed on a hot plate at 100°C for 2 h to release both methanol and water in a chemical hood. In the second method, the glycerol solution was placed in a vacuum distillation unit to extract the excess methanol at 60°C for 50 to 60 min at 95°C, and then to remove the water for 50–60 min at a pressure of 3 kPa. From an environmental standpoint, the latter method should perform better, as excess methanol can be recovered and reused in the biodiesel process.
Step 4: Activated carbon adsorption.
The glycerol solution obtained from the prior process steps was then passed through a column of commercially available activated carbon to remove odor, color, and other impurities and yield viscous, nearly colorless and odorless glycerol solution. Activated carbon can also be reused in the process at least a few times, provided the color and purity of the product glycerol are closely monitored.

6.5.3 Saponification

Soap is produced by the saponification of triglyceride from animal fat or vegetable oil (Fig. 6.12), wherein fatty acids react with NaOH or KOH to produce glycerol, fatty acid, and salts; 10%–15% of glycerol is produced during the saponification process. However, most commercial soap manufacturers remove and sell the glycerol, or convert it into more profitable beauty products. Glycerin serves as a humectant, attracting moisture from the environment to the surface layers of the skin and hair (Lodén and Maibach, 1999) and has cleansing, lubricating, and soothing properties. Glycerol by-product soap from biodiesel production would typically have more glycerin than traditional soap that has only 10%–20% glycerin content (Failor, 2000; Nicely, 2009). Different vegetable oils or animal fats, catalysts (NaOH/KOH), and glycerin grades can be used to produce soap (Fig. 6.13). The ingredients are chosen depending on the individual preference and the economic value of the desired product (Win et al., 2015).

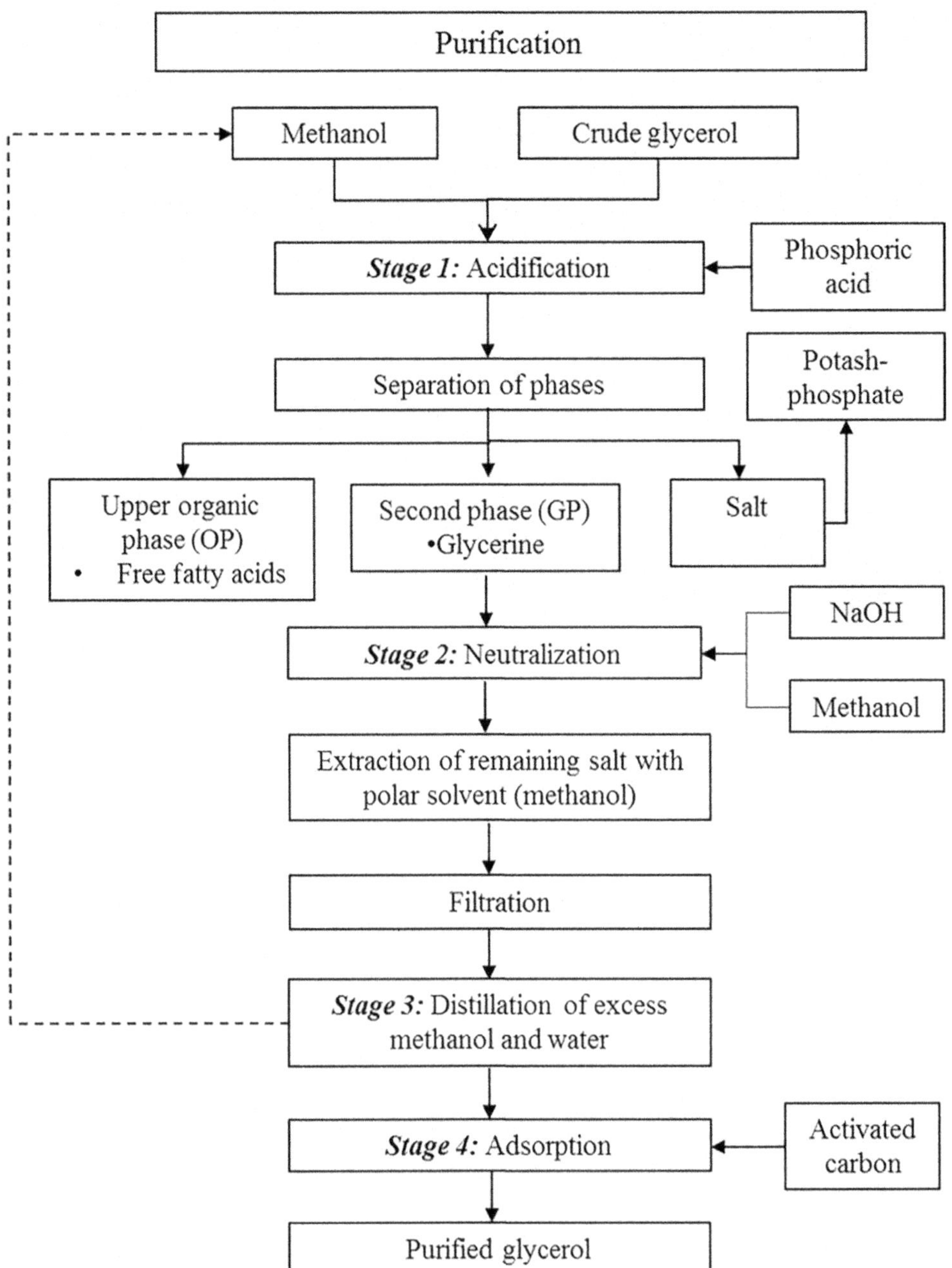

FIG. 6.11 Schematic diagram of glycerol purification process flow.

$$\text{Triglyceride} + \text{Caustic soda} \rightarrow \text{Glycerol} + \text{Soap}$$

$$\begin{array}{l} CH_2-O-\overset{O}{\overset{\|}{C}}-R_1 \\ \;| \\ CH-O-\overset{O}{\overset{\|}{C}}-R_2 \\ \;| \\ CH_2-O-\overset{O}{\overset{\|}{C}}-R_3 \end{array} + 3NaOH \rightarrow \begin{array}{l} CH_2-OH \\ \;| \\ CH-OH \\ \;| \\ CH_2-OH \end{array} + \begin{array}{l} Na^{+}\,{}^{-}O-\overset{O}{\overset{\|}{C}}-R_1 \\ Na^{+}\,{}^{-}O-\overset{O}{\overset{\|}{C}}-R_2 \\ Na^{+}\,{}^{-}O-\overset{O}{\overset{\|}{C}}-R_3 \end{array}$$

FIG. 6.12 Basic reaction of the saponification process (Tan et al., 2013).

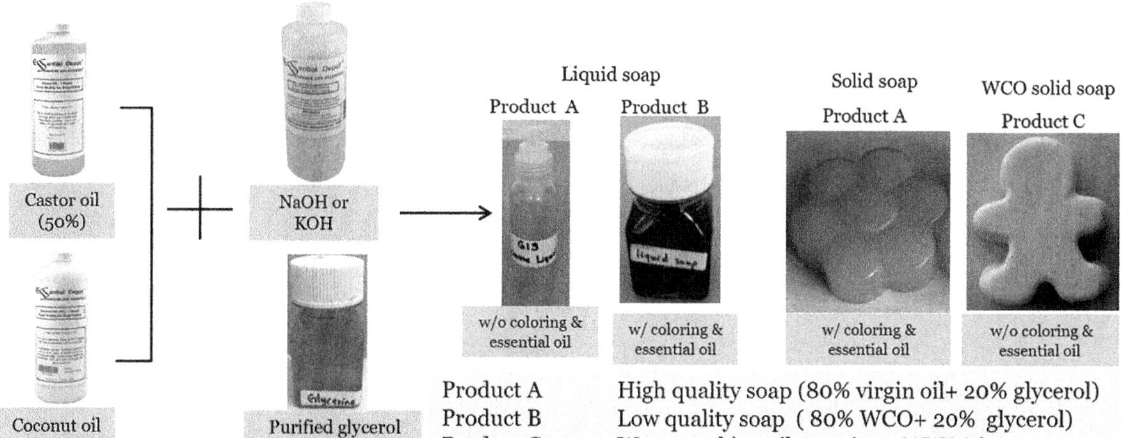

FIG. 6.13 General flowchart of the saponification process.

6.6 FUTURE PERSPECTIVE AND RESEARCH NEEDS

Waste cooking oil (WCO)-to-biodiesel conversion is not a new technology and has been implemented at the scale of individual WCO generating facilities, to community-scale plants, to large commercial enterprises. A wide diversity of research studies have been conducted in many different areas related to WCO-based biodiesel production. However, there are relatively few studies of WCO-to-biodiesel production in a constrained system at small scale (i.e., community scale), and few, if any, studies have considered the effect of utilizing by-product crude glycerol in saponification, to produce a secondary product that can enhance the economic and environmental performance of the overall system.

Crude glycerol from biodiesel production plants contains 80% glycerol in the best scenario and between 20% and 40% impurities. Purified through the costly refining process to cosmetic or pharmaceutical quality is generally not financially viable. Much research has been targeted on biodiesel production utilized for different applications, and other focused studies on conversion of low-value by-product crude glycerol into valued-added products. Many published papers have discussed ways to utilize crude glycerol from biodiesel production, and producing low-cost soap from edible or nonedible vegetable feedstocks in developing countries has already been implemented. Nonetheless, what has been minimally explored is the potential synergy between two waste products (WCO and crude glycerol) and converting the waste resources into two value-added products (biodiesel and soap), to determine the effect on economic development, resilience, energy independence, and waste for a constrained community-scale system.

Institutions generate waste cooking oil (WCO) from dining services and have few disposal options. The waste oil can properly be disposed at the facilities described previously, which can be costly, or sold for animal feed production for a small revenue. The development of a closed-loop biodiesel production system utilizing crude glycerol as an ingredient for soap production is compelling, especially in a constrained system where the WCO feedstock supply and biodiesel demand are in close proximity and controlled by a single entity. Biodiesel can be used in vehicles and other applications, while crude glycerol can be refined and used to produce soap. Potentially, the soap can be used in cafeterias and bathrooms across campus and dining services, to provide significant financial benefits beyond producing biodiesel fuel alone.

A case study of a community-based biodiesel production program in Atlanta, Georgia (GA) was conducted by Fyfe et al. (2006) to determine the opportunity to recycle yellow grease into biodiesel for urban fleets and other diesel users. The development of this yellow grease program in an urban area was likely to have potential based on the strong interest from the restaurant industry, with high grease volumes generated from nearby restaurants and Emory University, and diesel demand from the local public school buses. In another study, Skarlis et al. (2012) investigated the viability of a small-scale biodiesel production plant (10,000 ton/year) on the Greek island of Crete. They found that a small-scale plant would offer opportunities for decentralized rural development, which is important for the Greek economy. The investment of a small-scale project was found to be viable, as long as the biodiesel market value is consistently higher than the biodiesel production cost value. Schut et al. (2011) discussed the opportunities and constraints to implementing a sustainable community-based biofuel production and utilization in the Nhambita community in Mozambique. They developed three scenarios to understand how much jatropha oil production is needed to replace fossil fuels with pure plant oil (PPO) and to locally manufacture oil-based products such as soap. More rigorous economic and environment assessments should be

TABLE 6.8 Summary of U.S. Institutions With WCO-to-Biodiesel and Soap Programs

University	Location	Feedstock Source	Biodiesel Production Capacity	Facilities	Utilization and Benefits	Crude Glycerol Utilization
University of Arizona	Tucson, AZ	NA	NA	NA	NA	Produce soap
Appalachian State University	Boone, NC	NA	NA	NA	House heating systems and farm equipment	Produce soap and use in composting
Dickinson College	Carlisle, PA	WCO from dining services	5000 gal	Students manage production operation	NA	Produce soap for campus bathrooms and use in composting
Loyola University	Chicago, IL	WCO from Loyola, Northwestern University, and donation from neighborhood	10,000 gal	Student-run enterprise. First and only university operation licensed to sell biodiesel in the U.S. and a certified green business with the Illinois Green Business Association.	B100 (55–500 gal) is available for sale and also utilized in campus shuttle buses and a boiler in the Institute of Environmental Sustainability	Produce soap (marketed as BioSoap)
Clemson University	Clemson, SC	WCO from dining halls and local businesses	3000 gal per year in 90 gal batches	Student-run enterprise. Produce 20%–33% mix (i.e., 20%–33% biodiesel and 67%–80% diesel), with some vehicles running 100% biodiesel	Diesel trucks within fleet of vehicles on campus, landscaping, and utility trucks	Produce soap
Santa Fe College	Gainesville, FL	WCO from cafeteria	NA	Conduct biodiesel courses and provide education opportunities for students	Two fleet vehicles running with biodiesel	NA
Rochester Institute of Technology	Rochester, NY	WCO from cafeteria	5000 gal per year	Researchers produced biodiesel that met ASTM standards for vehicle use and heating fuel blends.	Performance of biodiesel was tested using campus vehicles and space heating appliances	Crude glycerol purified to produce soap of varying quality

conducted to determine the financial performance and environment impacts over the entire energy cycle, from WCO collection to biodiesel production and crude glycerol utilization. Table 6.8 provides a summary of universities in the United States that are currently producing biodiesel from waste cooking oil, and a few also have parallel soap production operations.

6.7 CONCLUSIONS

Today biodiesel offers one of the best substitutes for conventional transportation fuels because its physical and chemical properties are very similar to diesel fuel. The major disadvantages associated with biodiesel production are the high raw material cost and limited availability of vegetable oils resources that do not compete with food commodities. The high raw material cost is also due to high global biodiesel demand in blended mixtures with petroleum-based diesel, usually up to 20% in the U.S. (B20). In the future, vegetable oil prices are unlikely to be low enough for biodiesel produced from such feedstocks to compete with petroleum fuels, unless financial support from governments is made available (i.e., subsidies and tax credits). One of the widely accepted ways to reduce the raw material cost is the use of waste cooking oil (WCO) as a feedstock for biodiesel production.

Low-grade waste cooking oil is not as desirable as edible oils for biodiesel feedstock because of its generally high level of impurities. However, the growing volume of waste cooking oil generated by domestic and industrial operations due to the trend of increasing population growth has received a lot of attention. WCO has been facing serious disposal problems in many parts of the world, and proper management and utilization of this widely generated food supply chain waste material are urgently needed. Recycling of the waste cooking oil provides not only the benefit of producing a renewable energy resource that can displace high-GHG emitting transportation fuels, but also helps address a serious waste disposal problem. The importance of looking beyond the profitability and quality of oils for biodiesel production to address other benefits is essential. It has been shown that utilization of glycerol, widely considered a low-value output of the transesterification process, is essential to the economic viability of WCO-to-biodiesel conversion at small scale. The development of waste cooking oil-based biodiesel production in a community biorefinery system could contribute to sustainable waste management practices and also provide positive microeconomic outcomes for local populations.

REFERENCES

Ashokkumar, V., Agila, E., Sivakumar, P., Salam, Z., Rengasamy, R., Ani, F.N., 2014. Optimization and characterization of biodiesel production from microalgae *Botryococcus* grown at semi-continuous system. Energy Convers. Manag. 88, 936–946.

Atabani, A.E., Mahlia, T.M.I., Badruddin, I.A., Masjuki, H.H., Chong, W.T., Lee, K.T., 2013. Investigation of physical and chemical properties of potential edible and non-edible feedstocks for biodiesel production, a comparative analysis. Renew. Sust. Energ. Rev. 21, 749–755.

Babel, S., Arayawate, S., Faedsura, E., Sudrajat, H., 2018. Microwave-assisted transesterification of waste cooking oil for biodiesel production. In: Ghosh, S. (Ed.), Utilization and Management of Bioresources. Springer, Singapore.

Banani, R., Youssef, S., Bezzarga, M., Abderrabba, M., 2015. Waste frying oil with high levels of free fatty acids as one of the prominent sources of biodiesel production. J. Mater. Environ. Sci. 6 (4), 1178–1185.

Bart, J.C., Palmeri, N., Cavallaro, S., 2010. Biodiesel Science and Technology: From Soil to Oil. CRC Press, Woodhead Publishing Limited, Oxford.

Baskar, G., Aiswarya, R., 2016. Trends in catalytic production of biodiesel from various feedstocks. Renew. Sust. Energ. Rev. 57 (2016), 496–504.

Brorsen, W., 2015. Projections of US Production of Biodiesel Feedstock. Report prepared for Union of Concerned Scientists and The International Council on Clean Transportation.

Bruton, D.J., 2014. Waste Cooking Oil-To-Biodiesel Conversion for Space Heating Applications. M.S. Sustainable Systems thesis, Rochester Institute of Technology.

Chai, M., Tu, Q., Lu, M., Yang, Y.J., 2014. Esterification pretreatment of free fatty acid in biodiesel production, from laboratory to industry. Fuel Process. Technol. 125, 106–113.

Crocker, M. (Ed.), 2010. Thermochemical Conversion of Biomass to Liquid Fuels and Chemicals (No. 1). The Royal Society of Chemistry, Cambridge.

de Araújo, C.D.M., de Andrade, C.C., e Silva, E.D.S., Dupas, F.A., 2013. Biodiesel production from used cooking oil: a review. Renew. Sust. Energ. Rev. 27, 445–452.

Demirbas, A., 2009. Biodiesel from waste cooking oil via base-catalytic and supercritical methanol transesterification. Energy Convers. Manag. 50 (4), 923–927.

Deng, X., Fang, Z., Liu, Y.H., Yu, C.L., 2011. Production of biodiesel from Jatropha oil catalyzed by nanosized solid base catalyst. Energy 36 (2), 777–784.

DoE, U.S., 2014. Annual Energy Outlook 2014 With Projections to 2040. US Energy Information Administration, US Department of Energy, Washington, DC.

Energy Information Administration (EIA), 2017. Monthly biodiesel production report. In: Inputs to Biodiesel Production. Various Years.

Enweremadu, C.C., Mbarawa, M.M., 2009. Technical aspects of production and analysis of biodiesel from used cooking oil—a review. Renew. Sust. Energ. Rev. 13 (9), 2205–2224.

European Environment Agency (EEA), 2017. Landscapes in Transition—An Account of 25 Years of Land Cover Change in Europe.

Failor, C., 2000. Making Natural Liquid Soaps: Herbal Shower Gels, Conditioning Shampoos, Moisturizing Hand Soaps, Luxurious Bubble Baths, and More. Storey Publishing, North Adams, MA.

Frank, D.E., 2014. Waste Cooking Oil-To-Biodiesel Conversion for Institutional Vehicular Applications. M.S. Sustainable Systems thesis, Rochester Institute of Technology.

Fukuda, H., Kondo, A., Noda, H., 2001. Biodiesel fuel production by transesterification of oils. J. Biosci. Bioeng. 92 (5), 405–416.

Fyfe, E., Smith, B., Eisen, A., Pritchard, R., 2006. Community-Based Biodiesel Production from Restaurant Yellow Grease in Atlanta, Georgia. Emory University, Atlanta, Georgia.

Gan, S., Ng, H.K., Chan, P.H., Leong, F.L., 2012. Heterogeneous free fatty acids esterification in waste cooking oil using ion-exchange resins. Fuel Process. Technol. 102, 67–72.

Gholami, Z., Abdullah, A.Z., Lee, K.T., 2014. Dealing with the surplus of glycerol production from biodiesel industry through catalytic upgrading to polyglycerols and other value-added products. Renew. Sust. Energ. Rev. 39, 327–341.

Gnanaprakasam, A., Sivakumar, V.M., Surendhar, A., Thirumarimurugan, M., Kannadasan, T., 2013. Recent strategy of biodiesel production from waste cooking oil and process influencing parameters: a review. J. Energy 2013, 1–10. Article ID 926392.

Gui, M.M., Lee, K.T., Bhatia, S., 2008. Feasibility of edible oil vs. non-edible oil vs. waste edible oil as biodiesel feedstock. Energy 33 (11), 1646–1653.

Haile, M., 2014. Integrated volarization of spent coffee grounds to biofuels. Biofuel Res. J. 1 (2), 65–69.

Hajek, M., Skopal, F., 2010. Treatment of glycerol phase formed by biodiesel production. Bioresour. Technol. 101 (9), 3242–3245.

Hunsom, M., Saila, P., Chaiyakam, P., Kositnan, W., 2013. Comparison and combination of solvent extraction and adsorption for crude glycerol enrichment. Center Excell. Petrochem. Mater. Technol. 3 (2), 364–371.

Islam, M.A., Heimann, K., Brown, R.J., 2017. Microalgae biodiesel: current status and future needs for engine performance and emissions. Renew. Sust. Energ. Rev. 79, 1160–1170.

Kang, S.B., Oh, H.Y., Kim, J.J., Choi, K.S., 2017. Characteristics of spent coffee ground as a fuel and combustion test in a small boiler (6.5 kW). Renew. Energy 113, 1208–1214.

Karmakar, A., Karmakar, S., Mukherjee, S., 2010. Properties of various plants and animals feedstocks for biodiesel production. Bioresour. Technol. 101 (19), 7201–7210.

Kayode, B., Hart, A., 2017. An overview of transesterification methods for producing biodiesel from waste vegetable oils. Biofuels, 1–19.

Keera, S.T., El Sabagh, S.M., Taman, A.R., 2011. Transesterification of vegetable oil to biodiesel fuel using alkaline catalyst. Fuel 90 (1), 42–47.

Knothe, G., Razon, L.F., 2017. Biodiesel fuels. Prog. Energy Combust. Sci. 58, 36–59.

Kondamudi, N., Mohapatra, S.K., Misra, M., 2008. Spent coffee grounds as a versatile source of green energy. J. Agric. Food Chem. 56 (24), 11757–11760.

Kongjao, S., Damronglered, S., Hunsom, M., 2010. Purification of crude glycerol derived from waste used-oil methyl ester plant. Korean J. Chem. Eng. 27 (3), 944–949.

Lam, M.K., Lee, K.T., Mohamed, A.R., 2010. Homogeneous, heterogeneous and enzymatic catalysis for transesterification of high free fatty acid oil (waste cooking oil) to biodiesel: a review. Biotechnol. Adv. 28 (4), 500.

Leung, D.Y., Wu, X., Leung, M.K.H., 2010. A review on biodiesel production using catalyzed transesterification. Appl. Energy 87 (4), 1083–1095.

Lim, S., 2017. Post-Event Report. Inida, Oils and Fats International (OFI).

Lim, S., Teong, L.K., 2010. Recent trends, opportunities and challenges of biodiesel in Malaysia: an overview. Renew. Sust. Energ. Rev. 14 (3), 938–954.

Lin, L., Cunshan, Z., Vittayapadung, S., Xiangqian, S., Mingdong, D., 2011. Opportunities and challenges for biodiesel fuel. Appl. Energy 88 (4), 1020–1031.

Liu, Y., Tu, Q., Knothe, G., Lu, M., 2017. Direct transesterification of spent coffee grounds for biodiesel production. Fuel 199, 157–161.

Lodén, M., Maibach, H.I. (Eds.), 1999. Dry Skin and Moisturizers: Chemistry and Function. CRC Press, New York.

Luque, R., Melero, J.A. (Eds.), 2012. Advances in Biodiesel Production: Processes and Technologies. Woodhead Publishing Limited, Cambridge.

Manosak, R., Limpattayanate, S., Hunsom, M., 2011. Sequential-refining of crude glycerol derived from waste used-oil methyl ester plant via a combined process of chemical and adsorption. Fuel Process. Technol. 92, 92–99.

Marbun, B.T., Hutapea, P.A., Aristya, R., Husin, R., 2014. Purification and physio-chemical characterization of glycerol by-product of palm oil biodiesel industry as potential base fluid for synthetic oil-based drilling fluid system. Am. J. Oil Chem. Technol. 1(10).

Math, M.C., Kumar, S.P., Chetty, S.V., 2010. Technologies for biodiesel production from used cooking oil—a review. Energy Sustain. Dev. 14 (4), 339–345.

Mohammadshirazi, A., Akram, A., Rafiee, S., Kalhor, E.B., 2014. Energy and cost analyses of biodiesel production from waste cooking oil. Renew. Sust. Energ. Rev. 33, 44–49.

Mussgnug, J.H., Klassen, V., Schlüter, A., Kruse, O., 2010. Microalgae as substrates for fermentative biogas production in a combined biorefinery concept. J. Biotechnol. 150, 51–56.

Najdanovic-Visak, V., Lee, F.Y.L., Tavares, M.T., Armstrong, A., 2017. Kinetics of extraction and in situ transesterification of oils from spent coffee grounds. J. Environ. Chem. Eng. 5 (3), 2611–2616.

Naylor, R.L., Higgins, M.M., 2017. The political economy of biodiesel in an era of low oil prices. Renew. Sust. Energ. Rev. 77, 695–705.

Nicely, R., 2009. Making Biodiesel Soap. Knice-N-Clean Soap Company.

NREL, 2018. Alternative Fuels Data Center: Maps and Data. The Biofuels Atlas.

Quispe, C.A., Coronado, C.J., Carvalho Jr., J.A., 2013. Glycerol: production, consumption, prices, characterization and new trends in combustion. Renew. Sust. Energ. Rev. 27, 475–793.

Saifuddin, N., Refal, H., Kumaran, P., 2014. Rapid purification of glycerol by-product from biodiesel production through combined process of microwave assisted acidification and adsorption via chitosan immobilized with yeast. Res. J. Appl. Sci. Eng. Technol. 7, 593–602.

Sanli, H., Canakci, M., Alptekin, E., 2011. Characterization of waste frying oils obtained from different facilities. World Renewable Energy Congress-Sweden; 8–13 May; 2011; Linköping; Sweden. vol. 057. Linköping University Electronic Press, pp. 479–485.

Schut, M., van Paassen, A., Leeuwis, C., Bos, S., Leonardo, W., Lerner, A., 2011. Space for innovation for sustainable community-based biofuel production and use: lessons learned for policy from Nhambita community, Mozambique. Energy Policy 39 (9), 5116–5512.

Skarlis, S., Kondili, E., Kaldellis, J.K., 2012. Small-scale biodiesel production economics: a case study focus on Crete Island. J. Clean. Prod. 20 (1), 20–26.

Somnuk, K., Eawlex, P., Prateepchaikul, G., 2017. Optimization of coffee oil extraction from spent coffee grounds using four solvents and prototype-scale extraction using circulation process. Agric. Nat. Resour. 51 (3), 181–189.

Tan, H.W., Aziz, A.A., Aroua, M.K., 2013. Glycerol production and its applications as a raw material: a review. Renew. Sust. Energ. Rev. 27, 118–127.

Tiangfen, C., Huipen, L., Hua, Z., Kejian, L., 2013. Purification of crude glycerol from waste cooking oil based biodiesel production by orthogonal test method. Chin. Pet. Process Pe. Technol. 15 (1), 48–53.

Tsoutsos, T.D., Tournaki, S., Paraíba, O., Kaminaris, S.D., 2016. The used cooking oil-to-biodiesel chain in Europe assessment of best practices and environmental performance. Renew. Sust. Energ. Rev. 54, 74–83.

Tuntiwiwattanapun, N., Usapein, P., Tongcumpou, C., 2017. The energy usage and environmental impact assessment of spent coffee grounds biodiesel production by an in-situ transesterification process. Energy Sustain. Dev. 40, 50–58.

United States Department of Energy, 2014. Clean Cities Alternative Fuel Price Report.

Van Gerpen, J., 2005. Biodiesel processing and production. Fuel Process. Technol. 86 (10), 1097–1107.

Van Gerpen, J., Shanks, B., Pruszko, R., Clements, D., Knothe, G., 2004. Biodiesel Production Technology. Golden, Colorado. NREL/SR-510-36244.

Vardon, D.R., Moser, B.R., Zheng, W., Witkin, K., Evangelista, R.L., Strathmann, T.J., Rajagopalan, K., Sharma, B.K., 2013. Complete utilization of spent coffee grounds to produce biodiesel, bio-oil, and biochar. ACS Sustain. Chem. Eng. 1 (10), 1286–1294.

Verma, P., Sharma, M.P., 2016. Review of process parameters for biodiesel production from different feedstocks. Renew. Sust. Energ. Rev. 62, 1063–1071.

Wallace, T., Gibbons, D., O'Dwyer, M., Curran, T.P., 2017. International evolution of fat, oil and grease (FOG) waste management—a review. J. Environ. Manag. 187, 424–435.

Wan Isahak, W.N.R., Ramli, Z.A.C., Ismail, M., Mohd Jahim, J., Yarmo, M.A., 2014. Recovery and purification of crude glycerol from vegetable oil transesterification. Sep. Purif. Rev. 44, 250–267.

Wiltsee, G., 1999. Urban Waste Grease Resource Assessment (No. NREL/SR-570-26141). National Renewable Energy Lab, Golden, CO.

Win, S.S., Hedge, S., Trabold, T.A., 2015. Techno-economic assessment of different pathways for utilizing glycerol derived from waste cooking oil-based biodiesel.Proceedings of the ASME Power and Energy Conversion Conference, Paper Power Energy 2015–49563, San Diego, CA, June 28–July 2.

Yaakob, Z., Mohammad, M., Alherbawi, M., Alam, Z., Sopian, K., 2013. Overview of the production of biodiesel from waste cooking oil. Renew. Sust. Energ. Rev. 18, 184–193.

Yang, F., Hanna, M.A., Sun, R., 2013. Value-added uses for crude glycerol-a byproduct of biodiesel production. Biotechnol. Biofuels 5 (13), 1–10.

Chapter 7

Sustainable Waste-to-Energy Technologies: Bioelectrochemical Systems

Rami M.M. Ziara*, Bruce I. Dvorak†, Jeyamkondan Subbiah‡
*Department of Civil Engineering, University of Nebraska-Lincoln, Lincoln, NE, United States, †Department of Civil Engineering and Department of Biological Systems Engineering, University of Nebraska-Lincoln, Lincoln, NE, United States, ‡Department of Biological Systems Engineering and Department of Food Science and Technology, University of Nebraska-Lincoln, Lincoln, NE, United States

7.1 INTRODUCTION

Bioelectrochemical systems (BESs) are systems that use microorganisms to biochemically catalyze complex substrates into useful energy products, in which the catalytic reactions take place on electrodes. In other words, BESs are battery-like systems in which a biofilm grown on electrodes oxidizes substrates and generates energy. In wastewater treatment, a substrate refers to a contaminant that needs to be removed. For example, the major substrate removed from wastewater is organic matter which can be measured using different wastewater characteristics including chemical oxygen demand (COD) and biochemical oxygen demand (BOD). Wastewater characteristics could be represented as the total substrate concentration (e.g., Total BOD or Total COD) or concentration of the soluble substrates in the wastewater (e.g., soluble BOD or soluble COD). BESs are advantageous due to their ability to achieve a degree of substrate removal while generating energy. Typically, the energy generated from BESs is either in the form of electricity or energy-rich gasses. Therefore it is a promising technology toward energy positive or energy neutral treatment systems.

Potter (1911) was the first to report that electric potential can be produced in a cell using microorganisms; however, this technology did not gain much attention until the beginning of the 21st century. Over the past two decades, significant effort has been exerted in order to understand and develop BESs (Aghababaie et al., 2015; Wang et al., 2015). Many reactor configurations, architectures, and materials have been evaluated in efforts to optimize the technology. Thus different BESs types have emerged including:

- (i) microbial fuel cells (MFCs), which oxidize the substrate and generate electric power concurrently (Logan et al., 2006),
- (ii) microbial electrolysis cells (MECs) or biolectrochemically assisted microbial reactors (BEAMRs), for which an external power source is added to oxidize a substrate while generating useful by-products (Ditzig et al., 2007; Escapa et al., 2012),
- (iii) enzymatic biofuel cells, which use specific enzymes to oxidize the substrate and the enzymes are responsible for the transfer of electrons to the electrodes (Leech et al., 2012),
- (iv) microbial electrosynthesis cells which are used to synthesize organic chemicals from the substrate (Nevin et al., 2010), and
- (v) microbial desalination cells which can remove salinity from the substrate (Cao et al., 2009).

Logan et al. (2015) provided a comprehensive summary of additional secondary type MFCs and their relative performance. Most of BESs types were evaluated using nonfood source substrates. However, the studies that evaluated food waste focused mainly on MFCs and MECs, thus this chapter focuses on these two types of BESs.

The food industry produces a large amount of waste and wastewater. In the United States, fruits, vegetables, dairy, and grain products are the most common wasted foods, while in the UK, fruits, vegetables, bakery, and dairy are among the top wasted foods (Kosseva, 2013). Carbohydrates, proteins, lipids, and organic fibers constitute the majority of the waste mass, which makes food waste highly biodegradable and energy rich. Food production and processing are associated with the use of resources including water and energy. In addition, a large amount of waste and wastewater loads are generated during the production of food and must be treated before discharge. The constituents of the wastewater differ from industry to industry

Sustainable Food Waste-to-Energy Systems. https://doi.org/10.1016/B978-0-12-811157-4.00007-3

but generally, organic matter is the largest constituent of food industry wastewater. Summaries of waste and wastewater characteristics produced from food industries are provided through this chapter, as well as in Chapter 2. More detailed reviews of the wastes and wastewater produced from the food industry can also be found in earlier publications (e.g., ElMekawy et al., 2015; Kosseva, 2013).

7.2 THEORETICAL BACKGROUND AND PERFORMANCE INDICATORS

One of the advantages of the bioelectrochemical systems is that energy can be produced simultaneously while treating the wastewater through substrate degradation. In BESs, electrochemical reactions are carried out by a specific group of bacteria, *exoelectrogens*, which can transfer electrons outside the microbial cell (Kiely et al., 2011b; Liu et al., 2014; Logan, 2009; Sun et al., 2014). The fundamental principle behind BESs is redox potential. Gibbs free energy ($\Delta G°$) is the energy available in a chemical reaction to do useful work. Exergonic reactions produce energy ($\Delta G° < 0$), while endergonic reactions require energy to occur ($\Delta G° > 0$). Furthermore, Gibbs free energy can be converted into the electric potential using Nernst's law ($E° = -\Delta G°/nF$), where n is the number of electrons transferred in a chemical reactions and F is Faraday's constant (96,485 C/mol). Moreover, the electromotive force is the electrical potential available between an oxidizing reaction and a reduction reaction ($E°_{emf} = E°_{red} - E°_{oxi}$), where $E°_{red}$ and $E°_{oxi}$ are the electric potential for the reduction reaction and oxidation reaction, respectively.

7.2.1 Microbial Fuel Cells (MFCs)

Microbial fuel cells are a type of bioelectrochemical systems that oxidize substrates and generate electric current (i.e., $E°_{emf} > 0$) (Logan et al., 2006). A typical MFC contains two electrodes, anode and cathode, connected externally to a load or resistor and separated by a membrane. The oxidation of the substrate occurs at the anode and generates electrons (e^-) and protons (H^+). The electrons are transferred from the microorganisms into the electrode. Three means are reported in the literature by which electrons are shuttled from the microorganisms to the electrode; direct electron transfer, transfer through nanowire structures and through a mediator (Philips et al., 2016; Rabaey and Verstraete, 2005). The protons travel from the anode chamber to the cathode chamber through the liquid and ion exchange membrane, if applicable. The electrons and protons react with the terminal electron acceptor on the cathode. The terminal electron acceptor can theoretically be any chemical that has a redox potential less than that of the electron donor, for example, oxygen or nitrate. The transport of electrons through the external wire generates the electric current. The maximum voltage that can be produced by an MFC is limited by the thermodynamic relationships between the electron donor and the electron acceptor ($E°_{emf}$), as well as losses inside the cell. Electron losses are due to oxidation activation losses ($\eta_{oxi,\ act}$) and reduction activation losses ($\eta_{red,\ act}$); internal resistance (IR) of the cell due to losses in electrodes, electrolytes, membrane, and connections; and losses associated with mass transport and diffusion (η_{mt}) (Logan et al., 2006; Rabaey and Verstraete, 2005). The cell electrical potential is therefore $E_{cell} = E_{emf} - \eta_{oxi,\ act} - \eta_{red,\ act} - IR - \eta_{mt}$.

A typical microbial fuel cell design that contains two electrodes connected by a resistor (load) and separated by a membrane is illustrated in Fig. 7.1. The oxidation of substrate or wastewater is achieved by the biofilm that grows on the anode. Electrons produced from the oxidation of organic matter travel from the anode chamber to the cathode chamber where they are used in a reduction reaction at the cathode. In the shown case, the terminal electron acceptor is oxygen; however, other electron acceptors can be utilized including nitrate and sulfate. Several MFC architectures have been developed; the most commonly used are two-chamber MFC (TC-MFC) and single-chamber MFC (SC-MFC). Fig. 7.1 shows the architecture of a two-chamber microbial fuel cell, which has an anode and a cathode chamber separated by an ion exchange membrane. Single-chamber microbial fuel cells are MFCs that have a single chamber in which both electrodes are placed (Cheng et al., 2011). The use of ion exchange membrane in a SC-MFC is optional and when used it could be placed directly on the electrode. Tubular MFCs have a tube- or pipe-like architecture (Rabaey et al., 2005), while a three-chamber MFC has three chambers separated by ion exchange membranes (Zhang et al., 2013). Several sources are available for further reading on MFC architectures and materials used (Bajracharya et al., 2016; Du et al., 2007; Dumitru and Scott, 2016; Logan, 2008; Scott, 2016; Silver et al., 2014).

The performance of an MFC can be assessed based on several indicators, including power density, current density, coulombic efficiency (CE), and substrate reduction (e.g., ΔCOD). Power is the product of current and voltage, with current being an indication of electrons flow. Power and current densities are usually normalized by the anodic surface area (A_{an}) since the biofilm that oxidizes the substrate grows on the anode. Furthermore, power and current densities are sometimes normalized by the working volume of the cell. The current produced from a microbial fuel cell is usually small ($<1\ A/m^2$) since it is related to biochemical reactions that are limited by substrate utilization rate and electron production. Coulombic

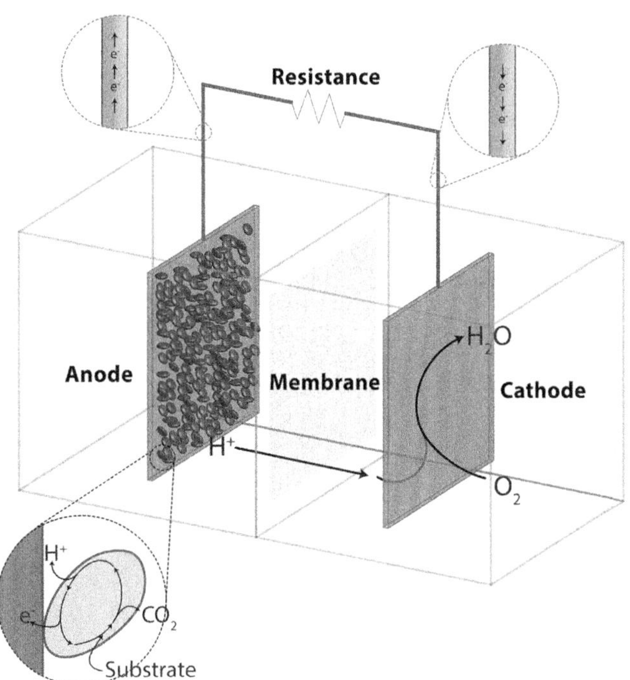

FIG. 7.1 A typical design of two-chamber microbial fuel cell (TC-MFC) which contains the anode and cathode chambers separated by ion exchange membrane.

efficiency (*CE*) is a parameter that indicates the fraction of electrons recovered as current, compared to that originally present in the organic matter. Therefore *CE* is an important indicator in mixed culture MFCs, where multiple microbial species compete for the substrate and it also can reflect electron loss in the cell. For more information on measurement and calculation methods, the reader is referred to Logan et al. (2006).

7.2.2 Microbial Electrolysis Cells (MECs)

Microbial electrolysis cells (MECs) are a type of bioelectrochemical systems that use an external power source to catalyze the substrate into by-products. This type of BES has been given many names, including BEAMR, biocatalyzed electrolysis cell (BEC), and microbial electrolysis cell (MEC) (Ditzig et al., 2007; Escapa et al., 2012). The latter, MEC, is the most commonly used. The external power is needed to force thermodynamically unfavorable reactions ($\Delta G° > 0$) to occur. Products ranging from methane (CH_4), to hydrogen gas (H_2), to hydrogen peroxide (H_2O_2) can be produced using MECs, depending on the redox reactions involved (Rozendal et al., 2009; Wagner et al., 2009).

Several MEC architectures have been evaluated, but generally most MFC architectures and materials are applicable for MECs, including TC-MEC, SC-MEC, and tubular MEC; the difference is that the cathode also operates under anaerobic condition. A typical two-chamber MEC is illustrated in Fig. 7.2. In MECs, electrons and protons are produced on the anode. The electrons travel through the electrode and the protons travel through the liquid to the cathode. The redox potential of the anodic and cathodic reactions is not enough to move these reactions forward, therefore an external power source is needed. Research studies have established that using a biocathode in MECs is more efficient than using an abiotic, microorganism-free, cathode (Rozendal et al., 2008; Wang et al., 2014; Xu et al., 2014). The theoretical voltage required to achieve a specific reaction can be calculated using (E°_{emf}), which will be negative in the case of MEC. Like MFCs, voltage losses occur within the cell, therefore the voltage needed to be added is usually slightly higher than the theoretical voltage.

The performance of an MEC can be assessed using multiple indicators, including *CE*, hydrogen yield (Y_{H_2}), cathodic hydrogen recovery (r_{cat}), overall hydrogen recovery (r_{H_2}), volumetric density, and hydrogen production rate (Call and Logan, 2008; Logan, 2008). Hydrogen yield is the mass fraction of the hydrogen produced to the substrate removed. The cathodic hydrogen recovery (r_{cat}) represents the fraction of hydrogen recovered to the estimated hydrogen produced based on measured current. The overall hydrogen recovery (r_{H_2}) is the efficiency of hydrogen production based on the total hydrogen moles recovered versus the theoretical possible production. The energy efficiency (η_W) is the efficiency based on the applied voltage. The volumetric hydrogen production rate (Q) represents how much hydrogen is produced per unit volume of reactor per unit time. For more information on measurement and calculation methods, the reader is referred to Call and Logan (2008) and Logan (2008).

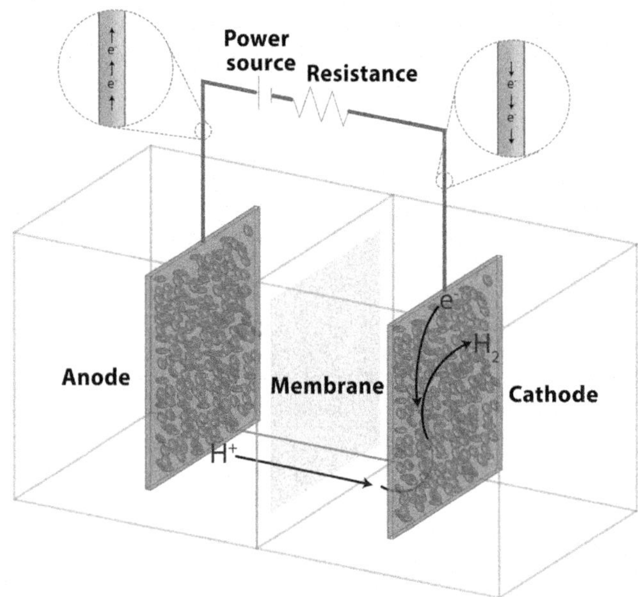

FIG. 7.2 A typical design of a two-chamber microbial electrolysis cell (TC-MEC) which has two chambers separated by ion exchange membrane.

7.3 ENERGY RECOVERY FROM FOOD INDUSTRY WASTES USING BESs

7.3.1 Microbial Fuel Cells

MFCs are bioelectrochemical systems that can achieve substrate removal and generate power simultaneously. Several architectures for MFCs exist; however, only a few have been evaluated using food industry wastes. Most of the food industry wastewater was evaluated using two-chamber MFCs and single-chamber MFCs at laboratory scales. The performance of MFCs using food wastes can be categorized according to the source of the waste as follows.

7.3.1.1 Brewery and Winery Wastewater

Brewery wastewater has high concentrations of carbohydrates and sugars which have high energy content and can be easily biodegraded (Wang et al., 2016). Due to wastewater generation patterns and variability among brewery and winery wastewater sources, traditionally biological wastewater treatment technologies are employed, including sequencing batch reactors (SBRs) and up-flow sludge blankets (USABs) systems (Simate et al., 2011). Aerobic and anaerobic biological treatment processes can achieve 70%–98% COD removal; however, the energy requirement for these processes is high (Feng et al., 2008). Therefore the use of MFC systems for brewery wastewater has been investigated extensively and has even been commercialized (Pandey et al., 2016).

Previous studies of MFCs to treat brewery and alcohol-based wastewaters have investigated parameters including substrate concentration, reactor configuration, electrode materials, and mixing with other substrates in batch and continuous operations modes. Table 7.1 provides a summary of the performance, reactor design, and materials used in 19 studies that evaluated the performance of MFCs using brewery and alcohol-based wastewaters. Most of the studies investigated cells with small working volume (<500 mL). Two studies investigated 4 L and 10 L MFC.

The highest power density using brewery wastewater was achieved using winery wastewater (6850 mg COD/L) in a tubular MFC with working volume of 170 mL (Penteado et al., 2016a). Their cell achieved a maximum power density of 890 mW/m^2, 10% COD removal, and maximum coulombic efficiency of 42.2%. Different solids retention times (SRT) were evaluated and Penteado et al. (2016a) concluded that SRT does not have a significant impact on biological treatment but has an effect on coulombic efficiency and power density. Feng et al. (2008) achieved the highest reported COD reduction of diluted brewery wastewater using single-chamber MFC (up to 98%), however lower power density (29–205 mW/m^2) was achieved; this study demonstrates that treating brewery wastewater with MFCs has the potential to be competitive with traditional energy-intensive biological processes.

A 4-L single-chamber MFC was investigated using diluted brewery wastewater (3707 mg COD/L); it produced 304 mW/m^2 and achieved >75% COD reduction (Wang et al., 2016). Despite this large COD removal, the coulombic efficiency was low which indicates that the organic matter might have been oxidized by fermentative and methanogenic microorganisms instead of exoelectrogens.

TABLE 7.1 Summary of Literature Studies Reporting Use of MFCs for Treating Brewery and Winery Wastewater

Wastewater Type	Cell Type[a]	Working Vol. (mL)	Anode Material	Cathode Material	Operation Mode	COD_{in} (mg/L)	ΔCOD (%)	Power Density mW/m^2 (mW/m^3)	Current Density mA/m^2 (mA/m^3)	CE (%)	Ref.
Alcohol	TC-MFC	84	Carbon cloth	Carbon paper-Pt		300		(627)	(3833)	<8	(Mohamed et al., 2016)
Alcohol	TC-MFC	84	Carbon cloth	Carbon paper-Pt		300		(164)	(833)	<1	(Mohamed et al., 2016)
Brewery	SC-MFC		Carbon cloth	Carbon cloth-Pt	Batch	84–2250	54–98	29–205		27–10	(Feng et al., 2008)
Brewery	SC-MFC		Carbon cloth	Carbon cloth-Pt		2239	85–87	435–483 (11 − 12)		21–38	(Wang et al., 2008)
Brewery	SC-MFC	4000	Carbon fiber brushes	Activated carbon	Continuous	3707 ± 220	75.4 ± 5.7	304 ± 31		1.5	(Wang et al., 2016)
Brewery	SC-MFC				Batch	3574	93	(<300)	(1100)		(Angosto et al., 2015)
Brewery	TC-MFC	200	Graphite felt with	Graphite cloth-Pt	Batch	2000	80	305	745		(Miran et al., 2015)
Brewery	SC-MFC	225	Graphite felt	Carbon cloth-Pt	Batch	510		251–552		31–41	(Yu et al., 2015)
Brewery	TC-MFC		Carbon paper	Carbon paper				1.68–38.34			(Mshoperi et al., 2014)
Brewery	3C-MFC	1200	Graphite plates	Graphite plates		850–4000	80–93	173.1	370		(Zhang et al., 2013)
Brewery	TC-MFC		Graphite felt	Graphite felt-Pt	Continuous	BOD: 125–1000	65		0.78		(Pisutpaisal and Sirisukpoca, 2012)
Brewery	SC-MFC	45	Carbon cloth anode	Carbon paper coated-Pt-PFTE	Batch	661	85		10 ± 1		(Velasquez-Orta et al., 2011)
Brewery	SC-MFC	100	Carbon fibers	Stainless steel-Activated carbon-PFTE	Continuous	1501	20.7	669 (24.1)		2.58	(Wen et al., 2010)

Continued

TABLE 7.1 Summary of Literature Studies Reporting Use of MFCs for Treating Brewery and Winery Wastewater—cont'd

Wastewater Type	Cell Type[a]	Working Vol. (mL)	Anode Material	Cathode Material	Operation Mode	COD_{in} (mg/L)	ΔCOD (%)	Power Density mW/m^2 (mW/m^3)	Current Density mA/m^2 (mA/m^3)	CE (%)	Ref.
Brewery	SC-MFC	100	Carbon fiber and graphite rods	Stainless steel-activated carbon-PFTE-Pt	Continuous	626.58	40.5–43	264 (9520)	1.79	19.75	(Wen et al., 2009)
Brewery; digester influent, effluent	TC-MFC	250	Copper mesh-Ti	Copper mesh-Ti	Continuous	2250 ± 80, 480 ± 20	<82	10.69–80.01, 12.36–18.43			(Çetinkaya et al., 2015)
Wine lees	TC-MFC	500	Graphite felt	Platinum mesh		10,843 ± 3904		0.8	6.6		(Cercado-Quezada et al., 2010a)
Winery	Tubular-MFC	170	Carbon felts	Carbon felts	Semicontinuous	6850	10	58–890		3.4–42.2	(Penteado et al., 2016a)
Winery	TC-MFC	70	Carbon felt	Carbon felt	Batch	6850	<17	105–465		2–15	(Penteado et al., 2016b)
Brewery; mixed with pig liquid manure	SC-MFC	100	Graphite granule and graphite rod	Carbon cloth-Pt	Batch	5028	53	(340)	(1200)	11	(Angosto et al., 2015)

[a] *TC-MFC stands for two-chamber microbial fuel cell, SC-MFC stands for single-chamber microbial fuel cell and 3C-MFC stands for three-chamber microbial fuel cell.*

Zhuang et al. (2012) scaled up an MFC to 10 L and operated it for 180 days continuously. A maximum power density of 4.1 W/m^3 was produced at 30 days of operation and power density dropped by 60% by the end of the experiment. The long-term COD removal rate was more stable than the power generation; the cell maintained COD removal larger than 85% throughout the experiment. The reported coulombic efficiency was low and >35% of the COD removed was estimated to be associated with nonexoelectrogenic microorganisms. However, high ammonia removal was concurrently achieved, which demonstrated the system's ability to treat multiple substrates. Zhuang et al.'s (2012) study demonstrates the MFC's general limitation: high electron loses in scaled-up systems.

Generally, two-chamber MFCs produced less current and achieved lower coulombic efficiency than single-chamber MFCs due to internal potential losses (Çetinkaya et al., 2015; Pisutpaisal and Sirisukpoca, 2012). Previous studies have shown that high COD and ammonia removal can be achieved using MFCs to treat brewery wastewater. However, proper methanogenic control should be employed to ensure that COD reduction is achieved by exoelectrogenic microorganisms and maximum coulombic efficiency is achieved. It is important to note that MFCs cannot achieve the required treatment for wastewater discharge, therefore they must be combined with a secondary process to further remove contaminants. The performance of MFC is similar to the performance of anaerobic technologies treating the same wastewater, and therefore MFC can compete with conventional anaerobic technologies.

7.3.1.2 Cafeteria and Canteen Wastes

Most cafeteria wastes are food leftovers that contain rice, bread, vegetables, oil, and meat products (Goud et al., 2011). Cafeteria and canteen wastes and wastewater were mostly investigated in a single-chamber or solid-phase MFC. Previous studies of MFCs to treat cafeteria and canteen wastes have investigated different parameters, including substrate concentration, reactor configuration, electrode materials, and pretreatment options in mostly batch operation modes. Table 7.2 provides a summary of the performance, reactor design, and materials used in 10 studies that evaluated the performance of MFCs using cafeteria and canteen wastes. Most of the studies investigated single- and two-chamber MFCs with small working volume (<500 mL).

Choi and Ahn (2015) fermented cafeteria waste and used the high strength leachate in a small single-chamber MFC. The cell power density was 1540 mW/m^2, the maximum reported from cafeteria waste. The cell also achieved high COD removal (85.1%) and high coulombic efficiency (88.8%). Sangeetha and Muthukumar (2011) investigated using canteen wastewater (COD; 7760 mg/L) and the cell achieved the highest COD removal reported for cafeteria and canteen waste, nearly 99%. However, the cell produced a maximum power density of only about 124 mW/m^2.

The studies reported for canteen and cafeteria-based waste show that MFCs can achieve high COD removal and high coulombic efficiencies. These studies also demonstrate that employing anaerobic fermentation as a waste pretreatment strategy for using high strength wastes is feasible. Similarly, higher power densities were reported by researchers who integrated fermentation of food wastes with MFC (Li et al., 2013; Rikame et al., 2012).

7.3.1.3 Dairy Industry and Cheese Whey

The dairy industry produces a large quantity of high-strength wastewater, with reported ranges of COD from 0.38 to 72.5 g/L, BOD from 0.19 to 68.6 mg/L, and up to 1462 mg TKN/L (Britz and van Schalkwyk, 2005). Several sources provide more specific dairy industry wastewater characterization (Britz and van Schalkwyk, 2005; Danalewich et al., 1998). Anaerobic biological treatment systems are usually used for the treatment of dairy industry wastewater, which includes UASB, up-flow anaerobic filters, and anaerobic suspended growth reactors, which can achieve 70%–99% COD reduction (Britz and van Schalkwyk, 2005; Demirel et al., 2005; Mohan et al., 2010b). Dairy industry wastewater has high concentrations of lipid, protein, and lactose content, and some of this may be emitted in the wastewater. These wastewater characteristics have encouraged researchers to investigate the performance of MFCs as a treatment and energy recovery technology. Table 7.3 provides a summary of the performance, reactor design, and materials used in 16 studies that evaluated the performance of MFCs using dairy industry waste and wastewater. Most of the studies investigated single- and two-chamber MFCs with working volume ranging between 28 and 2000 mL.

Dairy wastewater (3620 mg COD/L) was investigated by Mansoorian et al. (2016) in a two-chamber MFC and produced a maximum power density of 621 mW/m^2, which is the highest power density reported among studies listed in Table 7.3. The Mansoorian et al. (2016) study also reported >90% COD reduction and coulombic efficiency higher than 37%. Mohan et al. (2010b) investigated using diluted dairy wastewater in a single-chamber MFC under different organic loadings. The cell achieved the maximum COD removal reported for dairy industry wastewater (95%), however the cell achieved low coulombic efficiency and low power density. Mohan et al. (2010b) study also documented that high protein, turbidity, and carbohydrates removal can be achieved using MFC for the treatment of dairy wastewater. Kiely et al. (2011a)

TABLE 7.2 Summary of Literature Studies Reporting Use of MFCs for Treating Cafeteria and Canteen Wastes

Wastewater Type	Cell Type[a]	Working Vol. (mL)	Anode Material	Cathode Material	Operation Mode	COD_{in} (mg/L)	ΔCOD (%)	Power Density mW/m^2 (mW/m^3)	Current Density mA/m^2 (mA/m^3)	CE (%)	Ref.
Cafeteria waste; fermented	TC-MFC		Carbon felt	Carbon paper-Pt	Batch			15.3			(Choi et al., 2011)
Canteen	TC-MFC		Graphite felt	Graphite felt-Pt	Continuous	BOD; 125–1000	75		0.7		(Pisutpaisal and Sirisukpoca, 2012)
Canteen	TC-MFC	1500	Graphite plates	Graphite plates	Batch	7760	74.2–98.9	16.3–123.8	27.1–54.3		(Sangeetha and Muthukumar, 2011)
Canteen waste	TC-MFC	300	Graphite	Copper sheet		103.8–513.9	44	(19,151)			(Hou et al., 2016)
Canteen waste	SC SBES	300	Graphite	Graphite air-cathode	Batch	380	72	162.4	<4.5 mA		(Chandrasekhar et al., 2015)
Canteen waste	SC-MFC	22	Graphite fiber brush	Carbon cloth-Pt	Batch	2000–4900	77.2–86.4	371–556 (12–18)		23.5–27	(Jia et al., 2013)
Canteen waste	SC-MFC	430	Graphite plates	Graphite plates	Batch	sCOD 12,000	46.28–64.83	39.38–107.89	211–390		(Goud et al., 2011)
Canteen waste	Solid phase[b]	500	Graphite plates	Graphite plates			73–76	41.8–170.81			(Mohan and Chandrasekhar, 2011)
Cafeteria waste leachate	SC-MFC	24	Graphite brush	Carbon cloth-Pt	Batch	58,500 ± 3000	85.1	1540		88.8	(Choi and Ahn, 2015)
Canteen waste; Diluted	SC-MFC	120	Carbon cloth	Carbon cloth-Pt-PFTE	Batch	2700 ± 20	80.8	(5.6)	(15.3)		(Li et al., 2016)

[a] *TC-MFC stands for two-chamber microbial fuel cell, SC-MFC stands for single-chamber microbial fuel cell and 3C-MFC stands for three-chamber microbial fuel cell.*
[b] *This terminology was used because the waste was in solid phase. The design of the cell was conceptually similar to a single-chamber MFC.*

TABLE 7.3 Summary of Literature Studies Reporting Use of MFCs for Treating Dairy Wastewater

Wastewater Type	Cell type[a]	Working Vol. (mL)	Anode Material	Cathode Material	Operation Mode	COD_{in} (mg/L)	ΔCOD (%)	Power Density mW/m^2 (mW/m^3)	Current Density mA/m^2 (mA/m^3)	CE (%)	Ref.
Cheese whey	SC-MFC	28	Graphite fiber brush	Graphite fiber cloth-PTFE-Pt	Batch			(22.3)	10	49 ± 8	(Rago et al., 2017)
Cheese whey	TC-MFC	800	Graphite	Graphite	Batch			324.8 μW	1.19 mA		(Nasirahmadi and Safekordi, 2011)
Dairy	TC-MFC	84	Carbon cloth	Carbon paper-Pt		175.8		(503)	(1946)	<4	(Mohamed et al., 2016)
Dairy	TC-MFC	84	Carbon cloth	Carbon paper-Pt		175.8		(38)	(404)	<1	(Mohamed et al., 2016)
Dairy	TC-MFC	2000	Graphite plate	Graphite plate		3620	90.46	621.13	3.74 mA	37.16	(Mansoorian et al., 2016)
Dairy	TC-MFC		Carbon felt	Carbon-PFTE	Batch	2804	83.1	<450		32.4	(Pant et al., 2016)
Dairy	TC-MFC		Graphite felt	Platinum mesh		13,650 ± 3790			1009–1796		(Cercado et al., 2014)
Dairy	TC-MFC	30	Graphite plates	Graphite plates	Batch		<91	122–197 (2.7–3.2)		8–17	(Elakkiya and Matheswaran, 2013)
Dairy	SC-MFC	45	Carbon cloth anode	Carbon paper coated-Pt-PFTE	Batch	443–700	82		25 ± 1		(Velasquez-Orta et al., 2011)
Dairy	SC-MFC	480	Graphite plate	Graphite plate	Batch	45–444	67.79–95.49	0.366–1.28 (650–1100)		4.3–14.2	(Mohan et al., 2010b)
Dairy manure	3C-MFC	617	Graphite fiber brush	Graphite fiber brush and graphite granules			4434–8302 mg/L	(<300–14,000)		9.87–18.65	(Zhang et al., 2015)
Dairy manure	SC-MFC	28	Graphite fiber brushes	Carbon cloth-Pt-PFTE	Batch	4300	70	189		12	(Kiely et al., 2011a)

Continued

TABLE 7.3 Summary of Literature Studies Reporting Use of MFCs for Treating Dairy Wastewater—cont'd

Wastewater Type	Cell type	Working Vol. (mL)	Anode Material	Cathode Material	Operation Mode	COD_{in} (mg/L)	ΔCOD (%)	Power Density mW/m^2 (mW/m^3)	Current Density mA/m^2 (mA/m^3)	CE (%)	Ref.
Dairy wastewater; Synthetic	TC-MFC	480	Carbon Toray	Carbon Toray	Continuous	1513–3299	39–63	92.2 (1900)	665	2.2–24.2	(Faria et al., 2017)
Diary waste; activated sludge	TC-MFC	600	Graphite sheet	Graphite sheet	Batch			(0.5–0.715)			(Jayashree et al., 2014)
Yogurt waste	TC-MFC	500	Platinum mesh	Platinum mesh		91–594	87–91	38	<1450		(Cercado-Quezada et al., 2010b)
Yogurt waste	TC-MFC	500	Graphite felt	Platinum mesh		8169 ± 2568		2–53.8	14.5–231		(Cercado-Quezada et al., 2010a)

[a]*TC-MFC stands for two-chamber microbial fuel cell, SC-MFC stands for single-chamber microbial fuel cell and 3C-MFC stands for three-chamber microbial fuel cell.*

investigated using dairy manure (4300 mg COD/L) in a single-chamber small MFC operated in batch mode. The cell achieved a maximum power density of 189 mW/m^2, 70% COD reduction, and 12% coulombic efficiency. Zhang et al. (2015) investigated the performance of three-chamber MFC, two cathodes and one anode, in electricity production from dairy manure. The cell produced up to 14,000 mW/m^3 and reduced the COD by 4434–8302 mg/L.

Even though low power densities were achieved by MFCs using dairy wastewater, the COD removal indicates that if MFCs are better understood and optimized, they could be a viable alternative for current dairy wastewater treatment technologies.

7.3.1.4 Fruits, Vegetables, and Food Wastes

Fruit and vegetables constitute 20%–50% of household wastes. The percent of the fruit and vegetable in household waste is proportional to the proportion of vegetable and fruits in a country's diet (Bouallagui et al., 2003; Pekan et al., 2006). Further research concluded that the composition of fruit and vegetable wastes is related to the harvest period, demand for a product, handling requirements, and shelf life of the fruits and vegetables (Angulo et al., 2012; Kosseva, 2013). Thassitou and Arvanitoyannis (2001) collected the wastewater characteristics of fruit and vegetable processing industries including apples, carrots, cherries, corn, grapefruit, green peas, and tomatoes. The reported range of COD was 1.5–18.7 g/L, BOD was 0.8–9.6 g/L, and suspended solids was 0.21–4.12 g/L. Table 7.4 provides a summary of the performance, reactor design, and materials used in studies that evaluated the performance of MFCs using a variety of fruit, vegetable, and food wastes. Most of the studies investigated single- and two-chamber MFCs with working volume ranging between 25 and 1000 mL.

The highest power density reported for fruit and vegetable processing wastewater was achieved in a single-chamber MFC (Oh and Logan, 2005). However, the maximum power density achieved by a two-chamber MFC using the same wastewater dropped significantly, even though similar COD removal was achieved in both cells, as listed in Table 7.4. This demonstrates that two-chamber MFC has electron losses. Recently, Tian et al. (2017) evaluated the performance of small single-chamber MFC using potato pulp waste. The waste was diluted and the COD ranged between 2000 and 25,000 mg/L. The cell produced moderate power level that ranged between 20,400 and 32,100 mW/m^3 and achieved up to 68% COD reduction and up to 56% coulombic efficiency. However, Kiely et al. (2011a) was able to achieve higher COD removal using potato processing wastewater. Shrestha et al. (2016) investigated using tomato processing waste in two-chamber MFC. The tomato seeds and skin produced a maximum power density of 132 mW/m^2 while the tomato cull produced a maximum power density of 256 mW/m^2. Composited vegetable waste was investigated in a single-chamber, 430-mL MFC (Mohan et al., 2010a). The COD loading was varied from 0.70 to 2.08 kg/m^3/d and the cell achieved up to 63% COD reduction. The cell produced power density up to 216 mW/m^2.

The reported literature shows that 60%–85% COD removal can be achieved using MFC systems to treat fruit, vegetable, and food wastewater. In addition, lower power densities are generally achieved using this type of wastewater than that achieved by other wastewater. Further studies are needed particularly to evaluate the performance of scaled-up MFCs and how wastewater pretreatment, such as fermentation, may enhance the performance of MFC treating this wastewater. In addition, the economic feasibility of integrating MFC in vegetable and fruit waste treatment scheme must be evaluated since very high reduction COD cannot be achieved using MFC alone for this type of waste.

7.3.1.5 Animal Processing and Meat Industry

The global meat production was 280 million tonnes in 2008, with the production predicted to double by 2050 (Kosseva, 2013). To supply this global meat demand, livestock operations are intensified and thus produce large quantities of wastes and greenhouse gas emissions, which contribute to climate change (Caro et al., 2017; de Vries and de Boer, 2010; Naylor et al., 2005; Stehfest et al., 2013). The approximate edible mass portions of cows, sheep or goats, pigs, chicken, and turkey are 50%–54%, 52%, 60%–62%, 68%–72%, and 78%, respectively (Kosseva, 2013). Furthermore, meat processing in slaughterhouses and packing plants requires a large amount of water, for washing and cleaning, which is then discharged as wastewater. For example, the water used in a mid-size beef packing plant is approximately 3000 L/1000 kg live weight slaughtered (Ziara et al., 2016).

Meat processing wastewater is generally of high strength. Cattle slaughterhouse wastewaters have COD range of 3–12.9 g/L range, BOD of 0.9–7.24 g/L, average suspended solids (SS) of 3.6 g/L, average total nitrogen (TN) of 378 mg/L, and average total phosphorous (TP) around 79 mg/L (Banks and Wang, 2005; Kosseva, 2013). For hog slaughterhouses, the wastewater COD is about 3 g/L, BOD is in the 1.95–2.22 g/L range, average SS of 3.7 g/L, TN range of 14.3–253 mg/L, and TP range of 5.2–154 mg/L (Banks and Wang, 2005; Kosseva, 2013). The constituents of animal and meat-based industry are complex and not easily biodegradable, therefore anaerobic technologies are used the most

TABLE 7.4 Summary of Literature Studies Reporting Use of MFCs for Treating Fruits, Vegetables, and Food Waste and Wastewater

Wastewater Type	Cell Type[a]	Working Vol. (mL)	Anode Material	Cathode Material	Operation Mode	COD_{in} (mg/L)	ΔCOD (%)	Power Density mW/m^2 (mW/m^3)	Current Density mA/m^2 (mA/m^3)	CE (%)	Ref.
Baker's yeast	TC-MFC	100	Carbon felts	Carbon felts	Batch	3500–15,000	<40	9.75–18.41			(Liakos et al., 2017)
Bakery	SC-MFC	45	Carbon cloth	Carbon paper coated-Pt-PFTE	Batch	651	86		10 ± 1		(Velasquez-Orta et al., 2011)
Chilled ready-meal food production	Tubular-MFC	1000	Carbon veil	Carbon cloth	Continuous	843–1161	67–84	3.34–5.86			(Boghani et al., 2017)
Composite vegetable waste	SC-MFC	430	Graphite plates	Graphite plates	Batch	52,000	51.08–62.86	57.38–215.71			(Mohan et al., 2010a)
Fermented apple juice	TC-MFC	500	Graphite felt	Platinum mesh		3501 ± 2510		10.2–78	56.8–209		(Cercado-Quezada et al., 2010a)
Food	TC-MFC	84	Carbon cloth	Carbon paper-Pt	Batch	754		(1007)	(5524)	12	(Mohamed et al., 2016)
Food	TC-MFC	84	Carbon cloth	Carbon paper-Pt	Batch	754		(190.5)	(853)	7.6	(Mohamed et al., 2016)
Food industry	SC-MFC	250	Carbon cloth	Carbon cloth	Batch	810	64.2		0.78 mA		(Rasep et al., 2016)
Food industry	TC-MFC	250	Carbon cloth	Carbon cloth	Batch	810	62.96		0.72 mA		(Rasep et al., 2016)
Food processing	SC-MFC	250	Carbon paper	Carbon-Pt		sCOD; 595	95	371 ± 10			(Oh and Logan, 2005)
Food processing	TC-MFC	250	Carbon paper	Carbon-Pt		sCOD; 595	95	81 ± 7			(Oh and Logan, 2005)
Food waste leachate	TC-MFC	75.6	Carbon felt	Carbon felt	Batch	1000	74.1–85.4	(425.3–5591)		12.1–13.5	(Li et al., 2013)

TABLE 7.4 Summary of Literature Studies Reporting Use of MFCs for Treating Fruits, Vegetables, and Food Waste and Wastewater—cont'd

Wastewater Type	Cell Type	Working Vol. (mL)	Anode Material	Cathode Material	Operation Mode	COD_{in} (mg/L)	ΔCOD (%)	Power Density mW/m² (mW/m³)	Current Density mA/m² (mA/m³)	CE (%)	Ref.
Food waste leachate	TC-MFC	1200	Carbon electrode	Carbon electrode	Batch	5000	90	(15140)	(66750)		(Rikame et al., 2012)
Soy-based food	TC-MFC		Carbon felt	Carbon-PFTE	Batch	3107	71.4	<100		18.5	(Pant et al., 2016)
Tomato seeds and skin	TC-MFC		Graphite felt	Graphite felt	Batch	3000		132	456		(Shrestha et al., 2016)
Tomatoes Cull	TC-MFC		Graphite felt	Graphite felt	Batch	2000		256	1504		(Shrestha et al., 2016)
Vegetable waste	TC-MFC	35	Granular graphite and Graphite rod	Carbon Paper	Batch	sCOD; 1000–1500	87	(596–1019)		7.1–32.6	(Tao et al., 2013)
Potato	SC-MFC	28	Graphite fiber brushes	Carbon cloth-Pt-PFTE	Batch	7700	89	217		21	(Kiely et al., 2011a)
Potato processing	3C-MFC	800	Graphite particles	Graphite felt and graphite rods		1000	80		250–400 μA		(Durruty et al., 2012)
Potato pulp waste	SC-MFC	25	Graphite brush	Carbon Cloth	Batch	2000–25,000	55.4–68.4	(20,400–32,100)		18–56	(Tian et al., 2017)
Potato waste	TC-MFC	240	Carbon felts	Carbon felts	Batch	1569–4245	39.5–89.6	1.4–6.8	5–150	0.3–43.6	(Du and Li, 2016)

[a] *TC-MFC stands for two-chamber microbial fuel cell, SC-MFC stands for single-chamber microbial fuel cell and 3C-MFC stands for three-chamber microbial fuel cell.*

in the industry followed by secondary treatment for additional organics and nutrient removal (Banks and Wang, 2005). Table 7.5 provides a summary of the performance, reactor design, and materials used in 25 studies that evaluated the performance of MFCs using animal processing and meat industry waste and wastewaters. Most of the studies using this type of waste investigated cells with larger working volume (up to 2500 mL) than other wastewater sources discussed previously. Most of the studies focused on investigating two-chamber MFCs in both batch and continuous modes.

Using goat rumen fluid and hay in four two-chamber MFCs connected in series, Meignanalakshmi and Kumar (2016) reported the highest power density range of 34,390–42,110 mW/m^2 achieved using the waste type discussed in this section. However, further MFC performance indicators were not reported. Ismail and Mohammed (2016) reported the highest COD removal (99%) achieved using slaughterhouse wastewater in a tubular MFC operated in continuous mode. The highest coulombic efficiency of 47% was reported by Ichihashi and Hirooka (2012), who used swine slurry in a single-chamber MFC.

Swine waste produces significant greenhouse gas emissions during waste management operations, and therefore it has attracted special attention and is one of the most widely investigated wastes using BESs (Caro et al., 2017). Ma et al. (2016) achieved the highest power density (880–1056 mW/m^2) of the swine waste studies listed in Table 7.5 using swine farm wastewater in a two-chamber MFC. The anode was carbon fiber brush and the cathode was carbon cloth with Pt catalyst; other information was not reported. In a study aimed at evaluating the microbial dynamics in a continuous MFC, it was reported that that up to 5623 mW/m^3 was produced using swine slurry (Sotres et al., 2016). No other data was reported in that study regarding COD removal or coulombic efficiency. Zheng and Nirmalakhandan (2010) investigated the performance of two-chamber 1.85 L MFC using manure wash wastewater. The cell produced 216 mW/m^2 (2000 mW/m^3) and maximum coulombic efficiency of 5.2%.

Slaughterhouse and meat packing wastewater were also investigated using MFCs. Heilmann and Logan (2006) used diluted meat packing wastewater in a single-chamber MFC. The cell achieved >86% COD reduction, a maximum power density of 139 mW/m^2, and a maximum coulombic efficiency of 6%. The low coulombic efficiency indicates high internal resistance or that most of the COD reduction was achieved mainly by nonexoelectrogens.

Sulfur-based compounds can be present at higher concentrations in cattle and swine wastes. Hydrogen sulfide (H_2S) is the main sulfur-based emission from confined animal feedlot operations (CAFOs), which results from microbial degradation of sulfide (Rumsey and Aneja, 2014). Furthermore, sulfate-reducing bacteria such as *Desulfovibrio* and *Desulfotomaculum* use lactate as the main electron donor; lactic acid is one of the main organic acids used in slaughterhouses as an antimicrobial intervention (Algino et al., 2007; Ueki et al., 1986). Other bacteria such as *Desulfobacter postgatei*, *Desulfobulbus propionicus*, and *Desulfonema* can use acetate, proportionate and long-chain fatty acids as the main electron donor, which are the end products of anaerobic fermentation (Boone, 1982; Ueki et al., 1986, 1989, 1991). Three main sources of sulfur-based compounds have been identified: animal feed, degradation of animal proteins, and sulfate-based chemicals used for tanning hides (Abreu and Toffoli, 2009; Crawford, 2007; Miner, 1976; Sapkota et al., 2007; Sundar et al., 2002). Rabaey et al. (2006) and Zhao et al. (2009) showed that sulfur-based chemicals can be removed by MFCs. However, the performance of MFC in removing sulfur-based compounds from actual meat and animal-based wastewater has not been evaluated.

The studies reported in this section showed the potential of energy generation and treatment of animal waste and the meat processing wastewater. However, further research is needed to optimize the systems, evaluate pretreatment methods, and evaluate sulfur-based compound removal. Scaled-up systems still need to be developed and better methods for reduction of internal resistance and methanogenic control need to be researched.

7.3.1.6 Sugar-Based and Distillery Wastewater

Molasses wastewater is produced from sugar-based industry. Molasses wastewater is of high strength with COD ranging between 65,000 and 130,000 mg/L, low pH, and high concentrations of sugars and salts (Lee et al., 2016). The main by-product of distilleries is wastewater, with the wastewater volume being approximately 10 times larger than the volume of ethanol produced (Kosseva, 2013). The wastewater produced from distilleries is of high strength with COD range between 18,000 and 122,000 mg/L, high solids content, and low pH. The wastewater characteristics from distilleries depend on many factors including the feedstock, size, and capacity of plants, and wastewater utilization and biodegradation. Traditionally, molasses and distillery wastewater is treated using anaerobic processes, but has also been an attractive source for microbial fuel cells, due to the simplicity of the organic content which is primarily sugars (Pant and Adholeya, 2007). Table 7.6 provides a summary of the performance, reactor design, and materials used in studies that evaluated the performance of MFCs using molasses, distillery, and other sugar-based wastewater.

TABLE 7.5 Summary of Literature Studies Reporting Use of MFCs for Treating Animal Processing and Meat Industry Waste and Wastewater

Wastewater Type	Cell type[a]	Working Vol. (mL)	Anode Material	Cathode Material	Operation Mode	COD_{in} (mg/L)	ΔCOD (%)	Power Density mW/m^2 (mW/m^3)	Current Density mA/m^2 (mA/m^3)	CE (%)	Ref.
Cow manure, fruit waste and soil	TC-MFC	143	Graphite rod	Graphite rod			<71	31.92±4	190±9.1		(Vijay et al., 2016)
Cow's urine	TC-MFC	400	Carbon felt	Carbon felt	Batch	150–3000	45–82	(0.64–5.23)	(3.87–14.42)		(Jadhav et al., 2016)
Goat rumen fluid	TC-MFC	2500	Copper	Zinc				9700	0.24 A		(Meignanalakshmi and Kumar, 2016)
Goat rumen fluid and hay	4 TC-MFC in series	2500	Copper	Zinc				34,390–42,110	0.74–0.82 A		(Meignanalakshmi and Kumar, 2016)
Manure wash	TC-MFC	1850	Graphite fiber brush	Carbon cloth-Pt	Batch			216 (2000)	1380	1.3–5.2	(Zheng and Nirmalakhandan, 2010)
Manure; Diluted	TC-MFC	1850	Graphite fiber brush	Carbon cloth-Pt	Batch			46–93 (400–800)	370–780	1.3–5.2	(Zheng and Nirmalakhandan, 2010)
Meat packing	SC-MFC	28	Carbon paper	Carbon paper	Batch	6010	>86; diluted	139	1150	2.3–6.0	(Heilmann and Logan, 2006)
Slaughter house	TC-MFC	1000	Graphite	Zinc, graphite, and copper	Batch	10,815	67.9	700	318		(Christwardana et al., 2016)
Slaughter house	Tubular-MFC				Continuous	1000	99	165	472		(Ismail and Mohammed, 2016)
Swine	TC-MFC	1000	Carbon	Carbon rod		5400	85.92	3.55–88.45	0.14–0.49 mA		(Egbadon et al., 2016)
Swine	2 SC-MFC	100	Graphite fiber brushes	Activated carbon-PVDF-carbon black	Continuous	7000–7500	59±6	700–750 (2800–3000)	1400–1600		(Kim et al., 2016)
Protein food industry	TC-MFC	1500	Graphite sheets	Graphite sheets	Continuous	1900	86	230.3	527	5–21	(Mansoorian et al., 2013)
Swine	SC-MFC	70	Carbon felt	Carbon paper-Pt	Batch	60,000	76–91	1000–2300	6000–7000	37–47	(Ichihashi and Hirooka, 2012)
Swine	TC-MFC	450+350	Graphite granule and graphite rod	Carbon felt-Fe_2O_3	Batch	1652	62.2–76.7	(3.1–7.9)	1.7–2.8 mA		(Xu et al., 2011)

Continued

TABLE 7.5 Summary of Literature Studies Reporting Use of MFCs for Treating Animal Processing and Meat Industry Waste and Wastewater—cont'd

Wastewater Type	Cell type	Working Vol. (mL)	Anode Material	Cathode Material	Operation Mode	COD_{in} (mg/L)	ΔCOD (%)	Power Density mW/m^2 (mW/m^3)	Current Density mA/m^2 (mA/m^3)	CE (%)	Ref.
Swine	SC-MFC		Graphite brush		Batch		8–75				(Wagner et al., 2009)
Swine	TC-MFC SC-MFC	250	Carbon paper	Carbon-Pt	Batch	8320±190	88–92	261	1400	8	(Min et al., 2005)
Swine farm	TC-MFC		Carbon fiber brush	Carbon cloth-Pt	atch			880–1056			(Ma et al., 2016)
Swine farm	SC-MFC	128	Carbon fiber-Fe^{2+}	Carbon fiber-stainless steel mesh	Batch	6825±571	63.5–71.9	20–256	88–4000	0.9–39	(Estrada-Arriaga1 et al., 2015)
Swine manure	TC-MFC	420	Granular graphite and graphite rod	Granular graphite and graphite rod	Continuous	2200±665	2.02–2.09 kg/m^3/d	2–20		5–24	(Vilajeliu-Pons et al., 2015)
Swine manure	SC-MFC	28	Carbon paper	Carbon-Pt	Batch	8270±120	84	228			(Kim et al., 2008)
Swine manure; Diluted	SC-MFC	65	Carbon felt	Commercial Gas Diffusion Cathode-PFTE	Batch	2243 ± 25	15	28±20		24±3	(Vogl et al., 2016)
Swine slurry	TC-MFC	504	Carbon felt	Stainless steel mesh	Batch	6512	17–21		250		(Cerrillo et al., 2016)
Swine slurry	TC-MFC	269	Granular graphite and carbon felt	Stainless steel mesh in	Continuous	6908 mg/kg		(763–5623)			(Sotres et al., 2016)
Swine slurry liquid	TC-MFC	336	Carbon felt mesh	Stainless steel mesh	Continuous	3462	13.1–50.9	9.4–46.1	66.4–146.8	0.7–6.9	(Sotres et al., 2015)
Swine slurry; Digested	TC-MFC	504	Carbon felt	Stainless steel mesh	Batch	7951	7–12		225		(Cerrillo et al., 2016)

[a]*TC-MFC stands for two-chamber microbial fuel cell and SC-MFC stands for single-chamber microbial fuel.*

TABLE 7.6 Summary of Literature Studies Reporting Use of MFCs for Treating Sugar-Based and Distillery Wastewater

Wastewater Type	Cell type[a]	Working Vol. (mL)	Anode Material	Cathode Material	Operation Mode	COD_{in} (mg/L)	ΔCOD (%)	Power Density mW/m^2 (mW/m^3)	Current Density mA/m^2 (mA/m^3)	CE (%)	Ref.
Chitin solution	SC-MFC	300	Carbon brush	Carbon cloth-Pt	Batch			76–272		18–56	(Rezaei et al., 2009)
Chitin wastewater; fermented	TC-MFC	100	Carbon felt	Carbon felt	Batch				8.77 $\mu A/cm^2$		(Li et al., 2017)
Corn Stover Powder and solids	SC-MFC		Carbon paper	Carbon cloth-Pt	Batch			331–343			(Wang et al., 2009)
Distillery	TC-MFC	210	Graphite plate	Graphite plate	Batch	3200–6400	46.2–64.8	70–123.5	265–323.4	13.2–27	(Samsudeen et al., 2016)
Distillery	SC-MFC	28	Carbon cloth	Carbon cloth-Pt	Batch	125–3000	29.5–56.7	(5.46)	6.6–77.7		(Tanikkul and Pisutpaisal, 2015)
Distillery waste—Digested	TC-MFC	200	Graphite rods	Graphite rods	Batch		TOC; 60.78±0.95	(31490)			(Deval et al., 2017)
Molasses	SC-MFC	900	Carbon felt	Air diffusion electrode	Continuous	10,000	90.2±1.63	7.9±2.56	57.3±9.91		(Lee et al., 2016)
Molasses	SC-MFC	900	Carbon felt	MEET	Continuous	10,000	88.7±3.34	7.5±0.67	56.7±2.52		(Lee et al., 2016)
Molasses	TC-MFC	900	Carbon felt	Carbon felt	Continuous	10,000	50.3±5.06	17.0±10.15	80.2±29.11		(Lee et al., 2016)
Molasses	TC-MFC	300	Carbon cloth	Carbon cloth-Pt	Batch	130,000	67	2425	2600		(Ali et al., 2016)
Sugar mill	TC-MFC	500	Carbon felt	Carbon felt	Batch	7210	56	140	50	70	(Kumar et al., 2015)

[a] *TC-MFC stands for two-chamber microbial fuel cell and SC-MFC stands for single-chamber microbial fuel cell.*

Most of the studies investigated either single- or two-chamber MFCs with working volume ranging between 25 and 1000mL operated in batch mode. The highest power density among reported studies ranging between 331 and 343mW/m^2 using corn stover powder and solids was reported by Wang et al. (2009) in a single-chamber MFC. Lee et al. (2016) compared the performance of large two- and single-chamber MFCs using molasses wastewater (10,000mg COD/L) operating in a continuous mode, each with a working volume of 900mL. The single-chamber MFC achieved higher COD removal (90%) than two-chamber MFC (50%). However, the two-chamber MFC achieved a higher power density (17 $\pm$ 10.15mW/m^2) than the single-chamber MFC (7.9 $\pm$ 2.56mW/m^2). The performance of the single-chamber MFC was further evaluated for the effects of using a proton exchange membrane, and it was concluded that the membrane did not significantly impact COD removal and power generation. In addition, the study reported that methanogens existed in the reactors and contributed to 50%–90% of the COD removal. Therefore controlling methanogens in MFCs is an important operational parameter to ensure that the substrate is consumed during power production. Full-strength molasses wastewater (130,000mg COD/L) was used in a two-chamber MFC (Ali et al., 2016). The cell achieved 67% COD removal and produced a maximum power density of 242mW/m^3. These results show that MFCs can be adequate reactors for treatment of full strength molasses wastewater.

Distillery wastewater (3200–6400mg COD/L) was used in a two-chamber MFC operated in batch mode (Samsudeen et al., 2016). The cell achieved a maximum power density of 123.5mW/m^2, coulombic efficiency up to 27%, and COD removal up to 65%. Tanikkul and Pisutpaisal (2015) investigated the performance of a single-chamber MFC using distillery wastewater with varied COD range between 125 and 3000mg/L. The cell produced a maximum power density of 5.46mW/m^3 and up to 56.7% COD removal. Recently, Deval et al. (2017) evaluated the power production and carbon degradation of anaerobically digested distillery wastewater using a two-chamber MFC. Under optimum operating conditions, the cell produced a maximum power density of 31490mW/m^3 and achieved up to 61% TOC reduction. The COD reduction of these studies is considered lower than conventional anaerobic methods (Pant and Adholeya, 2007). Therefore further research is needed to understand current generation and substrate utilization in MFCs using distillery wastewater.

Generally, the performance of MFCs using molasses and distillery wastewater has been obtained from lab-scale, relatively small reactors, and the performance of pilot-scale MFCs in the treatment and energy recovery from molasses and distillery wastewater still needs to be evaluated. Power densities generated from this type of wastewater are relativity low, and methanogenic control is an essential parameter in operating MFCs using this wastewater. Employing anaerobic fermentation as a pretreatment may be a viable option which can produce energy-rich hydrogen gas and further break down organic substrates to fermentation products than can be consumed by the exoelectrogenic microorganisms.

7.3.1.7 Seafood Industry

The seafood industry is concentrated in coastal areas where seafood processing occurs. Processing seafood produces a large amount of waste and wastewater and may have a large impact on the local community (Kosseva, 2013). Processing seafood includes fish cleaning, cooling, equipment and floor cleaning, which produce wastewater with high organics, fats, oil and grease, and nitrogen content. Literature data on the characteristics of seafood wastewater is limited. However, it has been reported that the BOD load produced from seafood-processing operation ranges between 1 and 72.5kg per tonne of product (Kosseva, 2013).

The number of studies that evaluated the use of seafood wastewater in MFCs is limited, with only four studies identified in the literature, listed in Table 7.7. You et al. (2010) evaluated the performance of anoxic/oxic MFC in the power generation and treatment of seafood wastewater in continuous mode. The hydraulic retention time (HRT) was varied between 4.2 and 16.7h and the average COD varied between 2102 and 2522mg/L. The largest COD removal (80.2%) was achieved at HRT of 16.7h. However, the largest power density (16,200mW/m^3) was achieved at HRT of 4.2h. The performance of small single- and two-chamber MFCs was compared using seafood wastewater with COD about 1000mg/L (Sun, 2012). The study concluded that the single-chamber MFC produced higher power density (343.6–358.8mW/m^2) than the two-chamber MFC (258.7–291.6mW/m^2). Also, larger COD removal was achieved in the single-chamber MFC (85.1%) than the two-chamber MFC (64.7%). On the contrary, the two-chamber MFC achieved higher maximum coulombic efficiency (20.3%) than single-chamber MFC (14.2%). Jayashree et al. (2016) operated a continuous tubular MFC using seafood wastewater (4000mg COD/L). The study reported power density between 105 and 222mW/m^2 (221–886mW/m^3) and 83% COD removal. The power densities observed from using MFC to treat seafood wastewater are promising. However, treatment efficiencies of MFCs are comparable to the efficiencies achieved by fixed film filters treating seafood wastewater which are not sufficient to be employed as a stand-alone treatment technology (Tay et al., 2006). Further research is needed to evaluate the long-term and scaled-up performance and nutrient removal from this type of wastewater.

TABLE 7.7 Summary of Literature Studies Reporting Use of MFCs for Treating Seafood Wastewater

Wastewater Type	Cell Type[a]	Working Vol. (mL)	Anode Material	Cathode Material	Operation Mode	COD_{in} (mg/L)	ΔCOD (%)	Power Density mW/m^2 (mW/m^3)	Current Density mA/m^2 (mA/m^3)	CE (%)	Ref.
Seafood	Tubular-MFC	50	Activated carbon fiber felt	Activated carbon fiber felt	Continuous	700 ± 50	83	105–222 (221–886)		<30	(Jayashree et al., 2016)
Seafood	SC-MFC	26	Carbon cloth-steel mesh	Carbon cloth-Pt-PTFE	Batch	1015.6	85.1	343.6–358.8	360–1270	0.38–14.2	(Sun, 2012)
Seafood	TC-MFC	26	Carbon cloth-steel mesh	Carbon cloth-Pt-PTFE	Batch	1015.6	64.7	258.7–291.6	360–1270	0.65–20.3	(Sun, 2012)
Seafood	TC-MFC	98	Granular graphite and Graphite rod	Granular graphite and Graphite rod	Continuous	2102–2522	28.2–80.2	(8900–16,200)	(31,100–41,700)	2.11–15.2	(You et al., 2010)

[a]*TC-MFC stands for two-chamber microbial fuel cell and SC-MFC stands for single-chamber microbial fuel.*

7.3.1.8 Edible Oil Industry

The edible oil industry is seasonal and produces high strength wastes, with high COD (up to 220 g/L), solids (up to 102.6 g total solids/L), lipids (up to 30 g/L), sugars, nitrogen, and low pH that ranges between 5 and 5.9 (Hung et al., 2005; Kosseva, 2013). Olive oil and palm oil wastewater characterization and treatment are comprehensively reviewed in Hung et al. (2005) and Yacob et al. (2005), respectively.

Several researchers have studied using oil wastewater in MFCs as summarized in Table 7.8. Palm oil mill wastewater was investigated in a two-chamber MFC but the cell did not produce significant power densities ($<25\,mW/m^2$). Recently, Yu et al. (2017) investigated using soybean oil refinery wastewater (2900 mg COD/L) in a single-chamber MFC. The cell achieved a maximum power density of $746\,mW/m^2$, >96% COD removal, and up to 33.6% coulombic efficiency. An earlier study by Hamamoto et al. (2016) investigated full strength soybean oil wastewater (40 g COD/L) in a small single-chamber MFC. The cell achieved a maximum power density of $2240\,mW/m^2$, >77% COD removal, and up to 20% coulombic efficiency. This shows that increasing the strength of wastewater and the size of MFC can result in decreasing power density and the efficiency of the cell in removing COD.

7.3.2 Microbial Electrolysis Cells

Another type of BES is microbial electrolysis cell (MEC). Microbial electrolysis cells (MECs) use an external power source to catalyze the substrate into by-products, including methane (CH_4), hydrogen gas (H_2), and hydrogen peroxide (H_2O_2). While a considerable number of researchers investigated the use of MFCs to generate power from food industry waste and wastewater, the number of studies that investigated the use of MECs with this type of waste is limited. However, in some cases products produced by MEC may be more valuable than producing electricity, since these products be stored for later use or utilized in other processes. The food waste sources investigated in MECs include brewery and dairy wastewaters, molasses, animal waste, and winery wastewater, as shown in Table 7.9.

Methane production from brewery wastewater using MEC has recently been evaluated by Guo et al. (2017). The researchers used a single-chamber MEC and the average initial COD of the brewery wastewater was 1125 mg/L. The cell produced $0.14\,m^3$ CH_4/m^3/day and achieved a maximum COD removal of 80%. The maximum coulombic efficiency was low (32.7%) which suggests that most of the methane produced was not produced by the exoelectrogenic bacteria. Similar methane production rate was achieved in a small single-chamber MEC using soybean oil refinery wastewater of COD about 2900 mg/L (Yu et al., 2017). In their study, higher COD removal (95.8%) was achieved but the coulombic efficiency was not reported. Marone et al. (2016) evaluated different power inputs into an MEC using table olive oil processing brine wastewater as a substrate. The cell produced an average of 109 Normal mL CH_4 per g COD removed. The maximum COD removed was 29% and coulombic efficiency was 30%.

Several researchers evaluated biohydrogen production using MEC, since it is a cleaner fuel than methane which can be produced by conventional anaerobic wastewater treatment methods. The coulombic efficiencies achieved by hydrogen-producing MECs are generally higher than those used for methane production. This indicates that better MECs control might be achieved if methanogens are inhibited. The performance of MECs was evaluated using various food industry wastewaters, including molasses wastewater which achieved the highest hydrogen production rate and coulombic efficiency. Wang et al. (2014) used molasses wastewater in a single-chamber MEC, and reported up to 95% coulombic efficiency, >100% cathodic energy recovery, and produced up to $10.72\,m^3$ H_2/m^3/day of hydrogen. The results for molasses wastewater show that MECs can be used efficiently to treat wastewater and generate biohydrogen.

Several researchers evaluated swine waste as a substrate for MECs. Wagner et al. (2009) demonstrated that using high strength swine wastewater (12,825 mg COD/L), up to 70% coulombic efficiency can be achieved, with up to 75% COD removal, and up to $1\,m^3$ H_2/m^3/day hydrogen production. Using a continuous two-chamber MEC, Sotres et al. (2015) showed that up to 54% COD reduction and 57% coulombic efficiency can be achieved using swine slurry. Cerrillo et al. (2016) compared the performance of a two-chamber MEC using swine slurry and anaerobically digested swine slurry. The MEC with undigested slurry achieved higher COD removal but lower coulombic efficiency. It was also demonstrated that up to 40% ammonia removal from the slurry could be achieved using an MEC. However, Sotres et al. (2015) and Cerrillo et al. (2016) did not report the hydrogen production from the MECs.

Cusick et al. (2010) evaluated using a lab-scale single-chamber MEC with winery wastewater (2200 mg COD/L). The MEC achieved 47% COD removal, 50% coulombic efficiency and produced $0.17\,m^3$ H_2/m^3/d. Cusick et al. (2011) tested the first pilot-scale MEC operating on winery wastewater (1000 L with 144 electrode pairs). The anodes were made of graphite fiber brushes and the cathodes were made of stainless steel mesh. The operation period of the MEC was limited by the seasonal operation of the winery (around 100 days). The cell was operated at a voltage 0.9 V, with hydraulic retention

TABLE 7.8 Summary of Literature Studies Reporting Use of MFCs for Treating Oil Wastewater

Wastewater Type	Cell Type[a]	Working Vol. (mL)	Anode Material	Cathode Material	Operation Mode	COD_{in} (mg/L)	ΔCOD (%)	Power Density mW/m^2 (mW/m^3)	Current Density mA/m^2 (mA/m^3)	CE (%)	Ref.
Palm oil mill	TC-MFC	450	PACF carbon felt	PACF carbon felt		1000	70	22	~180	24	(Baranitharan et al., 2015)
Soybean oil	SC-MFC	18	Graphite felt	Carbon cloth-PTFE-Pt		40,000	77.9	2240 (31,600)	658	20.1	(Hamamoto et al., 2016)
Soybean oil refinery	SC-MFC	2	Graphite fiber Brush	Stainless steel mesh	Batch	2900±100	~96.4	746 (~24,100)		9.3–33.6	(Yu et al., 2017)
Vegetable oil	TC-MFC	500	Ti wire	Carbon cloth	Batch	925	86				(Abbasi et al., 2016)

[a] *TC-MFC stands for two-chamber microbial fuel cell and SC-MFC stands for single-chamber microbial fuel cell.*

time of 1 day. The cell enrichment and inoculation took ~60 days and the wastewater was diluted to enhance the inoculations and reduce the start-up time. The start-up time was affected by temperature, pH, and VFA content of the wastewater. The maximum gas production was $0.19\,m^3/m^3$/day and the majority of the gas produced was methane which was against the intent of the study. The cell provided favorable conditions for methanogens growth and no inhibition or methanogen control was employed. This study demonstrated some of the challenges of scaling up MECs, which included longer start-up time than lab-scale cells and methanogenic control. Extended continuous operation could enrich methanogens, as was also reported by other studies (Rader and Logan, 2010).

7.4 LIMITATIONS AND CHALLENGES OF BESs

Bioelectrochemical systems are unique systems that have the potential to recover energy and treat wastes. Over the past two decades, the growth of published research on BESs has been exponential (Aghababaie et al., 2015; Wang et al., 2015). The efforts of recovering energy while treating food waste have been dominated by microbial fuel cells as compared to other BESs. Despite the significant efforts in developing bioelectrochemical systems, there are still key limitations and challenges facing bioelectrochemical systems, as presented in this chapter for both microbial fuel cells and microbial electrolysis cells.

The cathode is the limiting electrode in a microbial fuels cell, and to enhance the performance a catalyst is traditionally used. There has been significant research effort applied in testing materials that are suitable for MFCs, with platinum (Pt) catalyst being among the most widely used. However, Pt is an expensive metal which increases the cost of constructing MFCs. Furthermore, the use of ion membrane in actual wastewater makes it susceptible to fouling which greatly increases the internal resistance of the cell and reduces the electric current. Similar challenges and limitations have been identified for MECs. Electron losses increase the required power input to the system, and methanogenic inhibition is also essential for controlling the system during operation.

Microbial fuel cells are devices that produce power while treating waste. The power produced in microbial fuels cells is lower than the theoretical power due to electron losses. Many factors contribute to electron losses in the cell including resistance to electron flow through the electrodes, connections, and membrane; activation energy needed to for redox reactions; losses in the bacterium; and losses due to concentration gradient (Logan et al., 2006). The sum of these losses in an MFC contributes to limiting the current produced. Furthermore, the power production is limited by microbial growth, substrate diffusion into the biofilm, and conversion of substrates in the cell environment. The coulombic efficiency of the cell is limited by the microbial culture in the cell and substrate. Substrate conversion is also limited by the substrate concentrations; high concentrations of substrate and low pH levels may inhibit exoelectrogenic activity (Kim and Logan, 2011; Lin et al., 2016; Rikame et al., 2012). In addition, higher concentrations of metals and toxins may inhibit microbial activity.

The optimum operation conditions of microbial fuels cells are close to the optimum conditions of methanogenesis. Therefore in continuous long-term operation of microbial fuel cells, methanogenic control is essential. In batch operation of MFC, methanogens are inhibited by the aeration of the electrodes between the batches. In continuous operation of MFC, some cells achieved high substrate reduction while low coulombic efficiency was achieved. This indicated that the substrate went through the fermentation and methanogenetic pathways and the electrons from the redox reactions were not transferred into the electrodes. Some substrates may provide inhibitory conditions to methanogens, like winery and brewery wastewaters, which have low pH levels.

Most studies using microbial fuel cells have been conducted at lab-scale and the number of scaled-up systems is limited. As demonstrated in the latter studies, the performance of the lab-scaled units cannot readily be extrapolated to commercially relevant sizes (Cusick et al., 2011; Hiegemann et al., 2016). Conventional anaerobic processes are sized according to the HRT needed to achieve the required degree of treatment, while BESs are also limited by the power production in addition to HRT. Low electric current and coulombic efficiencies are achieved in scaled-up systems which make the footprint large, even to power a small electronic device (Sun et al., 2016). Voltage reversal is also one of the factors that contributes to the electric current reduction in scaled-up systems. The substrate removal of scaled-up MFCs is not sufficient to operate MFCs as a sole unit process for waste treatment. It is envisioned that BESs can be a unit process within a waste treatment scheme. Furthermore, the internal resistance of scaled-up systems is increased which results in reducing electric current. Scaling up BESs also increases the start-up time of the systems, and COD reduction in scaled-up BESs generally takes longer.

TABLE 7.9 Summary of Literature Studies Reporting Use of MECs for Treating Food Waste and Food Wastewater

Wastewater Type	Cell Type[a]	Working Vol. (mL)	Anode Material	Cathode Material	Operation Mode	$E_{app.}$ (V)	COD_{in} (mg/L)	ΔCOD (%)	r_{cat} (%)	r_{H2} (%)	Q (m^3 H_2/m^3/day)	CE (%)	Ref.
Beer wastewater	SC-MEC	2100	Graphite fiber brushes	Circular stainless steel mesh	Semicontinuous	0.5–0.9	1125±66	65–80		32.1–91.2	0.14 (CH4)	5–32.7	(Guo et al., 2017)
Cheese whey	SC-MEC	32			Batch	0.8	2000		49±2				(Rago et al., 2017)
Cheese whey; Diluted and Fermented	MEC	50	Carbon felt	Gas diffusion electrode-Ni	Continuous	1	15.26	82			0.5		(Moreno et al., 2015)
Glycerol, starch and milk	SC-MEC	28	Graphite fiber brush	Graphite fiber cloth-Pt-PFTE	Batch	0.8		74–100	91		0–0.94	13–29	(Montpart et al., 2015)
Milk	SC-MEC	28	Graphite fiber brush	Graphite fiber cloth-Pt-PFTE	Batch	0.8	1000	73.5	14		0.086	36–52	(Montpart et al., 2015)
Molasses	SC-MEC	25	Graphite-fiber brush anodes	Carbon cloth—with and without Pt	Batch	0.6–0.8	2000		54.3–102	45.5–94	2.27–10.72	91–93	(Wang et al., 2014)
Potato	SC-MEC	28	Graphite fiber brushes	Carbon cloth-Pt-PFTE	Batch	0.8	7700	79			0.74	80	(Kiely et al., 2011a)
Soybean oil refinery	SC-MEC	22	Graphite fiber brush	Stainless steel mesh	Batch	1.2	2900±100	95.8			0.133±0.005 CH_4		(Yu et al., 2017)
Starch	SC-MEC	28	Graphite fiber brush	Graphite fiber cloth-Pt-PFTE	Batch	0.8	1185	85.1				15–28	(Montpart et al., 2015)
Swine	3c-MEC	2000	Carbon graphite	Carbon graphite	Continuous	0–2	10,136.9±850.5	59.7–67					(Lim et al., 2012)
Swine	TC-MEC				Batch	0.2–1	1298	45–52			0.061	9–30	(Jia et al., 2010)
Swine	SC-MEC				Batch		12,825	69–75	29–61	17–20	0.9–1	29–70	(Wagner et al., 2009)
Swine slurry	TC-MEC	504	Carbon felt	Stainless steel mesh	Batch	0–0.2	6512	29–35				7–9	(Cerrillo et al., 2016)
Swine slurry liquid	TC-MEC		Carbon felt mesh	Stainless steel mesh	Continuous	0.1–0.8	3462	13.5–53.8				3.2–56.9	(Sotres et al., 2015)
Swine slurry; Digested	TC-MEC	504	Carbon felt	Stainless steel mesh	Batch	0–0.2	7951	17–25				11–18	(Cerrillo et al., 2016)
Table olive oil brine processing	MEC	336	Graphite plates	Pt-radium grid	Batch	0.2–0.8		29			109±21 NmL CH_4/g COD_{rem}	30	(Marone et al., 2016)
Winery	SC-MEC				Batch	0.9	2200	47			0.17	50	(Cusick et al., 2010)

[a] *TC-MEC stands for two-chamber microbial electrolysis cell and SC-MEC stands for single-chamber microbial electrolysis cell.*

7.5 FUTURE PERSPECTIVE AND RESEARCH NEEDS

Bioelectrochemical systems is a promising technology that has the potential to recover resources, energy, and treat waste. With the expanding need to recover resources, secure the food supply, and maintain a clean and healthy environment, development of systems like BESs is essential. While in the food industry, anaerobic digestion is mostly used to recover energy and treat food wastes, BESs can be advantageous since they can be operated at the ambient wastewater temperature and do not require precise temperature control. In addition, BESs can be compacted and costumed to different shapes that can be installed inside buildings.

Over the past two decades, there have been great efforts to understand and optimize the performance of these systems. The effort has focused on optimizing lab-scale architecture, materials, and performance using both synthetic and actual wastewater. However, there is still need for more research to optimize scaled-up systems, increase power and current, reduce internal resistance for the entire system, increase efficiencies, and reduce the system footprint.

The investigation of new materials and reactor configuration is likely to continue, especially to reduce costs and discover cheaper catalysts. Further research is needed to better understand the electron transport from the microbes to the electrodes. Optimizing BESs by enhancing transport through the ion exchange membrane, improved fundamental understanding through mathematical modeling, and discovery of cheaper materials with comparable performance are needed. Like anaerobic biological processes, BESs operate optimally at around neutral pH and are sensitive to shock loadings. Therefore an equalization basin with pH adjustment might be needed in the process stream before the BES. However, there is still research needed to evaluate the long-term performance of continuous systems and their tolerance to changing environments.

Application of the BESs for purposes other than waste treatment requires further investigation. MFCs produce electrical current from organic substrates, and so they can be used as real-time sensors for substrates in various environments. MECs are suitable for generation of products on-site and can be incorporated in different industrial applications. The use of hybrid systems, which synergistically use multiple groups of microorganisms such as microalgae and bacteria, to optimize performance and treatment is also a promising approach. The performance of BESs in catalyzing organic carbon-based substrates is limited by the volatile fatty acids concentrations in the substrate. Therefore coupling BESs with anaerobic fermentation is still a promising strategy, especially for high strength wastewater.

The cost of membrane filters has been decreasing in recent years, and the combination of membrane filters with BESs in a continuous and recycling operation scheme may allow for the use of pure cultures that are known to produce higher current. Concentrations of nitrogen, phosphorus, and sulfur-based chemicals can be high in food wastes, especially meat-based wastes. Studies have shown that BESs have a promising ability to remove nitrogen and phosphorus from wastewater. Further research is still needed to optimize this approach and achieve a better understanding of the kinetics and pathways of nitrogen and phosphorus removal.

7.6 CONCLUSIONS

The efficiency of microbial fuel cells and microbial electrolysis cells in treating food wastes was reviewed. Bioelectrochemical systems are still in their infancy and further research is needed to better understand the systems and optimize their performance. Microbial fuel cells have been the focus of researchers for food industry waste and wastewater treatment due to their capability to produce electricity. Fewer researchers have investigated microbial electrolysis cells. Among the food waste investigated, brewery and sugar-based wastewater hold the most promise for higher power density generation from MFC. Other waste sources may have better performance if coupled with fermentation as a pretreatment process. Scaled-up systems using food waste have not been extensively evaluated. Several limitations and challenges are discussed including reduction of performance in scaled-up systems, treatment efficiency, electron loss, and internal resistance of the systems. The control of methanogenic microbes is essential, especially for continuous long-term operation. Further research on the removal of sulfur-based compounds from actual food wastewater using bioelectrochemical systems is needed.

REFERENCES

Abbasi, U., Jin, W., Pervez, A., Bhatti, Z.A., Tariq, M., Shaheen, S., Iqbal, A., Mahmood, Q., 2016. Anaerobic microbial fuel cell treating combined industrial wastewater: correlation of electricity generation with pollutants. Bioresour. Technol. 200, 1–7. https://doi.org/10.1016/j.biortech.2015.09.088.

Abreu, M.A., Toffoli, S.M., 2009. Characterization of a chromium-rich tannery waste and its potential use in ceramics. Ceram. Int. 35, 2225–2234. https://doi.org/10.1016/j.ceramint.2008.12.011.

Aghababaie, M., Farhadian, M., Jeihanipour, A., Biria, D., 2015. Effective factors on the performance of microbial fuel cells in wastewater treatment—a review. Environ. Technol. Rev. 4, 71–89. https://doi.org/10.1080/09593330.2015.1077896.

Algino, R.J., Ingham, S.C., Zhu, J., 2007. Survey of antimicrobial effects of beef carcass intervention treatments in very small state-inspected slaughter plants. J. Food Sci. 72, M173–M179. https://doi.org/10.1111/j.1750-3841.2007.00386.x.

Ali, N., Yousaf, S., Anam, M., Bangash, Z., Maleeha, S., 2016. Evaluating the efficiency of a mixed culture biofilm for the treatment of black liquor and molasses in a mediator-less microbial fuel cell. Environ. Technol. 37, 2815–2822. https://doi.org/10.1080/09593330.2016.1166267.

Angosto, J.M., Fernández-López, J.A., Godínez, C., 2015. Brewery and liquid manure wastewaters as potential feedstocks for microbial fuel cells: a performance study. Environ. Technol. 36, 68–78. https://doi.org/10.1080/09593330.2014.937769.

Angulo, J., Mahecha, L., Yepes, S.A., Yepes, A.M., Bustamante, G., Jaramillo, H., Valencia, E., Villamil, T., Gallo, J., 2012. Quantitative and nutritional characterization of fruit and vegetable waste from marketplace: a potential use as bovine feedstuff? J. Environ. Manag. 95, 203–209. https://doi.org/10.1016/j.jenvman.2010.09.022.

Bajracharya, S., ElMekawy, A., Srikanth, S., Pant, D., 2016. Cathodes for microbial fuel cells. In: Scott, K., Yu, E.H. (Eds.), Microbial Electrochemical and Fuel Cells: Fundamentals and Applications. Woodhead Publishing is an imprint of Elsevier, Cambridge, UK, pp. 179–213.

Banks, C., Wang, Z., 2005. Treatment of meat wastes. In: Wang, L.K., Hung, Y.-T., Lo, H.H., Yapijakis, C. (Eds.), Waste Treatment in the Food Processing Industry. CRC Press, Boca Raton, FL, pp. 67–100.

Baranitharan, E., Khan, M.R., Prasad, D.M.R., Teo, W.F.A., Tan, G.Y.A., Jose, R., 2015. Effect of biofilm formation on the performance of microbial fuel cell for the treatment of palm oil mill effluent. Bioprocess Biosyst. Eng. 38, 15–24. https://doi.org/10.1007/s00449-014-1239-9.

Boghani, H., Kim, J.R., Dinsdale, R.M., Guwy, A.J., Premier, G.C., 2017. Reducing the burden of food processing washdown wastewaters using microbial fuel cells. Biochem. Eng. J. 117, 210–217. https://doi.org/10.1016/j.bej.2016.10.017.

Boone, D.R., 1982. Terminal reactions in the anaerobic digestion of animal waste. Appl. Environ. Microbiol. 43, 57–64.

Bouallagui, H., Cheikh, R.B., Marouani, L., Hamdi, M., 2003. Mesophilic biogas production from fruit and vegetable waste in a tubular digester. Bioresour. Technol. 86, 85–89. https://doi.org/10.1016/s0960-8524(02)00097-4.

Britz, T.J., van Schalkwyk, C., 2005. Treatment of dairy processing wastewaters. In: Wang, L.K., Hung, Y.-T., Lo, H.H., Yapijakis, C. (Eds.), Waste Treatment in the Food Processing Industry. CRC Press, Boca Raton, FL, pp. 1–24.

Call, D., Logan, B.E., 2008. Hydrogen production in a single chamber microbial electrolysis cell lacking a membrane. Environ. Sci. Technol. 42, 3401–3406. https://doi.org/10.1021/es8001822.

Cao, X., Huang, X., Liang, P., Xiao, K., Zhou, Y., Zhang, X., Logan, B.E., 2009. A new method for water desalination using microbial desalination cells. Environ. Sci. Technol. 43, 7148–7152. https://doi.org/10.1021/es901950j.

Caro, D., Davis, S.J., Bastianoni, S., Caldeira, K., 2017. Greenhouse gas emissions due to meat production in the last fifty years. In: Ahmed, M., Stockle, C.O. (Eds.), Quantification of Climate Variability, Adaptation and Mitigation for Agricultural Sustainability. Springer International Publishing, Switzerland, pp. 27–37.

Cercado, B., Vega-Guerrero, A.L., Rodríguez-Valadez, F., Hernández-López, J.L., Cházaro-Ruiz, L.F., Délia, M.L., Bergel, A., 2014. Carbonaceous and protein constituents in dairy wastewater lead to a differentiated current generation in microbial fuel cells (MFCs). J. Mex. Chem. Soc. 58, 309–314.

Cercado-Quezada, B., Delia, M.-L., Bergel, A., 2010a. Testing various food-industry wastes for electricity production in microbial fuel cell. Bioresour. Technol. 101, 2748–2754. https://doi.org/10.1016/j.biortech.2009.11.076.

Cercado-Quezada, B., Delia, M.-L., Bergel, A., 2010b. Treatment of dairy wastes with a microbial anode formed from garden compost. J. Appl. Electrochem. 40, 225–232. https://doi.org/10.1007/s10800-009-0001-5.

Cerrillo, M., Oliveras, J., Viñas, M., Bonmatí, A., 2016. Comparative assessment of raw and digested pig slurry treatment in bioelectrochemical systems. Bioelectrochemistry 110, 69–78. https://doi.org/10.1016/j.bioelechem.2016.03.004.

Çetinkaya, A.Y., Köroğlu, E.O., Demir, N.M., Baysoy, D.Y., Özkaya, B., Çakmakçı, M., 2015. Electricity production by a microbial fuel cell fueled by brewery wastewater and the factors in its membrane deterioration. Chin. J. Catal. 36, 1068–1076. https://doi.org/10.1016/s1872-2067(15)60833-6.

Chandrasekhar, K., Amulya, K., Mohan, S.V., 2015. Solid phase bio-electrofermentation of food waste to harvest value-added products associated with waste remediation. Waste Manag. 45, 57–65. https://doi.org/10.1016/j.wasman.2015.06.001.

Cheng, S., Xing, D., Logan, B.E., 2011. Electricity generation of single-chamber microbial fuel cells at low temperatures. Biosens. Bioelectron. 26, 1913–1917. https://doi.org/10.1016/j.bios.2010.05.016.

Choi, J., Ahn, Y., 2015. Enhanced bioelectricity harvesting in microbial fuel cells treating food waste leachate produced from biohydrogen fermentation. Bioresour. Technol. 183, 53–60. https://doi.org/10.1016/j.biortech.2015.01.109.

Choi, J., Chang, H.N., Han, J.-I., 2011. Performance of microbial fuel cell with volatile fatty acids from food wastes. Biotechnol. Lett. 33, 705–714. https://doi.org/10.1007/s10529-010-0507-2.

Christwardana, M., Prabowo, A.K., Tiarasukma, A.P., Ariyanti, D., 2016. Microbial fuel cells for simultaneous electricity generation and organic degradation from slaughterhouse wastewater. Int. J. Renew. Energy Dev. 5, 107–112. https://doi.org/10.14710/ijred.5.2.107-112.

Crawford, G., 2007. In: Managing sulfur concentrations in feed and water.Minnesota Nutrition Conference.

Cusick, R.D., Kiely, P.D., Logan, B.E., 2010. A monetary comparison of energy recovered from microbial fuel cells and microbial electrolysis cells fed winery or domestic wastewaters. Int. J. Hydrog. Energy 35, 8855–8861. https://doi.org/10.1016/j.ijhydene.2010.06.077.

Cusick, R.D., Bryan, B., Parker, D.S., Merrill, M.D., Mehanna, M., Kiely, P.D., Liu, G., Logan, B.E., 2011. Performance of a pilot-scale continuous flow microbial electrolysis cell fed winery wastewater. Appl. Microbiol. Biotechnol. 89, 2053–2063. https://doi.org/10.1007/s00253-011-3130-9.

Danalewich, J.R., Papagiannis, T.G., Belyea, R.L., Tumbleson, M.E., Raskin, L., 1998. Characterization of dairy waste streams, current treatment practices, and potential for biological nutrient removal. Water Res. 32, 3555–3568. https://doi.org/10.1016/s0043-1354(98)00160-2.

de Vries, M., de Boer, I.J.M., 2010. Comparing environmental impacts for livestock products: a review of life cycle assessments. Livest. Sci. 128, 1–11. https://doi.org/10.1016/j.livsci.2009.11.007.

Demirel, B., Yenigun, O., Onay, T.T., 2005. Anaerobic treatment of dairy wastewaters: a review. Process Biochem. 40, 2583–2595. https://doi.org/10.1016/j.procbio.2004.12.015.

Deval, A.S., Parikh, H.A., Kadier, A., Chandrasekhar, K., Bhagwat, A.M., Dikshit, A.K., 2017. Sequential microbial activities mediated bioelectricity production from distillery wastewater using bio-electrochemical system with simultaneous waste remediation. Int. J. Hydrog. Energy 42, 1130–1141. https://doi.org/10.1016/j.ijhydene.2016.11.114.

Ditzig, J., Liua, H., Logan, B.E., 2007. Production of hydrogen from domestic wastewater using a bioelectrochemically assisted microbial reactor (BEAMR). Int. J. Hydrog. Energy 32, 2296–2304. https://doi.org/10.1016/j.ijhydene.2007.02.035.

Du, H., Li, F., 2016. Size effects of potato waste on its treatment by microbial fuel cell. Environ. Technol. 37, 1305–1313. https://doi.org/10.1080/09593330.2015.1114027.

Du, Z., Li, H., Gu, T., 2007. A state of the art review on microbial fuel cells: a promising technology for wastewater treatment and bioenergy. Biotechnol. Adv. 25, 464–482. https://doi.org/10.1016/j.biotechadv.2007.05.004.

Dumitru, A., Scott, K., 2016. Anode materials for microbial fuel cells. In: Scott, K., Yu, E.H. (Eds.), Microbial Electrochemical and Fuel Cells: Fundamentals and Applications. Woodhead Publishing is an imprint of Elsevier, Cambridge, UK, pp. 117–152.

Durruty, I., Bonanni, P.S., González, J.F., Busalmen, J.P., 2012. Evaluation of potato-processing wastewater treatment in a microbial fuel cell. Bioresour. Technol. 105, 81–87. https://doi.org/10.1016/j.biortech.2011.11.095.

Egbadon, E.O., Akujobi, C.O., Nweke, C.O., Braide, W., Akaluka, C.K., Adeleye, S.A., 2016. Simultaneous generation of bioelectricity and treatment of swine wastewater in a microbial fuel cell. Int. Lett. Nat. Sci. 54, 100–107. https://doi.org/10.18052/www.scipress.com/ilns.54.100.

Elakkiya, E., Matheswaran, M., 2013. Comparison of anodic metabolisms in bioelectricity production during treatment of dairy wastewater in microbial fuel cell. Bioresour. Technol. 136, 407–412. https://doi.org/10.1016/j.biortech.2013.02.113.

ElMekawy, A., Srikanth, S., Bajracharya, S., Hegab, H.M., Nigam, P.S., Singh, A., Mohan, S.V., Pant, D., 2015. Food and agricultural wastes as substrates for bioelectrochemical system (BES): the synchronized recovery of sustainable energy and waste treatment. Food Res. Int. 73, 213–225. https://doi.org/10.1016/j.foodres.2014.11.045.

Escapa, A., Gil-Carrera, L., García, V., Morán, A., 2012. Performance of a continuous flow microbial electrolysis cell (MEC) fed with domestic wastewater. Bioresour. Technol. 117, 55–62. https://doi.org/10.1016/j.biortech.2012.04.060.

Estrada-Arriagal, E.B., García-Sánchez, L., Garzón-Zuñiga, M.A., González-Rodríguez, J.G., 2015. Utilization of microbial fuel cells for the treatment of wastewater from a pig farm. Fresenius Environ. Bull. 24, 2512–2518.

Faria, A., Gonçalves, L., Peixoto, J.M., Peixoto, L., Brito, A.G., Martins, G., 2017. Resources recovery in the dairy industry: bioelectricity production using a continuous microbial fuel cell. J. Clean. Prod. 140, 971–976. https://doi.org/10.1016/j.jclepro.2016.04.027.

Feng, Y., Wang, X., Logan, B.E., Lee, H., 2008. Brewery wastewater treatment using air-cathode microbial fuel cells. Appl. Microbiol. Biotechnol. 78, 873–880. https://doi.org/10.1007/s00253-008-1360-2.

Goud, R.K., Babu, P.S., Mohan, S.V., 2011. Canteen based composite food waste as potential anodic fuel for bioelectricity generation in single chambered microbial fuel cell (MFC): bio-electrochemical evaluation under increasing substrate loading condition. Int. J. Hydrog. Energy 36, 6210–6218. https://doi.org/10.1016/j.ijhydene.2011.02.056.

Guo, Z., Thangavel, S., Wang, L., He, Z., Cai, W., Wang, A., Liu, W., 2017. Efficient methane production from beer wastewater in a membraneless microbial electrolysis cell with a stacked cathode: the effect of the cathode/anode ratio on bioenergy recovery. Energy Fuel 31, 615–620. https://doi.org/10.1021/acs.energyfuels.6b02375.

Hamamoto, K., Miyahara, M., Kouzuma, A., Matsumoto, A., Yoda, M., Ishiguro, T., Watanabe, K., 2016. Evaluation of microbial fuel cells for electricity generation from oil-contaminated wastewater. J. Biosci. Bioeng. 122, 589–593. https://doi.org/10.1016/j.jbiosc.2016.03.025.

Heilmann, J., Logan, B.E., 2006. Production of electricity from proteins using a microbial fuel cell. Water Environ. Res. 78, 531–537. https://doi.org/10.2175/106143005x73046.

Hiegemann, H., Herzer, D., Nettmann, E., Lübken, M., Schulte, P., Schmelz, K.-G., Gredigk-Hoffmann, S., Wichern, M., 2016. An integrated 45L pilot microbial fuel cell system at a full-scale wastewater treatment plant. Bioresour. Technol. 218, 115–122. https://doi.org/10.1016/j.biortech.2016.06.052.

Hou, Q., Pei, H., Hu, W., Jiang, L., Yu, Z., 2016. Mutual facilitations of food waste treatment, microbial fuel cell bioelectricity generation and Chlorella vulgaris lipid production. Bioresour. Technol. 203, 50–55. https://doi.org/10.1016/j.biortech.2015.12.049.

Hung, Y.-T., Salman, H., Awad, A., 2005. Olive oil waste treatment. In: Wang, L.K., Hung, Y.-T., Lo, H.H., Yapijakis, C. (Eds.), Waste Treatment in the Food Processing Industry. CRC Press, Boca Raton, FL, pp. 119–192.

Ichihashi, O., Hirooka, K., 2012. Removal and recovery of phosphorus as struvite from swine wastewater using microbial fuel cell. Bioresour. Technol. 114, 303–307. https://doi.org/10.1016/j.biortech.2012.02.124.

Ismail, Z.Z., Mohammed, A.J., 2016. Biotreatment of slaughterhouse wastewater accompanied with sustainable electricity generation in microbial fuel cell. Syst. Cybertnetics Informatics 14, 30–35.

Jadhav, D.A., Jain, S.C., Ghangrekar, M.M., 2016. Cow's urine as a yellow gold for bioelectricity generation in low cost clayware microbial fuel cell. Energy 113, 76–84. https://doi.org/10.1016/j.energy.2016.07.025.

Jayashree, C., Janshi, G., Yeom, I.T., Kumar, S.A., Banu, J.R., 2014. Effect of low temperature thermo-chemical pretreatment of dairy waste activated sludge on the performance of microbial fuel cell. Int. J. Electrochem. Sci. 9, 5732–5742.

Jayashree, C., Tamilarasan, K., Rajkumar, M., Arulazhagan, P., Yogalakshmi, K.N., Srikanth, M., Banu, J.R., 2016. Treatment of seafood processing wastewater using upflow microbial fuel cell for power generation and identification of bacterial community in anodic biofilm. J. Environ. Manag. 180, 351–358. https://doi.org/10.1016/j.jenvman.2016.05.050.

Jia, Y.H., Choi, J.Y., Ryu, J.H., Kim, C.H., Lee, W.K., Tran, H.T., Zhang, R.H., Ahn, D.H., 2010. Hydrogen production from wastewater using a microbial electrolysis cell. Korean J. Chem. Eng. 27, 1854–1859. https://doi.org/10.1007/s11814-010-0310-8.

Jia, J., Tang, Y., Liu, B., Wu, D., Ren, N., Xing, D., 2013. Electricity generation from food wastes and microbial community structure in microbial fuel cells. Bioresour. Technol. 144, 94–99. https://doi.org/10.1016/j.biortech.2013.06.072.

Kiely, P.D., Cusick, R., Call, D.F., Selembo, P.A., Regan, J.M., Logan, B.E., 2011a. Anode microbial communities produced by changing from microbial fuel cell to microbial electrolysis cell operation using two different wastewaters. Bioresour. Technol. 102, 388–394. https://doi.org/10.1016/j.biortech.2010.05.019.

Kiely, P.D., Regan, J.M., Logan, B.E., 2011b. The electric picnic: synergistic requirements for exoelectrogenic microbial communities. Curr. Opin. Biotechnol. 22, 378–385. https://doi.org/10.1016/j.copbio.2011.03.003.

Kim, Y., Logan, B.E., 2011. Hydrogen production from inexhaustible supplies of fresh and salt water using microbial reverse-electrodialysis electrolysis cells. Proc. Natl. Acad. Sci. 108, 16176–16181. https://doi.org/10.1073/pnas.1106335108.

Kim, J.R., Dec, J., Bruns, M.A., Logan, B.E., 2008. Removal of odors from swine wastewater by using microbial fuel cells. Appl. Environ. Microbiol. 74, 2540–2543. https://doi.org/10.1128/aem.02268-07.

Kim, K.-Y., Yang, W., Evans, P.J., Logan, B.E., 2016. Continuous treatment of high strength wastewaters using air-cathode microbial fuel cells. Bioresour. Technol. 221, 96–101. https://doi.org/10.1016/j.biortech.2016.09.031.

Kosseva, M.R., 2013. Sources, characterization, and composition of food industry wastes. In: Kosseva, M., Webb, C. (Eds.), Food Industry Wastes—Assessment and Recuperation of Commodities. Elsevier, London, UK, pp. 37–60. https://doi.org/10.1016/B978-0-12-391921-2.00003-2.

Kumar, R., Singh, L., Zularisam, A.W., 2015. Bioelectricity generation and treatment of sugar mill effluent using a microbial fuel cell. J. Clean Energy Technol. 4, 249–252. https://doi.org/10.7763/jocet.2016.v4.291.

Lee, Y.-Y., Kim, T.G., Cho, K.-S., 2016. Characterization of the COD removal, electricity generation, and bacterial communities in microbial fuel cells treating molasses wastewater. J. Environ. Sci. Health Part A 51, 1131–1138. https://doi.org/10.1080/10934529.2016.1199926.

Leech, D., Kavanagh, P., Schuhmann, W., 2012. Enzymatic fuel cells: recent progress. Electrochim. Acta 84, 223–234. https://doi.org/10.1016/j.electacta.2012.02.087.

Li, X.M., Cheng, K.Y., Wong, J.W.C., 2013. Bioelectricity production from food waste leachate using microbial fuel cells: effect of NaCl and pH. Bioresour. Technol. 149, 452–458. https://doi.org/10.1016/j.biortech.2013.09.037.

Li, H., Tian, Y., Zuo, W., Zhang, J., Pan, X., Li, L., Su, X., 2016. Electricity generation from food wastes and characteristics of organic matters in microbial fuel cell. Bioresour. Technol. 205, 104–110. https://doi.org/10.1016/j.biortech.2016.01.042.

Li, S.-W., He, H., Zeng, R.J., Sheng, G.-P., 2017. Chitin degradation and electricity generation by *Aeromonas hydrophila* in microbial fuel cells. Chemosphere 168, 293–299. https://doi.org/10.1016/j.chemosphere.2016.10.080.

Liakos, T.I., Sotiropoulos, S., Lazaridis, N.K., 2017. Electrochemical and bio-electrochemical treatment of baker's yeast effluents. J. Environ. Chem. Eng. 5, 699–708. https://doi.org/10.1016/j.jece.2016.12.048.

Lim, S.J., Park, W., Kim, T.-H., Shin, I.H., 2012. Swine wastewater treatment using a unique sequence of ion exchange membranes and bioelectrochemical system. Bioresour. Technol. 118, 163–169. https://doi.org/10.1016/j.biortech.2012.05.021.

Lin, H., Wu, X., Nelson, C., Miller, C., Zhu, J., 2016. Electricity generation and nutrients removal from high-strength liquid manure by air-cathode microbial fuel cells. J. Environ. Sci. Health Part A 51, 240–250. https://doi.org/10.1080/10934529.2015.1094342.

Liu, J., Zhang, F., He, W., Zhang, X., Feng, Y., Logan, B.E., 2014. Intermittent contact of fluidized anode particles containing exoelectrogenic biofilms for continuous power generation in microbial fuel cells. J. Power Sources 261, 278–284. https://doi.org/10.1016/j.jpowsour.2014.03.071.

Logan, B., 2008. Microbial Fuel Cells. John Wiley & Sons, Inc., Hoboken, NJ.

Logan, B.E., 2009. Exoelectrogenic bacteria that power microbial fuel cells. Nat. Rev. Microbiol. 7, 375–381. https://doi.org/10.1038/nrmicro2113.

Logan, B.E., Hamelers, B., Rozendal, R., Schröder, U., Keller, J., Freguia, S., Aelterman, P., Verstraete, W., Rabaey, K., 2006. Microbial fuel cells: methodology and technology. Environ. Sci. Technol. 40, 5181–5192. https://doi.org/10.1021/es0605016.

Logan, B.E., Wallack, M.J., Kim, K.-Y., He, W., Feng, Y., Saikaly, P.E., 2015. Assessment of microbial fuel cell configurations and power densities. Environ. Sci. Technol. Lett. 2, 206–214. https://doi.org/10.1021/acs.estlett.5b00180.

Ma, J., Ni, H., Su, D., Meng, X., 2016. Bioelectricity generation from pig farm wastewater in microbial fuel cell using carbon brush as electrode. Int. J. Hydrog. Energy 41, 16191–16195. https://doi.org/10.1016/j.ijhydene.2016.05.255.

Mansoorian, H.J., Mahvi, A.H., Jafari, A.J., Amin, M.M., Rajabizadeh, A., Khanjani, N., 2013. Bioelectricity generation using two chamber microbial fuel cell treating wastewater from food processing. Enzym. Microb. Technol. 52, 352–357. https://doi.org/10.1016/j.enzmictec.2013.03.004.

Mansoorian, H.J., Mahvi, A.H., Jafari, A.J., Khanjani, N., 2016. Evaluation of dairy industry wastewater treatment and simultaneous bioelectricity generation in a catalyst-less and mediator-less membrane microbial fuel cell. J. Saudi Chem. Soc. 20, 88–100. https://doi.org/10.1016/j.jscs.2014.08.002.

Marone, A., Carmona-Martínez, A.A., Sire, Y., Meudec, E., Steyer, J.P., Bernet, N., Trably, E., 2016. Bioelectrochemical treatment of table olive brine processing wastewater for biogas production and phenolic compounds removal. Water Res. 100, 316–325. https://doi.org/10.1016/j.watres.2016.05.008.

Meignanalakshmi, S., Kumar, S.V., 2016. Bioelectricity production by using goat (*Capra hircus*) rumen fluid from slaughterhouse waste in mediator-less microbial fuel cells. Energy Sources, Part A Recover. Util. Environ. Eff. 38, 1364–1369. https://doi.org/10.1080/15567036.2014.907846.

Min, B., Kim, J., Oh, S., Regan, J.M., Logan, B.E., 2005. Electricity generation from swine wastewater using microbial fuel cells. Water Res. 39, 4961–4968. https://doi.org/10.1016/j.watres.2005.09.039.

Miner, J.R., 1976. Production and Transport of Geseous NH3 and H2S Associated with Livestock Production. Environmental Protection Agency, Office of Research and Development, Robert S. Kerr Environmental Research Laboratory.

Miran, W., Nawaz, M., Kadam, A., Shin, S., Heo, J., Jang, J., Lee, D.S., 2015. Microbial community structure in a dual chamber microbial fuel cell fed with brewery waste for azo dye degradation and electricity generation. Environ. Sci. Pollut. Res. 22, 13477–13485. https://doi.org/10.1007/s11356-015-4582-8.

Mohamed, H.O., Obaid, M., Khalil, K.A., Barakat, N.A.M., 2016. Power generation from unconditioned industrial wastewaters using commercial membranes-based microbial fuel cells. Int. J. Hydrog. Energy 41, 4251–4263. https://doi.org/10.1016/j.ijhydene.2016.01.022.

Mohan, S.V., Chandrasekhar, K., 2011. Solid phase microbial fuel cell (SMFC) for harnessing bioelectricity from composite food waste fermentation: influence of electrode assembly and buffering capacity. Bioresour. Technol. 102, 7077–7085. https://doi.org/10.1016/j.biortech.2011.04.039.

Mohan, S.V., Mohanakrishna, G., Sarma, P.N., 2010a. Composite vegetable waste as renewable resource for bioelectricity generation through non-catalyzed open-air cathode microbial fuel cell. Bioresour. Technol. 101, 970–976. https://doi.org/10.1016/j.biortech.2009.09.005.

Mohan, S.V., Mohanakrishna, G., Velvizhi, G., Babu, V.L., Sarma, P.N., 2010b. Bio-catalyzed electrochemical treatment of real field dairy wastewater with simultaneous power generation. Biochem. Eng. J. 51, 32–39. https://doi.org/10.1016/j.bej.2010.04.012.

Montpart, N., Rago, L., Baeza, J.A., Guisasola, A., 2015. Hydrogen production in single chamber microbial electrolysis cells with different complex substrates. Water Res. 68, 601–615. https://doi.org/10.1016/j.watres.2014.10.026.

Moreno, R., Escapa, A., Cara, J., Carracedo, B., Gómez, X., 2015. A two-stage process for hydrogen production from cheese whey: integration of dark fermentation and biocatalyzed electrolysis. Int. J. Hydrog. Energy 40, 168–175. https://doi.org/10.1016/j.ijhydene.2014.10.120.

Mshoperi, E., Fogel, R., Limson, J., 2014. Application of carbon black and iron phthalocyanine composites in bioelectricity production at a brewery wastewater fed microbial fuel cell. Electrochim. Acta 128, 311–317. https://doi.org/10.1016/j.electacta.2013.11.016.

Nasirahmadi, S., Safekordi, A.A., 2011. Whey as a substrate for generation of bioelectricity in microbial fuel cell using *E. coli*. Int. J. Environ. Sci. Technol. 8, 823–830. https://doi.org/10.1007/bf03326265.

Naylor, R., Steinfeld, H., Falcon, W., Galloway, J., Smil, V., Bradford, E., Alder, J., Mooney, H., 2005. Agriculture: losing the links between livestock and land. Science 310, 1621–1622. https://doi.org/10.1126/science.1117856.

Nevin, K.P., Woodard, T.L., Franks, A.E., Summers, Z.M., Lovley, D.R., 2010. Microbial electrosynthesis: feeding microbes electricity to convert carbon dioxide and water to multicarbon extracellular organic compounds. MBio. 1. https://doi.org/10.1128/mbio.00103-10.

Oh, S., Logan, B.E., 2005. Hydrogen and electricity production from a food processing wastewater using fermentation and microbial fuel cell technologies. Water Res. 39, 4673–4682. https://doi.org/10.1016/j.watres.2005.09.019.

Pandey, P., Shinde, V.N., Deopurkar, R.L., Kale, S.P., Patil, S.A., Pant, D., 2016. Recent advances in the use of different substrates in microbial fuel cells toward wastewater treatment and simultaneous energy recovery. Appl. Energy 168, 706–723. https://doi.org/10.1016/j.apenergy.2016.01.056.

Pant, D., Adholeya, A., 2007. Biological approaches for treatment of distillery wastewater: a review. Bioresour. Technol. 98, 2321–2334. https://doi.org/10.1016/j.biortech.2006.09.027.

Pant, D., Van Bogaert, G., Alvarez-gallego, Y., Diels, L., Vanbroekhoven, K., 2016. Evaluation of bioelectrogenic potential of four industrial effluents as substrate for low cost microbial fuel cells operation. Environ. Eng. Manag. J. 15, 1897–1904.

Pekan, G., Koksal, E., Küçükerdönmez, Ö., Özel, H., 2006. Household Food Wastage in Turkey. Food and Agriculture Organization of the United Nations.

Penteado, E.D., Fernandez-Marchante, C.M., Zaiat, M., Cañizares, P., Gonzalez, E.R., Rodrigo, M.A., 2016a. Influence of sludge age on the performance of MFC treating winery wastewater. Chemosphere 151, 163–170. https://doi.org/10.1016/j.chemosphere.2016.01.030.

Penteado, E.D., Fernandez-Marchante, C.M., Zaiat, M., Cañizares, P., Gonzalez, E.R., Rodrigo, M.A., 2016b. Energy recovery from winery wastewater using a dual chamber microbial fuel cell. J. Chem. Technol. Biotechnol. 91, 1802–1808. https://doi.org/10.1002/jctb.4771.

Philips, J., Verbeeck, K., Rabaey, K., Arends, J.B.A., 2016. Electron transfer mechanisms in biofilms. In: Scott, K., Yu, E.H. (Eds.), Microbial Electrochemical and Fuel Cells: Fundamentals and Applications. Woodhead Publishing is an Imprint of Elsevier, pp. 67–113.

Pisutpaisal, N., Sirisukpoca, U., 2012. Bioelectricity generation from wastewaters in microbial fuel cells. Adv. Mater. Res. 512–515, 1456–1460. https://doi.org/10.4028/www.scientific.net/amr.512-515.1456.

Potter, M.C., 1911. Electrical effects accompanying the decomposition of organic compounds. Proc. R. Soc. B Biol. Sci. 84, 260–276. https://doi.org/10.1098/rspb.1911.0073.

Rabaey, K., Verstraete, W., 2005. Microbial fuel cells: novel biotechnology for energy generation. Trends Biotechnol. 23, 291–298. https://doi.org/10.1016/j.tibtech.2005.04.008.

Rabaey, K., Clauwaert, P., Aelterman, P., Verstraete, W., 2005. Tubular microbial fuel cells for efficient electricity generation. Environ. Sci. Technol. 39, 8077–8082. https://doi.org/10.1021/es050986.

Rabaey, K., Van de Sompel, K., Maignien, L., Boon, N., Aelterman, P., Clauwaert, P., De Schamphelaire, L., Pham, H.T., Vermeulen, J., Verhaege, M., Lens, P., Verstraete, W., 2006. Microbial fuel cells for sulfide removal. Environ. Sci. Technol. 40, 5218–5224. https://doi.org/10.1021/es060382u.

Rader, G.K., Logan, B.E., 2010. Multi-electrode continuous flow microbial electrolysis cell for biogas production from acetate. Int. J. Hydrog. Energy 35, 8848–8854. https://doi.org/10.1016/j.ijhydene.2010.06.033.

Rago, L., Baeza, J.A., Guisasola, A., 2017. Bioelectrochemical hydrogen production with cheese whey as sole substrate. J. Chem. Technol. Biotechnol. 92, 173–179. https://doi.org/10.1002/jctb.4987.

Rasep, Z., Aripen, N.S.M., Ghazali, M.S.M., Yahya, N., Arida, A.S., Som, A.M., Mustaza, M.F., 2016. Microbial fuel cell for conversion of chemical energy to electrical energy from food industry wastewater. J. Environ. Sci. Technol. 9, 481–485. https://doi.org/10.3923/jest.2016.481.485.

Rezaei, F., Richard, T.L., Logan, B.E., 2009. Analysis of chitin particle size on maximum power generation, power longevity, and Coulombic efficiency in solid-substrate microbial fuel cells. J. Power Sources 192, 304–309. https://doi.org/10.1016/j.jpowsour.2009.03.023.

Rikame, S.S., Mungray, A.A., Mungray, A.K., 2012. Electricity generation from acidogenic food waste leachate using dual chamber mediator less microbial fuel cell. Int. Biodeterior. Biodegrad. 75, 131–137. https://doi.org/10.1016/j.ibiod.2012.09.006.

Rozendal, R., Jeremiasse, A.W., Hamelers, H.V.M., Buisman, C.J.N., 2008. Hydrogen production with a microbial biocathode. Environ. Sci. Technol. 42, 629–634. https://doi.org/10.1021/es071720+.

Rozendal, R.A., Leone, E., Keller, J., Rabaey, K., 2009. Efficient hydrogen peroxide generation from organic matter in a bioelectrochemical system. Electrochem. Commun. 11, 1752–1755. https://doi.org/10.1016/j.elecom.2009.07.008.

Rumsey, I.C., Aneja, V.P., 2014. Measurement and modeling of hydrogen sulfide lagoon emissions from a swine concentrated animal feeding operation. Environ. Sci. Technol. 48, 1609–1617. https://doi.org/10.1021/es403716w.

Samsudeen, N., Radhakrishnan, T.K., Matheswaran, M., 2016. Effect of isolated bacterial strains from distillery wastewater on power generation in microbial fuel cell. Process Biochem. 51, 1876–1884. https://doi.org/10.1016/j.procbio.2016.06.007.

Sangeetha, T., Muthukumar, M., 2011. Catholyte performance as an influencing factor on electricity production in a dual-chambered microbial fuel cell employing food processing wastewater. Energy Sources, Part A Recover. Util. Environ. Eff. 33, 1514–1522. https://doi.org/10.1080/15567030903397966.

Sapkota, A.R., Lefferts, L.Y., McKenzie, S., Walker, P., 2007. What do we feed to food-production animals? A review of animal feed ingredients and their potential impacts on human health. Environ. Health Perspect. 115, 663–670. https://doi.org/10.1289/ehp.9760.

Scott, K., 2016. Membranes and separators for microbial fuel cells. In: Scott, K., Yu, E.H. (Eds.), Microbial Electrochemical and Fuel Cells: Fundamentals and Applications. Woodhead Publishing is an imprint of Elsevier, Cambridge, UK, pp. 153–178.

Shrestha, N., Fogg, A., Wilder, J., Franco, D., Komisar, S., Gadhamshetty, V., 2016. Electricity generation from defective tomatoes. Bioelectrochemistry 112, 67–76. https://doi.org/10.1016/j.bioelechem.2016.07.005.

Silver, M., Buck, J., Kiely, P., Guzman, J.J., 2014. Bio-electrochemical Systems. IWA Publishing, London, UK.

Simate, G.S., Cluett, J., Iyuke, S.E., Musapatika, E.T., Ndlovu, S., Walubita, L.F., Alvarez, A.E., 2011. The treatment of brewery wastewater for reuse: state of the art. Desalination 273, 235–247. https://doi.org/10.1016/j.desal.2011.02.035.

Sotres, A., Cerrillo, M., Viñas, M., Bonmatí, A., 2015. Nitrogen recovery from pig slurry in a two-chambered bioelectrochemical system. Bioresour. Technol. 194, 373–382. https://doi.org/10.1016/j.biortech.2015.07.036.

Sotres, A., Tey, L., Bonmatí, A., Viñas, M., 2016. Microbial community dynamics in continuous microbial fuel cells fed with synthetic wastewater and pig slurry. Bioelectrochemistry 111, 70–82. https://doi.org/10.1016/j.bioelechem.2016.04.007.

Stehfest, E., van den Berg, M., Woltjer, G., Msangi, S., Westhoek, H., 2013. Options to reduce the environmental effects of livestock production—comparison of two economic models. Agric. Syst. 114, 38–53. https://doi.org/10.1016/j.agsy.2012.07.002.

Sun, H.-L., 2012. Electricity generation from seafood wastewater in a single- and dual-chamber microbial fuel cell with CoTMPP oxygen-reduction electrocatalyst. J. Chem. Technol. Biotechnol. 87, 1167–1172. https://doi.org/10.1002/jctb.3741.

Sun, D., Wang, A., Cheng, S., Yates, M.D., Logan, B., 2014. *Geobacter anodireducens* sp. nov., a novel exoelectrogenic microbe in bioelectrochemical systems. Int. J. Syst. Evol. Microbiol 3485–3491. https://doi.org/10.1099/ijs.0.061598-0.

Sun, M., Zhai, L.-F., Li, W.-W., Yu, H.-Q., 2016. Harvest and utilization of chemical energy in wastes by microbial fuel cells. Chem. Soc. Rev. 45, 2847–2870. https://doi.org/10.1039/c5cs00903k.

Sundar, V., Raghava Rao, J., Muralidharan, C., 2002. Cleaner chrome tanning—emerging options. J. Clean. Prod. 10, 69–74. https://doi.org/10.1016/s0959-6526(01)00015-4.

Tanikkul, P., Pisutpaisal, N., 2015. Performance of a membrane-less air-cathode single chamber microbial fuel cell in electricity generation from distillery wastewater. Energy Procedia 79, 646–650. https://doi.org/10.1016/j.egypro.2015.11.548.

Tao, K., Quan, X., Quan, Y., 2013. Composite vegetable degradation and electricity generation in microbial fuel cell with ultrasonic pretreatment. Environ. Eng. Manag. J. 12, 1423–1427.

Tay, J.-H., Show, K.-Y., Hung, Y.-T., 2006. Seafood processing wastewater treatment. ChemInform 37, 29–66. https://doi.org/10.1002/chin.200613273.

Thassitou, P., Arvanitoyannis, I., 2001. Bioremediation: a novel approach to food waste management. Trends Food Sci. Technol. 12, 185–196. https://doi.org/10.1016/s0924-2244(01)00081-4.

Tian, Y., Mei, X., Liang, Q., Wu, D., Ren, N., Xing, D., 2017. Biological degradation of potato pulp waste and microbial community structure in microbial fuel cells. RSC Adv. 7, 8376–8380. https://doi.org/10.1039/c6ra27385h.

Ueki, A., Matsuda, K., Ohtsuki, C., 1986. Sulfate-reduction in the anaerobic digestion of animal waste. J. Gen. Appl. Microbiol. 32, 111–123. https://doi.org/10.2323/jgam.32.111.

Ueki, A., Ueki, K., Oguma, A., O'htsuki, C., 1989. Partition of electrons between methanogenesis and sulfate reduction in the anaerobic digestion of animal waste. J. Gen. Appl. Microbiol. 35, 151–162. https://doi.org/10.2323/jgam.35.151.

Ueki, K., Ueki, A., Itoh, K., Tanaka, T., Satoh, A., 1991. Removal of sulfate and heavy metals from acid mine water by anaerobic treatment with cattle waste: effects of heavy metals on sulfate reduction. J. Environ. Sci. Health 26, 1471–1489. https://doi.org/10.1080/10934529109375708.

Velasquez-Orta, S.B., Head, I.M., Curtis, T.P., Scott, K., 2011. Factors affecting current production in microbial fuel cells using different industrial wastewaters. Bioresour. Technol. 102, 5105–5112. https://doi.org/10.1016/j.biortech.2011.01.059.

Vijay, A., Vaishnava, M., Chhabra, M., 2016. Microbial fuel cell assisted nitrate nitrogen removal using cow manure and soil. Environ. Sci. Pollut. Res. 23, 7744–7756. https://doi.org/10.1007/s11356-015-5934-0.

Vilajeliu-Pons, A., Puig, S., Pous, N., Salcedo-Dávila, I., Bañeras, L., Balaguer, M.D., Colprim, J., 2015. Microbiome characterization of MFCs used for the treatment of swine manure. J. Hazard. Mater. 288, 60–68. https://doi.org/10.1016/j.jhazmat.2015.02.014.

Vogl, A., Bischof, F., Wichern, M., 2016. Single chamber microbial fuel cells for high strength wastewater and blackwater treatment—a comparison of idealized wastewater, synthetic human blackwater, and diluted pig manure. Biochem. Eng. J. 115, 64–71. https://doi.org/10.1016/j.bej.2016.08.007.

Wagner, R.C., Regan, J.M., Oh, S.-E., Zuo, Y., Logan, B.E., 2009. Hydrogen and methane production from swine wastewater using microbial electrolysis cells. Water Res. 43, 1480–1488. https://doi.org/10.1016/j.watres.2008.12.037.

Wang, X., Feng, Y.J., Lee, H., 2008. Electricity production from beer brewery wastewater using single chamber microbial fuel cell. Water Sci. Technol. 57, 1117–1121. https://doi.org/10.2166/wst.2008.064.

Wang, X., Feng, Y., Wang, H., Qu, Y., Yu, Y., Ren, N., Li, N., Wang, E., Lee, H., Logan, B.E., 2009. Bioaugmentation for electricity generation from corn stover biomass using microbial fuel cells. Environ. Sci. Technol. 43, 6088–6093. https://doi.org/10.1021/es900391b.

Wang, Y., Guo, W.-Q., Xing, D.-F., Chang, J.-S., Ren, N.-Q., 2014. Hydrogen production using biocathode single-chamber microbial electrolysis cells fed by molasses wastewater at low temperature. Int. J. Hydrog. Energy 39, 19369–19375. https://doi.org/10.1016/j.ijhydene.2014.07.071.

Wang, J., Zheng, T., Wang, Q., Xu, B., Wang, L., 2015. A bibliometric review of research trends on bioelectrochemical systems. Curr. Sci. 109, 2204–2211. https://doi.org/10.18520/v109/i12/2204-2211.

Wang, H., Qu, Y., Li, D., Ambuchi, J.J., He, W., Zhou, X., Liu, J., Feng, Y., 2016. Cascade degradation of organic matters in brewery wastewater using a continuous stirred microbial electrochemical reactor and analysis of microbial communities. Sci. Rep. 6. https://doi.org/10.1038/srep27023.

Wen, Q., Wu, Y., Cao, D., Zhao, L., Sun, Q., 2009. Electricity generation and modeling of microbial fuel cell from continuous beer brewery wastewater. Bioresour. Technol. 100, 4171–4175. https://doi.org/10.1016/j.biortech.2009.02.058.

Wen, Q., Wu, Y., Zhao, L., Sun, Q., 2010. Production of electricity from the treatment of continuous brewery wastewater using a microbial fuel cell. Fuel 89, 1381–1385. https://doi.org/10.1016/j.fuel.2009.11.004.

Xu, N., Zhou, S., Yuan, Y., Qin, H., Zheng, Y., Shu, C., 2011. Coupling of anodic biooxidation and cathodic bioelectro-Fenton for enhanced swine wastewater treatment. Bioresour. Technol. 102, 7777–7783. https://doi.org/10.1016/j.biortech.2011.06.030.

Xu, Y., Jiang, Y., Chen, Y., Zhu, S., Shen, S., 2014. Hydrogen production and wastewater treatment in a microbial electrolysis cell with a biocathode. Water Environ. Res. 86, 649–653. https://doi.org/10.2175/106143014x13975035525500.

Yacob, S., Hung, Y., Shirai, Y., Ali Hassan, M., 2005. Treatment of palm oil wastewaters. In: Wang, L.K., Hung, Y.-T., Lo, H.H., Yapijakis, C. (Eds.), Waste Treatment in the Food Processing Industry. CRC Press, Boca Raton, FL, pp. 101–117.

You, S.-J., Zhang, J.-N., Yuan, Y.-X., Ren, N.-Q., Wang, X.-H., 2010. Development of microbial fuel cell with anoxic/oxic design for treatment of saline seafood wastewater and biological electricity generation. J. Chem. Technol. Biotechnol. 85, 1077–1083. https://doi.org/10.1002/jctb.2400.

Yu, J., Park, Y., Kim, B., Lee, T., 2015. Power densities and microbial communities of brewery wastewater-fed microbial fuel cells according to the initial substrates. Bioprocess Biosyst. Eng. 38, 85–92. https://doi.org/10.1007/s00449-014-1246-x.

Yu, N., Xing, D., Li, W., Yang, Y., Li, Z., Li, Y., Ren, N., 2017. Electricity and methane production from soybean edible oil refinery wastewater using microbial electrochemical systems. Int. J. Hydrog. Energy 42, 96–102. https://doi.org/10.1016/j.ijhydene.2016.11.116.

Zhang, X., Zhu, F., Chen, L., Zhao, Q., Tao, G., 2013. Removal of ammonia nitrogen from wastewater using an aerobic cathode microbial fuel cell. Bioresour. Technol. 146, 161–168. https://doi.org/10.1016/j.biortech.2013.07.024.

Zhang, G., Zhao, Q., Jiao, Y., Lee, D.J., 2015. Long-term operation of manure-microbial fuel cell. Bioresour. Technol. 180, 365–369. https://doi.org/10.1016/j.biortech.2015.01.002.

Zhao, F., Rahunen, N., Varcoe, J.R., Roberts, A.J., Avignone-Rossa, C., Thumser, A.E., Slade, R.C.T., 2009. Factors affecting the performance of microbial fuel cells for sulfur pollutants removal. Biosens. Bioelectron. 24, 1931–1936. https://doi.org/10.1016/j.bios.2008.09.030.

Zheng, X., Nirmalakhandan, N., 2010. Cattle wastes as substrates for bioelectricity production via microbial fuel cells. Biotechnol. Lett. 32, 1809–1814. https://doi.org/10.1007/s10529-010-0360-3.

Zhuang, L., Yuan, Y., Wang, Y., Zhou, S., 2012. Long-term evaluation of a 10-liter serpentine-type microbial fuel cell stack treating brewery wastewater. Bioresour. Technol. 123, 406–412. https://doi.org/10.1016/j.biortech.2012.07.038.

Ziara, R.M.M., Li, S., Dvorak, B.I., Subbiah, J., 2016. Water and energy use of antimicrobial interventions in a mid-size beef packing plant. Appl. Eng. Agric. 32, 873–879. https://doi.org/10.13031/aea.32.11615.

Chapter 8

Sustainable Waste-to-Energy Technologies: Gasification and Pyrolysis

Serpil Guran
Rutgers University EcoComplex "Clean Energy Innovation Center", Bordentown, NJ, United States

8.1 INTRODUCTION

The world population is estimated to reach 9.7 billion by 2050 (United Nations, 2014) and 11 billion by the end of this century (United Nations, 2015). This rapid growth in population will bring increased demand for energy, water, and food resources. The demand for food, in particular, is expected to increase by 59%–98% by 2050 (Valin et al., 2014). The increased population will also dramatically increase waste generation along with the systems needed to manage it. Recent research states that across the farm-to-fork spectrum, food consumes 10% of the energy budget and 80% of all freshwater used in the United States and accounts for 50% of land use (Gunders, 2012; Webber, 2012). The amount of uneaten and wasted food approaches 40% of total food production, valued at $218 billion annually, and also wastes embodied resources such as freshwater (21%), chemicals (19%), and cropland (18%) (USDA, 2006; Buzby and Hyman, 2012; ReFED, 2016). Food waste, which accounts for over 21% of landfill volume, represents a unique opportunity for resource recovery (energy, nutrients, chemicals, water) while preserving landfill space. Therefore more efficient food systems with efficient disposal and valorization options are needed. In order to create a closed-loop sustainable bioeconomy, meet increased demand for these resources, and reduce the amount of food waste, the utilization and valorization of pre- and postconsumer food waste and food processing by-products across the food processing supply chain is necessary.

8.2 COUPLING FOOD WASTE WITH SUITABLE CONVERSION TECHNOLOGIES

Food waste can be classified into four major groups by source generation as *residential*, *institutional*, *commercial*, and *industrial* waste. From these, *commercial* (i.e., agricultural waste, supermarket waste) and *industrial* (i.e., food processing industry) food waste can also be classified as preconsumption food waste, whereas *residential* and *institutional* (i.e., cafeteria, hospital) wastes are considered to be postconsumer food waste. Mixed food waste sources from postconsumer groups are characterized by high moisture content (60%–90%), high organic content (more than 95% of dry matter), high salt content, and rich nutrition, which are very valuable for recycling and valorization (Qu et al., 2015; Arancon et al., 2013). The mixed characteristics of postconsumer waste make it more challenging to convert into energy, bio-based materials, and high value chemicals. In addition, postconsumer institutional waste generation from cafeterias and hospitals is often contaminated by plastic utensils, while residential food waste may be contaminated by plastic packaging. By comparison, preconsumer food waste is more homogenous than postconsumer mixed waste. Literature reports that *commercial* and *industrial* food waste are less susceptible to quick deterioration compared to mixed food waste sourced from *residential* and *institutional* waste (Galanakis, 2012).

As mentioned previously, food waste has numerous components, with moisture and chemical contents that vary, such as the mixture of carbohydrates, lipids, and proteins (Girotto et al., 2015). Food waste conversion into power, heat, fuels, and bioproducts varies based on the specific feedstock and is generally categorized into two major conversion pathways—biochemical and thermochemical. Thermochemical conversion pathways can be further categorized as follows:

- Combustion
- Gasification
- Pyrolysis
- Hydrothermal liquefaction (Chapter 9).

Sustainable Food Waste-to-Energy Systems. https://doi.org/10.1016/B978-0-12-811157-4.00008-5

Combustion is an exothermic chemical reaction of a fuel with oxygen and common products are carbon dioxide and water with the release of heat (Naik et al., 2010). Direct combustion systems oxidize viable biomass or municipal solid waste to generate hot flue gas that can be used for heat or fed into a boiler to generate steam. The steam can then be used in a turbine system to generate electricity (Naik et al., 2010; Peterson and Haase, 2009). Few studies have focused on "mixed food waste" combustion because of the high moisture content of postconsumer food waste. This will cause energy loss in evaporating the high moisture content of the waste and will generally not be energy efficient (Pham et al., 2015; Caton et al., 2010; Kim et al., 2013). Dehydration of high moisture content wastes could be costly, relative to the energy released during combustion, due to the high enthalpy of vaporization of water. Also, high moisture content lowers the gravimetric energy density, making waste transportation expensive and necessitating local resource use (Caton et al., 2010). Therefore, instead of direct combustion, converting food waste by suitable conversion technologies would provide better and more efficient food valorization approaches (Lai et al., 2009). Additionally, food industries are not in favor of combustion technology due to the potential negative impacts of existing controversial municipal solid waste incinerators on the environmental and human health, as well as economic considerations (Murugan et al., 2013). Instead of just disposing their wastes, food industries are now more interested in deriving useful products from these materials (Murugan et al., 2013). To achieve environmentally and economically sustainable systems (Pham et al., 2015), thermochemical conversion technologies are generally suitable for low moisture biomass feedstocks. The focus of this chapter is on efficient food waste valorization via gasification and pyrolysis technologies.

8.2.1 Gasification Technology Summary

Gasification is a thermochemical conversion process in which carbon- and hydrogen-containing substances, such as biomass and municipal solid waste, are partially oxidized at high temperatures (800–1100°C) in the presence of a gasifying agent (air, steam, and oxygen) and converted into gaseous products (Alauddin et al., 2010; Ruiz et al., 2013; Devi et al., 2003). Heating can be applied directly or indirectly, depending on the system's configuration (USEPA, 2014). The reaction mechanism of biomass gasification consists of four consecutive substages which are dehydration, pyrolysis, oxidation, and gasification. The product gas, commonly referred to as producer gas or syngas, consists mainly of hydrogen (H_2), carbon monoxide (CO), carbon dioxide (CO_2), water (H_2O), and small amounts of hydrocarbons such as methane (CH_4) (Alauddin et al., 2010; Baxter, 2005; Devi et al., 2003). Based on the reaction conditions, the H_2 and CO content of the gasification reactions can be modified. To enhance hydrogen production, catalytic treatments have been widely researched, with Ni-based catalysts reported to be especially effective for hydrogen production from biomass gasification (Wu et al., 2011; Lu et al., 2010). The gas leaving the gasifier also contains ashes, tars, and a small amount of unburnt char, alkali, metal compounds, sulfur, nitrogen, carbonyl sulfide, ammonia, and hydrogen cyanide (HCN). Following gasification, the syngas needs to be cleaned prior to further utilization. Syngas is best used either as fuel for stationary power and heat generation, or with catalytic conversion to manufacture a range of liquid fuels, chemical intermediates, and end products (USEPA, 2014). The other emerging syngas utilization option is biomass-to-liquid fuels production via Fischer-Tropsch synthesis which requires a H_2:CO ratio of around 2.15. Biomass-generated syngas H_2:CO ratio is typically between 1.0 and 2.2 and may require further adjustment, which can be achieved by either water gas shift or reverse shift reactions (Pala et al., 2017; Warnecke, 2000; van Steen and Claeys, 2008; Govender et al., 2006; Sharma et al., 2014).

An important challenge of biomass gasification is the formation of tar which can lead to corrosion, fouling, and blocking. Biomass tar is a light hydrocarbon and phenolic mixture and it can be converted into gaseous products through steam reforming and various types of catalysts which can enhance the conversion. Catalysts are categorized into three groups: (1) naturally occurring catalysts (i.e., dolomite, olivine); (2) metal catalysts (i.e., nickel and alkali metals); and (3) alkalis (i.e., potassium, potassium hydroxide (KOH), potassium bicarbonate ($KHCO_3$)). In order to reduce the tar formation in syngas, the operating parameters should be optimized. In some cases, the char is also used in reforming the syngas (Ahrenfelt et al., 2013; Wang et al., 2008; Sutton et al., 2001). Cyclones, ceramic and textile bag filters, electrostatic precipitators, scrubbers, and rotating particle separators are considered to be efficient gas cleaning systems for the removal of dust, particles, and tars from syngas (Ruiz et al., 2013; Wang et al., 2008).

Gasifier designs can be categorized into three main types: fixed bed, fluidized bed, and entrained suspension bed gasifiers. Numerous research studies have determined that despite their simple operation, fixed-bed gasifiers result in low calorific value gas with high tar yield. Fluidized bed gasifiers operate with uniform temperature distribution, which achieves more efficient heat and mass transfer (Thorin et al., 2012; Ruiz et al., 2013; Pereira et al., 2012; Bridgwater, 2003; Puig-Arnavat et al., 2010). Before gasification, the feedstock should be preprocessed to the required particle size, through methods including sorting/separation, shredding, grinding, blending, drying, and pelletization (Matsakas et al., 2017; Beneroso et al., 2014).

Gasification is one of the promising technologies for converting sustainable biomass and waste biomass into clean energy, as it adds value to low- or negative-value feedstocks by converting them into marketable fuels and products. Waste biomass feedstocks include agricultural wastes, wastes from food processing industry, and postconsumer food waste (Ruiz et al., 2013; Guran et al., 2018). Gasification of suitable food waste reduces landfill problems and helps to recover energy (Ahmed and Gupta, 2010).

8.2.2 Pyrolysis Technology Summary

Biomass pyrolysis is an emerging thermochemical technology that has the potential to serve as a viable conversion pathway for sustainable biomass, including food waste into advanced fuels and chemicals to displace fossil-based counterparts. The pyrolysis process involves the thermal degradation of materials in the absence of an oxidizing agent, causing irreversible rupture of polymer structures into smaller molecules leading to the formation of solid (char), liquid (bio-oils), and noncondensable gaseous products (i.e., CO, CO_2, H_2, CH_4, C_2H_6) (Babu and Chaurasia, 2003; Bridgwater et al., 1999; Williams and Besler, 1996; Bridgwater and Bridge, 1991; Guran et al., 2018). Biomass pyrolysis reactions occur at lower temperatures of 400–700°C, whereas gasification reactions take place at higher temperatures in the range of 800–1100°C. Pyrolysis reactions are strongly influenced by the reactor design and operating parameters, including temperature, heating rate, biomass feed rate, particle size, and residence time (Wright et al., 2010; Couhert et al., 2009; Ringer et al., 2006; Lede, 2013; Mohan et al., 2006).

Pyrolysis processes can be categorized according to the biomass conversion rate as *slow pyrolysis* in fixed-bed reactors, or *fast/flash pyrolysis* in fluidized bed reactors (including bubbling bed and circulating bed), vacuum reactors, or transported beds. Slow heating rates and lower temperatures provide higher char yields because slow heating and longer residence times in the reactor result in slower decomposition. Conversely, in fast pyrolysis, the rapid heat transfer and short residence times coupled with medium and higher temperatures produce higher percentages of liquid products with decreased amount of biochar yields. Both types of reaction conditions also generate noncondensable gaseous products such as C_1-C_4 hydrocarbons (Williams and Besler, 1996; Bridgwater and Bridge, 1991; Demirbas et al., 2004). When compared to combustion and gasification, pyrolysis produces much lower amounts of gaseous products, potentially enabling elimination of a gas cleaning subsystem. Biochar from pyrolysis can be used either as activated carbon if its pore structure and surface areas are suitable, or for soil quality enhancement (Granatstein et al., 2009). Bio-oils can be upgraded by lowering the oxygen content via catalytic treatment and hydrogenation, while noncondensable gases can be used to supply energy for the pyrolysis system. Therefore emerging pyrolysis technology has the potential to be used for the production of a wide range of fuels, solvents, chemicals, and other products from biomass feedstocks (Thorin et al., 2012; Williams and Besler, 1996; Bridgwater and Bridge, 1991; Demirbas et al., 2004; Heschel and Klose, 1995; Hernandez et al., 2014; Shafizadeh, 1985; Guran et al., 2018).

8.3 THERMOCHEMICAL CONVERSION OF SOURCE-SPECIFIC FOOD WASTE AND RESIDUES

As stated previously, thermochemical conversion processes have significant potential as viable food waste valorization technologies, but they are best suited for relatively low moisture content materials, such as those listed in Table 8.1.

8.3.1 Thermochemical Conversion of Animal-Sourced Food Waste and Residues

Published literature reporting studies of experimental thermochemical conversion of animal-sourced food waste residues are summarized in Table 8.2.

8.3.1.1 Meat Rendering Process By-products

Meat and bone meal (MBM) is a by-product of meat rendering industries obtained after cooking carcasses and eliminating fat, followed by drying and crushing. MBM has been utilized as a form of cattle feed in the past. However, Canada, the United States, and the European Union have all adopted regulations banning MBM as a feed, not only for cattle but also for sheep and goats to avoid potential Bovine Spongiform Encephalopathy (BSE) outbreaks in cattle and chronic wasting disease (CWD) in small animals such as sheep and goats (Fedorowicz et al., 2007; Cascarosa et al., 2012a;

TABLE 8.1 Major Food Processing Wastes and Residue Generation Pathways Suitable for Thermochemical Conversion

Food Type	Major Processed Food Categories	Types of Residues
Animal sourced		
- Meat - Poultry - Fish - Dairy	- Processed meat (beef, pork, lamb) - Processed poultry products - Canned, filleted, smoked, salted, or processed fish - Milk, butter, yogurt, cheese, ice cream	Meat and bone meal, carcass, fat, bones, heads, horns, hoof, offal, feathers, feet, poultry offal, scrap meat, scales, fins, shells, bones, gut remains, residual solids of dairy industry
Plant sourced		
- Cereal, grains, rice, wheat, corn - Fruits, nuts, and vegetables - Edible oils - Sugars - Beverages	- Grain for flour, bread and bakery products, starch - Juice, preserved vegetables and fruits, starch, sugar - Nuts - Oils, fats - Coffee, tea, fruit and grain-based alcoholic beverages, cocoa	Straw, stem, leaves, husk, cobs, pits, shells, peel, pomace, nonedible fibers, rotten and dried fruits and vegetables, pods, pulp, solid waste from malting, brewer and distillers grains

Adapted and revised from de las Fuentes, L., Sanders, B., Lorenzo, A., Alber, S., 2004. AWARANET: agro-food wastes minimization and reduction network. In: Waldron, K., Faulds, C., Smith, A. (Eds.), Total Food Exploiting Co-products-Minimizing Waste. Institute of Food Research, Norwich, pp. 233–244; Murugan, K., Simon, V., Chandrasekaran, V.S., Karthikeyan, P., Al-Sohaibani, S., 2013. Current state-of-the-art of food processing by-products. In: Chandrasekaran, M. (Eds.), Valorization of Food Processing By-products, pp. 35–59; Galanakis, C.M., 2012, Recovery of high added-value components from food wastes: conventional, emerging technologies, and commercialized applications. Trends Food Sci. Technol. 26, 49–70.

Bujak, 2015; Conesa et al., 2003). BSE and CWD are forms of Transmissible Spongiform Encephalopathy (TSE) and literature states that TSEs also include Creutzfeldt-Jakob Disease (CJD) in humans (Fedorowicz et al., 2007; Bolton and Bendheim, 1998; Prusiner, 1982).

To avoid the MBM-sourced outbreaks, most meat rendering industry by-products are incinerated, generally providing no additional value to the facility. Also, power plants and cement kilns combust MBM along with conventional fuels (Fedorowicz et al., 2007). Research has demonstrated that the ash produced by incineration of animal by-products can be converted into highly effective mineral fertilizer called eco-phosphate (Bujak, 2015). A fresh carcass contains approximately 32% dry matter (MBM), of which 50%–52% is protein, 35%–41% is fat, and 6%–10% is ash (Auvermann et al., 2004; Fedorowicz et al., 2007). Several studies state that agglomeration problems occur in combustion and cocombustion of MBM with coal and olive bagasse, especially in fluidized bed combustion. The agglomeration problem generally is attributed to high fat and protein content of MBM (Fryda et al., 2006, 2007; Gulyurtlu et al., 2005; Miller et al., 2006; Cascarosa et al., 2012a). In order to find better utilization options than direct combustion of MBM, cow carcasses and other Specified Risk Materials (SRM), including skull, brain trigeminal ganglia, eyes, tonsils, spinal cord, spinal cord column, dorsal root ganglia, and a small portion of intestine, have been considered in numerous gasification and pyrolysis studies. MBM gasification or cogasification with other materials was especially considered to investigate the viability of MBM-to-H_2 pathways (Fedorowicz et al., 2007; Cascarosa et al., 2012b; Bujak, 2015; Conesa et al., 2003).

Gasification of meat processing industry waste by-products creates various gaseous components such as hydrogen (H_2), carbon monoxide (CO), methane (CH_4), ethane (C_2H_6), carbon dioxide (CO_2), ethylene (C_2H_4), and acetylene (C_2H_2). The analysis of gas yields from MBM, carcass, and cull-cow SRM gasification concluded that hydrogen was produced at rates ranging from 3 to 11 ft^3 per pound (0.19 to 0.68 m^3/kg) of product fed into the reactor. In addition, other combustible gases were generated at the rate of 6 to 12 ft^3 per pound (0.37 to 0.74 m^3/kg) of material fed into the reactor. Other studies have also concluded that the highest gas yields were observed from the gasification of soft tissue SRM, which is comprised mainly of carbon, hydrogen, oxygen, and nitrogen. The energy density of the product gas was estimated to reach 400 MJ/m^3 for oxygen-fired gasification. However, tar formation was observed as a challenging issue to be addressed during the gasification of meat rendering by-products (Fedorowicz et al., 2007; Rao et al., 2004). Research with fixed-bed gasification of MBM (with and without steam) indicated that MBM can effectively serve as a feedstock for production

TABLE 8.2 Summary of Literature Studies Reporting Thermochemical Conversion of Animal-Sourced Food Waste Residues

Waste type	Type of Conversion	Temp.	Analyzed Product	Reference
MBM	TGA-pyrolysis	80–800°C	Decomposition kinetics	Conesa et al., 2003
				Ayllon et al., 2005
MBM	Fluidized-bed pyrolysis	480–580°C	Liquid, gas	Cascarosa et al., 2011
		300–900°C	Liquid, gas char	Ayllon et al., 2006
MBM, coal	Cogasification	850–900°C	Gas	Cascarosa et al., 2012b
Animal bones	Pyrolysis	700°C	Liquid	Prevsuren et al., 2004
MBM	Vacuum pyrolysis	500°C	Liquid, gas, char	Chaala and Roy, 2003
MBM, cow carcass	Gasification	1000°C	Gas	Fedorowicz et al., 2007
MBM	Gasification	650–850°C	Syngas	Soni et al., 2011
Cow bones	Combustion	850–900°C	Ash as mineral fertilizer	Bujak, 2015
Bone char	Regeneration	100–700°C	Char for chemical removal	Nigri et al., 2017
Cow, fish bone	Pyrolysis	300–600°C	Char	Brunson and Sabatini, 2009
Meat waste and bone	Gasification	700–750°C	Gas, liquid	Rao, 2004;
		650–750°C	Liquid	Rao et al., 2004
Cow bone, waste meat	Pyrolysis	220–750°C	Char	Zwetsloot et al., 2015
MBM, meat, bone	Gasification	650–850°C	H_2	Soni et al., 2009
Animal fats	Pyrolysis	420°C	Liquid	Ito et al., 2012
	Pyrolysis	500°C	Liquid, char	Hassen-Trabelsi et al., 2014
Mixed poultry waste	Gasification	1200°C	Gas	Naveed et al., 2009
Poultry waste	TGA-pyrolysis	400–700°C	Decomposition kinetics	Smith, 2008
Poultry waste	Gasification	600°C	Syngas	Ramzan et al., 2011
Poultry waste	Pyrolysis	450–600°C	Liquid	Marculescu and Stan, 2012
Poultry waste, coal	Cogasification	1100°C	Gas	Priyadarsan et al., 2005
Fish fats and waste	Pyrolysis	500 °C	Liquid	Kraiem et al., 2015
	Pyrolysis	600 °C	Liquid	Fadhil et al., 2017
	Pyrolysis	525 °C	Liquid	Wiggers et al., 2009
Fish waste	Gasification	700 °C	Gas	Rowland et al., 2009
	Gasification	900 °C	Gas	Freitas and Soria, 2011

of syngas/hydrogen via gasification. Two-stage gasification improved the H_2 yield considerably from 7.3% to 22.3% (volume) while reducing the tar yield. If similar gasification is performed with steam, H_2 yields of 36.2%–49.2% (volume) were achieved, along with lower char and tar yields (Soni et al., 2009, 2011). Fluidized bed gasification of MBM resulted in syngas yields with a lower heating value when compared to gasification of olive residue (Campoy et al., 2014). Some studies cogasified MBM and other meat processing waste with coal and concluded that the addition of a low percentage of MBM, or other biomass waste, does not affect H_2 and CO concentration in the syngas. When the MBM concentration was increased in the fuel blend, a slight increase in hydrogen sulfide (H_2S) concentration was observed along with an increase in gas yield with lower heating value (Cascarosa et al., 2012a, 2012b; Priyadarsan et al., 2005).

In order to better utilize MBM and other meat rendering by-products, pyrolysis technology was also investigated by several researchers. Different pyrolysis technologies with various reaction conditions were tested by transforming these by-products into gas, liquid, and solid char components. The studies included vacuum pyrolysis, fixed-bed pyrolysis, and fluidized bed pyrolysis. Several studies were also carried out to understand the thermal decomposition kinetics of MBM by using thermogravimetric analysis (TGA) and differential thermogravimetric (DTG) approaches. TGA is generally used to determine the overall kinetic parameters for the pyrolysis of biomass, since biomass thermal decomposition is complicated and represents a large number of reactions in parallel and series. TGA analysis measures the overall weight loss due to these reactions and provides general information on the overall reaction kinetics rather than individual reactions. However, TGA is useful in providing a comparison of kinetic data generated with various reaction parameters, such as temperature and heating rate. Thermal decomposition of MBM with thermogravimetric analysis resulted in similar weight fractions of 0.311 and 0.391 in two separate studies (Ayllon et al., 2005; Chaala and Roy, 2003). One study observed three shoulders and one peak on the DTG curve of MBM (Fig. 8.1). The research concluded that 3.25% of the mass loss was represented by the first shoulder in the temperature range of 50–148°C and can be attributed to the dehydration of MBM. The second shoulder which was observed between 148 and 225°C was attributed to the evaporation of low molecular weight compounds with the scission of C—C bonds. The highest weight loss rate was observed in the temperature range of 225–400°C where major degradation reactions occurred. The third shoulder, which appeared in the 400–500°C range, was attributed to the degradation of the bone portion of the feedstock and complete decomposition. Vacuum pyrolysis of MBM at 500°C resulted in yields of 37.1% pyrolysis oil, 11.6% aqueous phase, 39.1% solid char, and 9.4% gaseous phase. The pyrolysis oil's low ash content and high calorific value indicate that it has the potential to be substituted for diesel fuel.

The fixed-bed pyrolysis study also concluded that the optimum temperature was 450°C with main yields of liquid and solid char. Both types of pyrolysis experiments yielded similar gas compositions of C_1-C_4 gases. The char yield had relatively low surface area of 45 m^2/g, indicating that char from these experimental parameters may not be suitable for direct use as activated carbon (Chaala and Roy, 2003; Ayllon et al., 2006; Conesa et al., 2003). Fluidized bed pyrolysis of MBM produced pyrolysis oil that could be used as fuel, which can also be considered for refining to recover various fractions and examine whether the oil can displace fossil fuel-based chemical compounds (Cascarosa et al., 2011).

In another study, pyrolysis of cattle bones at 500–700°C resulted in low yield of oil with high C content, also comprised of aromatic compounds, pyridines, and phenols (Prevsuren et al., 2004). Reaction parameters, including pyrolysis temperature, significantly alter the yield of bone char. Prior research has demonstrated that pyrolysis modifies the chemical and physical structure of bone and with increasing pyrolysis temperatures bone char shows higher calcium phosphate crystallinity, smaller ash particle size, and fewer carboxyl and amide functional groups. This research also demonstrated that copyrolyzing bones with other biomass at higher temperatures can be used to tailor phosphorous characteristics of bone-based fertilizer (Zwetsloot et al., 2015).

Various studies have considered utilizing char produced from pyrolysis of bone to remove excessive fluoride from drinking water. For example, fish bone char is approximately 75% hydroxyapatite [$Ca_{10}(PO_4)_6(OH)_2$] and was considered for removal of fluoride and heavy metals from water (Brunson and Sabatini, 2009). Prolonged ingestion of water containing

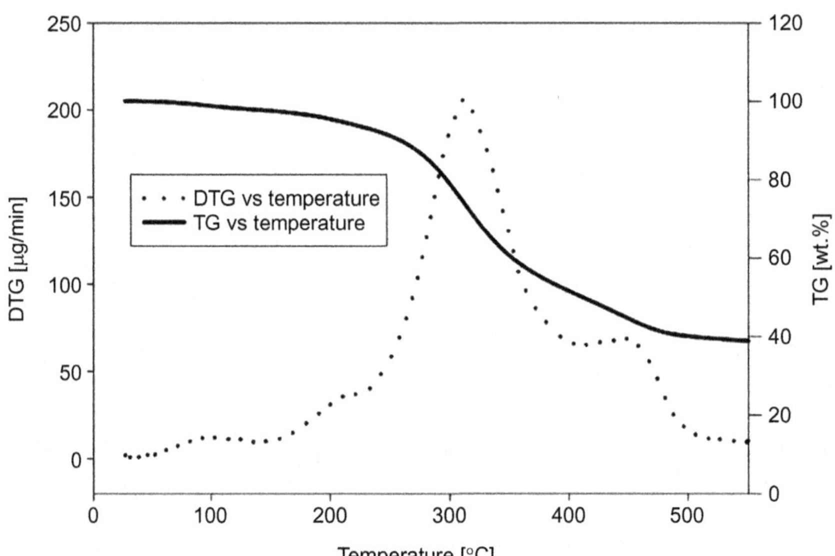

FIG. 8.1 TGA and DTG of MBM. *(Adapted from Chaala, A., Roy, C., 2003. Recycling of meat and bone meal animal feed by vacuum pyrolysis. Environ. Sci. Technol. 37 (19), 4517–4522.)*

more than 2 mg/L of fluoride can cause diseases such as fluorosis, osteoporosis, and arthritis (Nigri et al., 2017). The research demonstrated that the surface of bone char is porous and irregular, and fluoride adsorption capacity varies between 0.075 and 3.62 mg fluoride per gram of bone char. Bone char was reported to be a desirable technology because it may be an effective option to remove fluoride and arsenic depending on the local water chemistry and can be locally produced in many developing regions of the world (Mlilo et al., 2010).

Pyrolysis of waste animal fats (lamb, poultry, and swine) at 500°C indicated that the fatty waste can serve as viable feedstock for pyrolysis technology. The oils from the pyrolysis of animal fatty wastes ranged from 58% to 78%, with high aliphatic character with the presence of straight chain hydrocarbons. Highly oxygenated groups were also observed. When the oil fractions were examined, a large variety of components (alkanes, alkenes, cyclic hydrocarbons, carboxylic acids, aldehydes, ketones, alcohols) were identified. The presence of these components indicates that, with additional enhancement, they can serve as feedstocks or intermediaries in the chemical industry (Hassen-Trabelsi et al., 2014).

8.3.1.2 Poultry and Fish Processing By-products and Residues

Poultry Processing Residues

Similar to the meat processing industry, the poultry industry has carcass disposal challenges for routine and catastrophic mortality. A flock of 50,000 broilers grown to age of 49 days has a 0.1% daily mortality (4.9% total mortality) which results in 2.4 tons of carcasses. Similarly, a turkey flock of 30,000 birds averages 0.5% weekly mortality or 9% total mortality during an 18-week period and produces approximately 13.9 tons of carcasses. It has been estimated that the US poultry industry has to address the challenge of 650,000 tons of broiler and 195,100 tons of turkey carcasses yearly. These estimates do not include catastrophic losses from disease outbreaks, accidents, and natural disasters. Current disposal methods of carcasses and other by-products include landfilling, composting, rendering, and incineration. With all these practices, the spread of pathogenic microorganisms creates a significant concern that needs to be addressed (Blake et al., 2008).

A steady-state simulation model for the gasification of poultry waste investigated the effect of reaction parameters such as temperature and rate of steam injection. Also, the simulation model was validated using experimental data from the gasification of poultry waste. Increased temperature above 600°C caused an increase of CO mole fraction in the syngas. Also, up to 600°C, the H_2 fraction increased with increasing temperature, while with elevated steam injection rate, both H_2 and CO levels increased (Ramzan et al., 2011). Mixed poultry waste including blood, flesh, and chicken litter have also been gasified. Produced syngas was observed to have energy content as low as 9.24 MJ/kg. It was therefore suggested that poultry waste cannot be gasified alone and should be combined with charcoal and other high calorific value substances (Naveed et al., 2009).

Poultry feather and bone pyrolysis (at 450–600°C) produced biochar and bio-oils with lower heating value (LHV) of 23 and 28 MJ/kg, respectively. The waste products of the poultry industry can benefit from an optimized pyrolysis process (Marculescu and Stan, 2012). Pyrolysis of dissolved air flotation (DAF) skimmings from a poultry operation resulted in a bio-oil yield with high heating values, but it was also reported to be high in saturated fatty acids. The level of saturation of fatty acids causes the bio-oil to have a high cloud point and viscosity due to the high melting point of the saturated solvent extraction's fatty acids. However, this problem was addressed by solvent extraction to reduce fatty acids. This approach resulted in a bio-oil containing unsaturated fatty acids that could be esterified into biodiesel and fatty nitriles that could be a feedstock to produce bio-based surfactants (Smith, 2008).

Fish Processing Residues

Fisheries and aquaculture industries remain important sources of food, nutrition, income, and livelihood for hundreds of millions of people around the world. The global per capita seafood supply reached a new record high of 20 kg in 2014 (FAO, 2016). In addition to processing by filleting, freezing, and chilling for direct consumption, a major portion of the seafood is processed into various products by canning, mincing, and curing for later consumption. The nonedible parts of finfish include the head (14–20 wt%), gut (15–20 wt%), skin and scales (1–3 wt%), bones (10–16 wt%), and trimming (15 wt%) adding up to approximately 50% of the whole fish weight. The nonedible parts of shell fish (crab, lobster, molluscan) add up to 60%–70% of the whole shell fish weight. In addition, a large part of landed fresh catch gets discarded for various reasons. Research states that marine fishing operations in Alaska may be discarding up to 60% of their landed catch, by weight, as processing waste (Crapo and Bechtel, 2003). Total US fish processing waste generation adds up to 1 million metric tons of fish waste per year (Folador et al., 2006). Nonedible parts of seafood include valuable bioactive components such as collagen, gelatin, lipid, protein, protein hydrolysate, cartilage, calcium, chitin, chitosan pigments, enzymes, glucosamine (Suresh and Prabhu, 2013). Except for a small percentage, a major portion of the seafood by-products are usually either underutilized or discarded into landfills or the ocean because of their complex processing steps. In addition to serving as potential feedstocks for biochemical conversion, these materials can be converted into valuable end products via thermochemical conversion.

Gasification of red salmon by-products at 700°C resulted in syngas components including H_2, CO, CH_4, C_2H_2, C_2H_4, and C_2H_6. Average H_2:CO ratio was reported to be around 0.35. The research did not report any further improvement of the H_2:CO ratio; however, it stated that the syngas can be utilized at fisheries for energy needs. Also, gasification can serve better economically if the oil content of the fish by-products are extracted for fish oil or biodiesel production before gasification (Rowland et al., 2009). Cogasification of salmon processing slurry and alder in an updraft gasifier generated high quality syngas that could be utilized as an energy source at rural fishing communities. This research further identified modeling needed to optimize reactor design for efficient gasification (Freitas and Soria, 2011).

Pyrolysis of waste fish fats at 500°C with a 10°C/min heating rate resulted in a 54.6% liquid fraction, including a 37.5 wt% organic fraction and 17.1 wt% aqueous fraction. When a similar pyrolysis process was conducted in the presence of Al_2O_3/Na_2CO_3 catalyst, the liquid yield increased to 72%. The analysis of pyrolysis oil revealed the presence of cyclic hydrocarbons, ketones, aldehydes, carboxylic acids, and ester groups. These findings indicate that the oil can either be a good renewable energy source or it can be used as a feedstock for relevant chemical industries (Kraiem et al., 2015). Similarly, pyrolysis of waste fish oil at 525°C resulted in 72.83% liquid, 15.85% gas, and 11.32% solid yields. The study concluded that the oil product after upgrading can serve as an energy source (Wiggers et al., 2009). Pyrolysis of discarded fish parts was performed after oil removal from the samples. Oil from discarded fish parts is considered to be feedstock for biodiesel production via transesterification. The remaining fish waste can then be pyrolyzed at 500°C, resulting in 57.13% oil yield with properties similar to those reported in the literature. The biochar produced during the same process was also successfully converted into mesoporous-activated carbon (Fadhil et al., 2017).

Bone char is considered to be an effective and affordable option to remove fluoride and arsenic from potable water, especially in developing regions of the world. However, due to various religious and cultural beliefs, the use of cow or pig bones may be considered offensive and thus may not be suitable (Brunson and Sabatini, 2009). Therefore fish bone char may be an alternative option to serve the needs of certain areas in the world. Char produced by pyrolysis of fish bone at 500°C removed a sufficient amount of fluoride in 1 h to meet the World Health Organization standard of 1.5 mg/L for starting fluoride concentration of up to 10 mg/L. Pyrolysis temperature was found to be an important parameter for the surface area and porosity which affect the char's adsorption capacity. It was concluded that 500°C is the optimal pyrolysis condition for the fish bone charring, and it was shown that fish bone char is a viable technology to simultaneously remove fluoride and arsenic contaminants (Brunson and Sabatini, 2009).

8.3.2 Thermochemical Conversion of Plant-Sourced Food Waste and Residues

Published literature reporting studies of experimental thermochemical conversion of plant-sourced food waste and residues are summarized in Table 8.4.

8.3.2.1 Nutshells, Kernels, Seeds, Fruit, and Vegetable Skins

Nuts belong to the category of fruits that are dry rather than fleshy. Nuts do not split and the seed is only released when the fruit wall rots or is damaged. Nuts have hard and woody shells that need to be removed and can create a waste disposal problem. Over 78,000 tons of pecan shells and 266,000 tons of walnut shells are generated from US-produced nuts (Table 8.3) (Morecraft, 2015). In addition to nut shells, other agri-food processing wastes such as waste from olive oil, wine, jams and preserves, fruit juice manufacturing pulp and wastes can serve as valuable sources of energy and also raw chemical feedstock for thermochemical conversion technologies.

When thermochemical conversion of biomass is considered, the type of feedstock and the desired end products are essential to know for selecting the most efficient conversion technology. Biomass conversion conditions may differ

TABLE 8.3 Tree Nut Production of the United States

	Almonds	Walnuts		Pistachios	Pecans	
Crop (2012-thousand tons)	834.4[a]	183.3[a]	449.8[b]	251.9[b]	61.6[a]	140.0[b]

[a] *Kernel basis.*
[b] *In-shell basis.*
Adopted and revised from Morecraft, B., 2015. Growing Markets for U.S. Tree Nuts. https://www.usda.gov/oce/forum/2015_Speeches/BMorecraft.pdf
Source: Robobank- Riding the Tree Growth Curve.

due to the fact that biomass feedstock characteristics can be widely different. This includes many factors, such as type of biomass, species, and the climate in which they are grown, cultivated, and stored (Tsamba et al., 2006; Orfao et al., 1999). In order to understand the decomposition kinetics, numerous research projects on thermogravimetric decomposition were carried out to determine the kinetic parameters which can be useful to optimize reactor designs for lignocellulosic biomass pyrolysis and gasification processes (Manara et al., 2015; Zhang et al., 2016; Buessing, 2012; Tsamba et al., 2006). Decomposition kinetics research included grape pomace/seeds, peach pits, and olive pits. The internal and external heat and mass transport phenomena influence pyrolysis kinetics and yields. It is important to identify biomass components according to their response to changing temperature. Research has concluded that similar to other lignocellulosic biomass decomposition, cellulose decomposition has the highest activation energy of 143–175 kJ/mol, representing the largest contribution to the devolatilization process, while the lignin component had the lowest energy (29–37 kJ/mol; Manara et al., 2015). Coconut and cashew nut shell kinetic studies revealed that these species are different from ordinary woody biomass due to their higher hydrocarbon content which resulted in lower amounts of biochar. In addition, thermal decomposition with TGA analysis showed two distinct peaks for hemicellulose (first) and cellulose (second) decomposition steps which differ from the TGA analysis of other woody biomass species that generally result in one overlapping peak (Table 8.4) (Tsamba et al., 2006).

Studies with olive mill residues emphasized that the olive oil industry is in need of identifying alternatives to residue lagooning for fertilizer. Solutions should be economically and environmentally viable alternatives for managing the millions of tons of olive mill waste generated annually in the Mediterranean basin (Goldfarb et al., 2017; Buessing, 2012). Prior research has demonstrated that it is possible to recover lucrative concentrations of polypenols and polyunsaturated fatty acids via CO_2-supercritical fluid extraction, which was enhanced by using ethanol as a cosolvent to olive mill residues. The remaining solid waste also can be converted into high surface area activated carbon, which showed promising absorption ability, but further research is needed (Goldfarb et al., 2017; Buessing, 2012).

Vacuum pyrolysis of cashew nut shells at lower temperatures resulted in a 40% oil yield that indicated fuel-like properties (Das and Ganesh, 2003). Also, when almond shells and potato skins were copyrolyzed with high-density polyethylene, the properties of the oil yield resembled fossil diesel in terms of H:C ratio and heating values. In addition, less oxygen content was observed, which made the oil more stable (Onal et al., 2012, 2014). Studies of catalytic pyrolysis of cotton cocoon shell, olive husk, and tea factory waste concluded that total hydrogen-rich gaseous products increased with increasing pyrolysis temperature. Na_2CO_3, K_2CO_3, and $ZnCl_2$ were used as catalysts (Demirbas, 2001, 2002). The presence of date pyrolysis char in wood gasification improved the gasification performance by reducing tar formation (Al-Rahbi et al., 2016). Coconut pith and orange peel pyrolysis chars were shown to be highly effective in removing elemental mercury and cadmium ions, respectively, from aqueous solutions (Johari et al., 2016; Tran et al., 2016).

The gasification of walnut shells yielded promising results, with one study reporting that temperature had a significant effect on the total syngas yield, and H_2 content increased with increasing temperature (Safari et al., 2017). A viability assessment on the use of palm oil residue with supercritical water gasification and integration into a combined cycle power plant demonstrated that it has great potential as a renewable energy source (Lange and Pellegrini, 2013).

Palm shell pyrolysis at 500°C resulted in bio-oil with a high oxygen content and low heating value. To increase the heating value and improve the properties of the bio-oil, the undesirable oxygenated content needs to be reduced, and copyrolysis of palm oil with polystyrene resulted in higher grade bio-oil with higher heating values. The research demonstrated that the upgraded oil could potentially be used to displace fossil-based fuel (Abnisa et al., 2011, 2013).

8.3.2.2 *Rice and Corn Residues*

Worldwide annual rice production has reached 700 million tonnes. This level of production yields approximately 100 million tons of rice husk yearly, and some rice producing countries allow open burning or landfilling to address this significant disposal problem (Pinto et al., 2016; FAO, 2015).

Experiments on the thermochemical conversion of rice husks have been conducted widely for many years. A thermogravimetric decomposition study of rice husk showed that the decomposition starts at approximately 250°C, and there is a major loss of weight where the main devolatilization occurs, which is essentially complete by approximately 450°C. Further loss of weight occurs until 720°C, after which there is essentially no further loss of weight, indicating that the decomposition is complete. The amounts of residual char (29 wt%) and ash (11 wt%) were unaffected by the heating rate (Williams and Besler, 1993; Isa et al., 2011). Slow pyrolysis of rice husks in a fixed-bed reactor resulted in gaseous yields of CO, CO_2, H_2, CH_4, and C_2H_2, and lower concentrations of propane, propene, butane, and butene were also observed. The highest calorific value of the gas was reported to be 15.60 MJ/m^3. Fast pyrolysis of rice husks yielded higher

TABLE 8.4 Summary of Literature Studies Reporting Thermochemical Conversion of Plant-Sourced Food Waste and Residues

Waste type[a]	Type of Conversion	Temp.	Analyzed Product	Reference
Olive mill waste	Pyrolysis	600–800°C	Char	Goldfarb et al., 2017
	Pyrolysis	600°C	Decomposition	Buessing, 2012
Olive mill waste	Gasification	770–870°C	Gas	Campoy et al., 2014
Cashew nut shell	Pyrolysis	400–550°C	Liquid	Das and Ganesh, 2003
	Pyrolysis	250–900°C	Liquid	Tsamba et al., 2006
Pecan, almond shells	Pyrolysis	600°C	Liquid	Miranda et al., 2012
Almond shell, PE	Copyrolysis	500°C	Liquid	Onal et al., 2014
Potato skin, PE	Copyrolysis	440–550°C	Liquid	Onal et al., 2012
Palm shell, PS	Copyrolysis	600°C	Liquid	Abnisa et al., 2013
Palm shell	Pyrolysis	500°C	Liquid	Abnisa et al., 2011
Coconut pith	Pyrolysis	300–900°C	Liquid, char	Johari et al., 2016
Orange peel	Pyrolysis	400–800°C	Char	Tran et al., 2016
Walnut shell	Gasification	400–440°C	Gas	Safari et al., 2017
Walnut shell, coal	Copyrolysis, TGA	1100°C	Decomposition	Zhang et al., 2016
Grape pomace, peach kernels	TGA	200–800°C	Decomposition	Manara et al., 2015
Corncob, straw	Pyrolysis	500°C	Liquid, char	Yanik et al., 2007
Cotton stalk	Pyrolysis	480–530°C	Liquid, char	Zheng et al., 2008
Rice husk	TGA	600°C	Decomposition	Monteneiro et al., 2012
	TGA	800°C	Decomposition	Isa et al., 2011
Apple pomace	TGA	500°C	Decomposition	Yates et al., 2017
Rice straw, husk	Pyrolysis	500–850°C	Gas, liquid, char	Chen et al., 2003
	Pyrolysis	400–500°C	Gas, liquid, char	Heo et al., 2010
	Pyrolysis	420–540°C	Liquid	Zheng, 2007
	Catalytic pyrolysis	400–600°C	Gas, liquid, char	Williams and Nugranad, 2000
	Pyrolysis	400–800°C	Liquid, char	Tsai et al., 2006; Tsai et al., 2007
	Pyrolysis	720°C	Liquid, gas, char	Williams and Besler, 1993
Rice straw, husk, PE	Cogasification	850°C	Gas	Pinto et al., 2016
Sugarcane bagasse, coconut shell	Pyrolysis	400–700°C	Liquid, char	Tsai et al., 2006
Corncob	Catalytic pyrolysis	600–800°C	Liquid	Ates and Isikdag, 2009
Coffee hulls, wheat	Microwave pyrolysis	300–1100° C	Liquid, gas	Luque et al., 2012

PE: polyethylene; *PS*: polystyrene.

amounts of bio-oil, up to 60% (Heo et al., 2010; Zheng, 2007; Tsai et al., 2007). Catalytic pyrolysis of rice husks resulted in gaseous yields with higher H_2 content than a noncatalytic process (Chen et al., 2003).

Corncob pyrolysis also resulted in similar yields to other lignocellulosic biomass pyrolysis processes. Moderate temperatures provided the maximum oil yields, whereas higher temperatures favored cracking reactions, resulting in higher gas yields. Catalytic pyrolysis yielded an increase of aliphatic compounds in the oil (Ates and Isikdag, 2009). Catalytic pyrolysis can be considered when converting agricultural waste such as rice, wheat, and corncob into potential fuels and chemicals.

8.3.3 Thermochemical Conversion of Postconsumer Mixed Food Waste

Published literature reporting studies of experimental thermochemical conversion of postconsumer food waste and residues are summarized in Table 8.5.

Gasification

Postconsumer food waste from residential, institutional (e.g., cafeterias, hospitals, nursing homes, etc.) and some commercial sectors (e.g., mixed supermarket waste, fast-food restaurants) is represented by heterogeneous chemical characteristics, including carbohydrates, lipids, amino acids, phosphates, vitamins, and carbon but also containing other substances (Grycova et al., 2016a,b). In addition, postconsumer mixed waste may have a high moisture content. It has been reported that average proximate analysis of food waste is 80% volatile matter, 15% fixed carbon, and 5% ash. There is limited knowledge on whether or not the decomposition of different types of the main components of food waste would result in characteristics similar to the product of mixed food waste decomposition (Collins, 2015). As with preconsumer food waste, low moisture postconsumer waste can also be considered as a viable feedstock for gasification and pyrolysis conversion (Ahmed and Gupta, 2010). However, research on thermal decomposition of mixed food waste suggests that with a moisture content higher than 45%, a steam gasification approach would be viable because the water vapor liberated in the

TABLE 8.5 Summary of Literature Studies Reporting Thermochemical Conversion of Postconsumer Food Waste and Residues

Waste Type	Type of Conversion	Temp.	Analyzed Product	Reference
Tea waste	Gasification	450–850°C	Gas	Ayas and Esen, 2016
Tea waste	Pyrolysis	400–700°C	Liquid, char	Uzun et al., 2010
Coffee hulls	Microwave pyrolysis	300–1100°C	Liquid, gas	Luque et al., 2012
Food waste	Catalytic microwave pyrolysis	up to ~800°C @ 300–600 W	Liquid, char	Liu et al., 2014
Food waste	Steam gasification	300°C	Gas	Ko et al., 2001
Cereals, crisps	Pyrolysis	880°C	Liquid, gas, char	Grycova et al., 2016b
Waste fats, cellulose	Pyrolysis	800°C	Liquid, gas, char	Grycova et al., 2016a
Mixed food waste sludge	Pyrolysis	200–450°C	Activated carbon	Mahapatra et al., 2012
Food ind. waste	Pyrolysis and gasification	350–1100°C	Gas, energy	Marculescu, 2012
Cafeteria waste	Gasification	725°C	Gas	Caton et al., 2010;
	Gasification	800–1000°C	Char	Yang et al., 2016;
	Gasification	800–900°C	Gas, char	Ahmed and Gupta, 2010;
	TGA	50–950°C	Decomposition	Collins, 2015;
	Gasification	800°C	Syngas	Collins, 2015

pyrolysis stage can be used in the steam gasification stage and therefore the energy consumed to evaporate moisture can be recovered (Marculescu, 2012).

To understand the thermal decomposition and predict yields for biomass gasification decomposition, simulation models have been developed using Aspen Plus. However, relatively little research has focused on Aspen simulation of gasification and syngas upgrading processes using water gas shift and reverse water gas shift reactions. Pala et al. (2017) developed an integrated Aspen Plus model for food waste steam gasification and needed adjustments for syngas using shift and reverse shift reactions. Their model showed that hydrogen content remains almost constant with increasing gasification temperature, but it increases with increasing steam-to-biomass ratio. The model's prediction of hydrogen production at higher temperatures was found to be in good agreement with experimental data. The model was especially tailored for gasifying municipal solid waste (MSW), mixed food wastes, and coffee bean husks. It estimated that syngas adjustment can be achieved by the reverse water gas shift reaction to result in a H_2:CO molar ratio close to 2.15, which is desirable as feedstock for Fisher-Tropsch synthesis to produce liquid fuels and chemicals (Pala et al., 2017). Experimental data for steam gasification of food waste showed that increasing the amount of water added not only enhanced the reactivity of gasification by increasing the rate of the steam reforming reaction, but also by diminishing the level of carbon monoxide by water gas shift reaction. The reaction temperature affected the conversion of carbonized solids into gas and the composition of the gas (Ko et al., 2001).

As in other biomass gasification processes, after the dehydration stage, food waste gasification also contains a pyrolysis stage which is followed by a char gasification stage. Generally, the char gasification stage is slower than the prior pyrolysis stage and is considered to be the rate-limiting step. Understanding the catalytic effect of char on the overall efficiency of gasification is essential. Because of the presence of inorganic constituents in the food waste, char from the initial pyrolysis step was found to have a catalytic effect. Char reactivity increased with the degree of conversion (Ahmed and Gupta, 2010). Research on postconsumer mixed food waste considered preprocessing the food waste by dehydrating, grinding, and pelletizing due to the heterogeneous properties of the waste. Comparison of simulated gasification results with direct combustion indicated that, although combustion of pelletized food waste is energetically comparable to wood combustion, gasification results are also in good agreement with other biomass gasification literature (Caton et al., 2010).

Research on the gasification of various foodstuffs and specific components of food waste examined several compounds such as dextrose, phenylalanine and palmitic acid, as well as potatoes, beef, and peanuts, and found that H_2 and CO yields obtained from pure samples were at the highest levels from palmitic acid, followed by phenylalanine with the lowest amount from dextrose. Also, among the food samples, beef resulted in the highest yields, followed by peanuts and the lowest from potato waste gasification (Collins, 2015). Other findings suggested that although component analysis may provide a good understanding of physical and chemical properties, mixed food waste should still be thermally converted as received in order to understand conversion characteristics in detail (Kothari et al., 2010). Additional research has been conducted on the gasification and cogasification of wood with mixed food waste with moisture levels of 54%, 66%, and 83%, and concluded that food waste gasification and cogasification can provide viable valorization pathways. Biochars from the gasification experiments were tested to examine their ability to improve nutrient-poor acidic soil. The initial results showed that biochar can be used to reverse the conditions of poor soil and consequently allow agriculture and reforestation on degraded land in South Asia (Yang et al., 2016).

Pyrolysis

Pyrolysis of postconsumer mixed solid waste may provide a promising valorization approach to generate liquid, biochar, and gaseous coproducts. Similar to gasification, mixed food waste has its challenges because of the heterogeneous properties of the feedstock material. Several studies on mixed food waste pyrolysis concluded that food waste can be converted into useful forms of energy (Liu et al., 2014; Grycova et al., 2016a,b; Mahapatra et al., 2012). All these studies demonstrated that properties of the resulting pyrolysis products are dependent on the composition of the feedstock and reaction conditions. To understand mixed food waste thermal decomposition characteristics via pyrolysis, some studies have individually converted food waste components (carbohydrates, lipids, and proteins) and demonstrated that carbohydrates produced furan- and sugar-based products, whereas pyrolysis of proteins resulted mainly in hydrocarbon-based products in the aromatic form. The pyrolysis of lipids produced high amounts of acids and low concentrations of hydrocarbon and alkene products (Collins, 2015). The results also showed that pyrolysis oil from mixed waste is a multicomponent mixture depending on the feedstock. Substances that can serve as feedstocks for other industries can be isolated from the pyrolysis oil and recycled. Some studies showed that pyrolysis of waste cellulose from edible oil production can yield pyrolysis oil-containing substances such as capronaldehyde, valeric acid, caprylic acid, and caprolic acid that can be used in the flavors industry (Grycova et al., 2016a), and the combustible components of the product gas phase can be used for process energy

(Grycova et al., 2016b). Microwave pyrolysis is also considered to be a viable technology to convert mixed food waste into upcycled products. Microwave-assisted pyrolysis of coffee hulls resulted in higher oil yields compared to conventional pyrolysis (Luque et al., 2012). Liu et al. (2014) examined the catalytic effect of metals oxides such as MgO, Fe_2O_3, and MnO_2 and chloride salts such as $CuCl_2$ and NaCl during microwave pyrolysis. Their results showed that the presence of metal salts negatively impacted oil yield and enhanced gas yield, whereas chloride salts provided the opposite effect.

Pyrolysis research has especially concentrated on producing biochar to be used as activated carbon, and Mahapatra et al. (2012) demonstrated that $ZnCl_2$ was a better activation agent for the biochar produced from pyrolysis of mixed sludge from the food industry. Pyrolysis of postconsumer fats and oils, such as waste cooking oil with animal fat content, resulted in triacylglycerols that were decomposed to fatty acids. Unsaturated fatty acids decomposed to hydrocarbons at temperatures up to 390°C. At higher temperatures, hydrocarbons were formed by decarboxylation of ester groups. The presence of a Pd/C catalyst during the pyrolysis allowed selective decarboxylation to produce light oil fractions (Ito et al., 2012).

8.4 FUTURE PERSPECTIVE AND RESEARCH NEEDS

Pre- and postconsumer food wastes have numerous components with varying chemical and physical properties. With the new trends of diverting organic waste from landfills, food waste will be more available as a feedstock. It is expected that with more careful approaches postconsumer food waste may be minimized; however, preconsumer food waste is still expected to increase with the increased population and increased demand for food. Thermochemical conversion of food waste is an emerging research area that needs further research both for gasification and pyrolysis pathways. It is most important to consider coupling specific feedstocks with a suitable conversion technology and optimizing process conditions for successful commercialization. Further research is needed, especially from bench-scale to pilot-scale, for any chosen specific food waste-technology combination to avoid scale-up problems. In addition, it is recommended that detailed modeling should accompany the thermochemical conversion of pre- and postconsumer food waste. Because of the complex nature of mixed food waste, carrying out component-based research conditions should also be repeated for the mixed food waste to achieve repeatability and avoid design problems. Thermochemical conversion pathways should be researched further with and without existing catalyst materials and to develop new catalysts to achieve even higher conversion efficiency. If gasification is performed, the generated syngas H2:CO ratio is important to converting syngas catalytically into clean fuels and chemicals. Further research is also needed to achieve the needed H2:CO ratios for downstream conversion to fuels or chemicals. Similarly, pyrolysis oil and biochar yield properties for any specific feedstock still need further research for improvement and successful scale-up.

8.5 CONCLUSIONS

Thermochemical conversion technologies, namely, gasification and pyrolysis, are suitable pathways to valorize low moisture content food wastes into desired end products such as biofuels and chemicals. This paper concludes that gasification is well suited to convert preconsumer food processing industry wastes with low moisture content, such as nut shells, animal bones, and carcass, into syngas. Achieving syngas H2:CO ratio improvements still needs further research. Similarly, despite the technology still being in its infancy and the need for further scale-up research, pyrolysis technology appears to be able to convert viable feedstocks onto bio-oil and biochar. Increased temperature results in increased oil and reduced char yields, similar to other biomass pyrolysis research in the literature. Bio-oil from pyrolysis was shown to be suitable for upgrading to biofuels or compounds that can be isolated and serve as feedstocks or intermediaries for the chemical and flavor industry. Mixed food waste pyrolysis oil analysis revealed that the bio-oil is a multicomponent mixture and highly feedstock dependent. Postconsumer food waste has components such as carbohydrates, lipids, and proteins, and pyrolysis of these individual components will result in furan and sugar-based, hydrocarbon-based and acid compounds, respectively. Substances that can serve as feedstocks for other industries can be isolated from the pyrolysis oil and recycled. Some studies showed that pyrolysis of waste cellulose from edible oil production can yield pyrolysis oil-containing substances such as capronaldehyde, valeric acid, caprylic acid, and caprolic acid that can be used in the flavors industry. When designing larger scale operations, consistent feedstock characteristics will be essential to achieve successful conversion.

Nut shell kinetic studies revealed that these species are different from ordinary woody biomass due to their higher hydrocarbon content which resulted in lower amounts of biochar. However animal-based biochars, especially cattle and fish bone pyrolysis, generated biochars that can be utilized as an affordable potable water cleaning medium, especially in developing parts of the world.

REFERENCES

Abnisa, F., Wan Daud, W.M.A., Sahu, J.N., 2011. Optimization and characterization studies on bio-oil production from palm shell by pyrolysis using response surface methodology. Biomass Bioenergy 35, 3604–3616.

Abnisa, F., Wan Daud, W.M.A., Ramalingam, S., Azemi, M.N.B.M., Sahu, J.N., 2013. Co-pyrolysis of palm shall and polystyrene waste mixtures to synthesis liquid fuel. Fuel 108, 311–318.

Ahmed, I.I., Gupta, A.K., 2010. Pyrolysis and gasification of food waste: syngas characteristics and char gasification kinetics. Appl. Energy 87 (1), 101–108.

Ahrenfelt, J., Thomsen, T., Henriksen, U.B., Rongaard, L., 2013. Biomass gasification cogeneration—a review of state of the art technology and near future perspectives. Appl. Therm. Eng. 50, 1407–1417.

Alauddin, Z.A.B.Z., Lahijani, P., Mohammadi, M., Mohamed, A.R., 2010. Gasification of lignocellulosic biomass in fluidized beds for renewable energy development: a review. Renew. Sust. Energ. Rev. 14, 2852–2862.

Al-Rahbi, A., Onwudili, J.A., Williams, P.T., 2016. Thermal decomposition and gasification of biomass pyrolysis gases using hot bed of waste derived pyrolysis char. Bioresour. Technol. 2014, 71–79 (ISSN 0960-8524).

Arancon, R.A.D., Lin, C.S., Chan, K.M., Kwan, T.H., Luque, R., 2013. Advances on waste valorization: new horizons for a more sustainable society. Energy, Science & Engineering 1 (2), 53–71.

Ates, F., Isikdag, M.A., 2009. Influence of temperature and alumina catalyst on pyrolysis of corncob. Fuel 88, 1991–1997.

Auvermann, B., Kalbasi, A., Ahmed, A., 2004. Rendering. In: Carcass Disposal: A Comprehensive Review. Carcass Disposal Working Group, USDA-APHIS Cooperative Agreement Project, National Agricultural Biosecurity Center Consortium. Kansas State University.

Ayas, N., Esen, T., 2016. Hydrogen production from tea waste. Int. J. Hydrog. Energy 41 (19), 8067–8072.

Ayllon, M., Gea, G., Murillo, M.B., Sanchez, J.L., Arauzo, J., 2005. Kinetic study of meat and bone meal pyrolysis: an evaluation and comparison of different possible kinetic models. J. Anal. Appl. Pyrolysis 74, 445–453.

Ayllon, M., Aznar, M., Sanchez, J.L., Gea, G., Arauzo, J., 2006. Influence of temperature and heating rate on the fixed bed pyrolysis of meat and bone meal. Chem. Eng. J. 121, 85–96.

Babu, B.V., Chaurasia, A.S., 2003. Modeling for pyrolysis of solid particle: kinetics and heat transfer effects. Energy Convers. Manag. 44, 2251–2275.

Baxter, L., 2005. Biomass-coal co-combustion: opportunity for affordable renewable energy. Fuel 2 (84), 2141–2162.

Beneroso, D., Bermudez, J.M., Arenillas, A., Menedez, J.A., 2014. Integrated microwave drying, pyrolysis and gasification valorization of organic wastes to syngas. Fuel 132, 20–26.

Blake, J.P., Carey, J.B., Haque, A.K.M., Malone, G.W., Patterson, P.H., Tablante, N.L., Zimmermann, N.G., 2008. Poultry carcass disposal options for routine and catastrophic mortality. Council Agric. Sci. Technol. Issue Paper 40, 1–20.

Bolton, D.C., Bendheim, P.E., 1998. A modified host protein model of scrapie. Ciba Found. Symp. 135, 164–181.

Bridgwater, A.V., 2003. Renewable fuels and chemicals by thermal processing of biomass. Chem. Eng. J. 91, 87–102.

Bridgwater, A.V., Bridge, S.A., 1991. A review of biomass pyrolysis and pyrolysis technologies. In: Bridgewater, A.V. et al., (Ed.), Biomass Pyrolysis Liquids Upgrading and Utilization. vol. 2. ECSC, EEC, EAEC, Brussels and Luxembourg, pp. 11–91.

Bridgwater, A.V., Meier, D., Radlein, D., 1999. An overview of fast pyrolysis of biomass. Org. Geochem. 30, 1479–1493.

Brunson, L.R., Sabatini, D.A., 2009. An evaluation of fish bone char as an appropriate arsenic and fluoride removal technology for emerging regions. Environ. Eng. Sci. 26 (12), 1777–1784.

Buessing, L., 2012. Energy From Olive Mill Waste: Pyrolysis and Oxidation Kinetics of Olive Mill Waste. Masters of Science Thesis, University of New Hampshire. UMI: 1518018.

Bujak, J.W., 2015. New insights into waste management: meat industry. Renew. Energy 83, 1174–1186.

Buzby, J., Hyman, J., 2012. Total and per capita value of food loss in the United States. Food Policy 37, 561–570.

Campoy, M., Gomez-Barea, A., Ollero, P., Nilsson, S., 2014. Gasification of wastes in a pilot fluidized bed gasifier. Fuel Process. Technol. 121, 63–69.

Cascarosa, E., Fonts, I., Mesa, J.M., Sanchez, J.L., Arauzo, J., 2011. Characterization of the liquid and solid products obtained from the oxidative pyrolysis of meat and bone meal in a pilot-scale fluidized bed plant. Fuel Process. Technol. 99, 1954–1962.

Cascarosa, E., Gea, G., Arauzo, J., 2012a. Thermochemical processing of meat and bone meal: a review. Renew. Sust. Energ. Rev. 16 (1), 942–957.

Cascarosa, E., Gasco, L., Garcia, G., Arauzo, J., 2012b. Meat and bone meal and coal co-gasification: environmental advantages. Resour. Conserv. Recycl. 59, 32–37.

Caton, P.A., Carr, M.A., Kim, S.S., Beautyman, M.J., 2010. Energy recovery from waste food by combustion or gasification with the potential for regenerative dehydration: a case study. Energy Convers. Manag. 51, 1157–1169.

Chaala, A., Roy, C., 2003. Recycling of meat and bone meal animal feed by vacuum pyrolysis. Environ. Sci. Technol. 37 (19), 4517–4522.

Chen, G., Andries, J., Spliethoff, H., 2003. Catalytic pyrolysis of biomass for hydrogen rich gas production. Energy Convers. Manag. 44, 2289–2296.

Collins, J.C., 2015. Thermal Decomposition and Gasification of Carbohydrates, Lipids, Proteins and Food Waste. Master of Science Thesis, Oklahoma State University.

Conesa, J.A., Fullana, A., Font, R., 2003. Thermal decomposition of meat and bone meal. J. Anal. Appl. Pyrolysis 70, 619–630.

Couhert, C., Commandre, J.M., Salvador, S., 2009. Is it possible to predict gas yields of any biomass after rapid pyrolysis at high temperature from its composition in cellulose, hemicellulose and lignin? Fuel 88 (3), 408–417.

Crapo, C., Bechtel, P., 2003. In: Bechtel, P. (Ed.), Utilization of Alaska's seafood processing byproducts. Advances in Seafood Byproducts, 2002 Conference Proceedings. Alaska Sea Grant College Program, Univ. of Alaska Fairbanks, Fairbanks, Alaska, pp. 105–119.

Das, P., Ganesh, A., 2003. Bio-oil from pyrolysis of cashew nut shell-a near fuel. Biomass Bioenergy 25, 113–117.

Demirbas, A., 2001. Yields of hydrogen-rich gaseous products via pyrolysis from selected biomass samples. Fuel 80, 1885–1891.

Demirbas, A., 2002. Gaseous products from biomass by pyrolysis and gasification: effect of catalyst on hydrogen yield. Energy Convers. Manag. 43, 897–909.

Demirbas, A., et al., 2004. Effects of temperature and particle size on bio-char yield from pyrolysis of agricultural residues. J. Anal. Appl. Pyrolysis 72, 243–248.

Devi, L., Ptasinski, K.J., Janssen, F.J.J.G., 2003. A review of the primary measures for tar elimination in biomass gasification processes. Biomass Bioenergy 24, 125–140.

Fadhil, A.B., Ahmed, A.I., Salih, H.A., 2017. Production of liquid fuels and activated carbons from fish waste. Fuel 187, 435–445.

FAO, 2015. FAO Rice Market Monitor. vol. XVIII (2) Food and Agriculture Organization of the United Nations.

FAO, 2016. The state of world fisheries and aquaculture. In: Contributing to Food Security and Nutrition for All. Rome. ISBN 978-92-5-109185-2. 200 pp.

Fedorowicz, E.M., Miller, S.F., Miller, B.G., 2007. Biomass gasification as a means of carcass and specified risk materials disposal and energy production in the beef rendering and meatpacking industries. Energy Fuel 21, 3225–3232.

Folador, J.F., Karr-Lilienthal, L.K., Parsons, C.M., Bauer, L.L., Utterback, P.L., Schasteen, C.S., Bechtel, P.J., Fahey Jr., G.C., 2006. Fish meals, fish components, and fish protein hydrolysates as potential ingredients in pet foods. J. Anim. Sci. 84, 2752–2765. https://doi.org/10.2527/jas.2005-560.

Freitas, S.R., Soria, J.A., 2011. Development of a gasification system for utilizing fish processing waste and coastal small diameter wood in rural areas. Energy Fuel 25, 2292–2300.

Fryda, L., Panopoulos, K., Vourliotis, P., Pavlidou, E., Kakaras, E., 2006. Experimental investigation of fluidized bed co-combustion of meat and bone meal with coals and olive bagasses. Fuel 85, 1685–1699.

Fryda, L., Panopoulos, K., Vourliotis, P., Kakaras, E., Pavlidou, E., 2007. Mean and bone meal as secondary fuel in fluidized bed combustion. Proc. Combust. Inst. 31, 2829–2837.

Galanakis, C.M., 2012. Recovery of high added-value components from food wastes: conventional, emerging technologies, and commercialized applications. Trends Food Sci. Technol. 26, 49–70.

Girotto, F., Alibardi, L., Cossu, R., 2015. Food waste generation and industrial uses: a review. Waste Manag. 45, 32–41.

Goldfarb, J.L., Buessing, L., Gunn, E., Lever, M., Billias, A., Casoliba, E., Schievano, A., Adani, F., 2017. Novel integrated biorefinery for olive mill waste management: utilization of secondary waste for water treatment. ACS Sustain. Chem. Eng. 5, 876 – 884.

Govender, N.S., van Vuuren, M.J., Claeys, M., van Steen, E., 2006. Importance of the usage ratio in iron-based fisher-tropsch sysnthesis with recycle. Ind. Eng. Chem. Res. 45 (25), 8629–8633.

Granatstein, D., Kruger, C., Collins, H., Garcia-Perez, M., Yoder, J., 2009. Use of biochar from the pyrolysis of waste organic material as a soil amendment. In: Washington State University, Sustaining Agriculture and Natural Resources. C0800248.

Grycova, B., Koutnik, I., Pryszcz, A., Kaloc, M., 2016a. Application of pyrolysis process in processing of mixed food wastes. Pol. J. Chem. Technol. 18 (1), 19–23.

Grycova, B., Koutnik, I., Pryszcz, A., 2016b. Pyrolysis process for the treatment of food waste. Bioresour. Technol. 218, 1203–1207.

Gulyurtlu, I., Boavida, D., Abelha, P., Lopes, M.H., Cabrita, I., 2005. Co-combustion of coal and meat and bone meal. Fuel 84, 2137–2148.

Gunders, D., August 2012. Wasted: How America Is Losing up to 40 Percent of its Food from Farm to Fork to Landfill. NRDC Issue Paper, IP: 12-06-B.

Guran, S., Agblevor, F.A., Brennan-Tonetta, M., 2018. Bio-Power, and Bio-Products from Sustainable Biomass: Coupling Energy Crops and Waste with Clean Energy Technologies. In: Chang, H.N. (Ed.), Wiley Biotechnology Series, https://doi.org/10.1002/9783527803293.ch8.

Hassen-Trabelsi, A.B., Kraiem, T., Naoui, S., Belayouni, H., 2014. Pyrolysis of waste animal fats in a fixed-bed reactor: production and characterization of bio-oil and bio-char. Waste Manag. 34, 210–218.

Heo, H.S., Park, H.J., Dong, J., Park, S.H., Kim, S., Suh, D.J., Suh, Y., Kim, S., Park, Y., 2010. Fast pyrolysis of rice husk under different reaction conditions. J. Ind. Eng. Chem. 16, 27–31.

Hernandez, A.M., Labady, M., Laine, J., 2014. Granular activated carbon from wood originated from tropical virgin Forest. Open J. For. 4, 208–211.

Heschel, W., Klose, E., 1995. On the suitability of agricultural byproducts for the manufacture of granular activated carbon. Fuel 74, 1786–1791.

Isa, K.M., Daud, S., Hamidin, N., Ismail, K., Saad, S.A., Kasim, F.H., 2011. Thermogravimetric analysis and the optimization of bio-oil yield from fixed-bed pyrolysis of rice husk using response surface methodology (RSM). Ind. Crop. Prod. 33, 481–487.

Ito, T., Sakurai, Y., Sugano, M., Hirano, K., 2012. Biodiesel production from waste animal fats using pyrolysis method. Fuel Process. Technol. 94, 47–52.

Johari, K., Saman, N., Song, S.T., Cheu, S.C., Kong, H., Mat, H., 2016. Development of coconut pith chars towards high elemental mercury adsorption performance-effect of pyrolysis temperatures. Chemosphere 156, 56–68.

Kim, M.H., Song, H.B., Song, Y., Jeong, I.T., Kim, J.W., 2013. Evaluation of food waste disposal options in terms of global warming and energy recovery: Korea. Int. J. Energy Environ. Eng. 4 (1), 12.

Ko, M.K., Lee, W., Kim, S., Lee, K., Chun, H., 2001. Gasification of food waste with steam in fluidized bed. Korean J. Chem. Eng. 18, 961–964.

Kothari, R., Tyagi, V.V., Pathak, A., 2010. Waste-to-energy: a way from renewable energy sources to sustainable development. Renew. Sust. Energ. Rev. 14, 3164–3170.

Kraiem, T., Hassen-Trabelsi, A.B., Naoui, S., Belayouni, H., Jeguirim, M., 2015. Characterization of the liquid products obtained from Tunisian waste fish fats using pyrolysis process. Fuel Process. Technol. 138, 404–412.

Lai, C., Ke, G., Chung, M., 2009. Potential of food wastes for power generation and energy conservation in Taiwan. Renew. Energy 34, 1913–1915.

Lange, S., Pellegrini, L.A., 2013. Economic analysis of a combined production of hydrogen-energy from empty fruit bunches. Biomass Bioenergy 59, 520–531.

Lede, J., 2013. Biomass fast pyrolysis reactors: a review of a few scientific challenges and of related recommended research topics. Oil Gas Sci. Technol. 68, 801–814.

Liu, H., Ma, X., Hu, Z., Guo, P., Jiang, Y., 2014. The catalytic pyrolysis of food waste by microwave heating. Bioresour. Technol. 166, 45–50.

Lu, Y.J., Li, S., Guo, L.J., Zhang, X.M., 2010. Hydrogen production by biomass gasification in supercritical water over Ni/γAl_2O_3 and Ni/CeO_2-γAl_2O_3 catalysts. Int. J. Hydrog. Energy 35, 7161–7168.

Luque, R., Manendez, J.A., Arenillas, A., Cot, J., 2012. Microwave-assited pyrolysis of biomass feedstocks: the way forward? Energy Environ. Sci. 5, 5481–5488.

Mahapatra, K., Ramteke, D.S., Paliwal, L.J., 2012. Production of activated carbon from sludge of food processing industry under controlled pyrolysis and its application for methylene blue removal. J. Anal. Appl. Pyrolysis 95, 79–86.

Manara, P., Vamvuka, D., Sfakiotakis, S., Vanderghem, C., Richel, A., Zabaniotou, A., 2015. Mediterranean agri-food processing wastes pyrolysis after pre-treatment and recovery of precursor materials: a TGA-based kinetic modeling study. Food Res. Int. 73, 44–51.

Marculescu, C., 2012. Comparative analysis on waste to energy conversion chains using thermal-chemical processes. Energy Procedia 18, 604–611.

Marculescu, C., Stan, C., 2012. Non-oxidant thermal treatment for organic waste neutralization. Energy Procedia 18, 545–551.

Matsakas, L., Gao, Q., Jansson, S., Rova, U., Christakopoulos, P., 2017. Green conversion of municpal solid wastes into fuels and chemicals. Electron. J. Biotechnol. 26, 69–83.

Miller, B.G., Falcone Miller, S., Fedorowicz, E.M., Harlan, D.W., Detwiler, L.A., Rossman, M.L., 2006. Pilot-scale fluidized-bed combustor testing co-firing animal tissue biomass with coal as a carcass disposal option. Energy Fuel 20, 1828–1835.

Miranda, R., Sosa, C., Bustos, D., Carillo, E., Rodriguez-Cantu, M., 2012. Characterization of pyrolysis products obtained during the preparation of bio-oil and activated carbon. In: Montoya, V.H. (Ed.), Lignocellulosic Precursors Used in the Synthesis of Activated Carbon-Characterization Techniques in the Wastewater treatment. InTech, London.

Mlilo, T.B., Brunson, L.R., Sabatini, D.A., 2010. Arsenic and fluoride removal using simple materials. J. Environ. Eng. ASCE 391–398.

Mohan, D., Pittman, C.U., Steele, P.H., 2006. Pyrolysis of wood/biomass for bio-oil: a critical review. Energy Fuel 20, 848–889.

Monteneiro, S.N., Calado, V., Margem, F.M., Rodriguez, R.J.S., 2012. Thermogravimetric stability behavior of less common lignocellulosic fibers—a review. J. Mater. Res. Technol. 1 (3), 189–199.

Morecraft, B., 2015. Growing Markets for U.S. Tree Nuts. https://www.usda.gov/oce/forum/2015_Speeches/BMorecraft.pdf.

Murugan, K., Simon, V., Chandrasekaran, V.S., Karthikeyan, P., Al-Sohaibani, S., 2013. Current state-of-the-art of food processing by-products, In: Chandrasekaran, M. (Ed.), Valorization of Food Processing By-products.pp. 35–59.

Naik, S.N., Goud, V.V., Rout, P.K., Dalai, A.K., 2010. Production of first and second generation biofuels: a comprehensive review. Renew. Sust. Energ. Rev. 14, 578–597.

Naveed, S., Malik, A., Ramzan, N., Akram, M., 2009. A comparative study of gasification of food waste (FW), poultry waste (PW), municipal solid waste (MSW) and used tires (UT). Nucleus 46 (3), 1–5.

Nigri, E.B., Bhatnagar, A., Rocha, S.D.F., 2017. Thermal regeneration process of bone char used in the fluoride removal from aqueous solution. J. Clean. Prod. *142*,pp, 3558–3570.

Onal, E., Uzun, B.U., Putun, A.E., 2012. An experimental study on bio-oil production from co-pyrolysis with potato skin and high-density polyethylene (HDPE). Fuel Process. Technol. 104, 365–370.

Onal, E., Uzun, B.U., Putun, A.E., 2014. Bio-oil production via co-pyrolysis of almond shell as biomass and high density polyethylene. Energy Convers. Manag. 78, 704–710.

Orfao, J.J.M., Antunes, F.J.A., Figueiredo, J.L., 1999. Pyrolysis kinetics of lignocellulosic materials—three independent reactions model. Fuel 78, 349–358.

Pala, L.P.R., Wang, Q., Kolb, G., Hessel, V., 2017. Steam gasification of biomass with subsequent adjustment using shift reaction for syngas production an aspen plus model. Renew. Energy 101, 484–492.

Pereira, E.G., Silva, J.N., Oliveira, J.L., Machado, C.S., 2012. Sustainable energy: a review of gasification technologies. Renew. Sust. Energ. Rev. 16, 4753–4762.

Peterson, D., Haase, S., 2009. Market Assessment of Biomass Gasification and Combustion Technology for Small- and Medium Scale Applications. Technical report. NREL/TP-7 A2-46190.

Pham, T.P.T., Kaushik, R., Parshetti, G.K., Mahmood, R., Balasubramanian, R., 2015. Food waste-to-energy conversion technologies: current status and future directions. Waste Manag. 38, 399–408.

Pinto, F., Andre, R., Miranda, M., Neves, D., Varela, F., Santos, J., 2016. Effect of gasification agent on co-gasification of rice production wastes mixtures. Fuel 180, 407–416.

Prevsuren, B., Avid, B., Gerelmaa, T., Davaajav, Y., Morgan, T.J., Herod, A.A., Kandiyoti, R., 2004. The characterization of tar from the pyrolysis of animal bones. Fuel 83, 799–805.

Priyadarsan, S., Annamalai, K., Sweeten, J.M., Holtzapple, M.T., Mukhtar, S., 2005. Co-gasification of blended coal with feedlot and chicken litter biomass. Proc. Combust. Inst. 30, 2973–2980.

Prusiner, S.B., 1982. Novel proteinaceous infectious particles cause scrapie. Science 216, 135–140.

Puig-Arnavat, M., Bruno, J.C., Coronas, A., 2010. Review and analysis of biomass gasification models. Renew. Sust. Energ. Rev. 14, 2841–2851.

Qu, W., Song, A., Shan, Y., Wang, R., 2015. Component analysis of odor components in food waste treatment. J. Environ. Prot. 6, 671–674.

Ramzan, N., Ashraf, A., Naveed, S., Malik, A., 2011. Simulation of hybrid biomass gasification using Aspen plus: a comparative performance analysis for food, municipal solid and poultry waste. Biomass Bioenergy 35, 3962–3969.

Rao, B., 2004. Gasification of Food Processing Byproducts—An Economic Waste Handling Alternative. Master of Science Thesis, Oklahoma State University.

Rao, B.R., Weckler, P., Bowser, T., 2004. In: Gasification of food and agricultural processing byproducts- an economic waste handling alternative.ASAE/CSAE Meeting Presentation. Paper Number: 046143.

ReFED, Rethink Food Waste through Economics and Data, 2016. A Roadmap to Reduce U.S. Food Waste by 20 Percent.

Ringer, M., Putsche, V., Scahill, J., 2006. Large-Scale Pyrolysis Oil Production: A Technology Assessment and Economic Analysis. Technical Report, National Renewable Energy Laboratory. NREL/TP-510-37779.

Rowland, S., Bower, C.K., Patil, K.N., DeWitt, C.A.M., 2009. Updraft gasification of salmon processing waste. J. Food Sci. 74 (8), 426–431.

Ruiz, J.A., Juarez, M.C., Morales, M.P., Munoz, P., Mendivil, M.A., 2013. Biomass gasification for electricity generation: review of current technology barriers. Renew. Sust. Energ. Rev. 18, 174–183.

Safari, F., Tavasoli, A., Ataci, A., 2017. Gasification of Iranian walnut shell as a bio-renewable resource for hydrogen-rich gas production using supercritical water technology. Ind. J. Ind. Chem. 8, 29–36.

Shafizadeh, F., 1985. Pyrolytic reactions and products of biomass. In: Overend, R.P., Milne, T.A., Mudge, L.K. (Eds.), Fundamentals of Thermochemicals Biomass Conversion. Applied Science, London.

Sharma, A.M., Kumar, A., Madihally, S., Whiteley, J.R., Huhnke, R.I., 2014. Prediction of biomass-generated syngas using extents of major reactions in a continuous stirred-tank reactor. Energy 72, 222–232.

Smith, J.G., 2008. Producing Fuel and Specialty Chemicals from the Slow Pyrolysis of Poultry DAF Skimmings. Thesis, Master of Science, University of Georgia, GA.

Soni, C.G., Wang, Z., Dalai, A.K., Pugsley, T., Fonstad, T., 2009. Hydrogen production via gasification of meat and bone meal in two-stage fixed bed reactor system. Fuel 88, 920–925.

Soni, C.G., Dalai, A.K., Pugsley, T., Fonstad, T., 2011. Steam gasification of meat and bone meal in a two-stage fixed-bed reactor system. Asia Pac. J. Chem. Eng. 6, 71–77.

Suresh, P.V., Prabhu, N., 2013. In: Chandrasekaran, M. (Ed.), Seafood, Valorization of Food Processing By-products. CRC Press, Taylor & Francis Group LLC, Boca Raton, FL. ISBN: 978-1-4398-4885-2, pp. 685–736.

Sutton, D., Kelleher, B., Ross, J.R.H., 2001. Review of literature on catalysts for biomass gasification. Fuel Process. Technol. 73, 155–173.

Thorin, E.V., Boer, E.D., Belous, O., Song, H., 2012. In: Waste to energy—a review.International Conference on Applied Energy, ICAE, Suzhou. China Paper ID: ICAE2012-A10544.

Tran, H.N., You, S.J., Chao, H.P., 2016. Effect of pyrolysis temperatures and times on the adsorption of cadmium onto orange peel derived biochar. Waste Manag. Res. 34 (2), 129–138.

Tsai, W.T., Lee, M.K., Chang, Y.M., 2006. Fast pyrolysis of rice straw, sugarcane bagasse and coconut shell in an induction-heating reactor. J. Anal. Appl. Pyrol. 76 (1–2), 230–237.

Tsai, W.T., Lee, M.K., Chang, Y.M., 2007. Fast pyrolysis of rice husk: product yields and compositions. Bioresour. Technol. 98, 22–28.

Tsamba, A.J., Yang, W., Blasiak, W., 2006. Pyrolysis characteristics and global kinetics of coconut and cashew nut shells. Fuel Process. Technol. 87, 523–530.

United Nations, 2014. World Urbanization Prospects Revision Report. ISBN: 978-92-1-151517-6.

United Nations, 2015. World Population Prospects—The 2015 Revision. ESA/P/WP.241, Department of Economic and Social Affairs, New York.

USDA, United States Department of Agriculture, Economic Research Service Economic Bulletin No. (EIB-16), July 2006. Agricultural Resources and Environmental Indicators. Chapter 2.1.

USEPA, 2014. Food Waste Management Scoping Study. Office of Conservation and Recovery.

Uzun, B.B., Apaydin-Varol, E., Ates, F., Ozbay, N., Putun, A.E., 2010. Synthetic fuel production from tea waste: characterization of bio-oil and bio-char. Fuel 89, 176–184.

Valin, H., Sands, R.D., Mensbrugghe, D.V., Nelson, G.C., Ahammad, H., Blanc, E., Bodirsky, B., Fujimori, S., Hasegawa, T., Havlik, P., Heyhoe, E., Kyle, P., D'Croz, D.M., Paltsev, S., Rolinski, S., Tabeau, A., Meijl, H.V., Lampe, M.V., Willenbockel, D., 2014. The future of food demand: understanding differences in global economic models. Agric. Econ. 45, 51–67. https://doi.org/10.1111/agec.12089.

van Steen, E., Claeys, M., 2008. Fisher-Tropsh catalysts for biomass-to-liquid (BTL)-process. Chem. Eng. Technol. 31 (5), 655–666.

Wang, L., Weller, C., Jones, D., Hanna, M., 2008. Contemporary issues in thermal gasification of biomass and its application to electricity and fuel production. Biomass Bioenergy 32, 573–581.

Warnecke, R., 2000. Gasification of biomass: comparison of fixed-bed and fluidized bed gasifier. Biomass Bioenergy 18 (6), 489–497.

Webber, M.E., 2012. More food less energy. Sci. Am. 306(1).

Wiggers, V.R., Wisniewski Jr., A., Madureira, L.A.S., Barros, A.A.C., Meier, H.F., 2009. Biofuels from waste fish oil pyrolysis: continuous production in a pilot plant. Fuel 88, 2135–2141.

Williams, P.T., Besler, S., 1993. The pyrolysis of rice husks in thermogravimetric analyzer and static batch reactor. Fuel 72 (2), 151–159.

Williams, P.T., Besler, S., 1996. The influence of temperature and heating rate on the slow pyrolysis of biomass. Renew. Energy 7, 233–250.

Williams, P.T., Nugranad, N., 2000. Comparison of products from the pyrolysis and catalytic pyrolysis of rice husks. Energy 25 (6), 493–513.

Wright, M.M., Satrio, J.A., Brown, R.C., Daugaard, D.E., Hsu, D.D., 2010. Techno-Economic Analysis of Biomass Fast Pyrolysis to Transportation Fuels. Technical Report, NREL/TP-6A20-46586.

Wu, C., Wang, L., Williams, P.T., Shi, J., Huang, J., 2011. Hydrogen production from biomass gasification with Ni/MCM-41 catalysts: influence of Ni content. Appl. Catal. B Environ. 108-109, 6–13.

Yang, Z., Koh, S.K., Ng, W.C., Lim, R.C.J., Tan, H.T.W., Tong, T.W., Dai, Y., Chong, C., Wang, C., 2016. Potential application of gasification to recycle food waste and rehabilitate acidic soil from secondary forests on degraded land in South Asia. J. Environ. Manag. 172, 40–48.

Yanik, J., Kornmayer, C., Saglam, M., Yüksel, M., 2007. Fast pyrolysis of agricultural wastes: characterization of pyrolysis products. Fuel Process. Technol. 88 (10), 942–947.

Yates, M., Gomez, M.R., Martin-Luengo, M.A., Ibanez, V.Z., Serrano, A.M.M., 2017. Multivalorization of apple pomace towards materials and chemicals. Waste to wealth. J. Clean. Prod. 143, 847–853.

Zhang, Y., Fan, D., Zheng, Y., 2016. Comparative study on combined co-pyrolysis of walnut shell and bituminous coal by conventional and congruent-mass thermogravimetric analysis (TGA) methods. Bioresour. Technol. 199, 382385.

Zheng, J.-L., 2007. Bio-oil from fast pyrolysis of rice husk: yields and related properties and improvement of the pyrolysis system. J. Anal. Appl. Pyrol. 80 (1), 30–35.

Zheng, J.-L., Yi, W.M., Wang, N.N., 2008. Bio-oil production from cotton stalk. Energy Convers. Manage 49 (6), 1724–1730.

Zwetsloot, M.J., Lehmann, J., Solomon, D., 2015. Recycling slaughterhouse waste into fertilizer: how do pyrolysis temperature and biomass additions affect phosphorus availability and chemistry. J. Sci. Food Agric. 95, 281–288.

FURTHER READING

Basu, P., 2010. Gasification theory and modelling gasifiers. In: Gasification Design Handbook. Academic Press, Boston.

de las Fuentes, L., Sanders, B., Lorenzo, A., Alber, S., 2004. AWARANET: agro-food wastes minimization and reduction network. In: Waldron, K., Faulds, C., Smith, A. (Eds.), Total Food Exploiting Co-products-Minimizing Waste. Institute of Food Research, Norwich, pp. 233–244.

Food Waste Management Scoping Study, 2014. USEPA Office of Conservation and Recovery.

Chapter 9

Sustainable Waste-to-Energy Technologies: Hydrothermal Liquefaction

Serpil Guran
Rutgers University EcoComplex "Clean Energy Innovation Center", Bordentown, NJ, United States

9.1 INTRODUCTION

Despite worldwide efforts on valorization of food waste that have received increased attention in recent years, pre- and postconsumer food wastes are currently not fully utilized. Food wastage is considered a serious social, economic, and environmental problem. Underutilized food waste represents a missed opportunity to mitigate climate change, produce cleaner chemicals and fuels, spur local economies, and generate jobs (FAO, 2013; Vigano et al., 2015). Currently, a large fraction of pre- and postconsumer organic waste either ends up in landfills and incinerators or is discarded into water bodies. According to recent data, approximately one-third of all food produced for human consumption is lost or wasted, equivalent to 1.6 billion tons per year. Approximately 54% of this loss is caused during food production steps, including postharvest handling and storage, and 46% of the loss occurs during the downstream steps such as processing, distribution, and consumption. On a weight basis, approximately 30% of cereals; 40%–50% of root crops, fruits, and vegetables; 20% of oilseeds, meat, and dairy products; and 35% of fish are lost (Vigano et al., 2015). To address current and future efficiency improvements in the food industry, increased amounts of preconsumer food waste (agricultural, food processing, distribution, and retail) and postconsumer food waste (institutions, restaurants, residential) must be productively utilized to achieve sustainable practices and a closed-loop bioeconomy. Because pre- and postconsumer food waste is essentially biomass, technologies designed for biomass conversion can help in valorizing food waste for energy and producing clean chemicals (Karmee, 2016).

Biomass feedstocks that do not follow food-to-fuel pathways and do not result in forest conversion, and/or land clearing for biomass production are considered as sustainable biomass. Sustainable biomass includes organic matter, that is, agricultural crop residues (i.e., straw, husks, corn cobs, leaves, and brunches), dedicated and noninvasive fuel crops (fast-growing grasses and trees such as aspen, poplar, willow, switchgrass), native vegetation, forest residues, animal manure (i.e., equine, dairy and poultry), aquatic species (i.e., algae, duckweed), and unrecycled biomass in waste streams (food waste, yard waste, unrecycled paper, card boards, untreated waste wood). Preconsumer food waste, including agricultural crop residues and food processing waste, and postconsumer food waste including cafeteria waste are also categorized as sustainable biomass that can serve as feedstock for conversion. Food waste has numerous valuable components such as carbohydrates, lipids, and proteins (Girotto et al., 2015; Bardhan et al., 2015). Food waste conversion into power, heat, fuels and bio-products varies based on the specific feedstock and is generally categorized into two major conversion pathways: biochemical and thermochemical. The primary thermochemical conversion pathways can be categorized as combustion, gasification, pyrolysis, and liquefaction. Gasification and pyrolysis technologies and their applications to food waste valorization were discussed in Chapter 8 and this chapter focuses on liquefaction, with a specific emphasis on hydrothermal liquefaction (HTL) of high moisture content food waste.

9.2 LIQUEFACTION TECHNOLOGIES AND CONVERSION MECHANISMS

Because biomass is the only carbon-containing renewable energy source, conversion of biomass into biofuels, bio-based chemicals, and bio-based products has long been considered and researched. Biomass, based on its physical and chemical properties, can be converted into fuels and chemicals by various conversion technologies. These technologies may be applied under various temperature and pressure conditions, and using various reactor design parameters. Gasification and pyrolysis-based conversion processes are usually conducted in the higher temperature regimes, about 800–1100°C

Sustainable Food Waste-to-Energy Systems. https://doi.org/10.1016/B978-0-12-811157-4.00009-7

(Alauddin et al., 2010; Ruiz et al., 2013a; Devi et al., 2003) and 400–700°C, respectively (Wright et al., 2010; Couhert et al., 2009; Ringer et al., 2006; Lede, 2013; Mohan et al., 2006). Higher temperature conversions can cause severe depolymerization and potential repolymerization reactions that may not be suitable for conversion of some biomass species into value-added end products. In some cases, high operating temperatures under pyrolysis conditions (i.e., zero oxygen) can cause cross-linking repolymerization reactions between hydrocarbons and aromatics, and consequently result in tar formation (Liu and Zhang, 2008). Tar formation is not desired because tars are more difficult to decompose and refine than the bio-oil product of the pyrolysis reactions. The presence of solvents in the liquefaction reactions prevents the occurrences of cross-linking reactions and reverse reactions. Also, lower temperature reaction conditions of liquefaction result in lower energy consumption compared to pyrolysis and gasification. Numerous catalysts have been tested for liquefaction, including alkali (alkaline oxides, carbonates, and bicarbonate), metal and metal compounds (zinc, copper, nickel formate, iodine, cobalt sulfide), and widely used heterogeneous catalysts such as nickel and ruthenium.

Liquefaction of biomass occurs in three main steps: (i) depolymerization of biomass into monomers; (ii) decomposition of monomers through cleavage, dehydration, decarboxylation, and deamination resulting in the formation of unstable and very active light fragments of small molecules; (iii) condensation, cyclization, and polymerization reactions resulting in recombination of light fragments into new compounds (Fig. 9.1) (Huang and Yuan, 2015; Zhang et al., 2010; Toor et al., 2011).

Biomass conversion via liquefaction can be tailored and optimized to produce biofuels and value-added chemicals if suitable liquefaction solvents along with the other reaction conditions (i.e., temperature, pressure, catalysts) are chosen. Liquefaction reactions are performed at lower temperatures (250–450°C) and higher pressure (5–20 MPa) than other thermochemical processes in the presence of either water or other solvent fluids as the reaction medium. Liquefaction conversions, in the presence of a fluid, control the reaction rate and reaction mechanisms by pressure, liquefaction medium, and catalysts to produce oil and in some cases char and gas (Huber et al., 2006). Liquefaction fluids represent high degradation ability at hydrothermal conditions, and their properties under hydrothermal conditions are suitable for biomass liquefaction. However, under extreme operating conditions, corrosion and scaling are major limitations of sub/supercritical fluid operations (Akhtar and Amin, 2011).

Liquefaction processes can be referred to as "hydrothermal liquefaction" when water or solvent-water mixtures are employed, "solvolysis" with solvents, and "hydro-pyrolysis" when no carrier liquid solvent is used (Huber et al., 2006). Liquefaction fluids are classified as supercritical when the temperature and pressure are both higher than the critical temperature (T_c) and critical pressure (P_c). Under higher T_c and P_c conditions, fluids are considered as neither liquid nor gaseous, and this phase is referred to as the supercritical fluid phase (Mazaheri et al., 2010a,b). Supercritical fluids possess both liquid- and gas-like properties. Liquid-like properties such as high density can promote solvation of compounds in the supercritical fluids, while gas-like properties such as high diffusivity and low viscosity can promote higher mass transfer since the reactants can more easily penetrate into the polymer structure of the biomass feedstock. Therefore, the liquefaction reaction rate is usually controlled by the rate of diffusion. Some supercritical fluid properties, including viscosity and dielectric constant, are functions of density which of course can be adjusted by changing temperature or pressure. It has been reported that small changes in pressure can alter the density of a fluid significantly, and consequently controlling density-dependent properties such as dielectric constant and thermal conductivity, and the associated reaction conditions, is possible by adjusting pressure or temperature of the fluid (Jessop, 2006; Mazaheri et al., 2010a,b). Brand et al. (2014) investigated the influence of heating and cooling rate on liquefaction of lignocellulosic biomass in subcritical water and supercritical ethanol and they identified heating rate as an important rate controlling parameter for subcritical water liquefaction, however, nonessential for supercritical ethanol liquefaction.

Solvents, based on their polarity, can be classified into three categories: (i) polar protic, (ii) dipolar aprotic, and (iii) nonpolar. Polar solvents are compounds with a hydrogen atom attached to an electronegative atom like oxygen, whereas dipolar solvents are compounds that do not contain O—H bond. Dipolar solvents contain a multiple bond between

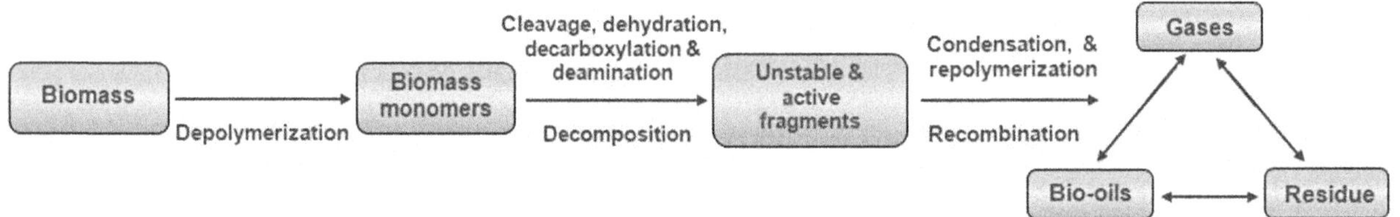

FIG. 9.1 Biomass liquefaction pathways. *(Adapted and revised from Huang, H., Yuan, X., 2015. Recent progress in the direct liquefaction of typical biomass. Prog. Energy Combust. Sci., 49, 59–80.)*

TABLE 9.1 Properties of Common Liquefaction Solvents (Mazaheri et al., 2010a,b; Huang and Yuan, 2015)

Solvent	Formula	T_c (°C)	P_c (MPa)	ρ_c (g/cm^3)	Polarity[a]	Dielectric Constant[b]
Water	H_2O	374	22.1	0.3320	100	79.7
Methanol	CH_4O	240	7.96	0.2720	76.2	32.6
Ethanol	C_2H_6O	243	6.39	0.2760	65.4	22.4
Acetone	C_3H_6O	235	4.8	0.2779	35.5	20.4
1,4-Dioxane	$C_4H_8O_2$	315	5.21	0.3702	16.4	2.21

[a]*Water as taken at 100°C.*
[b]*Measured at 20°C.*

carbon and either oxygen or nitrogen. Water and alcohols (R-OH) represent polar solvents while acetone (C_3H_6O) and acetic acid ($C_2H_4O_2$) can represent dipolar solvents. Solvents with low dielectric constants are defined as nonpolar solvents. These compounds contain bonds between atoms with similar electronegativities, such as carbon and hydrogen. Similar electronegativities will cause no partial charges that will make molecules nonpolar. These compounds are not miscible with water. Benzene (C_6H_6), toluene (C_6H_5-CH_3), and chloroform ($CHCl_3$) represent nonpolar solvents (Mazaheri et al., 2010a,b; Huang and Yuan, 2015; Liu and Zhang, 2008). Properties of common liquefaction solvents are listed in Table 9.1.

Biomass liquefaction with various solvents including water has been extensively researched. Review of the existing literature indicates that although liquefaction of biomass has been widely studied under various conditions, the findings generally cannot be directly compared because of the numerous types of biomass and solvents tested, and variation of other parameters including separation methods for the liquid products (Liu and Zhang, 2008; Karagoz et al., 2005). A comparison of the pros and cons of hydrothermal liquefaction and solvent liquefaction (solvolysis) is shown in Fig. 9.2 (Huang and Yuan, 2015).

Hydrothermal liquefaction (HTL) is a promising biomass conversion technology (Hardi et al., 2017). HTL is unique among thermochemical conversion methods in that it relies on water, with the primary goal of converting biomass with moisture content higher than 30%. High moisture-containing aquatic biomass like algae conversion have been also tested (Shuping et al., 2010; Ruiz et al., 2013b; Durak and Aysu, 2016; Yu et al., 2011; Liu et al., 2014; Dimitriadis and Bezergianni, 2017). Agriculture and food processing sectors of the food supply chain generate wastes that often contain large amounts of water, in some cases reaching 95% by mass (e.g., animal blood and dairy, cheese, and yogurt whey). The high moisture content of these food wastes make them well suited for conversion by HTL. The organic-phase products obtained can include high-value platform chemicals and fuels. Additionally, wastes from agriculture, food processing,

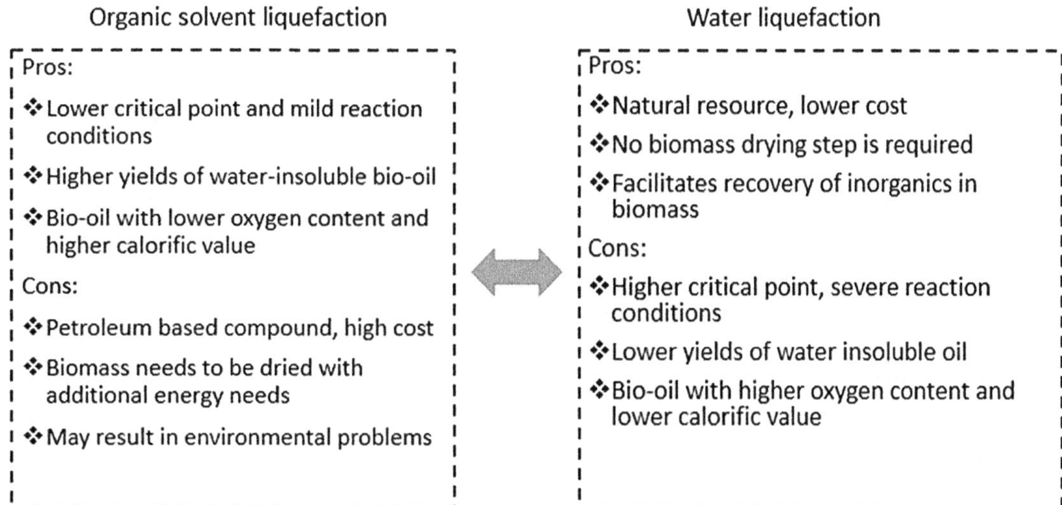

FIG. 9.2 Comparison of water and solvent liquefaction of biomass. *(Adapted and revised from Huang, H., Yuan, X., 2015. Recent progress in the direct liquefaction of typical biomass. Prog. Energy Combust. Sci., 49, 59–80.)*

TABLE 9.2 Comparison of Bio-Oils Produced From Via HTL and Fast Pyrolysis

	Hydrothermal Liquefaction	Fast Pyrolysis
Moisture content (wt%)	5	25
Elemental analysis (dry basis, wt%)		
C	77	58
H	8	6
O	12	36
Heating value (MJ kg^{-1})	37.5	22.6
Viscosity (cps)	15,000 @ 61C	59@ 40C

Adapted from Peterson, A.A., Vogel, F., Lachance, R.P., Froling, M., Antal, Jr., A.J., Tester, J.W., 2008, Thermochemical biofuel production in hydrothermal media: a review of sub- and supercritical water technologies. Energy Environ. Sci., 1, 32–65; Elliott, D.C., Schiefelbein, G.F., 1989. Liquid hydrocarbon fuels from biomass. Am. Chem. Soc., Div. Fuel Chem. Preprints 34 (4), 1160–1166.

and postconsumer sectors contain chemical compounds such as carbohydrates (sugar, cellulose, starch), lignin, proteins, oils, and fats that should be valorized in lieu of conventional disposal or treatment methods (Pavlovic et al., 2013; Gowthaman et al., 2012). Liu et al. (2014) also reported that mild acid and alkaline pretreatments of biomass were effective in increasing the bio-oil yield and reducing the liquefaction temperature.

As in other liquefaction processes, typical HTL processing conditions are in the temperature range of 250–375°C and operating pressures from 4 to 22 MPa. As compared to pyrolysis, HTL's low operating temperature allows energy efficient operations. In addition, HTL does not require predrying of biomass and results in lower tar than pyrolysis pathways (Tekin et al., 2014; Gollakota et al., 2018; Savage et al., 2010).

Previous research has indicated that bio-oils produced by HTL have much lower oxygen and moisture content and much higher C content than bio-oils produced via pyrolysis (Table 9.2). Typical HTL bio-oil has a heating value of about 37 MJ/kg as compared to pyrolysis bio-oil with a heating value generally in the range 22–23 MJ/kg. Zhu et al. (2014a) concluded that the HTL bio-oil heating value is more comparable to the heating value of 40–45 MJ/kg for petroleum-based fuels. Hydrothermal liquefaction of wet distiller's grain at subcritical conditions yielded similar oil qualities with a heating value of 36 MJ/kg. Over 75% of the product oil's content was in the boiling point range of diesel fuel (Toor et al., 2011).

Catalysts can enhance the hydrothermal liquefaction process efficiency, and thus suppress char formation while increasing oil yield and quality (Xu et al., 2014; Singh et al., 2013; Tortosa et al., 2014; Tu et al., 2016). A review of catalyst impacts on HTL bio-oils reported that alkaline solutions including Na_2CO_3, K_2CO_3, KOH, $Ca(OH)_2$, $Ba(OH)_2$, RbOH, and CsOH are widely employed and resulted in higher oil yields; see Table 9.3 (Xu et al., 2014; Karagoz et al., 2005).

Because pre- and postconsumer food waste is essentially biomass with a wide range of physical and chemical properties, understanding the fundamental mechanisms of biomass hydrothermal liquefaction will be beneficial in employing HTL to

TABLE 9.3 Summary of Catalysts Reported for Biomass HTL

Catalyst	Feedstock	T_c (°C)	Process Impact
K_2CO_3	Woody biomass	280	Reduced solid residue
Rb_2CO_3, Cs_2CO_3	Woody biomass	280	Increased oil yield
Na_2CO_3	Corn stalk	276–376	Increased oil yield
K_2CO_3	Woody biomass	280–360	Reduced solid residue
ZnCl2, Na_2CO_3, NaoH	Oil palm fruit fiber	210–330	Increased gas yields
Na_2CO_3, Ni	Cellulose	200–350	Char reduction
KOH and K_2CO_3	Wheat husk	280	Increased oil yield

Adapted and revised from Xu, C.C., Shao, Y., Yuan, Z., Cheng, S., Feng, S., Nazari, L., Tymchyshyn, M., 2014. Hydrothermal liquefaction of biomass in hot-compressed water, alcohols, alcohol-water co-solvents for biocrude production. In: Jin, F. (Ed.), Application of Hydrothermal Reactions to Biomass Conversion. Green Chemistry and Sustainable Technology.

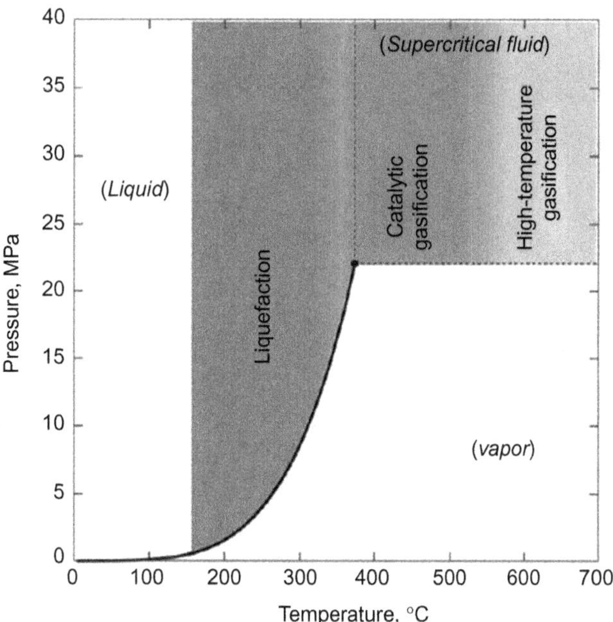

FIG. 9.3 Hydrothermal processing regions referenced to the pressure and temperature phase diagram of water. *(Adapted from Peterson, A.A., Vogel, F., Lachance, R.P., Froling, M., Antal, Jr., A.J., Tester, J.W., 2008, Thermochemical biofuel production in hydrothermal media: a review of sub- and supercritical water technologies. Energy Environ. Sci., 1, 32–65; Cansell, F., Beslin, P., Berdeu, B., 1998. Hydrothermal oxidation of model molecules and industrial wastes. Environ. Prog. 17 (4), 240–245.)*

valorize food waste. Biomass liquefaction processes are recognized as efficient pathways in reducing the oxygen content of the biomass that generally makes up 40%–50% by mass of the raw feedstock.

Hydrothermal processes are divided into two reaction conditions as subcritical and supercritical water conditions. The conditions are determined relative to the critical point conditions of water at 374°C and 22.1 MPa, and water behavior changes for each state. Subcritical and supercritical water have various advantages related to their properties. Water's solvent property can be changed as a function of temperature, and water can serve as an effective medium for acid-base-catalyzed organic reactions under critical conditions. It has been reported that the low viscosity of water is the rate controlling parameter, and with increased temperature lower viscosity is achieved, thus providing higher diffusion and reaction rates (Tekin et al., 2014) (Fig. 9.3).

Numerous complex reactions occur during the bio-oil formation due to the complex structure of biomass. Biomass components, including cellulose, hemicellulose, lignin, starch, proteins, fats, and oils, decompose under subcritical conditions resulting in various products. Prior research has indicated that the main hydrothermal liquefaction degradation steps are similar to other liquefaction reactions as shown in Fig. 9.1.

Generally, dehydration and decarboxylation reactions are the major reactions to remove the oxygen heteroatom from biomass structure, in the form of H_2O and CO_2, respectively (Akhtar and Amin, 2011). Removal of the amino acid content also may occur through deamination reactions. Removing oxygen as CO_2 is desirable since this will cause the oil H:C ratio to increase. The nitrogen heteroatom present in biomass also is converted into N_2O. Akhtar and Amin (2011) and Peterson et al. (2008) have reported that sulfur, chlorine, and phosphorus are oxidized to their respective inorganic acids (H_2SO_4, HCl, and H_3PO_4) that can be neutralized to salt by adding a suitable base. By the end of this series of reactions, complex chemicals may be synthesized by repolymerization of fragmented components. The end product bio-oil contains acids, alcohols, aldehydes, esters, ketones, phenols, and potentially other aromatic compounds (Gollakota et al., 2018; Chornet and Overend, 1985).

9.2.1 Conversion of Carbohydrates

Cellulose, hemicellulose, and starch are the most abundant carbohydrates in biomass and food waste. These carbohydrates are also classified as polysaccharides or sugar polymers that can be transformed into mono-sugars under hydrothermal treatment conditions (Toor et al., 2011; Brunner, 2009), followed by further reactions yielding important compounds such

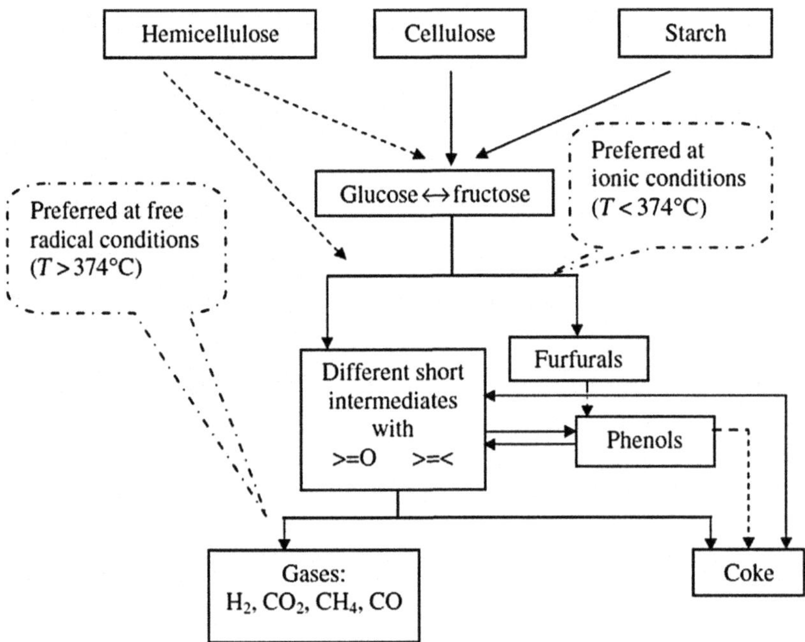

FIG. 9.4 Carbohydrate degradation at subcritical and supercritical conditions. *(Adapted from Toor, S.S., Rosendahl, L., Rudolf, A., 2011. Hydrothermal liquefaction of biomass: a review of subcritical water technologies. Energy, 36, 2328–2342.)*

as 5-hydroxymethylfurfural (5-HMF) that can serve as a building block for bio-based chemicals such as nylon 6 and nylon 6.6 (Fig. 9.4) (Toor et al., 2011; Kruse et al., 2005; Aida et al., 2007a).

One of the major components of biomass is cellulose, a polymer consisting of glucose units linked by β-(1→4) glycosidic bonds, and with a high crystallinity that makes cellulose insoluble in ambient water and resistant to enzyme conversions (Toor et al., 2011; Delmer and Amor, 1995). Based on the biomass species, cellulose may have both crystalline and chemical structure that can affect its decomposition behavior (Peterson et al., 2008). Subcritical water conversion of cellulose has been demonstrated to result in degradation to oligomer (i.e., cellobiose) and monomer sugars (i.e., glucose). Prior research has demonstrated that the crystallinity of cellulose disappears at 320°C and 25 MPa. The cellulose degradation reaction rate increases as the reaction temperature increases, until the supercritical water regime is reached, yielding 75% glucose. This research also demonstrated that cellulose destruction proceeds faster than glucose degradation at temperatures roughly at and above the critical point of water, indicating that high temperature and short processing time are needed to reform cellulose with maximized glucose yield (Peterson et al., 2008).

Monomer sugars such as glucose and fructose forming at high yields are important since dehydration of these compounds will yield formation of furfurals such as 5-hydroxymethylfuran (HMF) which is an important intermediary for production of levulinic acid and other chemicals (Fig. 9.5). Several prior studies concluded that dehydration of D-fructose yielded higher amounts of HMF than D-glucose; however, higher yields of furfural were observed from D-glusose than D-fructose (Brunner, 2009; Aida et al., 2007a,b).

The second important carbohydrate group present in biomass, including many food wastes, is hemicellulose which is composed of various monosaccharides such as xylose, mannose, glucose, and galactose. Hemicellulose composition, similar to cellulose, varies based on the type of biomass. However, due to the presence of side groups, it has less crystallinity than cellulose and can be solubilized and hydrolyzed in water at 180°C. Toor et al. (2011) reported glyceraldehyde, glycol aldehyde, and dihydroxyacetone as the main degradation products of HTL of hemicellulose.

Starch also is a main carbohydrate present in biomass species consisting of glucose monomers. Starches can be grouped in two different categories as amylose with a linear structure and amylopectin with a branched structure. Prior studies reported 5-hydroxymethylfurfural (HMF) as the main degradation product from HTL of starch and 1,6-anhydroglucose (Nagamori and Funazukuri, 2004; Toor et al., 2011). Sugars and starches are abundant in the food supply chain, especially as postconsumer waste. Therefore recovery of these sugars and converting them into value-added chemicals and potentially biofuels will help in developing a broad food waste valorization program (Table 9.4).

9.2.2 Conversion of Lignin

Lignin is the third main component of biomass with high molecular weight. Certain food supply chain by products may include biomass with high lignin concentration (e.g., nut shells) and hydrothermal conversion into acids, aldehydes,

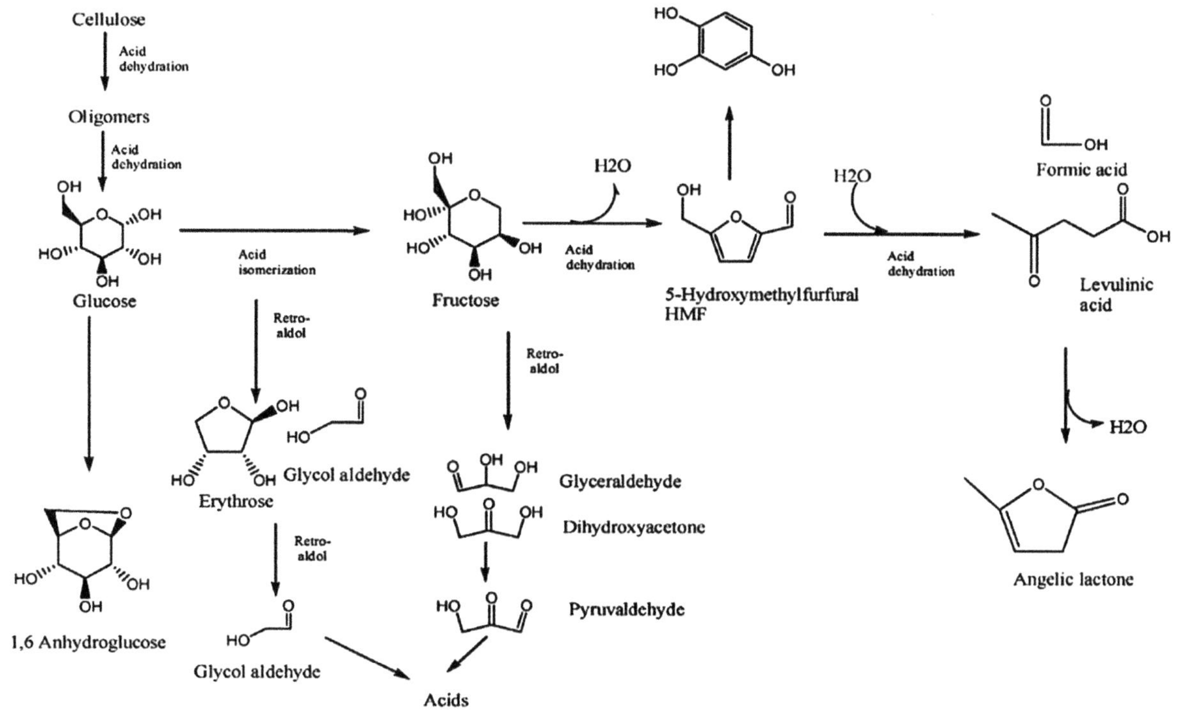

FIG. 9.5 Cellulose decomposition pathways in supercritical water. *(Adapted from Huber, G.W., Iborra, S., Corma, A., 2006. Synthesis of transportation fuels from biomass: chemistry, catalysts and engineering. Chem. Rev. 106, 4044–4098.)*

TABLE 9.4 Components of Agricultural and Food Wastes and Their Main Sources

Component	Main Sources
Cellulose	Cereals straw, wine shoots, rice husk
	Sunflower stalks, sugarcane bagasse, corn stalks, and cobs
Hemicellulose	Sugarcane bagasse, corn cobs, sunflower seed hulls, rice husk
Starch	Potato, cereal grains
Lignin	Coconut shell, walnut shell, olive pits
Lipids & fats	Oilseed cakes, slaughterhouse waste, meat food waste (poultry, beef), used vegetable oil.
Proteins	Meat waste (blood and fats), fish waste, oil seeds

Adapted and revised from Pavlovic, I., Knez, Z., Skerget, M., 2013. Hydrothermal reactions of agricultural and food processing wastes in sub- and supercritical water: a review of fundamentals, mechanisms, and state of research. J. Agric. Food Chem. 61, 8003–8025.

alcohols, and phenols is important for valorization of food waste (Fang et al., 2008; Peterson et al., 2008). It has a highly stable polymer structure built from three cross-linked phenylpropane (C_6-C_3) units of p-coumaryl, coniferyl, and sinapyl alcohol that are bonded with ether (C—O—C) and C—C bonds. Its structure is more heterogeneous than hemicellulose. Research has demonstrated that increased water density will increase the degradation into monomers and improve oil and gas yields. Supercritical water (SCW; 374.2°C and 22.1 MPa), as a weak polar solvent with a high value of ion product, will serve as a solvent that can dissolve and hydrolyze lignin for potential production of phenolic chemicals, or for upgrading lignin to fuels. Experimental research also concluded that after initial dissolution, at above 377°C, lignin undergoes hydrolysis and pyrolysis to phenolics, which were further changed to oil in the aqueous phase (Fang et al., 2008).

9.2.3 Conversion of Oils, Fats, and Proteins

Vegetable oils and fats are insoluble in water, but with increased temperature at supercritical conditions they become miscible with water. During HTL conversion, lipids hydrolyze to fatty acids and glycerol which further decomposes (without catalyst) into acetaldehyde, acrolein allyl alcohol and with catalysts, into alkanes (Pavlovic et al., 2013). Fatty acids have been reported to be stable below 300°C, and above this temperature decarboxylation of fatty acids generates alkanes, alkenes, or ketones (Deniel et al., 2016a).

Food waste including beef, poultry, swine, and seafood by-products consists of important levels of proteins. Proteins are combinations of amino acids connected with peptide bond which provide linkage between the carbonyl and amine groups. Hydrothermal liquefaction can serve an important pathway to convert proteins to amino acids as a precursor for bio-oil and chemical production (Pavlovic et al., 2013; Sato et al., 2004).

9.3 HYDROTHERMAL LIQUEFACTION OF SOURCE-SPECIFIC FOOD WASTES AND RESIDUES

In Chapter 8, we summarized the major food processing pathways that generate residues and wastes that can further be valorized via pyrolysis and gasification. Food processing industry (FPI) residues and wastes that are plant and animal-based can also potentially be converted via hydrothermal liquefaction, as described below.

9.3.1 Plant-Based FPI Waste Valorization With HTL

Plant-based FPI waste includes crop harvesting residues from the field (i.e., vegetables, grains residues, and some fruits), and trees (fruits), processing residues after the food is prepared (i.e., pulp and pomaces, peelings, seeds), and distribution and warehouse residues including lesser quality foods. Recent concerns regarding food-to-fuel pathways of first-generation biofuels, especially corn ethanol, diverted interests into lignocellulosic fuels including the recovery of sugars in agricultural residues such as corn stalks, corn stover, corn cob, and sugarcane bagasse. The main focus of prior research has been on achieving the successful release of sugars locked in the cell walls of biomass in order to produce lignocellulosic ethanol. In addition to acid and enzyme hydrolysis conversions, hot pressurized water at sub- and supercritical conditions has also been used to recover fermentable sugars. A recent comprehensive review demonstrated that the high cellulose and hemicellulose content of sugarcane bagasse makes it a great source of fermentable sugars via subcritical water treatment (Table 9.5). Corn stover liquefaction showed better liquid yield at lower temperatures (Zhang et al., 2008) and hydrothermal conversion of corn stalks demonstrated that catalysts greatly influenced the efficiency of the conversion process (Peng et al., 2014). Recommendations for process optimization have been provided by a number of studies (Prado et al., 2016; Lachos-Perez et al., 2016; Ruiz et al., 2013b; Allen et al., 1996; Capecchi et al., 2015; Sasaki et al., 2003; Faba et al., 2015).

Literature studies reporting experimental HTL conversion of plant sourced food waste residues are summarized in Table 9.6. Hydrothermal liquefaction of wheat straw with water and water-ethanol mixtures revealed that wheat straw could be effectively converted into value-added liquid products (Patil et al., 2014; Wang et al., 2012). Considering that world

TABLE 9.5 Composition of Selected Agricultural Food Residues

Agricultural Residues	Cellulose (%)	Hemicellulose (%)	Lignin (%)
Corn cobs	38.8–44	33–36.4	13.1–18
Corn stover	34.3–36.5	20–31.3	11.9–13.6
Wheat straw	33–40	20–33.8	15–26.8
Rice straw	35–36.6	16.1–22	12–14.9
Sugarcane bagasse	34.1–49	15.8–29.6	19.4–27.2
Barley straw	37.5	25.1–37.1	15.8–16.9
Rice husk	33.43	20.99	18.25
Rye straw	41.1–42.1	23.8–24.4	19.5–22.9
Rapeseed straw	36.6–37	19.6–24.2	15.5–18
Sunflower stalks	33.8	20.2–24.3	14.6–19.9
Sweet sorghum bagasse	41.3–45.3	22–26.3	15.2–16

Adapted and revised from Ruiz, H.A., Rodriguez-Jasso, R.M., Fernandes, B.D., Vicente, A.A., Teixeira, J.A., 2013. Hydrothermal processing, as an alternative for upgrading agriculture residues and marine biomass according to the biorefinery concept: a review. Renew. Sust. Energ. Rev. 21, 35–51.

TABLE 9.6 Summary of Literature Studies Reporting Experimental HTL Conversion of Plant-Sourced Food Waste and Residues

Waste Type	Temp (°C)	Reference
Barley straw	280–400	Zhu et al., 2014b
Blackcurrant pomace	290–335	Deniel et al., 2016b
	220–300	Mackela et al., 2015
Cherry stone	200–300	Akalin et al., 2012
Coffee grounds	200–300	Yang et al., 2016
Coriander seeds	40–70	Zekovic et al., 2017
Corn stalk, stover	350–400	Zhang et al., 2008
	200–400	Peng et al., 2014
	120–200	Tu et al., 2016
Olive mill waste	240–300	Hadhoum et al., 2016
Palm fruit fiber	270	Akhtar et al., 2010
	210–330	Mazaheri et al., 2010a,b
Peanut shells, meal	120–200	Tu et al., 2016
Rice straw	280	Karagoz et al., 2005
	230–270	Phaiboonsilpa et al., 2013
	200–300	Younas et al., 2017
Rice, potato starch	50–140	Li et al., 2016
Sugarcane bagasse	100–250	Lachos-Perez et al., 2016
	195–205	Capecchi et al., 2015
	190–230	Allen et al., 1996
	200–230	Sasaki et al., 2003
Wheat straw	350	Patil et al., 2014
	200	Chandra et al., 2012
	280	Singh et al., 2013
	300–374	Wang et al., 2012

wheat production is about 740 millions tonnes/year (FAO, 2017), the agricultural residues of wheat farming appear to be an important feedstock if they can be efficiently converted into fuels and chemicals. Wang et al. (2012) tested the impact of reaction temperature and residence time between 300 and 374°C and 1–16 min, respectively. The research reported the optimum conditions as 340°C reaction temperature and 5-min residence time. Patil et al. (2014) demonstrated that HTL with a water-ethanol mixture resulted in higher oil yields with higher heating values when Ru/H-Beta catalysts were employed during the liquefaction process. The study reported two types of bio-oils as higher calorific value oil (HCO) and lower calorific value oil (LCO). Oil analyses revealed the presence of furan derivatives and phenol-based compounds in HCO, and carboxylic acids and carbohydrate-based products in LCO. These oils have characteristics that make them suitable to be refined into bio-based products such as fuels or chemicals. Wheat straw and husks consist of 44.5% cellulose, 24.3% hemicellulose, and 21.3% lignin (Toor et al., 2011). Catalytic hydrothermal liquefaction of this material at 280°C with alkaline catalysts (KOH and K_2CO_3) resulted in approximately 31 wt% bio-oil. Analysis of the oil confirmed that all the components of wheat husk including lignin experienced thermal degradation to yield low molecular weight fragments such as syringols, guaiacols, catechols, and phenolics (Singh et al., 2013).

Another important food product residue is barley straw, which is rich in cellulose, hemicellulose, and lignin with typical content of 46%, 23%, and 15%, respectively (Sander, 1997; Zhu et al., 2014b). Sub- and supercritical water liquefaction of barley straw with alkaline catalysts (K_2CO_3) demonstrated that lower temperature (300°C) conversion favored higher bio-oil yields. Bio-oil analysis revealed the presence of carboxylic acids, phenolic compounds, ketones, and aldehydes.

Rice is one of the main food sources around the world, especially in Asia (Liu et al., 2011). Worldwide annual rice production has reached 700 million tonnes. This level of production yields approximately 100 million tonnes of rice husk yearly. Some rice-producing countries allow open burning or landfilling to address the disposal problems (Pinto et al., 2016; FAO, 2015). Due to high energy content and moderate lignin fraction, rice straw can be valorized after the edible part is removed. Research on hydrothermal liquefaction of rice residues at maximum temperature of 300–350°C and pressure of 16 MPa revealed that NiO nano-catalysts significantly influence oil recovery, with 52.8% of carbon recovery and 46.6% of energy recovery (Younas et al., 2017). The oil analyses also showed the presence of 5-HMF in high concentration when produced at temperatures lower than 260°C. These results confirmed the findings of similar studies conducted at lower temperature with 47% carbon recovery from rice husk liquefaction (Karagoz et al., 2005). A review paper also reported several studies of rice husk treatment via sub- and supercritical water hydrolysis (SCWH) conversion to recover sugars as monosaccharides and oligosaccharides. Two-step SCWH of rice husk resulted in 96.1% liquefaction efficiency and it was reported that the liquid product's sugar content was 36.3% (Vardanega et al., 2015; Phaiboonsilpa et al., 2013).

Various studies have reported food processing industry by-products that have been hydrothermally liquefied. Blackcurrant pomace, which includes seeds, peels, and pulp, was converted via HTL with bio-oil yield of 24%–31% at temperatures of 290–335°C (Deniel et al., 2016b). Alkaline catalysts resulted in higher oil yield. Research with peanut shells and deoiled peanut meal with NaOH and KOH catalysts resulted in higher oil yields as compared to similar processes with Na_2CO_3 and K_2CO_3 catalysts (Tu et al., 2016). However, liquefaction of empty palm fruit under subcritical conditions with K_2CO_3 resulted in higher oil yields than the same liquefaction process with KOH (Akhtar et al., 2010). Subcritical liquefaction of olive mill waste also demonstrated that KOH concentration significantly affects the extraction performance of resulting polyphenols and carbohydrates (Tortosa et al., 2014). A similar study with olive mill wastewater hydrothermal liquefaction tested temperatures from 240 to 300°C and observed the highest oil yield of 58% at 280°C. The bio-oil higher heating value (HHV) was 38 MJ/kg and analyses showed the presence of saturated and unsaturated fatty acids and phenolic compounds (Hadhoum et al., 2016).

Hydrothermal liquefaction of palm fruit press fiber with alkaline catalysts resulted in higher oil yields than other catalysts such as zinc. Higher temperatures achieved solid conversions of almost 90% but did not produce higher oil yields due to cracking reactions that led to more gaseous products (Akhtar et al., 2010). Research with cornelian cherry stone HTL at 300°C demonstrated a resulting bio-oil yield of 28%, with the presence of furfurals, phenols, acetic acid, vanillin, and fatty acids (Akalin et al., 2012). Hydrothermal liquefaction of rice, potato, and sweet potato starches yielded best yields of bio-oils as 15% at 300°C, 30% at 260°C, and 33% at 200°C, respectively (Li et al., 2016). Spent coffee grounds were converted via HTL at subcritical conditions of 275°C, yielding 47.3% bio-oil with HHV of 31 MJ/kg (Yang et al., 2016).

9.3.2 Animal-Based FPI Residue Valorization With HTL

The United States produces 4.28 billion pounds of red meat and 4.23 billion pounds of poultry meat per month (USDA, 2017). Slaughterhouses and processing facilities for cattle, swine, poultry, and fish industries generate large amounts of by-product wastes that constitute the inedible parts of animals derived from the production of meat, as well as blood and other animal by-products. Inedible animal tissues (organs, integument, ligaments, tendons, blood vessels, feathers, bone) can comprise up to 45% or more of the slaughtered animal (Franke-Whittle and Insam, 2013; Vigano et al., 2015). Efficient valorization of these wastes can help in the creation of closed-loop bioeconomy systems across the food industry.

Literature studies reporting experimental HTL conversion of animal sourced food waste residues are summarized in Table 9.7. Hydrothermal liquefaction of swine carcasses with alkaline catalysts at temperatures from 150 to 400°C resulted in the highest bio-oil yield of 62.2% at 250°C. Swine carcasses were converted to bio-oil that could be upgraded by catalytic cracking or hydrodeoxygenation for production of liquid transport fuels and valuable chemicals, such as hydroxymethylfurfural (HMF) and levulinic acid. Also, carcasses with high fatty meat residues were converted to produce bio-oils with HHV of 39.3 MJ/kg as compared to bio-oils from carcasses with nonfatty meat residues with HHV of 8.8 MJ/kg (Zheng et al., 2015).

Another animal-based biomass study examined hydrothermal decomposition of shrimp shells which contain approximately 42% protein (Shadidi, 1995; Quitain et al., 2001). Proteins can be hydrolyzed to form amino acids which have wide uses in pharmaceuticals, food products, animal nutrition, and cosmetic industries (Quitain et al., 2001). Research observed

TABLE 9.7 Summary of Literature Studies Reporting Experimental HTL Conversion of Animal Sourced Food Waste and Residues

Waste Type	Temp (°C)	Reference
Shellfish	60	Shadidi, 1995
Shrimp shellfish	100–250	Quitain et al., 2001
Swine Carcass	250	Zheng et al., 2015

the effect of temperature on the yield of amino acids. Amino acid yields at 250°C were about 2.5 times higher than the yields at 100°C (Quitain et al., 2001). Hydrothermal liquefaction of high moisture animal-based by-products still needs to be explored further to understand their fundamental reaction pathways and potential end-product characteristics.

9.3.3 Mixed Food Waste Valorization With HTL

Hydrothermal liquefaction research conducted with model compounds provides important reaction pathway information; however, interactions of these compounds in real food waste should also be well understood to successfully commercialize postconsumer food waste valorization facilities. Literature studies reporting experimental HTL conversion of mixed food waste residues and model compounds are summarized in Table 9.8. Research with three model compounds of potato starch ($C_6H_{10}O_5n$), bovine serum albumin (BSA), and linoleic acid ($C_{18}H_{32}O_2$) was chosen to represent carbohydrates, proteins, and vegetable oils, respectively. In addition, their binary and ternary mixtures were converted at subcritical conditions of 250–350°C and 5–20 MPa. This study reported that HTL of the ternary mixture produced bio-oil at a maximum yield of 67% and calorific content of 38 MJ/kg. For the binary mixtures, lipid and starch combination increased the oil yield as compared to starch liquefaction alone. The results demonstrated that food waste is a complex mixture of sugars, proteins, and lipids, and hydrothermal decomposition will be highly dependent on the specific properties of the mixture (Posmanik et al., 2017a).

TABLE 9.8 Summary of Literature Studies Reporting Experimental HTL Conversion of Mixed Food Waste and Residues and Model Compounds

Waste Type	Temp (°C)	Reference
D-Glucose	350–400	Aida et al., 2007a;
		Aida et al., 2007b
Cellulose	374	Cansell et al., 1998
	250	Pala et al., 2014
Lignin	400	Fang et al., 2008
	280	Karagoz et al., 2005
Starch, albumin, acid	250–350	Posmanik et al., 2017a
Mixed waste	350	Elliott and Schiefelbein, 1989
	330	Hammerschmidt et al., 2011
	250	Kaushik et al., 2014
	150–350	Mahmood et al., 2016
	250	Parshetti et al., 2014
	120–200	Tu et al., 2016
	250–315	Zastrow and Jennings, 2013

Mixed waste comprised of American cheese (12.8%), chicken breast (14.9%), brown gravy (2.1%), mashed potatoes (10.6%), green beans (14.9%), white rice (19.1%), apple desert (22.3%), and butter (3.2%) was converted via HTL at subcritical conditions of 250, 280, and 315°C. This research also tested model compounds of starch, casein, and soybean oil. The highest bio-oil yield of 45% was observed with HTL of the food mixture at 315°C with sodium carbonate catalyst. At the same reaction conditions, model compounds of starch and casein yielded 25% and 14%, respectively. Soybean oil did not provide any significant conversion at this temperature (Zastrow and Jennings, 2013). A study was also performed to investigate hydrothermal oxidation experiments of mixed food waste consisting of chicken, seafood, potato chips, vegetables, and bread at subcritical conditions of 150, 250, and 350°C, with and without enzymes, and concluded that increasing temperatures and presence of enzymes positively influenced the bio-oil yields. The study also performed techno-economic analysis (TEA) of the decomposition pathways and identified three key factors that would affect the profitability of the future commercial operations: bio-oil yield, cost of enzymes, and selling price of bio-oil (Mahmood et al., 2016). Another "real" food waste mixture study considered enzyme treatment of the mixture before liquefaction. The mixture included a variety of cooked foods (chicken, seafood, potato fries, vegetables, rice, and gravy), uncooked food (fruit peels, parts of vegetables), and condiments (salad dressing, ketchup, and cocktail sauce). The study concluded that enzymatic pretreatment was able to selectively generate specific products from food waste and narrow the distribution of compounds such as 5-HMF at 150 and 250°C (Kaushik et al., 2014).

Mixed food waste hydrothermal liquefaction was also tested for unspecified food supply chain sludge. This research utilized both homogeneous (K_2CO_3) and heterogeneous catalysts (ZrO_2) at the same conditions during HTL treatment. Feed concentrations varied as 12% dry matter with organic carbon content of 51%, 6.47% dry matter with organic carbon content of 36%, and 7.7% dry matter with organic carbon content of 6%. The HTL treatments were carried out at 330°C and 25 MPa. The results demonstrated that conversion of organic matter in aqueous waste streams into light oil is feasible; however, optimization of the process was also recommended (Hammerschmidt et al., 2011).

In another related study, food waste hydrothermal carbonization (HTC) was carried out at 250°C to valorize the waste as biochar for textile industry wastewater cleaning. The biochars were tested to remove textile dyes from contaminated waste water. The findings concluded that food waste biochars can serve as effective adsorbents for removal of dyes from wastewaters (Parshetti et al., 2014). Grape pomace as vinery waste was hydrothermally carbonized over a temperature range of 175–275°C. HTC yielded a higher energy densification and energy yield than torrefaction. IR spectra of the hydrochars thus produced demonstrated that decomposition of hemicelluloses occurred at lower temperatures (175°C), whereas cellulose decomposition began at temperatures higher than 250°C (Pala et al., 2014). Treatment of mixed food waste with a combination of enzymes prior to hydrothermal carbonization improved the quality of solid hydrochars, with higher carbon contents and calorific values (Kaushik et al., 2014).

9.3.4 Valorization of Aqueous Phase Food Waste Via HTL

In addition to bio-oil production, hydrothermal liquefaction processes create considerable amounts of aqueous phase by-products. The design of HTL processes should therefore include reuse and recycle of this by-product. The aqueous phase (AP) consists of a significant fraction of the organic carbon, varying from 10% to 40% depending on the process conditions and feedstock properties (Van Doren et al., 2017; Tommaso et al., 2015). Several prior research efforts attempted to valorize the aqueous phase by various approaches. Some research considered utilization of the aqueous phase for algae cultivation. A recent review paper summarized algae production trials conducted utilizing HTL of the aqueous phase (Elliott et al., 2015). The level of potentially toxic compounds in the process water was also shown to be significantly reduced following cultivation of the algae (Pham et al., 2013).

HTL aqueous phase valorization has also been tested by catalytic hydrothermal gasification (CHG); Elliott et al., 2013, 2015. This process converts the organic material in the aqueous by-product into methane, which can be easily separated from the water, and carbon dioxide gas, some of which passes to the gas phase and much of which remains dissolved in the water along with ammonia generated from any nitrogen-containing organic by-products (Elliott et al., 2015). CHG was also performed at low temperature of 350°C and 21 MPa with metal catalyst, and the gas thus produced can be used in combined heat and power (CHP) applications.

Anaerobic digestion (AD) of the aqueous phase of HTL appears to be another viable pathway to valorize the residual carbon in the aqueous phase. Recent research combined HTL and AD processes to valorize food waste, and this application can increase the energy product recovery. This study concluded that the total energy balance of this coupled technology approach represented trade-off between oil and biomethane production. Higher HTL temperatures favored oil production (44%–52% of COD) with lower methane production. This research also concluded that reduced biomethane production is potentially due to the formation of recalcitrant or other inhibitory products, and their concentrations depend on the reaction temperature, feedstock composition, and the pH of the HTL media (Posmanik et al., 2017b). Another HTL/AD combination study considered hydrothermally liquefying municipal wastes, food industry wastes (grape pomaces and sugar beet tailings), and algae cultivated in wastewater. The study concluded that the biogenic carbon concentration of AP is dependent on the feedstock (Maddi et al., 2017).

Techno-economic analyses of AP valorization via AD or CHG have revealed that HTL of algae by-products can be effectively valorized. Integrating anaerobic digestion with combined heat and power to HTL resulted in better economic results with lower capital and operational costs as compared to CHG, because the latter requires expensive heat-resistant materials, large heat exchangers, and high maintenance (Van Doren et al., 2017). Similar studies with food waste would provide detailed understanding of potential process and maintenance costs.

9.3.5 Upgrading Bio-Oil From Food Waste HTL

The bio-oil fraction produced from food waste hydrothermal liquefaction processes can be considered a renewable fuel or bio-based chemical source. The bio-oil appears to be a complex and chemically unstable mixture, and consists of oxygen that needs to be removed. Moisture content of the bio-oil also represents a concern by reducing its energy value, causing potential corrosion problems, and favoring microbial growth activity during long-term storage. Catalytic upgrading involves contacting the bio-oil with hydrogen under elevated pressure and high temperature to remove oxygen and reduce the molecular weight via hydrodeoxygenation, hydrodenitrogenation, and hydrodesulfurization reactions (Zhu et al., 2014a; Elliott, 2007; Deniel et al., 2016a). Heterogeneous catalysts such as CoMo, NiMo, Pt, Pd, and Ru supported on alumina can be used for hydrogenation up to 400°C and 20 MPa (Deniel et al., 2016a,b).

9.4 FUTURE PERSPECTIVES AND RESEARCH NEEDS

Because valorization of organic waste, particularly pre- and postconsumer food waste, is receiving increased attention to achieve closed-loop, low-carbon economy, successful commercialization of conversion technologies is essential. Although HTL of biomass has been researched extensively, HTL of food waste is still emerging to produce bioenergy and biochemicals. Due to the complex structure of food waste, especially postconsumer mixed food waste, it is essential to understand the decomposition pathways for specific feedstocks and components and their behaviors throughout the HTL reactions. Achieving successful, repeatable conversion pathways for a given feedstock and understanding their responses to critical reaction conditions such as temperature, pressure, residence time, feedstock moisture content, and catalyst type, is critically important. Also, understanding not only the behavior of model compounds found in organic waste but also discovering their interactions in "real" food waste is a key to achieve successful commercialization, and further research will be useful. Extensive research is also needed, especially for animal waste and mixed food waste, on coupling HTL with bio-oil upgrading and valorization of aqueous fraction of liquid yield. In addition, research on successful scale-up of food waste HTL should be supported by theoretical and computational modeling efforts to forecast specific yields at particular reaction conditions.

9.5 CONCLUSIONS

Hydrothermal liquefaction (HTL) technology appears to be a viable emerging valorization pathway for high moisture pre- and postconsumer food waste. HTL technology usually is performed at a moderate temperature < 400°C and high pressure up to 15–20 MPa. Generally, catalysts improve the reaction efficiency by suppressing char formation and increasing bio-oil yield which has lower oxygen content and higher carbon content, and consequently higher calorific vale as compared to pyrolysis technology. Numerous complex reactions occur during the bio-oil formation due to the complex structure of organic waste which generally consists of cellulose, hemicellulose, lignin, starch, proteins, fats, and oils. These compounds decompose under subcritical conditions, resulting in various end products. Cellulose, hemicellulose, and starch are the most abundant carbohydrates in food waste. The rate of degradation of these carbohydrates was reported to increase with reaction temperature and increases until the supercritical water regime is reached, yielding 75% glucose. Glucose and other monomer sugars such as fructose will consequently form furfurals such as 5-hydroxymethylfuran which is an important intermediary for production of chemical industry. In addition, conversion of fats and oils yields fatty acids which generate alkanes, alkenes, and ketones, and conversion of proteins yields amino acids which are also intermediaries for the chemical industry. Although HTL of the model compounds found in organic waste provides important reaction pathways, interactions of these compounds in "real" food waste should also be well understood to commercialize postconsumer food waste valorization facilities. The bio-oil fraction of the food waste HTL can be considered a renewable fuel or bio-based chemical source, and catalytic upgrading has been reported to be a viable solution. To create a successful closed-loop bioeconomy, it is essential to valorize every components of the food waste. Therefore, the aqueous fraction of the liquid yield of the HTL also can be valorized by anaerobic digestion or catalytic hydrothermal gasification.

REFERENCES

Aida, T.M., Sato, Y., Watanabe, M., Tajima, K., Nonaka, T., Hattori, H., Arai, K., 2007a. Dehydration of D-glucose in high temperature water at pressures up to 80 MPa. J. Supercrit. Fluids 40, 381–388.

Aida, T.M., Tajima, K., Watanabe, M., Saito, Y., Kuroda, K., Nonaka, T., Hattori, H., Smith Jr., R.L., Arai, K., 2007b. Reactions of D-fructose in water at temperatures up to 400C and pressures up to 100 MPa. J. Supercrit. Fluids 42, 110–119.

Akalin, M.K., Tekin, K., Karagoz, S., 2012. Hydrothermal liquefaction of cornelian cherry stones for bio-oil production. Bioresour. Technol. 110, 682–687.

Akhtar, J., Amin, N.A.S., 2011. A review on process conditions for optimum bio-oil yield in hydrothermal liquefaction of biomass. Renew. Sust. Energ. Rev. 15, 1615–1624.

Akhtar, J., Kuang, S.K., Amin, N.A.S., 2010. Liquefaction of empty palm fruit bunch (EPFB) in alkaline hot compressed water. Renew. Energy 36 (6), 1120–1127.

Alauddin, Z.A.B.Z., Lahijani, P., Mohammadi, M., Mohamed, A.R., 2010. Gasification of lignocellulosic biomass in fluidized beds for renewable energy development: a review. Renew. Sust. Energ. Rev. 14, 2852–2862.

Allen, S.G., Kam, L.C., Zeman, A.J., Antal, M.J., 1996. Fractionation of sugar cane with hot, compressed liquid water. Ind. Eng. Chem. Res. 35, 2709–2715.

Bardhan, S.K., Gupta, S., Gorman, M.E., Haider, A.A., 2015. Biorenewable chemicals: feedstocks, technologies and the conflict with food production. Renew. Sust. Energ. Rev. 51, 506–520.

Brand, S., Hardi, F., Kim, J., Suh, D.J., 2014. Effect of heating rate on biomass liquefaction: differences between subcritical water and supercritical ethanol. Energy 68, 420–427.

Brunner, G., 2009. Near critical and supercritical water. Part I. Hydrolytic and hydrothermal processes. J. Supercrit. Fluids 47, 373–381.

Cansell, F., Beslin, P., Berdeu, B., 1998. Hydrothermal oxidation of model molecules and industrial wastes. Environ. Prog. 17 (4), 240–245.

Capecchi, L., Galbe, M., Barbanti, L., Wallberg, O., 2015. Combined ethanol and methanol production using steam pretreated sugarcane bagasse. Ind. Crop. Prod. 74, 255–262.

Chandra, R., Takeuchi, H., Hasegawa, T., Kumar, R., 2012. Improving biodegradability and biogas production of wheat straw substrates using sodium hydroxide and hydrothermal pretreatments. Energy 43, 273–282.

Chornet, E., Overend, R., 1985. In: Biomass liquefaction: an overview. Proceedings of the International Conference of Fundamentals of Thermochemical Biomass Conversion. Elsevier Applied Science, New York, NY, pp. 967–1002.

Couhert, C., Commandre, J.M., Salvador, S., 2009. Is it possible to predict gas yields of any biomass after rapid pyrolysis at high temperature from its composition in cellulose, hemicellulose and lignin? Fuel 88 (3), 408–417.

Delmer, D.P., Amor, Y., 1995. Cellulose biosynthesis. Plant Cell 7, 987–1000.

Deniel, M., Haarlemmer, G., Roubaud, A., Weiss-Hortala, E., Fages, J., 2016a. Energy valorization of food processing residues and model compounds by hydrothermal; liquefaction. Renew. Sust. Energ. Rev. 54, 1632–1652.

Deniel, M., Haarlemmer, G., Roubaud, A., Weiss-Hortala, E., 2016b. Optimization of bio-oil production by hydrothermal liquefaction of agro-industrial residues: blackcurrant pomace (*Ribes nigrum* L.) as an example. Biomass Bioenergy 95, 273–285.

Devi, L., Ptasinski, K.J., Janssen, F.J.J.G., 2003. A review of the primary measures for tar elimination in biomass gasification processes. Biomass Bioenergy 24, 125–140.

Dimitriadis, A., Bezergianni, S., 2017. Hydrothermal liquefaction of various biomass and waste feedstocks for biocrude production: a state of the art review. Renew. Sust. Energ. Rev. 68, 113–125.

Durak, H., Aysu, T., 2016. Structural analysis of bio-oils from subcritical and supercritical hydrothermal liquefaction of *Datura stramonuim* L. J. Supercrit. Fluids 108, 123–135.

Elliott, D.C., 2007. Historical developments in hydroprocessing bio-oils. Energy Fuel 21, 1792–1815.

Elliott, D.C., Schiefelbein, G.F., 1989. Liquid hydrocarbon fuels from biomass. Am. Chem. Soc., Div. Fuel Chem. Preprints 34 (4), 1160–1166.

Elliott, D.C., Hart, T.R., Schmidt, A.J., Neuenschwander, G.G., Rotness, L.J., Olarte, M.V., Zacher, A.H., Albrecht, K.O., Hallen, R.T., Holladay, J.E., 2013. Process development for hydrothermal liquefaction of algae feedstocks in a continuous flow reactor. Algal Res. 2 (4), 445–454.

Elliott, D.C., Biller, P., Ross, A.B., Schmidt, A.J., Jones, S.B., 2015. Hydrothermal liquefaction of biomass: development from batch to continuous process. Bioresour. Technol. 178, 147–156.

Faba, L., Diaz, E., Ordonez, S., 2015. Recent developments on the catalytic technologies for the transformation of biomass into biofuels: a patent survey. Renew. Sust. Energ. Rev. 51, 273–287.

Fang, Z., Sato, T., Smith Jr., R.L., Inomata, H., Arai, K., Kozinski, A.J., 2008. Reaction chemistry and phase behavior of lignin in high-temperature and supercritical water. Bioresour. Technol. 99, 3424–3430.

FAO, 2013. Food Wastage Footprint: Impacts on Natural Resources. FAO, Roma, ISBN: 978-92-5-107752-8, p. 63. http://www.fao.org/docrep/018/i3347e/i3347e.pdf.

FAO, 2015. FAO Rice Market monitor. vol. XVIII (2). Food and Agriculture Organization of the United Nations.

FAO, 2017. Crop Prospects and Food Situation. Quarterly Report, http://www.fao.org/worldfoodsituation/csdb/en/.

Franke-Whittle, I.H., Insam, H., 2013. Treatment alternatives of slaughterhouse wastes, and their effect on the inactivation of different pathogens: a review. Crit. Rev. Microbiol. 39 (2), 139–151.

Girotto, F., Alibardi, L., Cossu, R., 2015. Food waste generation and industrial uses: a review. Waste Manag. 45, 32–41.

Gollakota, A.R.K., Kishore, N., Gu, S., 2018. A review on hydrothermal liquefaction of biomass. Renew. Sust. Energ. Rev. 81, 1378–1392.

Gowthaman, M.K., Gowthaman, P., Chandrasekaran, M., 2012. Principles of food technology and types of food waste processing technologies. In: - Chandrasekaran, M. (Ed.), Valorization of Food Processing By-Products. CRC Press, Boca Raton, FL, pp. 109–138.

Hadhoum, L., Balistrou, M., Burnens, G., Loubar, K., Tazerout, M., 2016. Hydrothermal liquefaction of oil mill wastewater for bio-oil production in subcritical conditions. Bioresour. Technol. 218, 9–17.

Hammerschmidt, A., Boukis, N., Hauer, E., Galla, U., Dinjus, E., Hitzmann, B., Larsen, T., Nygaard, S.D., 2011. Catalytic conversion of waste biomass by hydrothermal treatment. Fuel 90, 555–562.

Hardi, F., Makela, M., Yoshikawa, K., 2017. Non-catalytic hydrothermal liquefaction of biomass: an experimental design approach. Energy Procedia 105, 75–81. https://doi.org/10.1016/j.egypro.2017.03.282.

Huang, H., Yuan, X., 2015. Recent progress in the direct liquefaction of typical biomass. Prog. Energy Combust. Sci. 49, 59–80.

Huber, G.W., Iborra, S., Corma, A., 2006. Synthesis of transportation fuels from biomass: chemistry, catalysts and engineering. Chem. Rev. 106, 4044–4098.

Jessop, P.G., 2006. Homogeneous catalysis using critical fluids: recent trends and systems studied. J. Supercrit. Fluids 38, 211–231.

Karagoz, S., Bhaskar, T., Muto, A., Sakata, Y., 2005. Comparative studies of oil compositions produced from sawdust, rice husk, lignin and cellulose by hydrothermal treatment. Fuel 84, 875–884.

Karmee, S.K., 2016. Liquid biofuels from food waste: current trends, prospect and limitation. Renew. Sust. Energ. Rev. 53, 945–953.

Kaushik, R., Parshetti, G.K., Liu, Z., Balasubramanian, R., 2014. Enzyme-assisted hydrothermal treatment of food waste for co-production of hydrochar and bio-oil. Bioresour. Technol. 168, 267–274.

Kruse, A., Krupka, A., Schwarzkopf, V., Gamard, C., Henningsen, T., 2005. Influence of proteins on the hydrothermal gasification and liquefaction of biomass. 1. Comparison of different feedstocks. Ind. Eng. Chem. Res. 44, 3013–3020.

Lachos-Perez, D., Martinez-Jimenez, F., Rezende, C.A., Tompsett, G., Timko, M., Forster-Carneiro, T., 2016. Subcritical water hydrolysis of sugarcane bagasse: an approach on solid residues characterization. J. Supercrit. Fluids 108, 69–78.

Lede, J., 2013. Biomass fast pyrolysis reactors: a review of a few scientific challenges and of related recommended research topics. Oil Gas Sci. Technol. 68, 801–814.

Li, F., Liu, L., An, Y., He, W., Themelis, N.J., Li, G., 2016. Hydrothermal liquefaction of three kinds of starches into reducing sugars. J. Clean. Prod. 112, 1049–1054.

Liu, Z., Zhang, F., 2008. Effects of various solvents on the liquefaction of biomass to produce fuels and chemical feedstocks. Energy Convers. Manag. 49, 3498–3504.

Liu, Z., Xu, A., Zhao, T., 2011. Energy from combustion of rice straw: status and challenges to China. Energy Power Eng. 3, 325–331.

Liu, H.M., Wang, F.Y., Liu, Y.L., 2014. Characterization of bio-oils from alkaline pretreatment and hydrothermal liquefaction (APHL) of cypress. Bioresources 9 (2), 2772–2781.

Mackela, I., Kraujalis, P., Baranauskiene, R., Venskutonis, P.R., 2015. Biorefining of blackcurrant (*Ribes nigrum* L.) buds into high value aroma and antioxidant fractions by supercritical carbon dioxide and pressurized liquid extraction. J. Supercrit. Fluids 104, 291–300.

Maddi, B., Panisko, E., Wietsma, T., Lemmon, T., Swita, M., Albrecht, K., Howe, D., 2017. Quantitative characterization of aqueous byproducts from hydrothermal liquefaction of municipal waste, food industry wastes, and biomass grown on waste. ACS Sustain. Chem. Eng. 5, 2205–2214.

Mahmood, R., Parshetti, G.K., Balasubramanian, R., 2016. Energy, exergy and techno-economic analyses of hydrothermal oxidation of food waste to produce hydro-char and bio-oil. Energy 102, 187–198.

Mazaheri, H., Lee, K.T., Bhatia, S., Mohamed, A.D., 2010a. Sub/supercritical liquefaction of oil palm fruit press fiber for the production of bio-oil: effect of solvents. Bioresour. Technol. 101, 7641–7647.

Mazaheri, H., Lee, K.T., Bhatia, S., Mohamed, A.R., 2010b. Sub/supercritical liquefaction of oil palm fruit press fiber for the production of bio-oil: effect of catalysts. Bioresour. Technol. 101, 745–751.

Mohan, D., Pittman, C.U., Steele, P.H., 2006. Pyrolysis of wood/biomass for bio-oil: a critical review. Energy Fuel 20, 848–889.

Nagamori, M., Funazukuri, T., 2004. Glucose production by hydrolysis of starch under hydrothermal conditions. J. Chem. Technol. Biotechnol. 79, 229–233. https://doi.org/10.1002/jctb.976.

Pala, M., Katarli, I.C., Buyukisik, H.B., Yanik, J., 2014. Hydrothermal carbonization and torrefaction of grape pomace: a comparative evaluation. Bioresour. Technol. 161, 255–262.

Parshetti, G.K., Chowdhury, S., Balasubramanian, R., 2014. Hydrothermal conversion of urban food waste to chars for removal of textile dyes from contaminated waters. Bioresour. Technol. 161, 310–319.

Patil, P.T., Armbruster, U., Martin, A., 2014. Hydrothermal liquefaction of wheat straw in hot compressed water and subcritical water-alcohol mixtures. J. Supercrit. Fluids 93, 121–129.

Pavlovic, I., Knez, Z., Skerget, M., 2013. Hydrothermal reactions of agricultural and food processing wastes in sub- and supercritical water: a review of fundamentals, mechanisms, and state of research. J. Agric. Food Chem. 61, 8003–8025.

Peng, W., Wu, C., Wu, S., Wu, Y., Gao, J., 2014. The effect of heterogeneous catalysis on hydrothermal liquefaction of corn stalk under CO atmosphere. Energy Sources, Part A 36, 1388–1394.

Peterson, A.A., Vogel, F., Lachance, R.P., Froling, M., Antal Jr., A.J., Tester, J.W., 2008. Thermochemical biofuel production in hydrothermal media: a review of sub- and supercritical water technologies. Energy Environ. Sci. 1, 32–65.

Phaiboonsilpa, N., Ogura, M., Yamauchi, K., Rabemanolontsoa, H., Saka, S., 2013. Two-step hydrolysis of rice (*oryza sativa*) husk as treated by semi-flow hot-compressed water. Ind. Crop. Prod. 49, 484–491.

Pham, M., Schideman, L., Scott, J., Rajagopalan, N., Plewa, M.J., 2013. Chemical and biological characterization of wastewater generated from hydrothermal liquefaction of *Spirulina*. Environ. Sci. Technol. 47 (4), 2131–2138. https://doi.org/10.1021/es30453.

Pinto, F., Andre, R., Miranda, M., Neves, D., Varela, F., Santos, J., 2016. Effect of gasification agent on co-gasification of rice production wastes mixtures. Fuel 180, 407–416.

Posmanik, R., Cantero, D.A., Malkani, A., Sills, D.L., Tester, J.W., 2017a. Biomass conversion to bio-oil using sub-critical water: study of model compounds for food processing waste. J. Supercrit. Fluids 119, 26–35.

Posmanik, R., Labatut, R.A., Kim, A.H., Usack, J.G., Tester, J.W., 2017b. Coupling hydrothermal liquefaction and anaerobic digestion for energy valorization from model biomass feedstocks. Bioresour. Technol. 233, 134–143.

Prado, J.M., Lachos-Perez, D., Forster-Carneiro, T., Rostagno, M.A., 2016. Sub- and supercritical water hydrolysis of agricultural and food industry residues for the production of fermentable sugars: a review. Food Bioprod. Process. 98, 95–123.

Quitain, A.T., Sato, N., Daimon, H., Fujie, K., 2001. Production of valuable materials by hydrothermal treatment of shrimp shells. Ind. Eng. Chem. Res. 40, 5885–5888.

Ringer, M., Putsche, V., Scahill, J., 2006. Large-Scale Pyrolysis Oil Production: A Technology Assessment and Economic Analysis. Technical Report, National Renewable Energy Laboratory. NREL/TP-510-37779.

Ruiz, J.A., Juarez, M.C., Morales, M.P., Munoz, P., Mendivil, M.A., 2013a. Biomass gasification for electricity generation: review of current technology barriers. Renew. Sust. Energ. Rev. 18, 174–183.

Ruiz, H.A., Rodriguez-Jasso, R.M., Fernandes, B.D., Vicente, A.A., Teixeira, J.A., 2013b. Hydrothermal processing, as an alternative for upgrading agriculture residues and marine biomass according to the biorefinery concept: a review. Renew. Sust. Energ. Rev. 21, 35–51.

Sander, B., 1997. Properties of Danish biofuels and the requirements for power generation. Biomass Bioenergy 12, 177–183.

Sasaki, M., Adschiri, T., Arai, K., 2003. Fractionation of sugarcane bagasse by hydrothermal treatment. Bioresour. Technol. 86, 301–304.

Sato, N., Quitain, A.T., Kang, K., Daimon, H., Fujie, K., 2004. Reaction kinetics of amino acid decomposition in high-temperature and high-pressure water. Ind. Eng. Chem. Res. 43 (13), 3217–3222.

Savage, P.E., Levine, R.B., Huelsman, C.M., 2010. Hydrothermal processing of biomass: thermochemical conversion of biomass to liquid. In: Crocker, M. (Ed.), Fuels and Chemicals. RSC Publishing, Cambridge, pp. 192–215.

Shadidi, F., 1995. Extraction of value-added components from shellfish processing discards. In: Charalambous, G. (Ed.), Food Flavor: Ganeration, Analysis and Process Influence. Elsevier Science B.V., Amsterdam, pp. 1427–1439.

Shuping, Z., Yulong, W., Mingde, Y., Kaleem, I., Chin, L., Tong, J., 2010. Production and characterization of bio-oil from hydrothermal liquefaction of microalgae *Dunaliella tertiolecta* cake. Energy 35, 5406–5411.

Singh, R., Bhaskar, T., Dora, S., Balagurumurthy, B., 2013. Catalytic hydrothermal upgrading of wheat husk. Bioresour. Technol. 149, 446–451.

Tekin, K., Karagoz, S., Bektas, S., 2014. A review of hydrothermal biomass processing. Renew. Sust. Energ. Rev. 40, 673–687.

Tommaso, G., Chen, W.T., Li, P., Schideman, L., Zhang, Y., 2015. Chemical characterization and anaerobic biodegradability of hydrothermal liquefaction aqueous products from mixed-culture wastewater algae. Bioresour. Technol. 178, 139–146.

Toor, S.S., Rosendahl, L., Rudolf, A., 2011. Hydrothermal liquefaction of biomass: a review of subcritical water technologies. Energy 36, 2328–2342.

Tortosa, G., Alburquerque, J.A., Bedmar, E.J., Ait-Baddi, G., Cegarra, J., 2014. Strategies to produce commercial liquid organic fertilisers from "alperujo" composts. J. Clean. Prod. 82, 37–44.

Tu, Y., Huang, J., Xu, P., Wu, X., Yang, L., Peng, Z., 2016. Subcritical water hydrolysis treatment of waste biomass for nutrient extraction. Bioresources 11 (2), 5389–5403.

USDA, 2017. Livestock Slaughter, National Agricultural Statistics Service (NASS). Agricultural Statistics Board, United States Department of Agriculture (USDA). ISSN: 0499-0544, http://usda.mannlib.cornell.edu/usda/current/LiveSlau/LiveSlau-06-22-2017.pdf.

Van Doren, L.G., Posmanik, R., Bicalho, F.A., Tester, J.W., Sills, D.L., 2017. Prospects for energy recovery during hydrothermal and biological processing of waste biomass. Bioresour. Technol. 225, 67–74.

Vardanega, R., Prado, J.M., Meireles, M.A.A., 2015. Adding value to agri-food residues by means of supercritical technology. J. Supercrit. Fluids 96, 216–227.

Vigano, J., Machado, A.P.d.F., Martinez, J., 2015. Sub- and supercritical fluid technology applied to food waste processing. J. Supercrit. Fluids 96, 272–286.

Wang, H., Liu, X., Liu, Y., Chen, P., Sun, J., 2012. Hydrothermal liquefaction of whet straws for bio-oil and related analyses. Adv. Mater. Res. 347-353, 2419–2422.

Wright, M.M., Satrio, J.A., Brown, R.C., Daugaard, D.E., Hsu, D.D., 2010. Techno-Economic Analysis of Biomass Fast Pyrolysis to Transportation Fuels. Technical Report, NREL/TP-6A20-46586.

Xu, C.C., Shao, Y., Yuan, Z., Cheng, S., Feng, S., Nazari, L., Tymchyshyn, M., 2014. Hydrothermal liquefaction of biomass in hot-compressed water, alcohols, alcohol-water co-solvents for biocrude production. In: Jin, F. (Ed.), Application of Hydrothermal Reactions to Biomass Conversion. Green Chemistry and Sustainable Technology, https://doi.org/10.1007/978-3-642-54458-3_8. © Springer-Verlag Berlin Heidelberg 2014.

Yang, L., Nazari, L., Yuan, Z., Corcadden, K., Xu, C., He, Q., 2016. Hydrothermal liquefaction of spent coffee grounds in water medium for bio-oil production. Biomass Bioenergy 86, 191–198.

Younas, R., Hao, S., Zhang, L., Zhang, S., 2017. Hydrothermal liquefaction of rice straw with NiO nanocatalysts for bio-oil production. Renew. Energy 113, 532–545.

Yu, G., Zhang, Y., Schideman, L., Funk, T., Wang, Z., 2011. Distributions of carbon and nitrogen in the products from hydrothermal liquefaction of low-lipid microalgae. Energy Environ. Sci. 4, 4587–4595.

Zastrow, D.J., Jennings, P.A., 2013. In: Hydrothermal liquefaction of food waste and model food waste compounds.AIChE Annual Meeting Paper # 336978, http://www3.aiche.org/Proceedings/content/Annual-2013/extended-abstracts/P336978.pdf.

Zekovic, Z., Bera, O., Durovic, S., Pavlic, B., 2017. Supercritical fluid extraction of coriander seeds: kinetic modelling and ANN optimization. J. Supercrit. Fluids 125, 88–95.

Zhang, B., von Keitz, M., Valentas, K., 2008. Maximizing the liquid fuel yield in a biorefining process. Biotechnol. Bioeng. 105 (1), 103–912.

Zhang, L.H., Xu, C.B., Champagne, P., 2010. Overview of recent advances in thermochemical conversion of biomass. Energy Convers. Manag. 51, 969–982.

Zheng, J., Zhu, M., Wu, H., 2015. Alkaline hydrothermal liquefaction of swine carcasses to bio-oil. Waste Manag. 43, 230–238.

Zhu, Y., Biddy, M.J., Jones, S.B., Elliott, D.C., Schmidt, A.J., 2014a. Techno-economic analysis of liquid fuel production from woody biomass via hydrothermal liquefaction (HTL) and upgrading. Appl. Energy 129, 384–394.

Zhu, Z., Toor, S.S., Rosendahl, L., Chen, G., 2014b. Analyis of product distribution and characteristics in hydrothermal liquefaction of barley straw in subcritical and supercritical water. Environ. Prog. Sustain. Energy 33(3).

Chapter 10

Environmental Aspects of Food Waste-to-Energy Conversion

Jacqueline H. Ebner*, Swati Hegde[†], Shwe Sin Win[†], Callie W. Babbitt[†] and Thomas A. Trabold[†]

*College of Liberal Arts, Rochester Institute of Technology, Rochester, NY, United States, [†]Golisano Institute for Sustainability, Rochester Institute of Technology, Rochester, NY, United States

10.1 INTRODUCTION

The conversion of food waste to energy (FWtE) holds significant promise for contributing to a sustainable economy. Food waste is an abundant and currently undervalued biomass resource. When compared to first-generation crop-based biomass (such as corn or soybeans) as an energy feedstock, food waste does not incur the negative impacts associated with land use change or those associated with fertilizer or energy inputs required for cultivation. Furthermore, utilizing food waste (FW) can also provide a dual benefit in avoiding the negative environmental and social impacts associated with conventional waste management processes such as landfilling. As a renewable source of energy, FW has the potential to displace fossil fuel use, further energy independence, create economic opportunities, and return valuable materials and nutrients through circular economic patterns.

Realizing the potential of FWtE requires careful implementation of conversion technologies, particularly with respect to their environmental trade-offs. One of the most widely accepted methods for assessing environmental impacts is life cycle assessment (LCA). The LCA process has been codified through the ISO 14040/44 standards, which specify four steps to implement this method: (1) defining the study goal and scope, (2) preparing a life cycle inventory of all energy and resource inputs and emissions and product outputs, (3) assessing the ultimate impact of these inputs and outputs to human and environmental health, and (4) interpreting the study results according to the stated goal, with particular attention to uncertainty analysis (ISO, 2006a, b).

This chapter provides an overview of LCA as applied to FWtE technologies. However, LCA is a complex methodology and interpretation, and extrapolation of its results can be challenging. Therefore, we begin this chapter with a discussion of issues that arise and have particular relevance to LCA of FWtE systems. These issues include key assumptions, boundaries of the analysis, and common modeling choices that are important when applying and interpreting LCA of FWtE technologies. Next, we compare the inputs and outputs that influence environmental impacts for several common FWtE conversion technologies. We conclude this chapter by presenting an overview of comparative results for FWtE systems relative to other energy or food waste treatment systems. The goal of this section is to provide information relevant to assessing, interpreting, and inferring information about the environmental impacts of the FWtE systems. It is assumed that the reader has some familiarity with the LCA methodology. Additional resources on the methodology can be found in the ISO 14040/44 standards cited previously, Finnveden et al. (2000), Manfredi (2011), or additional resources.

10.2 LCA METHODOLOGY AND KEY ASSUMPTIONS

LCA analyzes the environmental impacts of a product, process, or system across all stages of its life cycle, from extraction and refinement of natural resources, manufacturing of components and products, distribution and use, to final disposal. The LCA methodology provides a structure for conducting an LCA and facilitates transparency. However, understanding the assumptions made and data used in the LCA is critical to interpreting results or making comparative assertions for different technologies. The following sections discuss some specific concerns relevant to the application of LCA to FWtE systems and interpretation of results.

Sustainable Food Waste-to-Energy Systems. https://doi.org/10.1016/B978-0-12-811157-4.00010-3

10.2.1 Functional Unit

The functional unit of an LCA specifies the function or service being provided by the product or system under study, and thus provides a reference to which all the tabulated inputs, outputs, and impacts can be related. This enables alternative processes or systems to be compared on a fair basis.

Functional unit definition presents unique challenges in the case of FWtE, as these systems provide two key functions: waste treatment and energy generation. The functional unit chosen depends on the objective of the study. If the objective of the study is to compare the alternative pathways of energy production, then a functional unit may be related to energy or fuel produced or the use of that energy in a particular application. For example, LCAs comparing FW-derived biofuels to gasoline may use "gallon gasoline equivalent" (GGE) as a functional unit to represent the volume of biofuel needed to achieve the equivalent energy of a gallon of gasoline. If the primary objective of the study is comparing food waste treatment alternatives, then an appropriate functional unit would be the treatment of a known quantity of food waste (e.g., 1 t mixed food waste). The use of different functional units in published FWtE LCA studies makes comparison of these studies challenging.

A further complication is that unlike a unit of energy, a functional unit of food waste does not have a universal definition. Defining a unit of food waste as a basis for assessing environmental impacts carries implicit or explicit assumptions about the waste stream's physical or chemical properties. The composition of the food waste is influenced by its source (i.e., industrial, household, etc.), consumption patterns specific to the region, and seasonality. Food waste composition impacts critical characteristics including moisture content, sugar content, nutrients present, biodegradability, and energy potential. These characteristics in turn influence LCA results through factors such as the energy yield, inputs required, and coproducts produced by the waste conversion process. Also, many FWtE systems process a combination of food waste along with other substrates, such as municipal solid waste (MSW), which contains plastics, metals, and glass, as well as other organic fractions including textiles and paper. Many anaerobic digesters process manure or sewage sludge combined with industrial or commercial food waste. The inclusion of cosubstrates may require additional process steps, cause synergistic (or antagonistic) effects, and make allocating the results to individual cosubstrates nearly impossible. Therefore, comparing results between studies requires a clear understanding of functional unit definition and any assumptions therein.

10.2.2 Biological Processes

Another LCA challenge unique to analyzing FWtE is that these systems combine biological and technical processes. Uncontrolled biological processes can occur throughout the system, particularly during food waste (FW) transportation or storage steps. As Bernstad and la Cour Jansen (2012) note, resulting fugitive emissions or mass losses can create additional environmental impact and reduce the potential nutrient and energy recovery, which will also influence environmental impact results. Biological processes also result in weight change due to evaporation, which can impact LCA results reported on a mass-based functional unit. These impacts can be overlooked in FWtE LCAs unless care is taken by the LCA practitioner and the interpreter of LCA results.

10.2.3 System Boundaries

Another important consideration is the delimitation of LCA system boundaries that clearly define which materials and processes are included in the system under study. A generic FWtE system diagram based on a food waste treatment functional unit is shown in Fig. 10.1A and for an energy generation functional unit in Fig. 10.1B. When modeling waste systems by LCA, a "zero burden approach" is often taken where the food waste enters the system free of environmental impact (Ekvall et al., 2007). Therefore, the cultivation, production, and consumption of the food itself is not considered. This approach is consistent with the objectives of most FWtE LCAs and with the prevailing philosophy that processes common to all systems being compared can be disregarded (Finnveden et al., 2005). FW collection, sorting, transportation to a processing facility, pretreatment, or on-site storage may or may not be included within the system boundaries, depending on the study objectives, data availability, and the extent to which these processes contribute to the ultimate life cycle impact.

Defining where to end the system boundary is also a challenge in FWtE LCA modeling. In reality, it is very difficult to include all emissions and uses for every resource associated with FWtE systems. LCA studies will include some, exclude others, and use simplifications and cutoffs depending on study objective. In addition, FWtE systems often entail products that are returned to the environment in different forms where biological processes can continue for a long time. Most notably, in the conventional treatment of food waste in a landfill (with or without landfill gas recovery) emissions are commonly considered over a100-year time frame (Edwards et al., 2018).

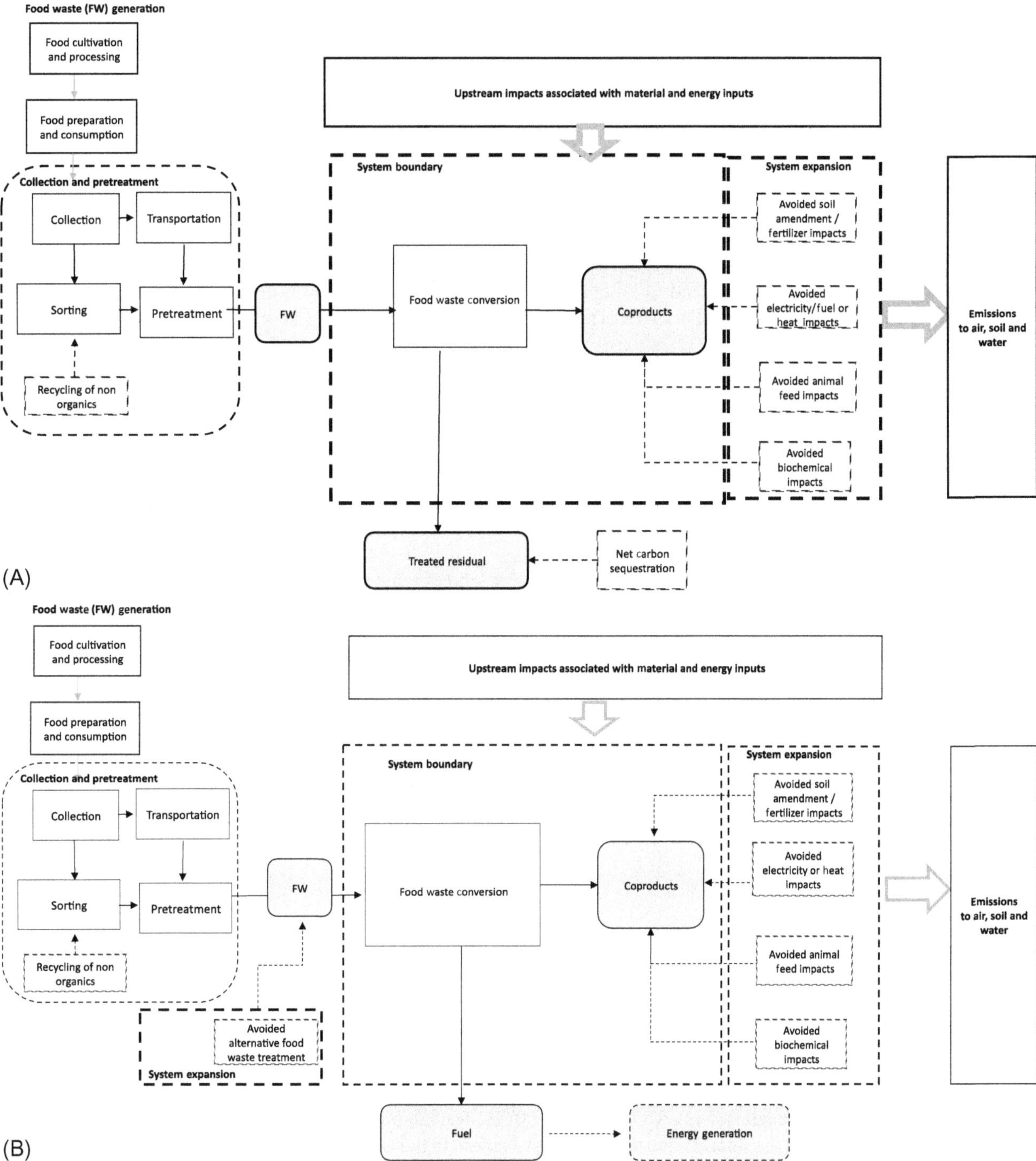

FIG. 10.1 (A) Generic system diagram for FWtE system based on a waste treatment functional unit. (B) Generic system diagram for FWtE system based on an energy (biofuel) production functional unit. Boldly dashed lines indicate the system boundaries. Processes shown with dashed borders are included in some studies and not others.

The downstream system boundary for biofuels or bioenergy can also vary depending upon the objective of the study. Some studies draw a system boundary at the production of the fuel. Due to the historical comparison to fossil-based fuels, these are called "well-to-gate" (sometimes "cradle-to-gate"). If the system boundary extends through fuel distribution, then it can be considered "well-to-tank" or "cradle-to-tank." When the objective is to compare use of the biofuel to use of fossil fuels, then the system boundary should extend to include combustion of the fuel to deliver a certain amount of energy or work. This is often referred to as "well-to-wheel" (or "cradle-to-grave").

10.2.4 Coproduct Credits

Many systems generate multiple products or coproducts. The treatment of coproducts is particularly relevant in LCAs pertaining to FWtE, because these systems inherently perform two functions: the treatment of food waste and the generation of energy or energy carriers. In addition, many FWtE systems operate as integrated biorefineries producing additional coproducts such as biofertilizers, soil amendments, animal feed, biochemicals, or animal bedding. Therefore the impacts incurred by the system must be allocated across the coproducts generated, or put another way, a credit must be given for the additional products generated given the impacts incurred.

The most common and preferred methodology is to use system expansion, although impacts may also be allocated to coproducts based upon a metric such as economic value or weight of the respective coproducts produced by the system. The choice of allocation methodology can have a significant effect on results and make the comparison of studies challenging. The system expansion methodology expands the studied boundaries to include processes that are displaced or avoided by generating a coproduct. In effect, a credit or offset is received for avoiding the production of the coproduct via an alternate process.

For example, consider the generic system shown in Fig. 10.1A where the objective is to understand or perhaps compare the impact of various food waste treatment options. All inputs and emissions associated with a FWtE pathway would be assessed relative to the reference or baseline treatment for the same food waste. However, when an energy coproduct such as electricity is also produced, it is assumed to displace an equivalent amount of electricity that would have otherwise been generated by conventional means, which in most cases is the regional electricity grid. The system can therefore be expanded to essentially provide a credit for avoiding the environmental impacts associated with that quantity of electricity via the grid (i.e., production and refinement of fossil fuels, combustion of fuels for electricity, etc.). Thus FWtE LCAs often contain region-specific assumptions about baseline coproducts or systems, which should be carefully scrutinized when generalizing results. In particular, the impact of grid mix can be significant (Schott et al., 2016).

Similarly, to account for other coproducts such as biofertilizers or animal feed, the system should be expanded to account for credit due to reference products that are hypothetically displaced. In cases where fossil-based fertilizers are displaced, this includes not only avoiding the production of the fertilizers but any differences in emissions that relate to the application of waste-derived vs. conventional fertilizers. In the case of animal feed coproducts, system expansion would include the avoidance of the crop cultivation and animal feed production, as well as impacts that result from feeding the coproduct to animals rather than the reference feed product.

Alternatively, if the functional unit for a FWtE system is a unit of fuel produced, then it may be appropriate to consider the treatment of food waste as a coproduct along with other energy, feed, or soil coproducts. In this case, system expansion would account for the avoided treatment of the food waste under a reference scenario such as landfilling. This system expansion is shown in Fig. 10.1B.

10.2.5 Impact Categories and Assessment Methods

Another significant set of assumptions to be clarified relates to which environmental impacts will be assessed. LCA can include a wide variety of impact categories, however, by far the most common in the literature is to assess climate change impact through the impact category of global warming potential (GWP), sometimes referred to as the "carbon footprint."

The standard methodology for assessing climate change impacts is to apply the Intergovernmental Panel on Climate Change (IPCC) characterization model which provides factors to assess the GWP of various greenhouse gas (GHG) emissions in kg carbon dioxide equivalents (CO_2e) (IPCC, 2014). The factors are provided for two time frames with different assumptions concerning the impacts due to methane. Advances in the underlying IPCC radiative heat forcing model result in periodic updates to characterization factors that can also be a source of variability when comparing older studies to newer studies.

With regard to assessing climate change impacts, an important assumption concerns the distinction between emissions that are considered part of the natural carbon cycle (biogenic emissions) and those that are attributed to man-made sources such as fossil fuel combustion (anthropogenic emissions). It is common in FWtE LCAs to exclude biogenic CO_2 emissions by selecting impact assessment methods that assign a characterization factor of "zero" to biogenic emissions. Generally, CO_2 resulting from the combustion of biofuels or biomass is considered biogenic along with CO_2 resulting from the composting process. Another consideration that becomes important in measuring comparative climate change impacts is the quantity and form of carbon in residuals. For example, carbon that remains in a stable form in a coproduct, landfill, or soil can be considered sequestered and thus provides an avoidance of (or negative) GWP impacts.

An alternative approach to GHG accounting will yield comparable results as long as the treatment is consistently applied Christensen et al. (2009). However, In a review of FWtE LCAs, Schott et al. (2016) found two studies that included biogenic CO_2 emission in the GWP balance, but neither applied a consistent treatment as recommended by Christensen et al. (2009). Thus when comparing FWtE climate change impacts, it is essential to ensure that the accounting of CO_2 emissions has been consistently applied. Other impact categories studied in FWtE systems include eutrophication potential (Guerrero and Muñoz, 2018); acidification potential (Guerrero and Muñoz, 2018; Schmitt et al., 2012); particulate matter (Hums et al., 2016; Schmitt et al., 2012); and cumulative energy demand (CED) (Dufour and Iribarren, 2012; Velásquez-Arredondo et al., 2010). A summary of major impact categories considered in prior LCA studies of FWtE systems (Tables 10.2–10.6) and the acronyms used to describe them is provided in Table 10.1.

TABLE 10.1 Acronyms for Impact Categories Used in Tables 10.2–10.6

Acronym	Acidification
ADP	Abiotic depletion potential
CC	Climate change
CED	Cumulative energy demand
ESD	Ecosystem damages
ESQ	Ecosystem quality
ET	Ecotoxicity
ETwc	Ecotoxicity in water chronic
FAETP, FEtox, FET	Freshwater ecotoxicity potential
FE	Freshwater eutrophication
FFD	Fossil fuel depletion
GHG	Greenhouse gas emissions
GWP	Global warming potential
HT	Human toxicity
HTa	Human toxicity via air
HTs	Human toxicity via soil
HTw, HTox	Human toxicity via water
LU	Land use
ME	Marine eutrophication
MEcox	Marine ecotoxicity
NE	Nutrient enrichment
ODP	Ozone depletion potential
POFP	Photochemical ozone formation potential
PM	Particulate matter
TA	Terrestrial acidification
TE	Terrestrial eutrophication
TEtox, TETP	Terrestrial ecotoxicity potential

10.3 LIFE CYCLE IMPACTS OF FOOD WASTE-TO-ENERGY CONVERSION

The following section discusses significant inputs and outputs across the life cycle of key FWtE technologies.

10.3.1 Anaerobic Digestion (AD)

A summary of studies reviewed concerning the environmental impacts of AD systems is provided in Table 10.2. Differences among AD systems make them complex to study from an environmental point of view and result in unique environmental impacts of each system (Börjesson and Berglund, 2006). System configuration is influenced by regional energy and agricultural policies, resource availability, and energy demand. The main factors relevant to defining an AD system are the types of food waste and other feedstock used, utilization of the biogas and coproducts generated, and the reactor technology implemented (Table 10.2).

The historical development of AD has been based on the treatment of wastewater or animal manure. Digestion of food waste can result in high volatile fatty acid (VFA) accumulation, leading to low biogas production and/or poor digestion process stability. However, the addition of food waste to wastewater or animal manure digestion is synergistic and can improve biogas production through mitigation of inhibition and the provision of nutrients. Thus codigestion of food waste with municipal wastewater or animal manure is common, although dedicated biogas systems to treat food waste are emerging.

Sources of food waste for AD vary and include industrial food wastes, such as brewery or food processing waste (Ebner et al., 2016; Poeschl et al., 2012a, b), retail food waste from supermarkets (Mondello et al., 2017; Jin et al., 2015), restaurant or cafeteria waste (Opatokun et al., 2017), household food waste or separated municipal food waste (Bernstad and la Cour Jansen, 2011; Opatokun et al., 2017). Characteristics of the food waste have an impact on the biomethane production potential as well as the nutrient composition in the residual effluent fertilizer (digestate). Ebner et al. (2016) noted that food waste with higher fat and carbohydrate content generally has higher biomethane potential, whereas foods with high water and lignin content have lower biomethane potential. However, it is worth noting that biomethane potential is a theoretical measure, as it represents biomethane production under ideal conditions. Inhibition, nutrient deficiencies, or pH imbalances can lead to lower biogas production in practice. Similarly, synergistic combinations can in some cases lead to even higher biogas production than single substrate biomethane potential. Environmental impact studies of AD are often based upon a specific feedstock. LCAs that consider codigestion of multiple feedstocks are difficult to disaggregate, and similarly, it is challenging to model the performance of codigestion systems by combining theoretical data on single substrates. There can be significant variation among all impact categories depending upon the food waste feedstock characteristics (Poeschl et al., 2012a, b).

AD technology can be implemented in a variety of system configurations, generally classified as wet or dry processes, and utilize either mesophilic or thermophilic microorganisms. Dry processes (sometimes called solid-state fermentation) utilize stackable feedstock generally with >30% solids, often in a batch style process. Although there are currently several commercial examples of dry digestion of the organic fraction of municipal solid food waste (OFMSW) in Europe and China, there are currently no studies of the environmental impacts associated with dry systems. While biogas production can be comparable between wet and dry systems, the latter generally processes less biodegradable feedstocks. Dry systems often require less water inputs than wet systems and have a less favorable energy balance, although it is difficult to draw comparable system boundaries (Angelonidi and Smith, 2015). Undigested residuals are often composted, and the offset value of the compost coproducts depends on the product assumed to be displaced. The lower nutrient composition in dry processes (which often include yard waste or other cellulosic feedstock) will provide less of an offset credit when compared to liquid fertilizer products; however, displacement of peat can provide significant environmental benefit due to the large impacts of traditional peat production.

Wet systems (<15% total solids) have been widely deployed for codigestion of food waste with wastewater or manure. Dedicated AD of industrial food waste, particularly brewery waste, is also becoming more common. Dedicated AD systems processing MSW or the OFMSW have received much attention, but we are just beginning to see commercial implementations. Therefore most of the studies available are modeled based on data from codigestion or manure-based systems (see Table 10.2).

Biogas utilization is also a highly significant factor that will influence net environmental impact. Biogas produced through AD can be utilized for on-site thermal applications, to generate electricity or both heat and electricity through cogeneration (also referred to as combined heat and power, CHP). Biogas can also be upgraded for use in natural gas applications or compressed/liquefied and distributed as compressed natural gas (CNG) or liquefied natural gas (LNG) for

TABLE 10.2 Summary of Selected Studies: Anaerobic Digestion

Reference	Feedstock	Functional Unit	System Boundaries	Impact Categories[a]	Key Findings
Artrip et al. (2013)	Manure	N/A	Cradle-to-grave	GHG (CH_4, N_2O, and CO_2)	Assessed an anaerobic lagoon only system (as a baseline) and a modified plug flow AD under real operational conditions and at the optimum potential of the AD. The results showed that AD system reduced the GHG emissions by 47.2% compared to lagoon
Bacenetti et al. (2016)	Energy crops, slurry, and manure from pigs and cattle; agricultural and agro-industry wastes	1 m^3, 1 MJ and 1 kWh of output energy or 1 ton of input feedstock	Cradle-to-grave and cradle-to-gate	GWP, AP, EP, CED	Reported that the energy crops production, the operation of anaerobic digesters, and digestate emission from open tanks are the main contributors to the environmental impacts
Börjesson and Berglund (2006, 2007)	Ley crops, straw, tops and leaves of sugar beet, liquid manure, food industry waste, and municipal organic waste	1 MJ of biogas	Cradle-to-grave	CO_2, CO, CH_4, NO_x, SO_2, N_2O and particles	Input feedstock, energy efficiency in biogas production, uncontrolled losses of methane, and the end-use application largely contributed to the environment impact of the biogas system. Emissions from the biogas production phase are higher than the end-use emissions for specific emissions
Chen et al. (2017)	Restaurant and hotel food waste in China	1 ton of food waste	Cradle-to-grave	GWP_{100}, AP, EP, $FAETP_{inf}$, and HTP_{inf}	The fuzzy model was used to evaluate the economic benefits and environmental impacts. The model indicated that when only methane production was considered, the total environment burden of AD technology was not significant
Ebner et al. (2016)	Codigestion of industrial food waste and cow manure	1 ton of influent (blended 27% IFW with manure)	Cradle-to-grave	GHG	Conventional treatment of liquid manure slurry and IFW as organic fertilizer was compared with AD with digestate management system. AD case reduces GHG emissions by 37.5 $kgCO_2eq$/ton influent. The largest reductions are from displacement of grid electricity
Evangelisti et al. (2014)	OFMSW	Total amount of OFMSW produced: 35,574 tons/year	Cradle-to-grave	GWP, AP, POCP, and NE	Environmental impacts of AD with energy and organic fertilizer production compared with incineration with CHP, and landfill with electricity production. AD with biogas production for CHP and digestate utilization was the best treatment option in terms of total GHG emissions and acidification which replaced nonrenewable electricity, heat, and inorganic fertilizer
Franchetti (2013)	Food waste	Annual management of food waste generated at University of Toledo	Cradle-to-gate	GWP	LCA analyzed the landfill disposal scenario with FWtE scenarios (e.g., AD with ultrasound pretreating, and one-stage AD). Study suggested that in terms of cost, energy, and GHG emissions, on-site organic FWtE systems are favorable

Continued

TABLE 10.2 Summary of Selected Studies: Anaerobic Digestion—cont'd

Reference	Feedstock	Functional Unit	System Boundaries	Impact Categories[a]	Key Findings
Jin et al. (2015)	Food waste from restaurants and hotels	1 ton of food waste	Cradle-to-grave	GWP_{100}, AP, EP, HTP_{inf}, $FAETP_{inf}$	System boundary includes pretreatment, AD, biogas utilization, disposal of digestate, and biological deodorization. The primary treatment and secondary pollution control systems are the biggest impact contributors and produce GWP of 96.97 $kgCO_2eq$/ton
Møller et al. (2009)	Source-separated municipal solid waste (MSW) and digestate use	1 ton of (wet) waste	Cradle-to-grave	GWP	Energy substation by biogas or substitution of natural gas in vehicles, N_2O emission from digestate in soil, fugitive emission of methane from AD system, unburned methane during combustion, carbon bound in soil, and fertilizer substitution influence the global warming potential
Opatokun et al. (2017)	Food waste-only AD in Australia with coal-based electricity	1 kg food waste	Cradle-to-grave	CC, OD, TA, FE, ME, HTox, POF, PMF, TEcox, FEcox, MEcox, WD, MD, FD	Food waste AD, pyrolysis, and integrated system all offer better environmental performance than landfilling, with AD and integrated system comparable. Food waste pretreatment (drying) minimized pyrolysis benefits, as did use of NaOH for wastewater treatment.
Poeschl et al. (2012a, 2012b)	Cattle manure (base case), several energy crops (straw, corn silage, grass silage), and several FW feedstocks (MSW, food residues, pomace, GTW, slaughter waste, grease separator sludge)	1 ton of organic material digested. System expanded for fertilizer displacement and electricity generated but not waste treatment	Cradle-to-grave	CO, NO_x, VOC, CO_2, CH_4, PM_{10}, N_2O	Emissions from the feedstock supply logistics are impacted by the input feedstocks used in biogas plant, which include fuel consumption and chemical fertilizer application during energy crop cultivation and fuel consumption during collection of MSW. Combined efficiency of energy generation (electrical and thermal), potential substitution of fossil fuels with biogas, and utilization of the heat by-product influence emissions from unit processes in biogas plant operation and biogas utilization. Recovered residual biogas from digestate storage areas has potential to reduce methane emission
Souza Filho et al. (2017)	Heat-treated potato liquor (HTPL) from potato starch production in Sweden	Treatment of 1 ton of HTPL residue	Cradle-to-grave	CC, AC, TE, ME, FE, FET	Cultivating filamentous fungus to produce a protein-rich biomass (fungus scenario) and production of biogas via AD were compared to the conventional method with HTPL as a fertilizer. Fungus cultivation demonstrated to be better economically and has the lowest impact in acidification; freshwater ecotoxicity; and the terrestrial, freshwater, and marine eutrophication categories

Tonini and Astrup (2012)	Residual municipal solid waste (MSW)	1 ton of Danish residual municipal solid (wet) waste	Cradle-to-grave	GW, AC, NE, ETwc, HTw, HTs, and HTa	Reported that the energy and environmental benefits of enzymatic refining of waste with utilization of products for recovery was a better alternative to incineration. Different energy utilization of the waste residue was assessed and the energy performance and environment impacts (i.e., global warming) of cocombustion in existing power plants and utilization of the liquid fraction for biogas production were the most favorable options
Woon et al. (2016)	Food waste	1 ton of food waste (wet base)	Bin-to-cradle	CC, PM, TA	Biogas fuel produced from AD for use as a petrol, diesel, and liquefied petroleum gas substitute for vehicle use performed better compared with biogas for electricity and heat and for city gas
Wulf et al. (2006)	Codigestion OFMSW with pig slurry	1 ton of OFMSW	Gate-to-grave (fermentation, storage and land application, and fertilizer displacement only considered)	CO_2, CH_4, NH_3, and N_2O	Codigestion of OFMSW with slurry increases biogas production and reduces the GHG emissions between 32 and 152 $kgCO_2/t$ depending on the application and storage techniques used for the fermentation residues. Majority of emission reduction comes from the management of fermentation residues. The system is also cost effective compared to composting
Xu et al. (2015)	Food waste	Management of 1 t volatile solid (VS)	Cradle-to-grave	18 midpoint categories	FW treatment at AD was compared with FW to landfill. The biggest potential impacts came from electricity consumption during AD and the transportation of FW to landfill

[a]*See Table 10.1.*

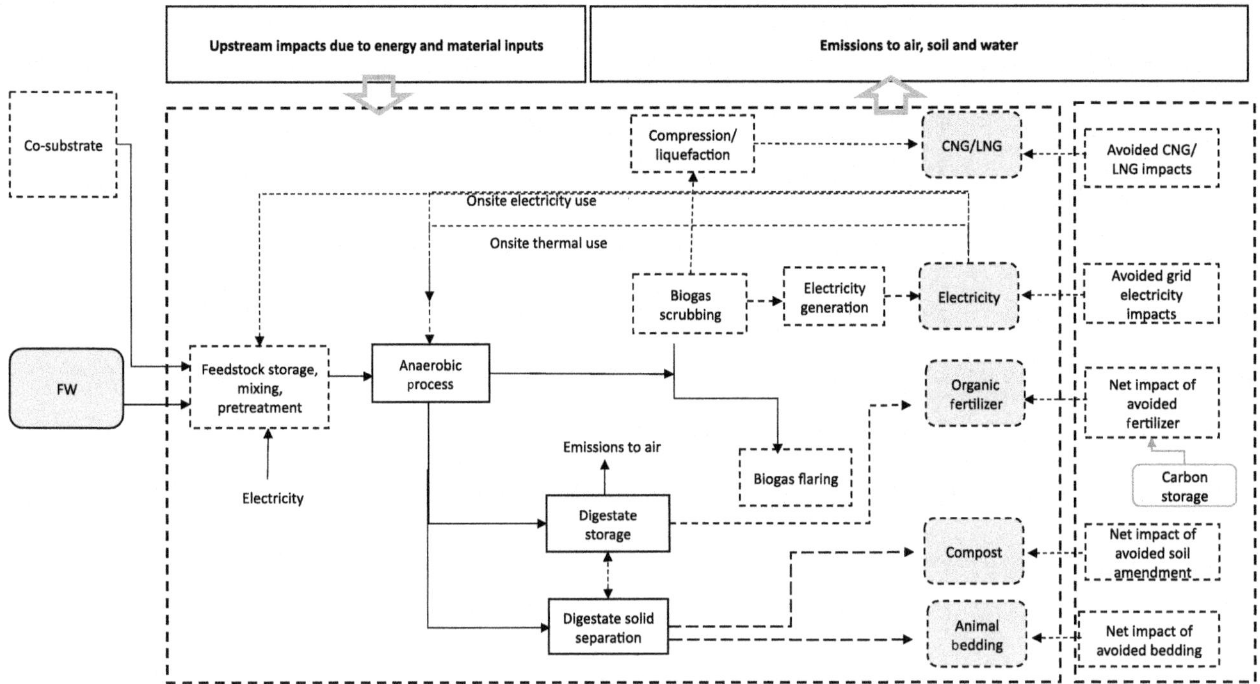

FIG. 10.2 System diagram for anaerobic digestion of food waste to biogas. Dashed processes indicate alternative system implementations.

transportation applications. In some cases, biogas can simply be flared, which converts most of the biomethane to biogenic CO_2. Additional coproducts include organic fertilizer, compost, or animal bedding, and the substitution of coproducts has a significant impact on LCA results. Fig. 10.2 shows the system diagram for AD of food waste to biogas.

The most significant sources of direct emissions due to AD are fugitive emissions from biogas leaks in tanks or piping, or due to uncontrolled releases of biogas through relief valves when flaring is inoperable (Ebner et al., 2016). Estimates for fugitive emissions range from 0% to 10% of the biogas volume produced (IPCC, 2006). However, according to Liebetrau et al. (2013), they can be close to 0% in plants where leaks are controlled and with proper flaring. The IPCC guidelines (2006) assume 3% with a range of 0%–10%. Due to the high GWP of methane, uncontrolled releases of methane can have significant impacts. For instance, methane losses equivalent to 1% of the biogas produced imply that the methane emissions from the production phase correspond to > 85% of the fuel-cycle emissions (Börjesson and Berglund, 2006). Other sources of direct emissions include combustion of fossil fuel for FW collection or pretreatment which is associated with respiratory problems in humans due to inorganic substances and dust (e.g., NO_x, SO_2, and PM10); and freshwater ecotoxicity (FET) due to toxic substances that are emitted to air, water, and soil (Poeschl et al., 2012a, b).

Biogas utilization can provide significant offsets through substitution of fossil energy products. Production of biogas from AD has been reported in the range of 92–165 m^3/ton treated food waste, with energy content of 2450–3890 MJ/ton (Bernstad and la Cour Jansen, 2012). Electricity generated beyond what is required onsite to operate pumps and mixers (around 12%) can be exported to the grid. Exported electricity is assumed to displace the generation of grid electricity, generally leading to a net benefit in terms of GWP for AD. The offset is based on the regional grid mix, thus implementation of AD in areas with fossil-heavy grid mixes provides a greater benefit. Produced biogas can also be upgraded, compressed/ liquefied, and distributed. Bernstad and la Cour Jansen (2012) reported large variation in reported values of energy required, chemical inputs, and fugitive CH_4 during the upgrading process which can have a significant impact on LCA results (Poeschl et al., 2012a, b; Table 10.2).

The digestate (effluent from the digester) can be treated and utilized in a variety of ways. A common utilization pathway is to apply the digestate to farmland as an organic fertilizer. Because land application of organic fertilizer can only be done when plants are growing and can take up the nutrients and when weather conditions permit, the digestate must usually be stored for some period of time. Depending on the local climatological conditions, CH_4, N_2O, and NH_3 can be emitted to air during storage when digestate is stored in open ponds or lagoons. Ebner et al. (2016) estimated digestate emissions for earthen pond storage to be in the range of 1.5 m^3 CH_4/t digestate based upon analysis of previous literature. When solids are removed (via solid separation) from the digestate prior to storage, the potential for emissions to air is reduced. Storage emissions are also impacted by temperature, with virtually no emissions occurring below 20°C and emissions increasing with temperature. They can be eliminated with covered storage systems.

Use of liquid digestate on land can serve as an organic fertilizer, where the nutrient content of the digestate is directly related to that of FW feedstock (Ebner et al., 2016). Using liquid digestate on land requires fuel inputs for transportation and spreading, and results in direct and indirect emissions to air including NH_3 (which can lead to potential changes in soil and water quality and is also associated with human health issues due to its role in the formation of particulates) and N_2O (which is a powerful GHG). Freshwater eutrophication (FE) and marine eutrophication (ME) impacts are associated with phosphorus concentration in freshwater and nitrogen enrichment in seawater, respectively, and are also directly linked to food waste composition. The use of digestate on land also results in indirect substitution of mineral fertilizer. Many studies find that the avoidance of mineral fertilizer provides a significant offset to GWP impacts (Bernstad and la Cour Jansen, 2012; Schott et al., 2016). However, when assessing the effect of mineral fertilizer substitution it is important to consider the net impact of organic fertilizer use vs. mineral fertilizer. This entails equivalent system boundaries to compare equivalent substitution rates; spreading or application of the fertilizers; emissions to the air, soil, and water; and net changes in remaining carbon soil content. The mineral content of the food waste greatly influences fertilizer displacement as does the crop, climate, and soil conditions. Data on actual displacement rates of mineral fertilizer are limited. Furthermore, some studies do not compare systems that include net impacts for both organic and mineral fertilizer application (Schott et al., 2016). In terms of other impacts, Fruergaard and Astrup (2011) report that land application of digestate incurs nutrient enrichment impacts due to nitrite runoff to surface water and heavy metals such as mercury contained in the treated waste may cause significant human toxicity impacts in water, while substitution of inorganic fertilizer can create large savings due to avoided emissions of Cd and Cu. Assumptions on carbon remaining in the soil after land application depend on the carbon complex of the food waste feedstock, climate, crops, and soil type and is based upon long-term studies which are limited. Therefore while some studies attribute an impact due to changes in carbon storage due to land application of organic fertilizer this modeling is highly complex and uncertain.

In summary, significant factors that influence the environmental impact of AD systems (Table 10.2) include FW composition, uncontrolled losses of methane, coproducts generated, and reference systems displaced. As Börjesson and Berglund (2006) concluded in their review of biogas fuel-cycle environmental impacts, it is not possible to specify a range of fuel-cycle emissions for biogas systems with reasonable reliability without defining the biogas system with high accuracy. In all of the studies assessed, the indirect benefits of AD systems outweigh the direct impacts with the substitution of grid electricity and of mineral fertilizer having the largest impacts.

10.3.2 Fermentation

A summary of selected literature on life cycle impacts of ethanol production from organic waste feedstocks is provided in Table 10.3. Environmental impact for conversion of food waste to bioethanol via fermentation is often based on a biofuel functional unit. When the objective is to compare production technologies for ethanol then a volume unit of ethanol (i.e., 1 L) is often chosen as the functional unit (Schmitt et al., 2012); however, since bioethanol is often blended with gasoline some studies use a blended unit such as E10 (a 10% blend of ethanol with 90% fossil-based gasoline) (Pourbafrani et al., 2013).

The system boundaries can include all upstream and production inputs and outputs to produce that unit of ethanol to the (i) factory gate, (ii) the distribution system, or (iii) combustion of fuel ethanol in vehicles depending upon the goal and scope of the study. Most food waste to bioethanol environmental impact studies to date consider either GWP or CED. A generic system diagram for a fermentation process to convert food waste to ethanol is shown in Fig. 10.3.

As with the other FWtE technologies, studies of the environmental impacts of bioethanol production from FW show varying results because of composition of the food waste and the system boundaries chosen (Borrion et al., 2012). As Ebner et al. (2014) note, a comparison of food waste-derived bioethanol to other sources of fuel should consider the net impact of avoiding alternative disposal of the FW. For example, avoiding treatment of rapidly biodegradable food waste in a landfill where a portion of methane releases are uncaptured (which is common even when landfill gas recovery systems are in place) results in a significant benefit greater than direct bioethanol process emissions. While diverting food waste from a beneficial use process (i.e., a carbon-negative process) such as composting or use as animal feed results in an impact to the system (Ebner et al., 2014).

The most common process for bioethanol converts sugars within the FW to ethanol via fermentation. The first step for most food wastes is a pretreatment to reduce particle size and increase surface area. The fuel inputs for grinding or milling and pumping carry associated environmental impacts. Several high sugar industrial food wastes, such as grape and beet pomaces (Rodríguez et al., 2010; Berłowska et al., 2016), banana or pineapple waste (Velásquez-Arredondo et al.,

TABLE 10.3 Summary of Selected Studies: Fermentation to Ethanol

Reference	Feedstock	Functional Unit	Process	System Boundaries	Impact Categories[a]	Key Impacts
Chester and Martin (2009)	MSW	15 dry MMT MSW available for converting to ethanol in California	Dilute acid prehydrolysis, enzymatic hydrolysis, fermentation	Cradle-to-grave	GHG, LU	Established that a complete MSW-to-ethanol facility in California would displace 110PJ of fossil energy and result in slight increase in GHG emissions. Landfilling of lignin residue is recommended over incineration to achieve improved GHG benefits
Ebner et al. (2014)	Food processing and retail waste	1 t combined	Simultaneous saccharification fermentation	Cradle-to-gate	GHG	Negative GHG emissions and approximately 500% improvement compared to corn ethanol production
Guerrero and Muñoz (2018)	Lignocellulosic waste from banana packaging plant	1 MJ of energy released during ethanol combustion in a passenger car	Simultaneous saccharification fermentation with steam explosion pretreatment	Well-to-wheels	GWP, AP, EP	Downstream wastewater treatment contributed significantly to GHG emissions. Increased acidification impact due to the use of chemicals in pretreatment. E65 blend was recommended for Ecuador to obtain net negative emissions
Kalogo et al. (2007)	MSW	1 ton of wet MSW treated; 1 km distance traveled	Selective hydrolysis of cellulose fraction of MSW, fermentation, and distillation	Cradle-to-grave	GHG	At an ethanol yield lower than 166 L/ton, MSW-to-ethanol conversion results in higher emissions than landfilling with LFG recovery. Well-to-wheels emissions for ethanol were higher than gasoline, corn ethanol, and lignocellulosic ethanol
Pourbafrani et al. (2013)	Citrus waste	The functional units are 1 MJ of E85, 1 kWh of generated electricity utilizing biomethane, 1 kg of limonene, and 1 kg of digestate	Acid hydrolysis and fermentation (removal of inhibitor compounds (limonene), AD of residuals	Well-to-wheels	GHG	Use of E85 in light duty vehicles resulted in 134% reduction in GHG compared to gasoline. The ethanol biorefinery integrated with biogas production would offer significant savings resulting from on-site electricity generation and fertilizer displacement
Schmitt et al. (2012)	MSW	1 L of denatured ethanol produced in Washington State	Dilute acid enzymatic hydrolysis and fermentation	Cradle-to-grave	GWP, AP, smog air, PM	Acid and enzyme production for pretreatment contributed significantly to energy consumption and acidification potential. Impacts from enzyme production have a high degree of uncertainty
Velásquez-Arredondo et al. (2010)	Banana pulp, fruit, flower stalk, and peel	Net energy analysis of a pilot plant capable of processing 4000 kg/day of banana fruit and its residual biomass	Dilute acid and enzyme hydrolysis, fermentation, and distillation	Cradle-to-grave	CED	Energy ratio of 1.9 for fruit and pulp estimated, slightly higher than ER for corn ethanol. Unsatisfactory energy ratio when fruit was cofermented with cellulosic residue

[a]See Table 10.1.

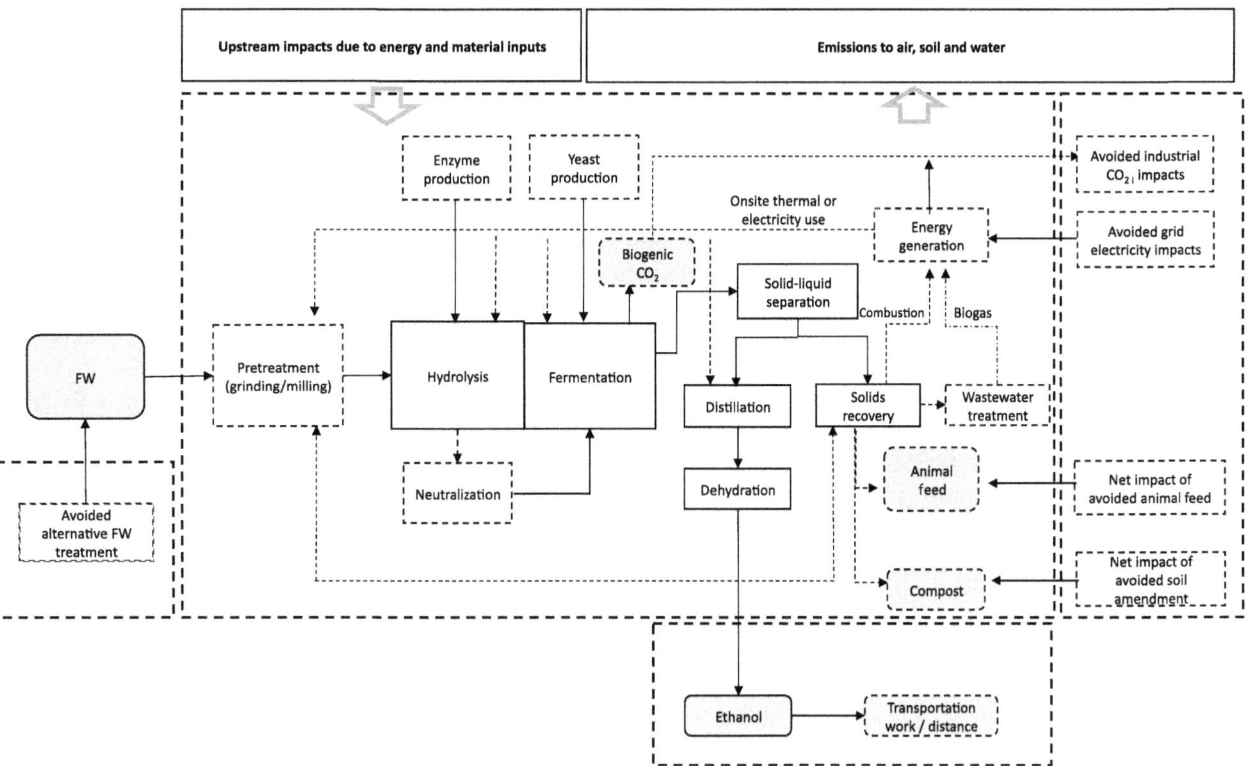

FIG. 10.3 System diagram for fermentation of food waste to ethanol.

2010; Upadhyay et al., 2013) and fruit syrups (Ebner et al., 2014), can be used to produce ethanol without energy-intensive pretreatment processes. High starch wastes such as bakery waste (Ujor et al., 2014), wasted bread (Pietrzak and Kawa-Rygielska, 2014), pasta and potato industry wastes (Walker et al., 2013), and several other food processing wastes like tofu processing waste, fruit pomaces, and beverage industry waste have been reviewed for ethanol production (Hegde et al., 2018). The simple carbohydrates in these food wastes can be readily converted to bioethanol via yeast. *Saccharomyces cerevisiae* is the most commonly used yeast for food wastes rich in glucose, and yeast belonging to *Kluyveromyces* sp. is used for lactose-based feedstocks (see Chapter 5 for a detailed description of the ethanol production process).

Depending on the composition of the FW, different additional pretreatments may be required to access the convertible sugars, including acid, alkali, organic solvents, or heat treatments. While commercial or pilot-scale processes currently convert high sugar industrial wastes to bioethanol, production from OFMSW or household waste is not yet found on a commercial scale. A variety of implementations are under development for lignocellulosic feedstock, making it difficult to predict environmental impacts at this time.

Gerbrandt et al. (2016) report that steam pretreatment processes (without added chemicals) for lignocellulosic ethanol production tend to be more energy efficient due to heat recovery, which can be used in downstream processes such as distillation. Processes using acids or bases require pH adjustment and are less likely to allow for heat recovery. The type of pretreatment used impacts the degree of cellulose breakdown and in turn enzyme use as well as the degree of hemicellulose converted to inhibitors which impacts yield (and in turn environmental impacts based on yield) (Gerbrandt et al., 2016). Kiran et al. (2014) report that harsh treatments may not be necessary to convert lignocellulosic food waste such as MSW or household waste to ethanol, but autoclave of FW is often required to improve product yield and purity. However, not only does thermal pretreatment incur energy and water inputs, but it may lead to reduction in useful sugars and amino acids and a decrease in the reaction efficiency between the enzymes and rewetted FW. In lieu of thermal sterilization, acidic conditions can prevent microbial contamination and putrefaction but require acid-tolerant microorganisms for fermentation. Since each pretreatment method incurs trade-offs in terms of process yield and downstream inputs, each will have different environmental implications.

In the hydrolysis step, enzymes are used to convert starch, cellulose, and hemicelluloses into simple sugars, which can be fermented by bacteria or yeast. Fermentation can follow enzymatic hydrolysis as a separate step, and this procedure is known as separate hydrolysis and fermentation (SHF). However, the most commonly used technique is called simultaneous saccharification and fermentation (SSF), which is carried out by combining fermentation and enzyme hydrolysis in the same step.

Saccharifying enzymes include α-amylase, β-amylase, glucoamylase, and cellulase. Previous studies on enzyme production have reported that glucoamylase uses the most energy-intensive production process. Therefore purely on a cradle-to-enzyme manufacturer gate perspective, glucoamylase production contributes the highest toward GHG emissions (Dunn et al., 2012; Knauf and Krauss, 2006). Novozymes, the largest industrial enzyme supplier in the world, published a cradle-to-gate environmental assessment for five enzymes including amylase and glucoamylase used in ethanol fermentation (Nielsen et al., 2007). The GHG emissions in glucoamylase production arise from electricity or energy requirements of the process, and for cellulase, the impacts are associated with chemicals and nutrients used in production. However, lignocellulosic ethanol production uses 25–250 times higher doses of cellulase compared to amylase and glucoamylase (Dunn et al., 2012) and thus can result in higher environmental impact. For example, studies comparing corn ethanol to cellulosic fermentation processes found yeast and enzyme production accounted for 27%–35% of life cycle GHG emissions in cellulosic ethanol vs. only 1.4%–3% GHG emissions for corn ethanol (Dunn et al., 2012; MacLean and Spatari, 2009). Hong et al. (2013) estimated that on-site production of cellulase could reduce 64% of total GHG emissions compared to off-site production. While GWP impacts are sensitive to cellulase consumption, in a review of lignocellulosic ethanol literature Gerbrandt et al. (2016) reported a wide range of values observed (1.24–37.4kg cellulase/dry ton feedstock). A trade-off can be observed between enzyme use and solids loading. Higher solid loading requires greater energy for mixing and higher enzyme inputs, but reduced energy for pumping, fermentation, and distillation downstream. Additional variables that impact the saccharification process include temperature, agitation speeds, and hydrolysis times. It is important to note that currently no large-scale commercial plants exist for cellulosic ethanol production and the LCA studies reported are based on future implementation assumptions. As cellulosic ethanol technology advances, a deeper understanding of commercial scale environmental impacts will emerge.

The fermentation process generates CO_2 that is considered biogenic in LCA. Many ethanol production processes capture this CO_2 as a coproduct for use in food grade or industrial applications. However since CO_2 is a common by-product of many industrial processes including natural gas production, the avoided impact is not significant.

Downstream processing of ethanol involves solid separation and purification that consists of distillation and dehydration. Distillation is highly energy intensive, for example, using waste banana pulp as the feedstock for ethanol production, distillation was estimated to use approximately 53% of the total energy encompassing hydrolysis, fermentation, and distillation. However, environmental impacts associated with distillation will depend on the sources of electricity used (grid mix) and emissions associated with them. Choices upstream on hydraulic load, enzyme pretreatment, and fermentation can have a significant impact on the purification stage.

Studies that consider distribution and storage of ethanol note that emissions of pollutants and GHGs do occur; however, these emissions are generally lower than those related to gasoline and largely influenced by the location of the biorefinery (Wang et al., 2012). The treatment of CO_2 emissions during combustion as biogenic CO_2 greatly reduces the GWP impact of bioethanol combustion relative to gasoline. A review-based analysis of tailpipe emissions indicated that combusting E85 resulted in decreased emissions of CO_2 and benzene (a known carcinogen) compared to gasoline, but increased emissions of acetaldehyde, which has human toxicity impacts.[1]

Leftover solids after food waste pretreatment and/or ethanol distillation can be converted into a variety of coproducts. In conventional grain-based ethanol production, these solids, called distiller's grains and solubles (DGS), are sold wet or dried as animal feed components. The offset impacts include avoidance of cultivating the equivalent amount of alternative feed (primarily corn) as well as any changes in the use of the feed including changes in emissions due to enteric fermentation. Commonly proposed lignocellulosic ethanol processes (Humbird et al., 2011) consist of a step to separate the solids with a high lignin content for use to generate energy as heat or in CHP. Lignin is a complex polymer characteristic of the cell wall of vascular plants and is highly recalcitrant toward both chemical and biological degradation. As lignin content is a characteristic of the feedstock it can vary depending on the type of biomass used. Food scraps contain lower amounts of lignin than other cellulosic biomass and hence offer less opportunity for energy coproduction. A final alternative for separated solids is composting to generate a soil amendment product. Depending on the characteristics of the finished soil amendment, this could displace conventional peat or mineral-fertilizer products (as discussed above in relation to AD).

In summary, feedstock with more free sugars available require less inputs and incur lower impacts due to ethanol production, compared to starch or cellulosic feedstocks. For example, the ethanol production phase (hydrolysis, fermentation,

1. https://www.afdc.energy.gov/vehicles/flexible_fuel_emissions.html.

and distillation) using sugar cane as feedstock contributed to only 7% of CO_2 equivalent emissions per MJ ethanol produced in a well-to-wheel life cycle, while for cellulosic and corn feedstocks it contributed 41%–50% (Wang et al., 2012). However, sugarcane incurred a significant impact during the farming stage. As food waste does not incur any cultivation or land use impacts, wastes rich in free sugars like dairy whey, brewery waste, and fruit processing wastes can result in net environmental benefits (Ebner et al., 2014). Enzyme production and distillation utilize the highest amount of energy and have been estimated as contributing up to 40%–80% of the life cycle GHG impact. Enzymes that are used to liquefy and saccharify starch have lower impacts compared to cellulase. Utilizing residual solids in energy generation via combustion or AD can displace onsite fossil fuel use and reduce environmental impacts. Coproducts such as compost or animal feed also have environmental benefits due to displacement of high impacts associated with peat or corn feed production.

10.3.3 Transesterification

The most common food waste-derived biodiesel utilizes waste cooking oil (WCO, also referred to as yellow grease), animal fats (e.g., beef or poultry), or grease trap waste (GTW, also referred to as brown grease) from commercial or industrial processes. Additional promising feedstocks include industrial food processing wastes with high lipid content such as pomace from olive production (Rajaeifar et al., 2016), spent coffee grounds from brewed coffee (Tuntiwiwattanapun et al., 2017), corn oil from ethanol production (Wang et al., 2015), and oil extracted from municipal food wastes (Ahamed et al., 2016). The main advantages of using waste feedstocks are that they are relatively low cost or even "free," and they avoid arable land and water use for dedicated energy crop production.

One of the drawbacks of using waste feedstocks is that they often contain impurities such as water, food residue, or high free fatty acids (FFA), which can inhibit the chemical process. Thus, depending on the composition of the waste feedstock, pretreatment may be required, which in turn incurs associated environmental burdens due to chemical and energy inputs. Waste from animal product manufacturing requires a rendering process to produce animal fat or tallow. Rendering requires a low level of constant heat and creates impacts due to energy inputs. For industrial and municipal wastes, a lipid extraction step, via mechanical disruption or solvent extraction method (often hexane), must be carried out prior to transesterification. Hexane recovery from the lipids can be achieved using a simple distillation method (Islam et al., 2017). Food wastes with high FFA content can be pretreated with acid esterification where methanol and sulfuric acid are mixed with oils to create methyl esters (Ezobor et al., 2014). A caustic stripping process can also be used to create a soap but results in significant yield loss and creates disposal issues. Pretreatment chemicals can also be recovered or removed but require additional chemical inputs. For example, Dufour and Iribarren (2012) report the use of calcium oxide and phosphoric acid to remove sulphuric acid and sodium hydroxide in a case study for WCO. Nonetheless, those processes are often costly, energy intensive, and require additional chemical inputs. When the waste contains high moisture, as in the cases of GTW, olive pomace, or sewage sludge, feedstock drying is often required along with waste water treatment. In this case, feedstock drying, extraction of oil, and wastewater treatment are the largest contribution to energy consumption and emissions (Hums et al., 2016; Dufour and Iribarren, 2012).

Biodiesel production via the transesterification reaction is one of the most common commercial processes, and a representative system diagram is provided in Fig. 10.4. During the transesterification reaction, oil or fats are reacted with methanol in the presence of a base catalyst (KOH or NaOH) at a desired temperature and atmospheric pressure to produce biodiesel and crude glycerol. The main impacts due to the transesterification process stem from thermal and electrical energy demands, which can contribute as high as 30% to the total system GWP in the case of biodiesel production from beef tallow and poultry fat (Dufour and Iribarren, 2012). In the case of WVO-based biodiesel production, thermal energy requirements for transesterification were responsible for >20% of the impact for GWP due to the use of steam-heated reactors.

The amount of methanol and catalysts required in transesterification is highly dependent on the purity of the feedstocks (de Araújo et al., 2013). Methanol can be recovered by distillation and reused in transesterification or crude glycerol purification processes. While this step avoids chemical inputs and associated upstream environmental impacts, it requires a trade-off with additional energy requirements. Tuntiwiwattanapun et al. (2017) reported that methanol recovery consumed 73% of the energy in the in situ transesterification process.

Crude glycerol is a coproduct of biodiesel production, and unlike many other FWtE conversion processes discussed in this chapter, system expansion is less commonly used in biodiesel LCAs. Instead, mass or economic allocation is commonly used to distribute the environmental burdens between the two coproducts (Dufour and Iribarren, 2012). For every 1 kg of biodiesel produced, approximately 100 g of crude glycerol is generated. Crude glycerol can be treated to produce different coproducts including feedstock for animal feeds or the production of value-added chemicals, depending on the glycerol purity (Yang et al., 2012). The impurities of crude glycerol derived from waste-based biodiesel (i.e., methanol, soap, water,

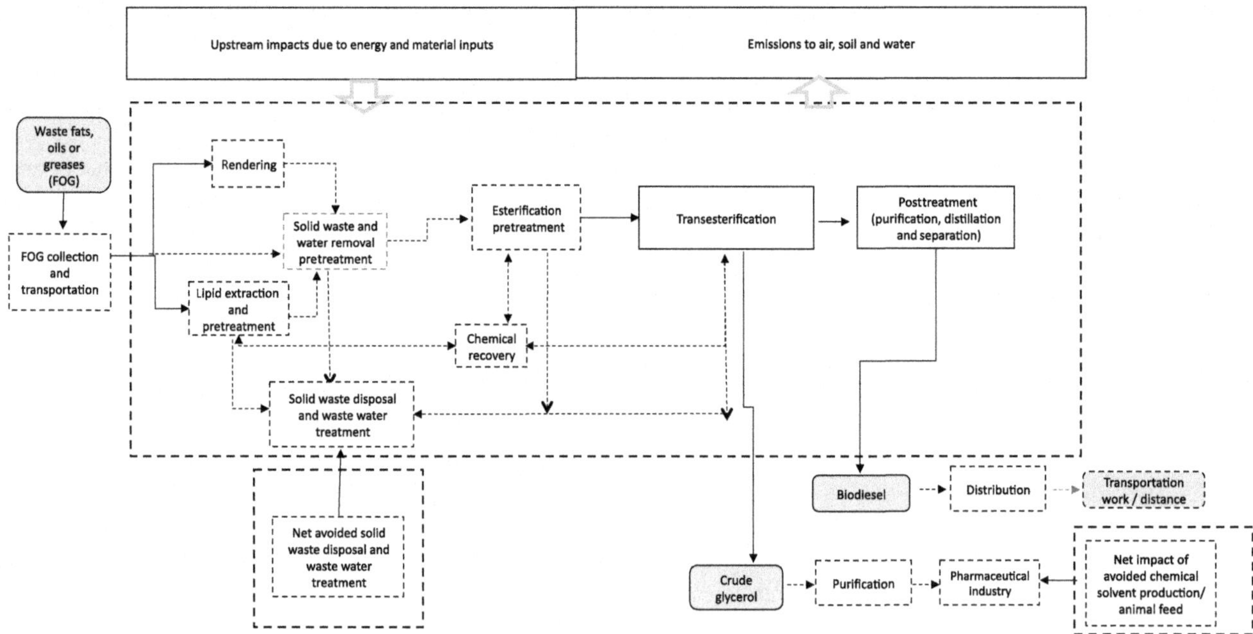

FIG. 10.4 System diagram for transesterification of food waste derived biodiesel.

excess catalyst, and esters) tend to be higher as it requires additional process steps and chemicals such as methanol and base catalysts (Thompson and He, 2006; Win et al., 2015). When a system expansion method is used, crude glycerol can avoid emissions associated with the cultivation and production of animal feeds or chemicals, but the final uses and associated benefits are very uncertain Janssen and Rutz (2008).

Prior studies have reported relative emissions impacts of biodiesel and conventional petroleum-based fuels. Argonne National Laboratory conducted an LCA of biodiesel (B100) and petroleum diesel and showed that GHG emissions of B100 are 74% lower than those from petroleum diesel.[2] Combustion of biodiesel in urban buses reduces the life cycle emissions of total PM, CO, and sulfur oxides compared to petroleum diesel (Sheehan et al., 1998). Cheung et al. (2015) also demonstrated that the addition of biodiesel resulted in decreased hydrocarbon and CO emissions and PM concentration; however, NO_x increased, which contributes to ground level ozone, smog, and acid rain impacts. A summary of literature related to the environmental impacts of food waste derived biodiesel production via transesterification is provided in Table 10.4.

10.3.4 Other Technologies

Numerous other FWtE technologies are being pursued via research, development, and commercialization. While the literature on these emerging technologies is less extensive, several LCAs have been carried out for technologies like bioelectrochemical systems (BES), pyrolysis, and gasification. Table 10.5 gives a summary of literature for FWtE technologies beyond AD, fermentation, and transesterification. Several studies involving thermal conversion technologies have used food waste functional units while those involving BES have used output energy as the functional unit. BES were reported to be environmentally efficient only when coproduct credits were allocated (Pant et al., 2011). When comparing BES with AD for energy generation or wastewater treatment, it is important to note that the energy generated and COD of the waste treated are an order of magnitude higher in AD compared to BES. Berge et al. (2015) assessed the impacts from energy generation using food waste-derived hydrochar via hydrothermal liquefaction. While hydrochar can offer benefits compared to coal, direct food waste combustion was more beneficial than hydrochar combustion. Also, the wastewater from hydrochar production had a significant environmental load. As certain FWtE technologies are still in their infancy, significant technological improvements are needed to realize greater environmental benefits. Ethanol production via gasification

2. https://www.afdc.energy.gov/vehicles/diesels_emissions.html.

TABLE 10.4 Summary of Selected Studies: Transesterification to Biodiesel

Reference	Feedstock	Functional Unit	System Boundaries	Impact Categories[a]	Key Findings
Ahamed et al. (2016)	Municipal food waste collected from households, food retail, and services in Singapore	1 ton of food waste was used to evaluate the potential waste treatment method	Waste-to-energy	GWP, AP, EP, CED	Treatment process (hydrothermal carbonization at 200° C) contributed the highest environment impacts (i.e., acidification potential and 100-year global warming potential) and cumulative energy demand (CED)
Bruton (2014)	Waste cooking oil (WCO) from university cafeteria	1600 MJ of energy output from a B5 blended heating fuel (5% biodiesel, 95% fuel oil)	Cradle-to-gate	GWP, CED	Biodiesel heating fuel blends of B75 and B45 were estimated to mark the critical point at which the production phase impacts in GWP and CED outweighed the impacts from the use phase
Dufour and Iribarren (2012)	Fatty acid-rich wastes (used cooking oil, beef tallow, poultry fat, and sewage sludge)	Production of 1 ton biodiesel from fatty acid-rich wastes and dispatched from biodiesel plants	Cradle-to-gate	GWP, AP, EP, ODP, POFP, CED	Feedstock drying and methanol production contributed significantly to overall energy consumption. Waste vegetable oil biodiesel was superior in environmental performance to low sulfur diesel and 1st generation soybean biodiesel
Frank (2014)	Waste cooking oil (WCO) from university cafeteria	100 km travel distance	Well-to-wheels	GWP, CED	By increasing the biodiesel component of the vehicle fuel blend, a decrease in life cycle impact was exhibited. The methanol solvent contributed the largest component to the energy and environmental impact
Hums et al. (2016)	Grease trap waste (GTW)	Production and combustion of 1 MJ of biodiesel from GTW	Well-to-wheels	GWP, CED, PM	Major impacts are associated with energy consumption in purification of GTW biodiesel to ASTM purity. GTW biodiesel increases GHG emission in waste management system, but reduces emissions significantly in the use phase
Rajaeifar et al. (2016)	Olive pomace	Combustion of 1 MJ of B20, B100 and petroleum diesel # 2	Well-to-wheels	HH, ESD, CC, Resources	B100 resulted in a 15% increase in NOx emissions compared to diesel #2. Olive agriculture contributed for >40% of the emissions and impacts. Authors recommended the use of B20 in Iran over B100 based on environmental performance
Sousa et al. (2017)	Beef tallow	1 MJ biodiesel plant	Cradle-to-gate	GWP, AP, EP	Biodiesel production offered significant reduction in GHG, AP, and EP compared to beef tallow production. Major contribution to GHG came from use of heavy oil to heat up the transesterification reactor
Tuntiwiwattanapun et al. (2017)	Spent coffee grounds (SCG)	1 kg of SCG biodiesel	Cradle-to-gate	HH, ESQ, CC, resources	A hexane-free production method resulted in higher energy requirement but improved human health benefits
Wang et al. (2015)	Corn oil from distillers dried grains with solubles (DDGS)	Production and combustion of 1 MJ of biodiesel from corn oil	Well-to-wheels	GWP, CED	Energy consumption in corn oil recovery process significantly influenced life cycle GHG emissions of corn oil biodiesel
Yang et al. (2017)	Yellow grease/Restaurant oil	100 km driven by buses in 2014 in Shanghai, China	Well-to-wheels	GWP, AP, urban air pollution	90% reduction in GHG and 46% reduction in particulates during use phase

[a]*See Table 10.1.*

TABLE 10.5 Summary of Selected Studies: Other Technologies Used in Food Waste-To-Energy Conversion

Reference	Feedstock	Functional Unit	Products/Process	Impact Categories[a]	Key Findings
Berge et al. (2015)	Food waste	1 kg of source separated food waste	Hydrochar for electricity generation	GWP, AP, EP, HT, ET, POF	Hydrochar offered environmental benefits compared with coal, but direct biomass combustion was superior to hydrochar combustion. If untreated, the liquid product resulting from carbonization imposes a significant environmental load
Ibarrola et al. (2012)	Several biodegradable wastes, including food waste	1 MWh electricity output	Energy from biochar and syngas. Comparison of slow pyrolysis, fast pyrolysis, and gasification with landfill and incineration reference cases	Carbon abatement	Slow pyrolysis offered the highest carbon abatement due to higher yield of biochar, irrespective of the feedstock, assuming biochar is used as either soil amendment or combusted for energy generation. Gasification of food waste performed the best regarding energy efficiency. In slow pyrolysis, carbon abatement was dominated by carbon storage offset and in fast pyrolysis and gasification by electricity offset. Higher carbon abatement when compared with landfill as reference case than with incineration
Pant et al. (2011)	Organic fraction of wastewater	1 kW/m3 of anode volume for energy production	Various co products/ bioelectrochemical conversion	GHG	Benefits can be realized when BES is integrated with energy generation and multiple product formation. Coproducts offer improved energy and environmental benefits.
Patterson et al. (2013)	Food waste and flour mill waste	The production of unit volume of biohydrogen/biomethane vehicle fuel	Hydrogen/biomethane for vehicle fuel application (one stage and two stage processes)	GWP	Combined methane and hydrogen production resulted in an increased environmental burden. Upgrading biogas from food waste to methane for vehicle fuel using a single-stage process was recommended over production/ upgrade to hydrogen
Stichnothe and Azapagic (2009)	Household waste and biodegradable municipal waste	Two FUs, treatment of 190,000 t/year of MSW and combustion of 1 MJ eqt fuel	Bioethanol/Gasification integrated with BAU	GHG	Significant emissions reductions during ethanol production compared to BAU (LF and incineration). Use of ethanol as gasoline substitute offers little or no advantage for GHG reduction
Yano et al. (2015)	Waste cooking oil	Treatment of 1142 kL/year and 1108 kL/year of waste cooking oil from households and businesses, 41.1 TJ of diesel fuel	Biodiesel/Hydrogenation process	GWP, FFD, AP, urban air pollution	A shift from FAME-type biodiesel to hydrogenated in the future would be more effective in reducing total environmental impacts in terms of GWP, fossil fuel consumption, urban air pollution, and AP

[a]*See Table 10.1.*

of household organic waste was reported to offer significant environmental benefits when integrated with "business as usual" waste management methods (Stichnothe and Azapagic, 2009). A shift from FAME-type biodiesel to hydrogenated biodiesel was found to be more effective in reducing total environmental impacts in terms of GWP, fossil fuel consumption, urban air pollution, and acidification (Yano et al., 2015).

10.4 COMPARISON OF TECHNOLOGIES

Because of the breadth of options for FWtE conversion, it is difficult to generalize results. Any comparison must consider the treatment of similar food waste feedstocks, and then the studies must carefully ensure comparable system boundaries and assumptions. However, comparison of the environmental impacts of FWtE technologies to incumbent processes for treatment of food waste or provision of energy carriers is critical to their successful implementation. Given these constraints, this section provides a review of comparative FWtE studies in an attempt to provide some bounded conclusions.

10.4.1 Food Waste Treatment Alternatives

A summary of comparative assessments in the literature is provided in Table 10.6. Bernstad and la Cour Jansen (2012) observed wide variation in results from a meta-analysis of 25 studies with the stated objective of evaluating different methods for treating the food waste fraction in MSW. Although some coherence was observed, variations in absolute values of GHG emissions from respective treatment alternatives as well as the ranking of different alternatives were striking, with incineration, AD, and composting each being ranked as the alternative most as well as the least contributing to GWP, while only landfilling in all cases was ranked as least beneficial alternative. Factors such as heterogeneity in food waste composition; emissions of carbon and loss of nutrients during the collection, storage, and pretreatment stages; potential energy recovery through combustion and AD; emissions from composting; emissions from storage and land use of biofertilizers; and avoided impacts from substituted goods have all been identified as highly relevant for the outcomes of comparative studies (Bernstad and la Cour Jansen, 2012). Therefore, reported differences in GWP may not be a result of actual differences in the environmental impacts from studied systems, but rather because of the differences in the how the LCA was conducted. Schott et al. (2016) provided an updated analysis reviewing 19 studies containing 103 different scenarios reaching a similar conclusion that GWP of food waste treatment alternatives can be mainly attributed to the choices made in relation to background systems (i.e., indirect emissions from energy, biofertilizer, and soil amendment substitutions), rather than emissions originating from the foreground system (direct emissions).

The majority of comparative studies concluded that AD offers the greatest GWP reductions compared to any other technology considered and is the preferable treatment alternative for food waste as long as energy is recovered in the process, as summarized in Table 10.6. It is important to note, however, that none of these studies have considered fermentation, pyrolysis, and gasification as food waste management systems. Land application of digestate where mineral fertilizer is displaced increases the GWP benefit, while assumptions concerning energy recovery in incineration, mineral fertilizer and peat displacement in composting, and landfill gas capture efficiency and carbon sequestration in landfilling are significant factors for alternative technologies. Fuel mix assumption for exported grid electricity is also a decisive factor for the GWP of AD. As grid mixes continue to lower emissions through increasing use of renewables, the intangible benefit of being able to store biomethane produced from AD as well as its potential for use in the transportation sector may make these end-use option more attractive in the future. Bernstad and la Cour Jansen (2011) asserted that AD offered greater benefits when biogas was used a vehicle fuel rather than for electricity.

10.4.2 Biofuel Comparisons

FW-derived biofuels are often compared to alternative biofuels or fossil fuels. In addition to avoiding the negative externalities of fossil fuel extraction and production, the CO_2 emissions resulting from combustion of biofuels are associated with the natural photosynthetic process and therefore considered biogenic. Waste-derived biofuels have the additional advantage of avoiding the environmental impacts of cultivation and land use associated with crop-based biofuels. In a review of biodiesel production from WCO, beef tallow, and chicken fat, Dufour and Iribarren (2012) concluded that these biofuels incurred lower impacts for every impact category relative to rapeseed biodiesel. The lipid content of the feedstock is important to the yield of biodiesel and therefore influences the environmental impacts per unit fuel produced. Hums et al. (2016) concluded that for all lipid contents studied, biodiesel made from grease trap waste (GTW) is environmentally preferred to low sulfur diesel and alternative treatment of GTW.

TABLE 10.6 Summary of Comparative Assessments of Food Waste Treatment Pathways

Reference	Feedstock	Functional Unit	FW Management Methods	Reference System	Impact Categories[a]	Key Findings
Ahamed et al. (2016)	Restaurant food waste	1 ton FW	Biodiesel from hydrothermal carbonization and AD	Incineration with energy production	AP, EP, GWP100, CED	For food waste with oil content >5%, biodiesel was the most preferred pathway and AD was preferred when oil <5%. AD was the most preferred option for system under study and incineration was the least preferred
Bernstad and la Cour Jansen (2011)	Household FW	24.9 kg organic waste per person per year	Anaerobic digestion and composting	Incineration with energy production	GWP, AP, NE, ODP, POF	The highest avoidance of environmental impact observed in a scenario with FW treatment using AD and use of biogas as vehicle fuel. The contribution to GWP was the largest in FW incineration and lowest in several scenarios involving anaerobic treatment
Edwards et al. (2018)	Municipal waste with a focus on FW treatment	Treatment of 1 years' municipal curb-side waste and 1 year' sewage sludge	AD, in-sink maceration, composting, mechanical treatment	Landfilling with energy recovery	ADP, AFFD, HT, ODP, POF, AP, EP, GWP	Anaerobic digestion-based scenarios outperformed centralized and home composting as well as BAU in terms of GWP. Mechanical biological treatment offered benefits in terms of ODP, EP, and POF. A negative GWP was estimated for AD with efficient household sorting of food waste
Eriksson et al. (2015)	Supermarket food waste	Treatment of 1 kg food waste, including packaging from super markets	Donation, composting, AD, incineration with energy recovery and animal feed	Landfilling without energy recovery	GHG emissions	Donation and AD were found to be the options with highest GHG reductions. The order of most preferred to least preferred alternatives were AD, donation, animal feed, incineration, composting, and landfill when solely considering GHG emissions
Gao et al. (2017)	University FW	2 ton FW per day	Composting, AD, incineration, heat moisture reaction (HMR)	Landfilling without energy recovery	CC, AP, EP, carcinogenicity	AD showed the lowest CC impact followed by composting, HMR, and incineration. Incineration performed the worst in terms of AP and EP. Composting showed highest carcinogenicity impact

Mondello et al. (2017)	Mass-retail FW	1 ton of FW	Composting, insect-based composting, AD, incineration	Landfilling with energy recovery	ADP, AP, EP, GWP100, ODP, HT, FAETP, TETP, LU	The results underscore the fact that the main environmental impacts are related to landfill and incineration. While insect-based composting showed the best performance in all impact categories except for GWP in which AD showed the lowest impacts
Opatokun et al. (2017)	Food waste	1 kg of FW	AD, pyrolysis, and integrated energy system (AD sequence with pyrolysis)	Landfilling without energy recovery	CC, OD, TA, FE, ME, HTox, POF, PMF, TEcox, FEcox, MEcox, WD, MD, FD	Environmental impacts and benefits of the integrated system are similar to AD alone
Thyberg and Tonjes (2017)	Municipal waste from suburban NY	1 ton Brookhaven residential residual MSW with a 100-year emissions time frame	Windrow composting, tunnel composting, and AD	Incineration with energy production	GWP, ODP, AP, EP, AFFD	Results indicated that overall environmental burdens can be reduced by source segregation of food waste and treating it by AD and then composting the AD residuals. However, in some impact categories, the business as usual scenario was a better choice than alternatives
Vandermeersch et al. (2014)	Food waste	1000 ton of food waste (with 100 ton of bread waste)	Anaerobic digestion and animal feed production	–	Exergetic analysis, 18 impact categories in mid-point and end-point, respectively	Showed that conversion of bread waste fraction into animal feed provides better environmental performance than valorization to produce electricity, heat, and digestate which would result from the avoided products from traditional animal feed produced from agricultural products. Also, exergy analysis is only focused on resource

[a]*See Table 10.1*

As part of the U.S. Renewable Fuel Standard (RFS), which mandates certain quantities of renewable fuels, the U.S. Environmental Protection Agency (EPA) has released life cycle GHG analyses for approved commercial pathways for biofuel production.[3] Biodiesel production from yellow grease using transesterifcation is reported to provide an 86% reduction in life cycle GHG emissions relative to the low sulfur petroleum diesel baseline, whereas soybean and palm oil biodiesel pathways are reported to deliver a reduction of 57% and 17% relative to the baseline, respectively. Eliminating upstream feedstock production is a large source of benefit where it contributes 24.8 kg CO_2e and 50.9 kg CO_2e per mmBtu for soybean and palm biodiesel, and can primarily be attributed to land use change impacts. The fuel production process for yellow grease also has lower impact (9.6 kg CO_2e per mmBtu) than alternative biodiesel pathways, including algal oil biodiesel (31.5kg CO_2e per mmBtu) which also incurs zero feedstock production impacts. Chen et al. (2017) analyzed high FFA tallow based biodiesel production with data from commercial plants and reported that both the rendering and transesterification processes contribute to greater GHG emissions compared to soy biodiesel production, because they are more energy and chemical intensive. However, because tallow biodiesel incurs no land use change from crop production, its overall emissions were still lower than those of soybean or canola based biodiesel (Chen et al., 2017).[3]

Data are limited for comparisons of waste derived bioethanol, but extensive research has been directed towards environmental impacts of corn starch and cellulosic ethanol. Wang et al. (2012) estimate that, on a life cycle basis, corn ethanol resulted in an average 34% reduction in GWP impact relative to gasoline. However, the ethanol production stage accounted for less than half of GHG emissions. Cultivation and land use impacts for corn ethanol account for 53% of GHG emissions; which would be eliminated with the use of FW feedstock. Thus, even with an increase in ethanol production impacts due to additional pre-treatment or inefficiencies, it is reasonable to assume that ethanol production from FW feedstock high in simple sugars or starch will result in significant reductions in impacts relative to gasoline. Similar insight into GWP impacts due to conversion of lignocellulosic FW can be extrapolated from cellulosic ethanol studies. For example, cellulosic ethanol could lower GHG emissions by more than 95% relative to gasoline (Wang et al., 2012). While cultivating cellulosic feedstock still incurs impacts from fertilizer production and application and equipment operation, land use change may result in sequestration of organic carbon in soil, reducing net GWP impacts for this stage. While the system configuration will vary depending upon FW characteristics, portions of the cellulosic conversion technology are likely to be similar in FWtE systems. However, modeling of cellulosic ethanol conversion processes typically includes an additional offset due to energy recovery from the combustion of lignin content from the feedstock. Because FW has lower lignin content, this offset may be reduced or replaced with an alternate coproduct in FWtE systems, pointing to a need for more comparative studies.

10.5 CONCLUSIONS

Sustainable food waste-to-energy systems, such as those presented in Chapters 4–9, are usually deployed with the expectation that they will minimize the environmental burden of conventional treatment methods, including landfilling, wastewater treatment, and incineration. LCA is a well-developed method that can be applied to assess the overall environmental impacts of FWtE systems and provide a consistent framework for comparing the environmental performance of different potential FWtE options. Appropriately defining the functional unit, often based on either a unit of energy output or unit of food waste treated, is a key initial step in executing an LCA. Equally critical to the process is unambiguous definition of the system boundary, which dictates the input and output materials flows and subprocesses contained within the overall FWtE system. Because of the dual objectives of most FWtE systems (i.e., food waste treatment combined with sustainable energy production), all of the various coproducts must be identified and represented within the LCA process. Many LCAs conducted to date have focused on GWP as the primary impact category, but others have considered eutrophication and acidification potential, particulate emissions, and CED, among others.

Among the technologies discussed in this chapter, AD is the most mature as a food waste conversion pathway, and many systems are already in operation worldwide. Because of the prevalence of commercial-scale system deployments, especially in Europe, there is a significant body of available data and a relatively large number of publications focused on applying LCA to quantify the associated environmental impacts. Most studies indicate that AD has the potential to reduce GWP and other impacts relative to incumbent systems, but composition of food waste feedstocks, uncontrolled methane leaks, coproducts generated (fertilizer, animal bedding, etc.), and reference systems displaced (grid electricity and synthetic fertilizer) all significantly influence the outcome. A number of LCA studies have also been published on fermentation and transesterification systems, but these are more limited in the types of feedstocks that have been considered, mostly

3. https://www.epa.gov/fuels-registration-reporting-and-compliance-help/lifecycle-greenhouse-gas-results.

sugar/starch-rich wastes and waste cooking oil, respectively. LCA has also been applied to emerging technologies, such as pyrolysis and BES, but with few commercial-scale deployments, representative model input data are lacking.

Although LCA is an effective tool to systematically compare environmental performance of different FWtE technologies, it should be recognized that the outcomes are strongly dependent on feedstock characteristics that can vary spatially and temporally, as well as many indirect effects such as grid electricity impact factors, distribution of transportation, water, natural gas and electricity infrastructures, etc. It is also difficult to compare biofuels (e.g., ethanol, biodiesel) which typically are based on a unit of fuel, to other food waste-to-energy systems (biogas, biothermal) for which the functional unit is based on a food waste treatment objective. To achieve true sustainability, food supply chain stakeholders also need to consider relative economics costs and benefits (Chapter 11) and the associated social impacts imposed in part by the existing policy and regulatory framework (Chapter 12).

REFERENCES

Ahamed, A., Yin, K., Ng, B.J.H., Ren, F., Chang, V.W.-C., Wang, J.-Y., 2016. Life cycle assessment of the present and proposed food waste management technologies from environmental and economic impact perspectives. J. Clean. Prod. 131, 607–614.

Angelonidi, E., Smith, S.R., 2015. A comparison of wet and dry anaerobic digestion processes for the treatment of municipal solid waste and food waste: Comparison of wet and dry anaerobic digestion processes. Water Environ. J. 29 (4), 549–557.

Artrip, K.G., Shrestha, D.S., Coats, E., Keiser, D., 2013. GHG emissions reduction from an anaerobic digester in a dairy farm: Theory and practice. Appl. Eng. Agric. 29 (5), 729–737.

Bacenetti, J., Sala, C., Fusi, A., Fiala, M., 2016. Agricultural anaerobic digestion plants: what LCA studies pointed out and what can be done to make them more environmentally sustainable. Appl. Energy 179, 669–686.

Berge, N.D., Li, L., Flora, J.R.V., Ro, K.S., 2015. Assessing the environmental impact of energy production from hydrochar generated via hydrothermal carbonization of food wastes. Waste Manag. 43, 203–217.

Berłowska, J., Pielech-Przybylska, K., Balcerek, M., Dziekońska-Kubczak, U., Patelski, P., Dziugan, P., Kręgiel, D., 2016. Simultaneous saccharification and fermentation of sugar beet pulp for efficient bioethanol production. Biomed. Res. Int. 2016, 1–10.

Bernstad, A., la Cour Jansen, J., 2011. A life cycle approach to the management of household food waste—a Swedish full-scale case study. Waste Manag. 31 (8), 1879–1896.

Bernstad, A., la Cour Jansen, J., 2012. Review of comparative LCAs of food waste management systems—Current status and potential improvements. Waste Manag. 32, 2439–2455.

Börjesson, P., Berglund, M., 2006. Environmental systems analysis of biogas systems. Part I. Fuel-cycle emissions. Biomass Bioenergy 30 (5), 469–485.

Börjesson, P., Berglund, M., 2007. Environmental systems analysis of biogas systems—Part II: the environmental impact of replacing various reference systems. Biomass Bioenergy 31 (5), 326–344.

Borrion, A.L., McManus, M.C., Hammond, G.P., 2012. Environmental life cycle assessment of lignocellulosic conversion to ethanol: a review. Renew. Sust. Energ. Rev. 16 (7), 4638–4650. https://doi.org/10.1016/j.rser.2012.04.016.

Bruton, D.J., 2014. Waste Cooking Oil-to-Biodiesel Conversion for Space Heating Applications. Rochester Institute of Technology.

Chen, T., Shen, D., Jin, Y., Li, H., Yu, Z., Feng, H., et al., 2017. Comprehensive evaluation of environ-economic benefits of anaerobic digestion technology in an integrated food waste-based methane plant using a fuzzy mathematical model. Appl. Energy 208, 666–677.

Cherubini, F., Strømman, A.H., 2011. Life cycle assessment of bioenergy systems: state of the art and future challenges. Bioresour. Technol. 102 (2), 437–451.

Chester, M., Martin, E., 2009. Cellulosic ethanol from municipal solid waste: a case study of the economic, energy, and greenhouse gas impacts in California. Environ. Sci. Technol. 43 (14), 5183–5189.

Cheung, C.S., Man, X.J., Fong, K.W., Tsang, O.K., 2015. Effect of waste cooking oil biodiesel on the emissions of a diesel engine. Energy Procedia 66, 93–96.

Christensen, T.H., Gentil, E., Boldrin, A., Larsen, A.W., Weidema, B.P., Hauschild, M., 2009. C balance, carbon dioxide emissions and global warming potentials in LCA-modelling of waste management systems. Waste Manag. Res. 27 (8), 707–715.

de Araújo, C.D.M., de Andrade, C.C., e Silva, E.D.S., Dupas, F.A., 2013. Biodiesel production from used cooking oil: a review. Renew. Sust. Energ. Rev. 27, 445–452.

Dufour, J., Iribarren, D., 2012. Life cycle assessment of biodiesel production from free fatty acid-rich wastes. Renew. Energy 38 (1), 155–162.

Dunn, J.B., Mueller, S., Wang, M., Han, J., 2012. Energy consumption and greenhouse gas emissions from enzyme and yeast manufacture for corn and cellulosic ethanol production. Biotechnol. Lett. 34 (12), 2259–2263. https://doi.org/10.1007/s10529-012-1057-6.

Ebner, J., Babbitt, C., Winer, M., Hilton, B., Williamson, A., 2014. Life cycle greenhouse gas (GHG) impacts of a novel process for converting food waste to ethanol and co-products. Appl. Energy 130, 86–93.

Ebner, J.H., et al., 2016. Anaerobic co-digestion of commercial food waste and dairy manure: characterizing biochemical parameters and synergistic effects. Waste Manag. 52 (2016), 286–294.

Edwards, J., Othman, M., Crossin, E., Burn, S., 2018. Life cycle assessment to compare the environmental impact of seven contemporary food waste management systems. Bioresour. Technol. 248, 156–173.

Ekvall, T., et al., 2007. What life-cycle assessment does and does not do in assessments of waste management. Waste Manag. 27 (8), 989–996.

Eriksson, M., Strid, I., Hansson, P.A., 2015. Carbon footprint of food waste management options in the waste hierarchy—a Swedish case study. J. Clean. Prod. 93, 115–125.

Evangelisti, S., Lettieri, P., Borello, D., Clift, R., 2014. Life cycle assessment of energy from waste via anaerobic digestion: a UK case study. Waste Manag. 34 (1), 226–237.

Finnveden, G., Jessica Johansson, and Lind, P., Life Cycle Assessments of Energy from Solid Waste, 2000, Stockholm University, ISBN 91-7056-103-6, http://citeseerx.ist.psu.edu/viewdoc/download?doi=10.1.1.197.4254&rep=rep1&type=pdf.

Finnveden, G., Johansson, J., Lind, P., Moberg, Å., 2005. Life cycle assessment of energy from solid waste. Part 1. General methodology and results. J. Clean. Prod. 13 (3), 213–229.

Franchetti, M., 2013. Economic and environmental analysis of four different configurations of anaerobic digestion for food waste to energy conversion using LCA for: a food service provider case study. J. Environ. Manag. 123, 42–48.

Frank, D.E., 2014. Waste Cooking Oil-To-Biodiesel Conversion for Institutional Vehicular Applications. Rochester Institute of Technology.

Fruergaard, T., Astrup, T., 2011. Optimal utilization of waste-to-energy in an LCA perspective. Waste Manag. 31 (3), 572–582.

Gao, A., Tian, Z., Wang, Z., Wennersten, R., Sun, Q., 2017. Comparison between the technologies for food waste treatment. Energy Procedia 105, 3915–3921.

Gerbrandt, K., Chu, P.L., Simmonds, A., Mullins, K.A., MacLean, H.L., Griffin, W.M., Saville, B.A., 2016. Life cycle assessment of lignocellulosic ethanol: a review of key factors and methods affecting calculated GHG emissions and energy use. Curr. Opin. Biotechnol. 38, 63–70.

Guerrero, A.B., Muñoz, E., 2018. Life cycle assessment of second generation ethanol derived from banana agricultural waste: environmental impacts and energy balance. J. Clean. Prod. 174, 710–717.

Hegde, S., Lodge, J.S.J.S., Trabold, T.A.T.A., 2018. Characteristics of food processing wastes and their use in sustainable alcohol production. Renew. Sust. Energ. Rev. 81, 510–523.

Hong, Y., Nizami, A.S., Pour Bafrani, M., Saville, B.A., Maclean, H.L., 2013. Impact of cellulase production on environmental and financial metrics for lignocellulosic ethanol. Biofuels Bioprod. Biorefin. 7 (3), 303–313.

Humbird, D., Davis, R., Tao, L., Kinchin, C., Hsu, D., Aden, A., Dudgeon, D., 2011. Process Design and Economics for Biochemical Conversion of Lignocellulosic Biomass to Ethanol: Dilute-Acid Pretreatment and Enzymatic Hydrolysis of Corn Stover. National Renewable Energy Laboratory (NREL).

Hums, M.E., Cairncross, R.A., Spatari, S., 2016. Life-cycle assessment of biodiesel produced from grease trap waste. Environ. Sci. Technol. 50 (5), 2718–2726.

Ibarrola, R., Shackley, S., Hammond, J., 2012. Pyrolysis biochar systems for recovering biodegradable materials: a life cycle carbon assessment. Waste Manag. 32 (5), 859–868.

IPCC, 2014. Climate change 2014: synthesis report. In: Core Writing Team,, Pachauri, R.K., Meyer, L.A. (Eds.), Contribution of Working Groups I, II and III to the Fifth Assessment Report of the Intergovernmental Panel on Climate Change. IPCC, Geneva, Switzerland (151 pp.).

IPCC, National Greenhouse Gas Inventories Programme, 2006. IPCC guidelines for national greenhouse gas inventories. In: Eggleston, H.S., Buendia, L., Miwa, K., Ngara, T., Tanabe, K. (Eds.), Intergovernmental Panel on Climate Change (IPCC). IGES, Japan.

Islam, M.A., Heimann, K., Brown, R.J., 2017. Microalgae biodiesel: current status and future needs for engine performance and emissions. Renew. Sust. Energ. Rev. 79, 1160–1170.

ISO, Organización Internacional de Normalización, 2006a. ISO 14044: Environmental Management, Life Cycle Assessment, Requirements and Guidelines.

ISO, Organización Internacional de Normalización, 2006b. ISO 14040: Environmental Management—Life Cycle Assessment—Principles and Framework.

Janssen, R., Rutz, D., 2008. Carbon Labelling in Europe. Doctoral Dissertation.

Jin, Y., Chen, T., Chen, X., Yu, Z., 2015. Life-cycle assessment of energy consumption and environmental impact of an integrated food waste-based biogas plant. Appl. Energy 151, 227–236.

Kalogo, Y., Habibi, S., MacLean, H., Joshi, S.V., 2007. Environmental implications of municipal solid waste-derived ethanol. Environ. Sci. Technol. Am. Chem. Soc. 41 (1), 35–41.

Kiran, E.U., et al., 2014. Bioconversion of food waste to energy: a review. Fuel 134, 389–399.

Knauf, M., Krauss, K., 2006. Specific yeasts developed for modern ethanol production. Zuckerindustrie 131 (11), 753–758.

Liebetrau, J., Reinelt, T., Clemens, J., Hafermann, C., Friehe, J., Weiland, P., 2013. Analysis of greenhouse gas emissions from 10 biogas plants within the agricultural sector. Water Sci. Technol. 67 (6), 1370–1379.

MacLean, H.L., Spatari, S., 2009. The contribution of enzymes and process chemicals to the life cycle of ethanol. Environ. Res. Lett. 4 (1), 014001.

Manfredi, S., 2011. ILCD, Supporting Environmentally Sound Decisions for Bio-waste Management—A practical guide to Life Cycle Thinking (LCT) and Life Cycle Assessment (LCA). https://ec.europa.eu/jrc/en/publication/eur-scientific-and-technical-research-reports/supporting-environmentally-sound-decisions-bio-waste-management-practical-guide-life-cycle.

Møller, J., Boldrin, A., & Christensen, T. H. 2009. Anaerobic digestion and digestate use: accounting of greenhouse gases and global warming contribution. Waste Manag. Res., 27(8), 813–824.

Mondello, G., Salomone, R., Ioppolo, G., Saija, G., Sparacia, S., Lucchetti, M.C., 2017. Comparative LCA of alternative scenarios for waste treatment: the case of food waste production by the mass-retail sector. Sustainability (Switzerland). 9(5).

Nielsen, P.H., Oxenbøll, K.M., Wenzel, H., 2007. Cradle-to-gate environmental assessment of enzyme products produced industrially in Denmark by Novozymes A/S. Int. J. Life Cycle Assess. 12 (6), 432.

Opatokun, S.A., Lopez-Sabiron, A., Ferreira, G., Strezov, V., 2017. Life cycle analysis of energy production from food waste through anaerobic digestion, pyrolysis and integrated energy system. Sustainability 9 (10), 1804.

Pant, D., Singh, A., Van Bogaert, G., Gallego, Y. A., Diels, L., & Vanbroekhoven, K. 2011. An introduction to the life cycle assessment (LCA) of bioelectrochemical systems (BES) for sustainable energy and product generation: Relevance and key aspects. Renew. Sust. Energ. Rev., 15(2), 1305–1313.

Patterson, T., Esteves, S., Dinsdale, R., Guwy, A., Maddy, J., 2013. Life cycle assessment of biohydrogen and biomethane production and utilisation as a vehicle fuel. Bioresour. Technol. 131, 235–245.

Pietrzak, W., Kawa-Rygielska, J., 2014. Ethanol fermentation of waste bread using granular starch hydrolyzing enzyme: effect of raw material pretreatment. Fuel 134, 250–256.

Poeschl, M., Ward, S., Owende, P., 2012a. Environmental impacts of biogas deployment. Part I. Life cycle inventory for evaluation of production process emissions to air. J. Clean. Prod. 24, 168–183.

Poeschl, M., Ward, S., Owende, P., 2012b. Environmental impacts of biogas deployment. Part II. Life cycle inventory for evaluation of production process emissions to air. J. Clean. Prod. 24, 184–201.

Pourbafrani, M., McKenchnie, J., MacLean, H., Bradley, S., 2013. Life cycle greenhouse gas impacts of ethanol, biomethane and limonene production from citrus waste. Environ. Res. Lett. 8 (1), 1–12.

Rodríguez, L.A., Toro, M.E., Vazquez, F., Correa-Daneri, M.L., Gouiric, S.C., Vallejo, M.D., 2010. Bioethanol production from grape and sugar beet pomaces by solid-state fermentation. Int. J. Hydrog. Energy 35 (11), 5914–5917.

Rajaeifar, M.A., Akram, A., Ghobadian, B., Rafiee, S., Heijungs, R., Tabatabaei, M., 2016. Environmental impact assessment of olive pomace oil biodiesel production and consumption: a comparative lifecycle assessment. Energy 106.

Schmitt, E., Bura, R., Gustafson, R., Cooper, J., Vajzovic, A., 2012. Converting lignocellulosic solid waste into ethanol for the state of Washington: An investigation of treatment technologies and environmental impacts. Bioresour. Technol. 104, 400–409.

Schott, A.B.S., Wenzel, H., la Cour Jansen, J., 2016. Identification of decisive factors for greenhouse gas emissions in comparative life cycle assessments of food waste management—an analytical review. J. Clean. Prod. 119, 13–24.

Sheehan, J., Camobreco, V., Duffield, J., Graboski, M., Graboski, M., Shapouri, H., 1998. Life Cycle Inventory of Biodiesel and Petroleum Diesel for Use in an Urban Bus (No. NREL/SR-580-24089). National Renewable Energy Lab (NREL), Golden, CO (United States).

Sousa, V.M.Z., Luz, S.M., Caldeira-Pires, A., Machado, F.S., Silveira, C.M., 2017. Life cycle assessment of biodiesel production from beef tallow in Brazil. Int. J. Life Cycle Assess. 22 (11), 1837–1850.

Souza Filho, P.F., Brancoli, P., Bolton, K., Zamani, A., Taherzadeh, M.J., 2017. Techno-economic and life cycle assessment of wastewater management from potato starch production: present status and alternative biotreatments. Fermentation 3 (4), 56.

Stichnothe, H., Azapagic, A., 2009. Bioethanol from waste: Life cycle estimation of the greenhouse gas saving potential. Resour. Conserv. Recycl. 53 (11), 624–630.

Thompson, J.C., He, B.B., 2006. Characterization of crude glycerol from biodiesel production from multiple feedstocks. Appl. Eng. Agric. 22 (2), 261–265.

Thyberg, K.L., Tonjes, D.J., 2017. The environmental impacts of alternative food waste treatment technologies in the U.S. J. Clean. Prod. 158, 101–108.

Tonini, D., Astrup, T., 2012. Life-cycle assessment of a waste refinery process for enzymatic treatment of municipal solid waste. Waste Manag. 32 (1), 165–176.

Tuntiwiwattanapun, N., Usapein, P., Tongcumpou, C., 2017. The energy usage and environmental impact assessment of spent coffee grounds biodiesel production by an in-situ transesterification process. Energy Sustain. Dev. 40, 50–58.

Ujor, V., Bharathidasan, A.K., Cornish, K., Ezeji, T.C., 2014. Feasibility of producing butanol from industrial starchy food wastes. Appl. Energy 136, 590–598.

Upadhyay, A., Lama, J.P., Tawata, S., 2013. Utilization of pineapple waste: a review. J. Food Sci. Technol. Nepal 6, 10–18.

Vandermeersch, T., Alvarenga, R.A.F., Ragaert, P., Dewulf, J., 2014. Environmental sustainability assessment of food waste valorization options. Resour. Conserv. Recycl. 87, 57–64.

Velásquez-Arredondo, H.I., Ruiz-Colorado, A.A., De Oliveira, S., 2010. Ethanol production process from banana fruit and its lignocellulosic residues: energy analysis. Energy 35 (7), 3081–3087.

Walker, K., Vadlani, P., Madl, R., Ugorowski, P., Hohn, K.L., 2013. Ethanol fermentation from food processing waste. Environ. Prog. Sustain. Energy 32 (4), 1280–1283.

Wang, Z., Dunn, J.B., Han, J., Wang, M.Q., 2015. Influence of corn oil recovery on life-cycle greenhouse gas emissions of corn ethanol and corn oil biodiesel. Biotechnol. Biofuels 8 (1), 178.

Wang, M., Han, J., Dunn, J.B., Cai, H., Elgowainy, A., 2012. Well-to-wheels energy use and greenhouse gas emissions of ethanol from corn, sugarcane and cellulosic biomass for US use. Environ. Res. Let. 7 (4), 045905.

Win, S.S., Hedge, S., Trabold, T.A., 2015. In: Techno-economic assessment of different pathways for utilizing glycerol derived from waste cooking oil-based biodiesel. Proceedings of the ASME Power and Energy Conversion Conference, Paper Power Energy 2015–49563, San Diego, CA, June 28–July 2.

Woon, K.S., Lo, I.M., Chiu, S.L., Yan, D.Y., 2016. Environmental assessment of food waste valorization in producing biogas for various types of energy use based on LCA approach. Waste Manag. 50, 290–299.

Wulf, S., Jäger, P., Döhler, H., 2006. Balancing of greenhouse gas emissions and economic efficiency for biogas-production through anaerobic co-fermentation of slurry with organic waste. Agric. Ecosyst. Environ. 112 (2–3), 178–185.

Xu, C., Shi, W., Hong, J., Zhang, F., Chen, W., 2015. Life cycle assessment of food waste-based biogas generation. Renew. Sust. Energ. Rev. 49, 169–177.

Yang, F., Hanna, M.A., Sun, R., 2012. Value-added uses for crude glycerol—a byproduct of biodiesel production. Biotechnol. Biofuels 5 (1), 13.

Yang, Y., Fu, T., Bao, W., Xie, G.H., 2017. Life cycle analysis of greenhouse gas and PM2.5 emissions from restaurant waste oil used for biodiesel production in China. Bioenergy Res. 10 (1), 199–207.

Yano, J., Aoki, T., Nakamura, K., Yamada, K., Sakai, S.S., 2015. Life cycle assessment of hydrogenated biodiesel production from waste cooking oil using the catalytic cracking and hydrogenation method. Waste Manag. 38 (1), 409–423.

Yanowitz, J., McCormick, R.L., 2009. Effect of E85 on tailpipe emissions from light-duty vehicles. J. Air Waste Manag. Assoc. 59 (2), 172–182.

Chapter 11

Economic Aspects of Food Waste-to-Energy System Deployment

Somik Ghose and Matthew J. Franchetti

Department of Mechanical, Industrial and Manufacturing Engineering, The University of Toledo, Toledo, OH, United States

11.1 INTRODUCTION

Like any business endeavor, a food waste-to-energy project should undergo a thorough economic feasibility analysis that includes a budget of all expenditures and revenues, sources of funds, and any associated debt service on borrowed funds. Project development and successful implementation of food waste-to-energy systems is dependent on several considerations, from the characteristics and quantity of the food waste to the choice of energy conversion system and the revenue streams from waste tipping fees, energy products, and by-products.

Many biomass power generation options that may be applied to food waste to energy are mature, commercially available technologies (e.g., anaerobic digestion, direct combustion in stoker boilers, low-percentage cofiring, municipal solid waste incineration, and combined heat and power). While others are less mature and only at the beginning of their deployment (e.g., atmospheric biomass gasification and pyrolysis), still others are only at the demonstration or R&D phases (e.g., integrated gasification combined cycle, biorefineries, biohydrogen) (International Renewable Energy Agency (IRENA), 2012).

Anaerobic digestion (AD) systems have been most successful for food waste-to-energy applications and are often suitable for liquid, wet, and high moisture-content organic wastes. AD systems codigesting food waste with other organic waste, such as livestock manure or wastewater sludge, are operational in large and medium scales at single-site livestock farms, centralized locations for multiple farms, and water resource recovery facilities (WRRF). AD systems operating solely on mixed food waste at centralized locations and at large food waste generators are increasing.

The cost for the commercially mature technologies, including AD, are high due to large capital equipment and infrastructure investments for large projects; further, many of these technologies including AD are today only commercially available and financially viable at large scales.

11.2 PROJECT FEASIBILITY CONSIDERATIONS

11.2.1 Technical Feasibility

11.2.1.1 Waste Quantity, Characteristics, and Energy Potential Estimation

An initial step in a waste-to-energy project development is the technical evaluation and characterization of the food waste and the energy generation potential. Waste quantity, generation location and characteristics such as percentage content of moisture, biodegradable organic matter, calorific or heating value, fats and oils, sugars and carbohydrate, transportability, particle size, any periodic variations, and potential contamination are important factors in the choice of conversion technology and the scale and location of the waste-to-energy system.

For example, low-solids wet anaerobic digesters can only handle pumpable feedstock with total solids concentration in the 5%–15% range while some high-solids dry anaerobic digester designs can handle stackable dry feedstock with solids concentration in the 25%–40% range; although there is no standard cutoff point. Wet waste feedstocks are generally suitable for biochemical conversion technologies like AD and fermentation while drier waste feedstock with low moisture content are generally essential for thermochemical conversion technologies like gasification and pyrolysis. Prepared feedstock for processes such as AD, fermentation, liquefaction, and hydrolysis may be diluted with process water to achieve the desirable solids content during the preparation stages, while drying of feedstock is usually cost prohibitive without

Sustainable Food Waste-to-Energy Systems. https://doi.org/10.1016/B978-0-12-811157-4.00011-5

inexpensive heat sources such as solar drying or waste heat recovery. Fermentation is essentially suitable for wet waste and wastewater feedstock with high sugar and carbohydrate content while transesterification requires feedstock with higher fat, oil and grease content. Postconsumer food waste from residential, institutional, and some commercial sources typically have high moisture content of 70% or more and heating value ranging from 1,500 to 3,000 Btu per pound.

Moisture and solids content further impact the material handling, transportability, and associated costs for the waste; liquid wastes may be pumpable but would require tanker trucks for off-site transport, while solid waste could be transported in garbage trucks but could require conveyors or augers for handling. Large particle sizes in the waste may require size reduction through grinders for use as suitable feedstock.

11.2.1.2 Reliability and Operational Complexity

The reliability of a food waste-to-energy project can depend on the quality, quantity, and consistent availability of food waste streams. The challenges of creating energy from food waste relate to the receiving, conditioning, and feeding of the food waste into the biochemical or thermochemical waste-to-energy conversion system. Unfortunately, food wastes can have significant quantities of contaminants such as plastic and metals, which could harm waste treatment plant mechanical equipment. In general, source-separated food wastes are most desirable since this type of waste requires minimal processing at the waste-to-energy plant. Once the food waste is sorted, it is important that it is metered into the system to prevent an upset of the process due to overfeeding.

A food waste-to-energy program should have a modest impact on operations. Impacts come from the construction, operations and maintenance of the receiving station, administration of the food waste program, and the O&M of the generation equipment. The receiving station could require an area for the hauler interconnection equipment and the drive-up pad. The receiving station and generation equipment pads would be modest in size. Careful consideration must be given to the design to minimize impacts on existing plant operations. The receiving and odor control equipment will require frequent cleaning and periodic maintenance. There also is a risk that haulers could bring in toxic or other undesirable materials to the facility that could harm the process. This risk can be mitigated either by clear rules, strict manifest requirements for waste haulers, and/or sampling of the waste received. Testing of the sampled waste usually is not done unless a hauler created a problem with the system operation. However, this sampling technique has been used by other agencies as an effective risk management tool.

11.2.2 Financial Feasibility

A financial feasibility assessment for a food waste-to-energy project should evaluate all the financial aspects of implementing and operating the project through its life time. This assessment would include items like project budget and start-up capital costs, operating expenses, revenues, sources of funds, loan and interest repayments on borrowed capital, and investor income and disbursements.

11.2.2.1 Project Budget

Project budget is a detailed estimate of all the costs required to complete project implementation and operation, and the revenues from the project. Project implementation costs could typically include design, engineering, and project management costs; land, building, and equipment including waste and by-products handling, and energy conversion and transport; and any regulatory permit and license to install and operate. Operating costs typically include administrative and technical personnel costs, costs in waste handling and system maintenance including equipment servicing and replacement.

11.2.2.2 Levelized Cost of Electricity (LCOE) Generation

The LCOE is the price of electricity required for a project where revenues would equal costs, including making a return on the capital invested equal to the discount rate. An electricity price above this would yield a greater return on capital while a price below it would yield a lower return on capital or even a loss.

In LCOE cost and performance analysis, the energy recovery is compared on the basis of electricity production for AD, conventional combustion and gasification. These processes can also produce heat, or heat combined with electricity (CHP), further enhancing their efficiency. In the case of AD, it is also possible to produce a pipeline quality biogas that can be used as fuel. Comparing on the basis of electricity, however, provides a good comparative view of how much energy can be recovered per tonne of input material.

The LCOE varies by technology, country, and project, based on the renewable energy resource, capital and operating costs, and the efficiency/performance of the technology. The approach used is based on a discounted cash flow (DCF) analysis (International Renewable Energy Agency (IRENA), 2012). The weighted average cost of capital (WACC), often also referred to as the discount rate, is an important part of the information required to evaluate biomass power generation projects and has an important impact on the LCOE.

11.2.2.3 Capital Cost

The total investment cost, also referred to as capital expenditure (CAPEX), for food waste-to-energy systems could be highly dependent on the technology and its mechanical system design and degree of sophistication. Many commercial-scale systems and most pilot-scale commercial demonstration systems with newer waste-to-energy technologies are owned and operated by private companies, and very little data are available on capital and operating costs. Among commercial-scale technologies, wet anaerobic digesters (codigesting food waste at livestock farms, municipal wastewater treatment plants, and some stand-alone bioenergy plants) and various incineration plants (cofiring of food waste with other biomass and waste feedstock and utilizing conventional combustion technologies like moving-grate mass burners, fluidized-bed burners, and modular close-coupled gasification systems) have longer operational histories. Technologies like fermentation and transesterification are in commercial-scale operations with conventional sugar and oil feedstocks, respectively, and their application to food waste currently is mostly in separate pilot-scale batch operations. Other newer waste-to-energy technologies such as gasification and pyrolysis, liquefaction and hydrolysis, as applied to food waste, are mostly in development and pilot-scale demonstration stages.

The International Renewable Energy Agency's (IRENA) publication on Biomass for Power Generation (2012) (International Renewable Energy Agency (IRENA), 2012) provides the following summary in Table 11.1 on capital costs and levelized cost of electricity for different biomass power technologies.

In Table 11.1, the total investment cost or capital expenditure (CAPEX) consists of the equipment (prime mover and fuel conversion system), fuel handling and preparation machinery, engineering and construction costs, and planning. It can also include grid connection, roads, and any new infrastructure or improvements to existing infrastructure required for the project. Assuming a cost of capital of 10%, the LCOE of biomass-fired electricity generation ranges from a low of USD 0.06/kWh to a high of USD 0.29/kWh.

The feedstock conversion system comprises boilers (stoker, fluidized bed, etc.), gasifiers, and anaerobic digesters with a gas collection system, as well as the gas cleaning systems for gasifiers and gas treatment systems for AD systems. The prime mover is the power generation technology and includes any inline elements, such as particulate matter filters, etc. A breakdown of the typical cost structure of different biomass power generation technologies indicates the prime mover, feedstock conversion technology, and feedstock preparation and handling machinery accounts for between 62% and 77% of the capital costs for the biomass power generation technologies.

TABLE 11.1 Typical Capital Costs and the Levelized Cost of Electricity (LCOE) of Biomass Power Technologies

	Investment Costs (USD/kW)[a]	LCOE Range (USD/kWh)[a]
Stoker boiler	1,880–4,260	0.06–0.21
Bubbling and circulating fluidized boilers	2,170–4,500	0.07–0.21
Fixed- and fluidized-bed gasifiers	2,140–5,700	0.07–0 24
Stoker CHP	3,550–6,820	0.07–0.29
Gasifier CHP	5,570–6,545	0.11–0.28
Landfill gas	1,917–2,436	0.09–0.12
Digesters	2,574–6,104	0.06–0.15
Cofiring	140–850	0.04–0.13

[a]*All costs are in 2010 USD.*
Adapted from International Renewable Energy Agency (IRENA), June 2012. Renewable Energy Technologies: Cost Analysis Series—Biomass for Power Generation, https://www.irena.org/DocumentDownloads/Publications/RE_Technologies_Cost_Analysis-BIOMASS.pdf (Retrieved June 2017).

Additionally, a 2016 study from Canada provides the following summary of capital cost for waste-to-energy technologies (Regional District of Nanaimo (Canada), 2016).

1. AD costs: The capital costs of AD facilities vary widely, depending on size, location, type of technology, and efficiency; capital costs from known Canadian facilities show an average of $500 (USD 2007) per metric tonne of installed annual capacity.
2. Conventional combustion/WTE costs: With a worldwide inventory of over 600 conventional combustion or WTE facilities, statistical information is readily available to characterize the costs of this technology. The average capital cost for a 50,000 metric tonne per year facility is about $1,200 (USD 2007) per tonne of installed capacity.
3. Gasification and pyrolysis costs: There are numerous firms that offer gasification and pyrolysis systems for MSW; however, many are at a demonstration or pilot scale, and very few plants have actually been built. For a capacity of 100,000 metric tonnes of feedstock per year (or about 14 tonnes per hour), the capital costs are projected at $100 million (USD 2007) for a plant in Edmonton by Enerkem, plus $40 million for converting feedstock into refuse derived fuel (RDF). This results in capital cost of about $1,400 (USD 2007) per metric tonne of installed annual capacity. Comparison of conventional WTE with the Enerkem facility costs indicates that gasification technologies may cost about 20% more than conventional WTE plants.
4. Mixed waste processing facilities (MWPF) costs: Most MWPF in North America are owned and operated by private companies and very little data are available on capital costs. One example that might be used is a single stream clean MRF recently built in London, ON. The MRF in London cost $22.4 million for a capacity of 75,000 metric tonnes per year or about $300 per metric tonne of installed capacity. Based on a 2015 study "Mixed Waste Processing Economic and Policy Study" by Burns & McDonnell, for a 300,000 tonnes per year facility, capital costs would be about $200 per tonne of installed capacity.

11.2.2.4 Operating Cost

Operation and maintenance (O&M) cost, also referred to as operational and maintenance expenditure (OPEX), includes the fixed and variable costs associated with operating a food waste-to-energy plant. According to IRENA Biomass for Power Generation (2012) (International Renewable Energy Agency (IRENA), 2012), the O&M cost for biomass power plants typically ranges from 1% to 6% of the initial capital expenditure (CAPEX) per year. Fixed O&M costs are incurred independent of energy generation and include labor, scheduled maintenance, routine equipment replacement, insurance, etc. As plant size increases, specific (per kW) fixed O&M costs typically decline, due to economies of scale. Variable O&M costs change with the system output and typically include nonbiomass fuel costs, residue disposal, chemicals, and unplanned maintenance (Table 11.2).

Additionally, the 2016 study from Canada provides the following summary of operating cost for waste-to-energy technologies (Regional District of Nanaimo (Canada), 2016).

1. AD: Most AD plants are privately operated, so reliable operating costs are difficult to obtain. A reasonable preliminary assumption for study purposes is approximately $90 (USD 2007) per metric tonne.
2. Conventional combustion/WTE: The operating costs would be in the range of $115 (USD 2007) per metric tonne.

TABLE 11.2 Fixed and Variable Operations and Maintenance Costs for Biomass Power

Technology	Fixed O&M (% of Installed Cost)	Variable O&M (2010 USD/MWh)
Stokers/fluidized bed boilers	3.2–4.2 3–6	3.8–4.7
Gasifiers	3 6	3.7
AD systems	2.1–3.2 2.3–7	4.2
Landfill gas systems	11–20	n.a.

Adapted from from International Renewable Energy Agency (IRENA), June 2012. Renewable Energy Technologies: Cost Analysis Series—Biomass for Power Generation, https://www.irena.org/DocumentDownloads/Publications/RE_Technologies_Cost_Analysis-BIOMASS.pdf (Retrieved June 2017).

3. Gasification and pyrolysis costs: Similar to capital costs, there are few reference facilities providing any kind of reliable costs. For gasification, $40 (USD 2007) per metric tonne for feedstock preparation should be assumed and added to conventional WTE operations costs.
4. Mixed waste processing facilities (MWPF) costs: Operation and maintenance costs for MWPF are directly related to throughput and complexity of equipment. Operating costs for MWPF are not generally public information. Burns & McDonnell (2015) calculated a cost of $36 per metric tonne for a 300,000 tonne per year facility. Generally net operating costs per tonne, after sales of recyclables, can be in the $30 to $50 range; although some reports indicate potential costs of over $100 per tonne.

11.2.2.5 Hauling Cost

The cost of transporting food waste from generators to landfills or WTE facilities is largely dependent on the distance between the two. To estimate hauling costs, the project developer could begin by calculating the distance (in miles) between each generator and its assigned disposal facility. The estimate involves calculating the number of truckloads of food waste produced by each generator in a given week, based on generator-specific estimates of weekly food waste tonnage and an assumed 20-tonne capacity of long-haul collection trucks. The calculation multiplies the number of truckloads of food waste by the hauling distance and the standard $4 per mile cost, based on a 2013 analysis of the economic benefits of food waste collection and composting in New York City (Global Green USA, 2013).

Generators may increase the frequency of food waste collection to help control odors. To account for the potential increase in collection frequency, the estimate could assume any generator producing up to 20 tonnes (one truckload) of food waste per week would increase food waste collection to two times (or two, partially empty 20-tonne truckloads) per week. For generators producing >20 tonnes of food waste per week, the estimate could assume collection frequency, as approximated by truckloads per week, remains unchanged.

11.2.2.6 Borrowed Capital and Debt Service (Capital Repayment and Interest Expense)

Borrowed capital consists of funds borrowed from either individuals such as private investors and venture capitalists, or institutions such as banks or other financial lending entities. Another way of raising capital in corporate finance is a debenture, a medium- to long-term debt instrument used by large companies to borrow money at a fixed rate of interest.

Debt service is the cash that is required to cover the repayment of interest and principal on a debt for a particular period. For a loan or a mortgage, companies must meet debt service requirements for loans and bonds issued to the public. Interest expense is the cost incurred by an entity for borrowed funds. Interest expense is a nonoperating expense shown on the income statement and represents interest payable on any borrowings—bonds, loans, convertible debt, or lines of credit.

Before a company approaches a banker for a commercial loan or considers what rate of interest to offer for a bond issue, the firm needs to compute the debt service coverage ratio. This ratio helps to determine the borrower's ability to make debt service payments because it compares the company's net income to the amount of principal and interest the firm must pay. The interest coverage ratio is defined as the ratio of a company's operating income (or EBIT—earnings before interest or taxes) to its interest expense. The ratio measures a company's ability to meet the interest expense on its debt with its operating income. A higher ratio indicates that a company has a better capacity to cover its interest expense.

11.2.2.7 Return on Investment

Return on investment (ROI) for project investor is a financial metric that is dependent on both a project's Internal Rate of Return (IRR) and its Weighted Average Cost of Capital (WACC). The IRR is extremely site specific and is developed initially by the project sponsor, which could be a private district energy company or private utility, or a public body such as a local authority or public utility. The IRR will depend on the costs and incomes of the project. WACC is the arithmetic average (mean) capital cost that weights the contribution of each capital source by the proportion of total funding it provides. Weighted average cost of capital usually appears as an annual percentage. The WACC depends on the project's risk profile and its current and future sponsors, as well as on the debt-to-equity ratio of its financial structuring (United Nations Environment Program (UNEP)—District Energy in Cities Initiative, 2015).

Typically, while private sector investors will focus primarily on the financial IRR of a given project, the public sector, either as a local authority or a public utility, will also account for additional socioeconomic costs and benefits that are external to standard project finance.

11.2.3 Market Feasibility

11.2.3.1 Operating Revenue

1. Tipping fees: The analysis of baseline tipping fees available to a waste-to-energy project considers the cost associated with alternative disposal of food waste at a landfill for food waste generators and waste haulers. Tipping fees at existing landfills facilities could range from approximately $30 to $70 per tonne; the national average landfill tipping fee in 2013 was $50.60 per tonne from the 2010 Nationwide Survey of Municipal Solid Waste Management in the United States (US EPA, 2016). The total potential revenue from tipping fees can be calculated as the estimated annual tonnage of food waste from each generator, multiplied by the tipping fee at their assigned disposal landfill facility. A 2014 US EPA case study (US EPA, 2014) on codigestion of food waste with sewage sludge at municipal Water Resource Recovery Facilities (WRRFs) mentions the following ranges of tipping fees: Septage $0.07/gal, FOG $0.11/gal nonconcentrated and $0.15/gal concentrated, liquid organic material $0.03/gal, protein material $0.06/gal up to 10% total solids and $0.08/gal over 10% total solids, solid organic material $30/tonne to $65/tonne.
2. Energy and by-products sales and commodity values: By generating electricity and other energy products, food waste-to-energy systems create the potential for sustainable revenue streams. The quantity and commodity value of the energy products would depend on the food waste feedstock quantities and the specifics of the waste-to-energy conversion technology. For example, the total potential revenue commodity value of electricity generated from anaerobic digestion of food waste could be based on the total amount of food waste sent to the facility, a factor of 550 kWh generated per tonne food waste, and the average wholesale electricity price of $0.04 per kWh (US EPA, 2014).

AD plants generate revenue from the sale of electricity (or gas, if it is upgraded to pipeline quality) produced with the biogas that is recovered through the process. A very efficient wet AD facility might be able to generate up to 260 kWh of electricity per metric tonne of feedstock and after internal usage sell 210 kWh per tonne of feedstock (Dry AD facilities generally produce about 20% less gas). Assuming the electricity is sold for $0.10 per kWh, this could result in revenue of $21 to $17 per tonne of feedstock, which is only adequate to offset some, but not all of the operating costs. It can be assumed that in conjunction with the AD plant, composting of the digestate could be carried out for $40 per tonne, so that after adding the cost of composting and subtracting the revenue from the sale of energy, the minimum break-even tipping fee (without profits or contingencies) would be approximately $110 per tonne (Regional District of Nanaimo (Canada), 2016).

An example of revenue from by-products from anaerobic digestion of food waste would be from compost sales. The model for estimating the commodity value of compost could be based on the total amount of food waste sent to compost facilities, a factor of 0.5 tonnes of compost produced per tonne of food waste, and an assumed market price of $20–25 per tonne (US EPA, 2014).

11.2.3.2 Project Financing and Sources of Funds

When designing a business model for a new food waste-to-energy system, it is important to consider site-specific circumstances, including the type of project finance that is available. In assembling the project finance for a project, two further considerations need to be made (United Nations Environment Program (UNEP)—District Energy in Cities Initiative, 2015):

1. The project needs to be built before it can begin to deliver revenues. This is referred to as the investment/revenue time lag. To reduce this lag, the network should be built from the generation plant outward, placing priority on any anchor loads.
2. The project will likely be developed in stages, requiring waves of capital investment. Taken together with the investment/revenue time lag, enough headroom needs to be built into the model to cope with these fluctuations in investment and to avoid cash flow problems. This headroom also must account for debt repayments relative to operating income to ensure that the project meets debt service requirements.

11.2.4 Ownership and Management Feasibility

11.2.4.1 Ownership Models

The appropriate structure for any food waste-to-energy facility is dependent on many factors, and no one structure will work for all projects. Nonetheless, there are certain types of entities that bear consideration in most situations. Generally, projects can be thought of as taking the form of private ownership or municipal ownership. However, private ownership need not exclude municipal involvement and municipal ownership need not exclude private involvement. Public and private partnership, using one or more entity types, may be appropriate to a project (Biomass Energy Resource Center (BERC), 2009).

11.2.4.2 Private Ownership Model

A private ownership model can take various forms; those most typically used are a for-profit corporation or a limited liability company. These forms of ownership rely on private capital and the success of any private ownership model would be largely dependent on the availability of private capital. However, private ownership could also take advantage of many of the federal and state tax incentives.

Private ownership is generally seen as the opposite of a municipal model; however, a municipality can participate in the development and operation of even a privately owned facility. The municipality may be a financial stakeholder or the owner of some of the components of the facility, such as the land or the means of distribution.

Corporations and LLC: The most common private ownership entity is the corporation. Corporations are generally private, as opposed to public entities. Another common private business model is the limited liability company. This type of entity provides limited liability protection like a corporation, but has pass-through tax characteristics like a general partnership. An LLC offers flexibility in attributing financial and management rights to members, thus allowing for novel structures that recognize the roles and expectations of different member owners.

Cooperatives: A cooperative is an entity often used to permit either the producers or consumers of a product (including energy) to jointly own the means of production or distribution of that product. Cooperatives are generally distinguished from other types of ownership in that they are owned by their members, they are not for profit, and their governance structure provides one vote per member. Participation by members is a desirable characteristic to many people and could work well in an appropriately sized district energy project.

Cooperative statutes need to be well structured for avoiding difficulties in resolving governance and operational issues. An example of existing cooperatives is an electric cooperative, organized under state statute for the purpose of creating or supplying energy. The concept should be applicable to both a food waste-to-energy facility, which can be formed as a consumer cooperative, with the consumers of energy being members with governance rights including the classic cooperative characteristic of 'one member, one vote.'

Community energy trust: The concept of a community energy trust is based on allowing a community to own and operate its own renewable energy project. The premise is that renewable resources belong to everyone and should be held in trust for current and future generations. While the concept is attractive, unfortunately there are very few examples of operating entities using this model and thus there are few answers to many of the fundamental structural questions about ownership, governance, and other responsibilities. The trust concept is employed by, and best understood by reference to, community land trusts, a concept widely used in the United States and typically organized as nonprofit corporations.

11.2.4.3 Municipal Ownership Model

A municipal ownership model of a food waste-to-energy facility means ownership by one or more municipal governments. Municipal ownership can take the form of an entire facility or portions thereof, such as the underlying land or the means of transmission. The extent or location, or the service area, of a facility may dictate a need for more than one municipality to participate in the facility. Multiple municipalities could own a facility pursuant to the existing authority of the municipalities or through an intermunicipal cooperation.

11.2.4.4 Public/Private Partnership Model

The municipal ownership model and the private ownership model are not mutually exclusive, and employing elements of both models could be beneficial in any number of projects. For example, private ownership of a facility may mesh well with public ownership of the means of transmission/distribution. While, in a given situation, it may be most beneficial to construct a facility through private means on privately owned land, it may still be valuable for a municipality to own the means of transmission/distribution. The municipality may be better suited to overcome the obstacles to ownership of a transmission/distribution system that overlaps a large number of publicly and/or privately owned properties.

If community ownership is a desirable goal for a food waste-to-energy project, it may be appropriate to consider a nonprofit corporation structured as a mutual benefit corporation as opposed to the commonly understood public benefit corporation. A mutual benefit nonprofit corporation is organized for the benefit of a limited number of owners as opposed to the public in general, but cannot obtain 501(c)(3) tax-advantaged status granted by the Internal Revenue Service. Consideration would need to be given to the goals of the project, such as whether it is intended to service a limited number of participating users or will sell energy beyond its ownership.

11.2.4.5 Ownership Challenges

Among the major obstacles to the use of food waste-to-energy systems is the reluctance of waste generators, farmers, and other prospective owners to incur the risks and responsibilities associated with owning the system, which include (Biomass Energy Resource Center (BERC), 2009):

- Whether the regulatory approvals can be awarded to allow the receipt and storage of the feedstocks, the operation of the system, the sale or use of the end products, access to the electricity transmission grid, etc.;
- Whether customers can be found for the sale or use of the end products, whether they will be interested in them, and whether they will need regulatory approval to use them (e.g., whole digestate from AD as an organic fertilizer); and,
- The costs and technical problems of purchasing and operating systems, especially for an operator/owner who is primarily focused on a different business activity (e.g., food processing or farming).

In most instances, these risks and responsibilities can be managed or mitigated in the design of an appropriate system ownership model. The ownership model may include the feedstock suppliers to ensure a reliable supply, a utility operator to ensure access to the electricity grid or regulatory compliance, a municipal partner to ensure a buyer exists for the heat produced, or a greenhouse partner to ensure a user for the by-products such as digestate or compost. With the risk/responsibility mitigation requirements identifying likely participants in the ownership model, there are several ways in which they can be brought together. Possibilities include farm ownership and operation, third party build-own-operate, utility company ownership, and farm cooperatives. There is a strong rationale for the public sector to partner in any of these models, giving rise to another alternative—the public/private partnership.

11.3 BARRIERS AND STRATEGIES FOR FOOD WASTE-TO-ENERGY DEPLOYMENT

There is a growing interest in food waste-to-energy management, driven by several factors such as increasing number of landfills reaching capacity and the need for municipalities to evaluate landfill extensions and associated costs to meet future needs, better understanding of environmental benefits of organics management and landfill diversion for alternative beneficial use, and improving economics and efficiencies of waste-to-energy facilities that have now successfully operated over a period of time along with proven performance of reducing contamination rates. Still, most cities and municipalities in the United States, as well as most small and medium food-related businesses in urban and semiurban areas, landfill the majority of food scraps and food-related waste.

Successful organic waste recycling programs, including anaerobic digestion (waste-to-energy and alternative biofuels along with recycled organic products) and composting (organics recycling only with no energy recovery), would require municipal decision makers, haulers, investors, and businesses to optimize across a wide variety of variables: cost of disposal, transportation and logistics costs, material supply assurance, packaging and contamination, access to finance, end-market development, and permitting and siting (ReFED (Rethink Food Waste Through Economics and Data), 2016).

Some barriers and challenges include the following.

- High cost of commercially available large-scale waste-to-energy systems with electricity generation, including incremental costs for advanced biogas to biofuel processing and food waste preprocessing.
- Low price or barriers for sale of electricity, such as utility interconnection agreements and costs, and absence of combined and heat and power (CHP) application opportunities for remote sites.
- Marginal or insufficient economic return that allows operating systems to barely break even and prevents financial feasibility for new projects.
- Requirements for management and maintenance with trained staff and often 24/7 supervision.
- Lack of reliable data and information on food waste characteristics and energy potential and operational parameters for food waste as feedstock in energy conversion systems.
- Distraction from core business activity or expertise and increased risk for disruption of core operation, such as for food-related businesses and livestock operations.

The 2016 report, "A Roadmap to Reduce U.S. Food Waste by 20 Percent," compiled by a multistakeholder nonprofit network, ReFED (Rethink Food Waste Through Economics and Data) (ReFED (Rethink Food Waste Through Economics and Data), 2016), identifies some barriers that apply to wider deployment and adoption of food waste-to-energy programs involving AD; these barriers could be considered as generally applicable for any other food waste-to-energy technologies. It should be further noted that not all of these barriers apply significantly to alternative organic waste recycling programs such as composting and animal feed, thereby enhancing the challenges in deploying food waste to energy.

To overcome the current status quo and improve food waste-to-energy deployment, multiple stakeholders need to engage collaboratively. First, "generators" (homes and businesses that create waste) must face a risk of penalty or receive an incentive to motivate them to sort food scraps into separate waste. Second, haulers must expect a higher profit from collecting food scraps and taking them to organics recycling facilities versus landfills. Finally, there must be available infrastructure in place to process the organics. Processing facilities have to carefully set the tipping fee—the disposal fee they receive for accepting waste—low enough to attract sufficient hauler volumes to keep their facilities at capacity but high enough to generate a profit margin (International Renewable Energy Agency (IRENA), 2012). The following sections overview these barriers and present potential strategies to make waste-to-energy systems more cost effective.

11.3.1 Cost of Disposal

Landfill disposal rates (i.e., tipping fees) have remained exceptionally low in the United States, especially beyond the Northwest and Northeast, relative to many other developed countries. Since tipping fees are typically the largest revenue streams for organic waste recycling operations involving anaerobic digester and composting, this has hurt the business case to expand organics recycling infrastructure.

Taxes and surcharges on land-filling cost per tonnage are effective measures to support long-term organic recycling and waste-to-energy project initiatives. Twenty states or more have implemented landfill taxes, an effective policy tool that can be expanded to directly change tipping fees. The income from the tax can be used in support of organic waste recycling efforts, including grants for new infrastructure or business and consumer education.

11.3.2 Transportation and Logistics Cost (Hauling Cost)

In efficient programs, the incremental vehicle, labor, and fuel costs from recycling generally add a 5%–10% net increase in collection costs versus landfill-only programs. Food waste for organic recycling through composting may lead to minimal extra hauling cost in certain scenarios, such as for residential programs when paired with existing yard waste programs where food scraps are simply added into existing collection routes. However, for food waste-to-energy through AD or any alternative technology, separate collection and hauling considerations and associated costs may have to be evaluated.

For inefficient programs, collection costs can be the driving factor for low cost-effectiveness. When commercial food businesses are spread out, lack of route density or inefficient scheduling of pickups leads to a high labor and fuel cost per volume collected. Once waste is collected and deposited to a transfer station, the distance that it must travel to a recycling facility versus a landfill becomes the key driver for system economics. If hauling costs to a recycling facility are high, haulers may not bring enough food scrap volumes to keep the facility at the high capacity rate needed to be profitable.

- Organic waste collection can benefit from integration with existing recycling and solid waste collection infrastructure. Reduced route redundancy through municipal organized collection or franchised service can bring efficiencies through fewer miles traveled per truck.
- Immense savings can be gained by siting recycling facilities much closer to urban centers than the landfill disposal alternative. A compost or AD facility located 50 miles closer to a city than a landfill would on average result in a $20 per tonne system cost savings stemming from lower truck depreciation costs and the bypass of transfer station fees, which can be shared with the hauler in the form of lower tipping fees.
- Retailers can use reverse logistics practices to transport food scraps from stores to distribution centers where they can be stored and collected for recycling in larger quantities.
- Organic waste haulers can use analytics and logistics software to optimize routes and reduce pickups of partially full loads to reduce fuel and labor costs; currently large solid waste hauling businesses with long-term municipal and business contracts are increasingly using web-based route mapping services.

11.3.3 Material Supply Assurance (Quantity)

A long-term guarantee of material is necessary for waste-to-energy facilities to access project finance and maintain long-term profitability. The cost of capital is both critical to project economics and correlated to uncertainty in the supply of feedstock for the life of the project. Waste-to-energy facilities, such as with anaerobic digesters, often need to secure a complex set of contracts with multiple points of waste generation. Large anchor generators are preferred, but most cities and businesses are reluctant to sign long-term waste supply contracts due to future price uncertainty and the perception that renegotiating contracts on a regular basis can lower costs for the generator. Also, higher management priorities in reducing food wastage throughout the entire supply chain, from farming to food processing and retail food businesses, poses uncertainties in the quantity of available food waste in any area over a longer period of time. Alternative management programs such as composting and animal feed could also compete with waste-to-energy technologies for food waste in an area and pose challenges to long-term stable supply of waste to a particular project.

- Material supply assurance begins by incentivizing businesses and homes to continue to sort their food scraps into separate waste bins for waste-to-energy projects. Local and state governments can incentivize this behavior by offering rebate and rewards program to reduce charges for solid waste collection, or sewer charges.
- Municipalities can encourage local generators to sign long-term contracts for food waste directly with project developers or haulers to lower the risks of maintaining long-term supply.
- Cities and states with organics bans or landfill diversion policies can provide staff to enforce the local policies. Enforcement can be a simple letter to all participants or a more time-intensive audit of waste loads at points of generation or transfer stations.
- Local planning commissions can publish data on planned prevention, recovery, and recycling activities to ensure "right-sizing" of recycling facilities across a wasteshed.

11.3.4 Packaging and Contamination (Quality)

Depackaging of food waste generated in the food processing and retail sectors from defective, expired, and unsellable products pose a major challenge in organic waste management, due to the nonbiodegradable or noncompostable nature of most packaging materials. Biodegradable and compostable packaging has helped reduce this problem in certain settings but it has not been widely adopted among retail grocers due to concern over costs, quality, and shortened product shelf life. Common contaminants include plastic, foam, or disposable food packaging that appears compostable but is not. The problem continues at the treatment facility. Compost or AD facilities that receive highly contaminated feedstock must spend more on pre- and postprocessing, which may hurt profitability.

Postconsumer food waste such as from households, restaurants, offices (kitchens and cafeterias), and institutions (hospitals and university dining services) are often comingled with nonbiodegradable contaminants such as plastic films and food wraps, plastic food packaging and containers including mini containers for single-serve items, and plastic disposable tableware items like forks and stirrers. Most existing waste-to-energy facilities involving anaerobic digestion are currently reluctant to include and manage postconsumer food waste in their operation due to contamination issues, though some collaborate efforts are underway at local levels by municipalities and small businesses, waste haulers, and waste-to-energy facilities to facilitate collection of source-separated organic waste for anaerobic digestion.

- Source separation of organic waste and an effective collection infrastructure through management initiatives and consumer education would be essential before postconsumer waste attracts significant interest from waste-to-energy projects.
- Municipalities can provide sufficient education to residents regarding which items can be separated for waste to energy and offer financial incentives to residences and businesses that demonstrate low contamination rates through random audits.
- Entrepreneurs can partner with the AD industry to bring innovative packaging solutions to market that can both improve shelf life performance and be readily processed by organics recycling infrastructure. Additionally, they can develop low-cost depackaging equipment that can be used by businesses and processors to reduce contamination.
- Businesses can ensure that packaging is clearly marked as biodegradable and meets standards for processing in local anaerobic digester facilities.
- Entrepreneurs and processors can continue to bring innovative, low-cost depackaging equipment to the market to be widely distributed to food waste generators and processors.

11.3.5 Access to Financing

Waste-to-energy projects involve equipment and systems for waste handling and preprocessing, waste-to-energy conversion and postprocessing of energy and by-products, and tend to be highly capital-intensive projects. Large projects, like centralized anaerobic digesters, are the most sensitive to variations in financing rates, which can be a differentiator for success. Projects can typically only secure low debt capital rates if they also secure long-term feedstock and product off-take agreements for energy and digestate products with credit-worthy counterparties to lend against. Long-term off-take agreements are more common in the energy sector and are challenging to achieve for compost or animal feed.

- Many projects proposed today would be able to move forward if federal and state programs or impact investors could supply 10% of all project capital in the form of grants.
- Renewable energy mandates that encourage long-term contracts can lower the cost of financing.
- Long-term material supply assurance allows project debt to offer low-cost financing.
- Government projects, such as USDA financing, lower the total blended cost of capital for a project by offering grants or low-cost capital at a rate of 4% or below.
- New impact investment funds from corporate stakeholders, high net worth families, or foundations can fill a critical project financing gap to help lift the profitability of a project with borderline economics; an example is the Closed Loop Fund that offers 0% financing to cities and low-cost debt for hard-to-finance municipal solid waste recycling projects.

11.3.6 End-Market Development

Recycling facilities must maintain certain price levels for their end products (compost, digestate, energy) to maintain financial viability over time. This is impacted by customer demand. Energy demand is massive and linked into regional markets. However, for AD projects, the sharp ups and downs of natural gas market prices make it challenging to finance projects that require stable, long-term cash flows. Recycled organics products such as digestate and compost are typically restricted to local markets and end-use due to limitations on long-distance transportability and higher transportation costs.

- Local governments and state agencies can promote digestate and compost products for greater agricultural applications as a way to improve soil health and mitigate the effects of drought, and for industrial applications such as highway right-of-way revegetation, slope stabilization, and wetland rejuvenation. Municipalities can include incentives for use into RFPs for construction and landscaping programs to create an end-use market for locally generated compost.
- Industry and impact investors can host competitions to spark innovation to build "high value" digestate and compost markets such as value-added products and applications.
- Government resources should be made available as a resource or clearinghouse to help connect generators of products with potential users.

11.3.7 Permitting and Siting

Organic recycling operations such as anaerobic digestion and compost facilities have trouble rallying local support because the benefits are often hidden to communities. Centralized AD faces issues of scale, are typically only economical when processing 50,000–250,000 tonnes of feedstock per facility, and must be sited near waste suppliers and energy users, a challenging requirement for both very rural and very dense urban areas. In addition, communities often object to organic waste processing facilities due to NIMBY (not-in-my-back-yard) concerns related to odors, pests, or increased truck traffic, impacts which vary greatly based on the management of the waste facility.

- Policymakers can reduce barriers for large-scale Centralized AD facilities by better coordinating existing state agency regulations so that they have clear standards to meet.
- Nonprofits can compile permitting best practices and provide assistance to galvanize community support for new projects undergoing community and environmental impact studies.
- Case studies of well-managed anaerobic digester and other waste-to-energy facilities can help disprove myths regarding odors and pests and provide benchmarks for future operators.
- Beyond the immediate economic benefits of food waste reduction, greater environmental and social impacts of waste diversion can be factored in locally to determine the total system cost-benefit analysis of food waste reduction. These include the cost of siting and building new landfills; health impacts related to incineration; societal cost of greenhouse gas emissions; infrastructure constraints; and benefits associated with removing trucks from roads, increased local job creation, and improved energy security.

11.4 SCALE FOR FOOD WASTE-TO-ENERGY SYSTEMS

Scale and size for a food waste-to-energy system would depend on several factors such as the type of conversion technology; design and engineering of the system; availability, transportability, and handling capacity of waste feedstock; energy, fuel, and by-products handling and marketing capability; fixed and variable cost of system components; and economic feasibility for a site-specific project, which would be subject to local factors like land, capital, labor, feedstock, and energy costs; and regulatory and permitting considerations. As an example, pressurized gasification vessels or completely mixed digester tanks would have engineering and manufacturing size limitations, and a gasification or AD plant has to be designed within those boundaries. Also, scale of a waste-to-energy conversion system could also be limited by the scale of ancillary and support systems, such as material handling and feeding equipment (grinders and pulpers, feed pumps, augers, etc.), mixing equipment (circulation pumps, blade mixers, etc.), and power and fuel-refining equipment (electricity generators, gas scrubbers, etc.).

1. AD Systems: A 2012 feasibility study by CH2MHILL summarizes the following scale and characteristics for three AD technologies (CH2MHILL, 2012)

   Wet low-solids anaerobic digestion
   - Capacities from 30,000 to 250,000 wet tonnes/year
   - Footprint of 5 ha for large end of range
   - 2%–20% solids-liquid
   - Good for feedstocks that are liquid upon arrival
   - Feedstock preparation: Size reduction to 5 cm or less
   - Processing times: 30–40 days (hydraulic retention time)
   - Wastewater production: Up to 500 L per tonne of waste; Characteristics: 1,500 mg/L BOD, 3,000 mg/L SS.
   - Digestate must be dewatered and can then be dried and used as fertilizer, or composted.

   Wet high-solids or dry pumpable AD
   - Capacities from 3,000 to 250,000 wet tonnes/yr
   - Footprint of 5 ha for large end of range
   - 20%–40% solids—Plug Flow—must be pumpable
   - Good for feedstocks that are liquid or slurry upon arrival
   - Feedstock preparation: Size reduction to 5 cm or less
   - Processing times: 14–28 day
   - Wastewater: Up to 300 L per tonne of waste, Characteristics: 1,500 mg/L BOD, 3,000 mg/L SS.
   - Digestate can be dried and used as fertilizer, or composted

   Dry anaerobic digestion
   - Capacities: 10,000–200,000 tpy
   - Footprint: 3,000 m^2 for 15,000 tpy system (1 ha/50 k tpy)
   - Solids content 40% or greater—must be "stackable"—Food waste mixed with shredded yard wastes to percolate.
   - Feedstock preparation: Size reduction to 5–8 in. Avoid aggressive size reduction
   - Feedstocks are loaded with front end loader
   - Processing times: 14–28 days
   - Wastewater: 20–30 L per tonne of waste; Characteristics: 2,000–5,000 mg/L BOD5, 0–5,000 mg/L SS
   - Digestate of 50%–60% moisture can be composted without dewatering.

   In developing countries, some small-scale manure and wastewater AD systems associated with electricity generation have been installed under Clean Development Mechanism projects—42 manure and 82 wastewater projects—most of them with capacities between 1 and 3 MW and investments between USD 500 and USD 5,000/kW (International Renewable Energy Agency (IRENA), 2012).

2. Conventional combustion/WTE systems: The 2016 study from Canada provides the following summary of scales for waste-to-energy technologies (Regional District of Nanaimo (Canada), 2016)
   - Mass Burners: Used for large applications, usually over 200 tonnes per day or 70,000 tonnes per year.
   - Controlled air, starved air, or modular systems (sometimes also called close coupled gasification systems): For applications up to 300 tonnes per day or 100,000 tonnes per year.
   - Fluidized bed technologies: For preprocessed waste with capacities up to about 200 tonnes per day (70,000 tonnes per year).
   - Rotary kilns: Usually used for specialty waste that requires a high degree of agitation and containment, such as hazardous or medical waste (these systems are highly specialized, costly, and not normally used for MSW).

3. Gasifiers: One of the key characteristics of gasifiers, in addition to the gas they produce, is the size range to which they are suited. Fixed-bed downdraft gasifiers do not scale well above around 1 MWth in size due to the difficulty in maintaining uniform reaction conditions. Fixed-bed updraft gasifiers have fewer restrictions on their scale while atmospheric and pressurized fluidized bed and circulating bed, and entrained flow gasifiers can provide large-scale gasification

solutions. Fixed-bed gasifiers are the preferred solution for small- to medium-scale applications with thermal requirements up to 1 MWth. Updraft gasifiers can scale up to as much as 40 MWth (International Renewable Energy Agency (IRENA), 2012)

Small biomass gasification is successfully applied in India, and rice-husk gasification is a widely deployed technology (International Renewable Energy Agency (IRENA), 2012). To produce electricity, piles of rice husks are fed into small biomass gasifiers, and the gas produced is used to fuel internal combustion engines. The operation's by-product is rice-husk ash, which can be sold for use in concrete. Several equipment suppliers are active and one, Husk Power Systems (HPS), has installed 60 mini-power plants that power around 25,000 households in >250 communities. Investment costs are low (USD 1,000 to USD 1,500/kW) and overall efficiencies are between 7% and 14%, but they are labor intensive as there is significant fouling. One of the keys to their success has been the recruitment of reliable staff with a vested interest in the ongoing operation of the plant to ensure this regular maintenance.

4. Biomass Feedstock: Other important cost considerations for biomass feedstocks include the preparation the biomass requires before it can be used to fuel the power plant. Analysis suggests that there are significant economies of scale in biomass feedstock preparation and handling. The capital costs fall from around USD 29,100/tonnes/day for systems with 90 tonnes/day throughput to USD 8,700/tonnes/day for systems with 800 tonnes/day. The capital costs for preparation and handling can represent around 6%–20% of total investment costs of the power plant for systems above 550 tonnes/day. Assuming a heat value of forest residue with 35% moisture content to be 11,500 kJ/kg, the handling capital costs could therefore range from a low of USD 772/GJ/day to as high as USD 2522/GJ/day.(International Renewable Energy Agency (IRENA), 2012)

11.5 ECONOMIC INCENTIVES AND OPPORTUNITIES FOR FOOD WASTE-TO-ENERGY PROJECTS

In the United States, several federal, state, and local government agencies provide a variety of financial incentives such as grants, loans, tax credits, and tax exemptions, to businesses, individuals, and institutions to facilitate renewable energy projects. Eligible technologies under these programs vary and need to be checked for a specific food waste-to-energy system. Programs that accommodate anaerobic digesters, waste to biofuels, biomass for solid fuel, biodiesel and ethanol, biorefinery and advanced biofuels, renewable electricity production, and carbon credits could be generally applicable for food waste-to-energy projects.

A useful search engine for renewable energy financial incentives is the Database of State Incentives for Renewable Energy (DSIRE) by U.S. DOE (US DOE Office of Energy Efficiency and Renewable Energy and North Carolina State University, n.d.). DSIRE allows search by zip codes for all available federal, state, and local programs and their incentive type, such as grant, loan, or tax credit.

Fig. 11.1 shows a partial list of the search results from 2017 for zip code 43606 (Toledo, Ohio) in DSIRE.

1. USDA Rural Energy for America Program (REAP) Renewable Energy Systems & Energy Efficiency Improvement Loans & Grants (USDA, n.d.-a)
    - Provides guaranteed loan financing and grant funding to (a) agricultural producers with at least 50% of gross income coming from agricultural operations, and (b) small businesses in eligible rural areas, to purchase or install systems, including anaerobic digesters, and biomass systems for solid fuel, biodiesel, and ethanol.
    - Funding availability: Loan guarantees on loans up to 75% of total eligible project costs; Grants for up to 25% of total eligible project costs; combined grant and loan guarantee funding up to 75% of total eligible project costs.
    - Loan guarantees: $5,000 minimum loan amount; $25 million maximum loan amount; Up to 85% loan guarantee; maximum term of 15 years, or useful life, for machinery and equipment; maximum term of 7 years for capital loans; maximum term of 30 years for combined real estate and equipment loans.
    - Renewable energy system grants: $2,500 minimum; $500,000 maximum.
    - Additional requirements: Applicants must provide at least 75% of the project cost if applying for a grant only; and at least 25% of the project cost if applying for loan, or loan and grant combination. Projects greater than $200,000 require a technical report.
2. USDA Biorefinery, Renewable Chemical, and Biobased Product Manufacturing Assistance Program (USDA, n.d.-b)
    - Provides loan guarantees to assist in the development of advanced biofuels, renewable chemicals, and biobased products manufacturing facilities.

Name	State/ Territory	Category	Policy/Incentive Type
Energy Conversion and Thermal Efficiency Sales Tax Exemption	OH	Financial Incentive	Sales Tax Incentive
Air-Quality Improvement Tax Incentives	OH	Financial Incentive	Other Incentive
Business Energy Investment Tax Credit (ITC)	US	Financial Incentive	Corporate Tax Credit
Residential Energy Conservation Subsidy Exclusion (Personal)	US	Financial Incentive	Personal Tax Exemption
Modified Accelerated Cost-Recovery System (MACRS)	US	Financial Incentive	Corporate Depreciation
Residential Energy Conservation Subsidy Exclusion (Corporate)	US	Financial Incentive	Corporate Tax Exemption
Renewable Electricity Production Tax Credit (PTC)	US	Financial Incentive	Corporate Tax Credit
Energy-Efficient Mortgages	US	Financial Incentive	Loan Program
USDA - Rural Energy for America Program (REAP) Grants	US	Financial Incentive	Grant Program

FIG. 11.1 List (partial) of programs and financial incentives for zip code 43606 in DSIRE.

- Project requirements: Must be for the development and construction or the retrofitting of a commercial-scale biorefinery using an eligible technology; must use an eligible feedstock for the production of advanced biofuels and biobased products.
- Applicable sectors: Commercial, construction, industrial, investor-owned utility, local government, municipal utilities, cooperative utilities, state government, federal government, tribal government, agricultural, institutional
- Eligible advanced biofuels:
  Biofuel derived from cellulose, hemicellulose, or lignin, or other fuels derived from cellulose.
  Biofuel derived from sugar, starch, excluding ethanol derived from corn kernel starch.
  Biofuel derived from waste material, including crop residue, vegetative waste material, animal waste, food waste, and yard waste.
  Diesel fuel derived from renewable biomass, including vegetable oil and animal fat.
  Biogas, including landfill gas and sewage waste treatment gas, produced through the conversion of organic matter from renewable biomass.
- Maximum loan: Maximum loan amount: 80% of project costs or $250 million
- Loan term: 20 years or the useful life of the project, whichever is less.
  Rate: Lender's customary commercial interest rate, fixed or variable.
  Fees vary with % guarantee and loan amount.

3. Federal Business Energy Investment Tax Credit (ITC) (US DOE, n.d.-a)
   - Provides federal corporate tax credit through IRS Form 5695, 30% for solar, fuel cells, wind; and 10% for geothermal, microturbines, and CHP.
   - Eligibility: Commercial, industrial, investor-owned utility, cooperative utilities, agricultural
   - Savings category: Solar water heat, solar space heat, geothermal electric, solar thermal electric solar thermal process heat, solar photovoltaics, wind (all), geothermal heat pumps, municipal solid waste, combined heat & power, fuel cells using non-renewable fuels, tidal wind (small), geothermal direct-use, fuel cells using renewable fuels, microturbines.
   - Maximum rebate: All other eligible technologies: no limit

Fuel cells: $1,500 per 0.5 kW; microturbines: $200 per kW; Small wind turbines placed in service 10/4/08–12/31/08: $4,000; small wind turbines placed in service after 12/31/08: no limit.

4. Federal Renewable Electricity Production Tax Credit (PTC) (US DOE, n.d.-b)
   - Provides an inflation-adjusted per-kilowatt-hour (kWh) tax credit for electricity generated by qualified energy resources and sold by the taxpayer to an unrelated person during the taxable year. The duration of the credit is 10 years after the date the facility is placed in service for all facilities placed in service after August 8, 2005.
   - Eligibility: Commercial, industrial
   - Savings category: Geothermal electric, solar thermal electric, solar photovoltaics, wind (all), biomass, hydroelectric, municipal solid waste, landfill gas, tidal wave, ocean thermal, wind (small), hydroelectric (small)
   - Rebate amount: Systems commencing construction after December 31, 2016—Wind: $0.0184/kWh for first 10 years of operation; All other technologies: Not eligible.

     Systems commencing construction prior to January 1, 2017—Wind, Geothermal, Closed-loop Biomass, and Solar Systems not claiming the ITC: $0.023/kWh; Other eligible technologies: $0.012/kWh. Applies to first 10 years of operation
   - Maximum rebate: None specified

5. U.S. Department of Energy—Loan Guarantee Program (US DOE, n.d.-c)
   - Provides loan guarantees for projects with high technology risks that "avoid, reduce or sequester air pollutants or anthropogenic emissions of greenhouse gases; and employ new or significantly improved technologies as compared to commercial technologies in service in the United States at the time the guarantee is issued." Loan guarantees are intended to encourage early commercial use of new or significantly improved technologies in energy projects. The loan guarantee program generally does not support research and development projects.
   - Applicable sectors: Commercial, industrial, local government, nonprofit, schools, state government, agricultural, institutional.
   - Eligible technologies: Geothermal electric, solar thermal electric, solar thermal process heat, solar photovoltaics, wind (all), biomass, hydroelectric, fuel cells using non-renewable fuels, landfill gas, tidal, wave, ocean thermal, daylighting, fuel cells using renewable fuels.
   - Maximum loan: Not specified
   - Loan term: Full repayment is required over a period not to exceed the lesser of 30 years or 90% of the projected useful life of the physical asset to be financed.

6. USDA Rural Energy for America Program Energy Audit & Renewable Energy Development Assistance Grants (USDA, n.d.-c)
   - Grantees assist rural small businesses and agricultural producers by conducting and promoting energy audits, and providing renewable energy development assistance (REDA).
   - Applicable sectors: Local government, schools, state government, federal government, agricultural, institutional.
   - Funds usage: The assistance must be provided to agricultural producers and rural small businesses. Rural small businesses must be located in eligible rural areas. This restriction does not apply to agricultural producers. Assistance provided must consist of Energy audits, Renewable energy technical assistance, Renewable energy site assessments.
   - Grant terms: Applicants must submit separate applications, limited to one energy audit and one REDA per fiscal year. The maximum aggregate amount of an energy audit and REDA grant in a Federal fiscal year is $100,000.

11.6 CASE STUDIES

Three case study sections presented here provide information on food waste-to-energy via AD systems: (1) codigester at a dairy farm for manure and food waste; (2) codigester at water resource recovery facilities (WRRFs) for wastewater solids and food waste; (3) digester at a food processing facility or codigestion of food processing waste at a centralized digester, and smaller-scale digester considerations. In the United States, roughly 73% of the recycling opportunity of organic waste is expected to come from creating centralized composting (organics recycling only with no energy recovery or waste to energy) and centralized AD facilities (waste-to-energy and alternative biofuels along with recycled organic products). Another 17% could come from new and upgraded digesters at water resource recovery facilities (WRRF with AD), also known as wastewater treatment plants. The remaining 10% could come from smaller scale decentralized solutions in homes and businesses; decentralized systems are more effective in rural areas or dense urban areas where collection infrastructure is cost prohibitive (ReFED (Rethink Food Waste Through Economics and Data), 2016).

11.7 CASE STUDY OF FOOD WASTE AND DAIRY MANURE CODIGESTION

11.7.1 Project Overview

The anaerobic codigester project with biogas and electricity generation from dairy manure and food waste was developed in northwest Ohio in 2007. The project involved a dairy farm expansion from 650 cows to 2,100 cows and was in final stages of implementation with loan approvals and a $500,000 grant from the State; however, the concurrent financial crisis in the United States postponed the dairy expansion and the project has not been implemented. Codigestion of food waste along with livestock waste at animal farms has successful operational history in the United States, with systems operating from the 1990s. The US EPA AgStar program maintains a detailed database of livestock waste anaerobic digesters, and lists about 90 farm-scale digesters that are codigesting food waste, and about another 140 farm-scale digesters without codigestion (US EPA, 2017).

Financial projections show that the project can generate sufficient cash flow to cover operating expenses and most of debt service (Table 11.5). However, the project was very capital intensive and the finances were marginal. There were also environmental and other benefits that were not included in the financial numbers. Following are the project highlights; Fig. 11.2 shows the process flow for the anaerobic codigester with food waste.

11.7.1.1 Project Highlights

Influent: Dairy manure, preconsumer waste
Herd size: 2,100 mature animals
Manure waste: 70,000 33 gal/day (8.5% solids)
Food waste: 14,000 gal/day (5%–20% solids)
Digester design: Hybrid two-stage mixed plug-flow
Description: Partially below-grade insulated concrete cell with insulated concrete roof
Mixing system: Biogas recirculation
Heating system: Engine waste heat capture and hot water recirculation
Biogas treatment: In-digester sulfur reduction
Biogas conditioning: Condensate trap and dryer
Digester size: 240′ × 72′ × 16′ (2.06 million gal, including 18″ head space for biogas)
Retention time: 22 days
Electrical gen-set: Two 450 kW
Additional equipment: Digester and engine control system, automatic flare, power transformer and transmission setup
Design and installation: An AD Co. from WI

Owner and operator: Cooperative (with dairy farm, a waste hauling business, several local crop farmers, and some agricultural and food processing businesses)
Location: Dairy Farm (Wood Co, Ohio)
Operation and maintenance: Wastewater-AD Management Co. from Ohio (subcontractor)
Total project cost: $2.75 million (2007 $)
Grant awarded: $500,000 (Ohio)
Total annual revenue: $447,000
Annual electricity generation: 4.7 million kWh
Biogas use: Electricity, digester heating, farm use
Receiving utility: Rural Electric CoOp
Solids separation: Screen, screw-press
Solids use: By dairy as stall bedding and compost
Liquid effluent use: Applied to cropland
Revenue streams: Electricity, RECs, manure and food waste tipping fees, carbon credits (for flaring only)
Additional environmental benefits: Odor and pathogen reduction, landfill diversion, and beneficial disposal means for food waste

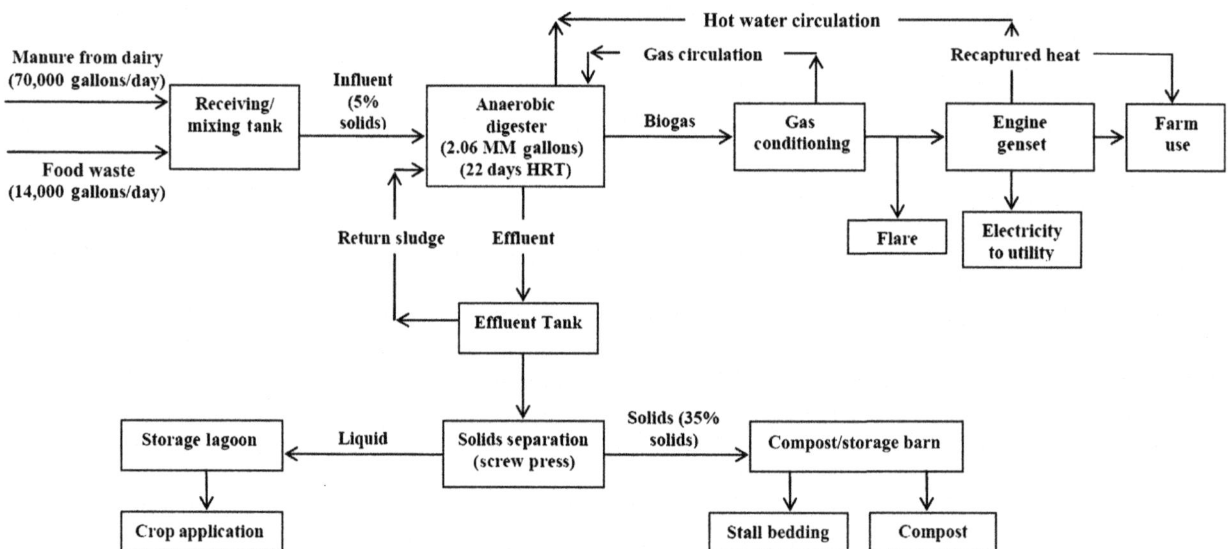

FIG. 11.2 Process flow for anaerobic codigester with food waste.

11.7.2 Project Development

A project development team, consisting of representatives from the cooperative members and some local engineering and government organizations, conducted a technical and financial feasibility study and noted the following.

- The team conducted visits to multiple dairy farms with codigesters receiving food waste and evaluated technical, operational, and financial considerations. These included (1) Vander Haak Dairy, WA (mixed plug-flow digester for 1,500 cows and food waste, operating from 2004, and research and operational support from WSU and multiple agencies) (Kruger et al., 2008; Washington State Department of Agriculture, 2011; Northwest CHP Application Center, 2005); (2) Ridgeline Farm, NY (formerly Matlink Dairy Farm, complete pump-mixed digester for 650 cows and food waste, operating from 2001, and installed with support from NYSERDA) (Cornell University, 2008; NYSERDA, n.d.); Bridgewater Dairy, OH (mixed plug-flow digester for 3350 cows and food waste, operating from 2008, and received USDA grant) (OSU Extension, 2017); Scenic View Dairy at Fennville MI (complete stirred tank digester for 3,800 cows and food waste, operating from 2006 with biogas cleanup and pipeline injection, and received USDA grant) (MI Department of Agriculture, n.d.).
- AD and biogas generation from livestock manure is well established, but reliable and operational information on food waste codigestion is subject to wide variations. Technical and research support on food waste characteristics, codigester operation, and biogas potential from food waste was supported by work done at OSU (USDA National Institute of Food and Agriculture Project Award to OSU, 2004–2009; Schanbacher, 2009).

11.7.3 Outcomes and Challenges

- Mixed plug-flow concrete tank digester with electricity gen-sets and sale to electric utility was found to be most cost effective and operationally simple option for a codigester at a dairy farm; information from system designer indicated that the plug-flow digester could handle only 20%–30% additional food waste to manure waste and boost biogas production and higher methane content in biogas to an extent but is not capable of handling higher food waste, such as some completely stirred tank digesters operating on 100% food waste. Further, given the variability in food waste, the plug-flow digester needs to be properly operated and maintained to prevent settling of solids and accumulating grit and gunk at the bottom; completely stirred tank reactors or pumped mixed reactors may have lesser issues with settling.
- The financial success of this project depended upon four sources of revenue. Two of these revenue streams required sales into external markets: the sale of renewable electricity to the local power utility and carbon credits on the Chicago Climate Exchange (CCX). The other two revenue streams would be generated from the manure tipping fee paid by the dairy and tipping fees for food waste paid by the waste generators and waste haulers.
- Electric generation from biogas would generate renewable credits, but with low heat recapture for digester and farm operation only and no CHP application for the large waste heat, the revenue potential is reduced. However, any system upgrades for pipeline gas, CNG, or gas-to-liquid fuel would be cost prohibitive with marginal finances. Excess biogas has to be flared with only small carbon credits available for flaring. Also, while the rural electric co-op agreed to a 10-year power purchase agreement (4.6 cents/kWh including renewable credits), additional costs of $200,000 in equipment for utility hookup and line upgrades have to be paid by the owners.
- Most codigesters at livestock farms are owned and operated by the farms with support from the system installer and with partnerships with food waste haulers. This project was developed with a cooperative model whose members included the dairy farm, an anchor waste hauling business, several local crop farmers, and some agricultural and food processing businesses. The cooperative model allowed multiple stakeholders to share the large financial risk and also helped receive a state grant and local and state support.
- In the cooperative model, the dairy farm would pay the system owner-operator a set fee of 0.7 cents per gallon of manure that flows into the digester. The operating of the anaerobic digester and energy system would be integrated to the dairy operation and the farm would manage the waste receiving and mixing and also the digester effluent including solids separation and liquid storage and land application. This agreement would allow the dairy to continue ongoing contracts with local farmers for liquid fertilizer and also to utilize the separated solids as stall bedding, reducing sand bedding costs. The digester and energy system would be integrated to their expansion plans with required permit modifications.
- Waste haulers and generators will pay the system owner-operator around 5 cents per gallon tipping fees for the food waste disposal. The food waste will be mixed with the manure before being fed to the digester. A local waste management company and member of the agricultural cooperative would provide a major portion of the food waste. Other potential sources of food processing waste from local companies were also identified, including a green bean processor, a meat processor, a canned soup manufacturer, a cookie manufacturer, a turkey processor, a snack food manufacturer, a canned vegetable processor, and a milk processor.

- The agricultural cooperative and owner of the system planned to operate and manage the system with a professional 3rd party O&M contract with a local water and wastewater management business with specific expertise in operating and maintaining wastewater treatment plants including anaerobic sludge digesters. This would allow the dairy farm to manage the influent waste and effluent digestate as part of its normal dairy business and not to be overburdened with the day-to-day operational requirements of the AD system.

11.7.4 Technical Feasibility

A detailed waste and energy estimation is provided in Table 11.3.

11.7.5 Financial Feasibility

Following are the project financial highlights. Detailed project budget from the digester company with some projections from the cooperative is shown in Table 11.4 and operating projections in Table 11.5.

TABLE 11.3 Waste and Energy Estimation

Dairy cow heads	*2,100*
Dairy manure waste estimates	
Waste volume per head per day (manure + wastewater) (gal/head/day)	33
% Total solids	8.5%
% Volatile solids of total solids	70.0%
Total manure waste volume for 2100 head (gal/day)	*69,300*
Food waste substrate estimates	
% Food waste substrate (% of manure volume)	20%
Total food waste volume (gal/day)	*13,860*
Digester volume estimates	
Design hydraulic retention time (days)	22
Total digester hydraulic volume (gal)	1,829,520
% Headspace for gas collection (% of hydraulic volume)	10%
Total digester volume required (gal)	*2,012,472*
Digester design size (240′ × 72′ × 16′) (gal)	*2,068,070*
Biogas production estimates	
Biogas yield from manure waste (ft^3/head/day)	110
Average Btu/ft3 (60% methane content in biogas)	600
Gross energy from manure waste (Btu/head/day)	66,000
Total gross energy from manure waste for 2100 head (Btu/day)	*138,600,000*
% Increase in biogas generation with food waste substrate	50%
Total gross energy from manure & food waste (Btu/day)	*207,900,000*
Electricity generation estimates	
Engine genset biogas-to-electricity conversion efficiency	30%
Total generator capacity (KW)	*761*
(3413 Btu/kWh @ 30% engine efficiency = 11,377 Btu/kWh, 24 h operation)	
Annual kWh generation @ 70% capacity factor	*4,669,070*

TABLE 11.4 Project Budget

Project Budget for Anaerobic Digester and Energy System for 2100 Head Dairy With Food Waste Substrate	
I. Reception pit[a]	
- Piping to and from digester	
II. Digester (16′ × 72′ × 240′)	
- Walls and floor (poured concrete)	$300,000
- Precast roof	$120,000
- Insulation/concrete coating/foam	$137,000
- Excavation (only for digester)	$55,000
III. Digester heating system	
- Heat exchanger header	$16,000
- Heat exchangers	$51,000
- Heat piping	$18,000
- Piping racks	$12,000
- HDPE draft wall	$17,000
- Circulation pumps/solenoids (5 digester/1 b)	$20,000
- Sludge recirculation pump	$15,500
- Misc piping/valves	$17,500
- Labor/mobilization/equipment rental	$85,000
IV. Gas mixing system	
- Diffuser heads/header	$27,500
- Blower/solenoids	$30,000
- Labor/equipment rental	$15,750
V. Building interior plumbing and electrical	
- Labor	$50,000
- Electrical (includes digester controls)	$82,500
- Automatic flare and flame arrestors	$19,000
VI. Electrical gen-set	
- Gen-Set and switch gear (2–450 kW Gen-set with 2-m connection)	$640,000
- Generator installation, plumbing/hookup	$75,000
VII. Building	
- Solids press/handling equipment	$160,000
- Electrical	$25,000
- Building shell	$80,000
- Concrete	$15,000
VIII. Utility hookup[b]	
- Transformer	$50,000
- Line upgrade	$150,000
- Utility impact study	$7500

Continued

TABLE 11.4 Project Budget—cont'd

Project Budget for Anaerobic Digester and Energy System for 2100 Head Dairy With Food Waste Substrate	
IX. Engineering, administrative and contingency	
Engineering/start-up	$200,000
Administrative	$65,000
Contingency	$200,000
Total	*$ 2,756,250*

Note: All dollar values are from 2007. Cumulative Rate of Inflation from 2007 to 2017 is about 20%; and $1,000 from 2007 has the same buying power as $1,200 in 2017.
[a]*By dairy farm.* [b]*Projections by cooperative system owner and operator.*
Source: US DOL, U.S. DOL Bureau of Labor Statistics—Consumer Price Index CPI Inflation Calculator. https://www.bls.gov/data/inflation_calculator.htm (accessed June 2017).

TABLE 11.5 Operating Projections

Operating Projections for Anaerobic Digester and Energy System		**Estimate**
I. Revenue		
Electricity sales (Co-op power utility)[a]	$215,000	
Manure tipping fees (dairy farm)[b]	$176,000	
Food waste tipping fees[c]	$51,000	
Carbon credits (credit for flaring only)[d]	$5,000	
Total revenue		$447,000
II. Operating expenses		
Administration & technical salary[e]	$30,000	
Sales & marketing	$8,000	
System maintenance[f]	$70,000	
Miscellaneous	$2,000	
Total operating expenses		−$110,000
III. Operation profit		$337,000
IV. Nonoperating expense		
Interest expenses (8%)[g]		−$160,000
V. Cash flow income		$177,000
VI. Capital repayment		
First Year principle (10-year amortization)[g]		−$138,000
VII. Net income		$39,000
VII. Without $500,000 state grant		
Additional annual amortization on $500,000		−$73,000
Net Income without $500,000 grant (loss)		($34,000)

Note: All dollar values are from 2007. Cumulative Rate of Inflation from 2007 to 2017 is about 20%; and $1,000 from 2007 has the same buying power as $1,200 in 2017.
[a]*4.67 million kWh/year, 70% capacity factor, 4.6 cents per kWh.* [b]*69,000 gal/day, 365 days, 0.7 cents/gal.* [c]*14,000 gal/day, 365 day, 5.0 cents/gal for 20% of total volume.* [d]*Projections.* [e]*$40/h, 2 h per day, 365 days (part-time skilled worker).* [f]*1.5 cents/kWh average system maintenance, 4.67 million kWh/year.* [g]*$2,750,000 facility; $500,000 grant, $250,000 capital; $2,000,000 loan, 10-year amortization; 8% interest.*
Source: US DOL, U.S. DOL Bureau of Labor Statistics—Consumer Price Index CPI Inflation Calculator. https://www.bls.gov/data/inflation_calculator.htm (accessed June 2017).

11.7.5.1 Financial Highlights

Project budget: **$2,750,000 (all design and engineering, the digester and its heating and mixing systems, the electric generator and controls, the automatic flare for biogas, power transformer and utility line upgrades, the solids separator and contingency, the dairy farm would incur additional cost for the waste receiving and mixing pits, waste piping, solids separation and compost building and effluent storage lagoon.)**

Operating revenue: $447,000/year (4.6 cents per KWh including renewable energy credits, 0.7 cents per gallon for the manure waste and 5 cents per gallon for the food waste).	*Sources and uses of funds:* Total $2.75 million ($2,000,000 debt financing, equity of $250,000 from various, $500,000 through grant funding.
Operating expenses: $110,000/year ($40/h operator salary for 2 h/day, 1.5 cents per kWh average for system maintenance.	*Nonoperating expenses:* $298,000/year (capital repayment and interest for loan amount of $2 million with 8% interest rate and a 10 year amortization period.

11.8 CASE STUDIES OF WATER RESOURCE RECOVERY FACILITY UPGRADES FOR CODIGESTION PROJECTS AT MUNICIPAL WWTP

An estimated 216 Wastewater Treatment Plants and Water Resource Recovery Facilities (WRRF) out of total 1240 with operating AD systems for sewage sludge or wastewater solids in the US haul in food waste (primarily fats, oils, and grease) for codigestion. This accounts for approximately 17% of WRRFs that process sewage sludge using anaerobic digestion, and with another 3600 WRRFs with no current AD systems, there is a huge potential to utilize existing WRRFs for food waste to energy (USDA-US EPA-US DOE, 2014).

Some of the advantages of WRRFs in operating AD systems are the availability of trained wastewater and AD staff and 24/7 operation; existing infrastructure for receiving and managing food waste and digester effluent; opportunities for combined heat and power (CHP) application; large electrical load at the WRRFs and opportunities for in-facility use of generated energy; and proximity to electric grid, gas pipeline, and urban centers for biofuel sale. However, it would take capital improvements in food waste receiving and storage, AD system expansion, and energy generation equipment at WRRFs where food waste codigestion and energy generation is found feasible. Waste tipping fees to food waste generators, long-term waste stable contacts with waste haulers and waste generators, and collection efficiency from small and medium waste generators would be critical for new AD projects and expansion to food waste codigestion at WRRFs.

An example is the Lucas Co WRRF (Lucas County, Ohio, 2017) in Ohio, which signed an agreement in 2017 for the improvements to the anaerobic digesters with Quasar Energy Group from Ohio, who operates several AD systems at WRRFs, centralized and farm locations (Quasar Energy Group, Ohio, 2017). The improvements include refurbishing the existing four digesters, construction of a dual purpose building for receiving and dewatering solids, construction and installation of the necessary tanks and heating equipment to produce Class A biosolids, installation of two one-megawatt generators, installation of two new centrifuges for solids dewatering, and construction of an arched roof system to cover the biosolids storage pad. The improvements will allow the WRRF to become more energy independent through cogeneration of electricity while producing a higher quality of biosolids (from Class B to Class A) for reuse.

Six additional examples of facilities are listed in Table 11.6 that are codigesting with food waste to boost biogas production (US EPA, 2014).

11.9 CASE STUDY OF PILOT-SCALE ANAEROBIC DIGESTION OF SNACK FOOD MANUFACTURING WASTE

11.9.1 Project Overview

The pilot was located at a major snack food manufacturing plant in Ohio during June 2004. The pilot program was a cooperative effort between the food processor, the Ohio Agricultural Research and Development Center (OARDC), Center for Innovative Food Technology (CIFT), and NewBio E Systems Inc. The project utilized systems including one 1,600-gal trailer-mounted and PLC-controlled AD system, one 50-gal mixed-tank PLC-controlled AD system, and multiple 5-gal bench-top PLC-controlled anaerobic biofermenters (USDA National Institute of Food and Agriculture Project Award to OSU, 2004–2009; Schanbacher, 2009).

Project goals included (1) determining the metabolic and nutritional requirements of digesters fed with food waste; (2) developing sensitive analytical technologies to monitor biodigestion and purity of biogas produced as a necessary guide in the development of anaerobic process strategies; and (3) evaluating scalability of anaerobic digesters of food waste to produce competitive quantities of clean biogas for reliable power for process heat, combustion or turbine engines, or solid-oxide fuel cells.

TABLE 11.6 Summary of Water Resource Recovery Facilities and Food Waste Codigestion Projects

Facility Name and Location	CMSA, San Rafael, CA	EBMUD, Oakland, CA	Hill Canyon, Thousand Oaks, CA	Sheboygan, Sheboygan, WI	West Lafayette, West Lafayette, IN	Janesville, Janesville, WI
Average dry weather flow/ treatment plant flow (MGD)	7/125	60/168	9/14	18.4/56.8	7.8/10.5	12.4/25
Anaerobic digester (MGD)	2	22	2.8	4.8	1.0	2.5
Types of codigested food waste	FOG, Postconsumer commercial	FOG, Winery waste, Industrial liquids and solids, Animal processing & rendering, Postconsumer commercial, Postconsumer residential (pilot)	FOG Industrial, including from fruit juice, frappe, beer, and cheese producers. Restaurant biodiesel waste, e.g., glycerin	FOG Industrial including: dairy, soda processing, and off-spec beverage Ethanol production waste: including thin stillage and corn syrup	FOG Purdue cafeteria food scraps Agricultural waste from Purdue's Ag. Research program Spoiled produce donations	Chocolate waste
Food waste (FW) and fats, oil and grease (FOG): Average quantity processed (GPD)	10,000 (FOG); 1100 (FW)	<600,000 total	>25,000 total	60,000 (FW); 500 (FOG	1 (FW) 142 (FOG)	857 (FW)
Food waste storage (gallons)	20,000	40,000 for solid wastes; 81,000 for liquid wastes	20,000 (expanding to 50,000)	500,000	16,000	7,000
% increase w/ codigestion	60%	Over 100%	250%	150–300%	N/A	40%
Biogas production (cubic feet/ day, averaged)	252,000	2,400,000	450,000	560,000	92,160	120,000
Biogas use	CHP ICE Boiler	CHP ICE Boiler Turbine	CHP ICE Boiler	CHP Microturbines Boilers	CHP Microturbines	CHP Microturbines CNG
Electricity (MWh/year)	3,460	52,000	4,600	2,300	679	1,717
Heat (therms/ year)	150,612	2,300,000	3,000,000	84,000	Not measured	>65,000
Biogas storage capacity (cubic feet)	200,000	200,000	None	Negligible	None	102,000
Biosolids quantity produced (dry metric tonnes per year)	1,302	14,716	2,011	3,278	370	1,277

TABLE 11.6 Summary of Water Resource Recovery Facilities and Food Waste Codigestion Projects—cont'd

Facility Name and Location	CMSA, San Rafael, CA	EBMUD, Oakland, CA	Hill Canyon, Thousand Oaks, CA	Sheboygan, Sheboygan, WI	West Lafayette, West Lafayette, IN	Janesville, Janesville, WI
Capital investment for upgrades (million $)	$7.65 million on digester upgrades, including, $1.9 million for waste receiving facility.	> $35 million ($5 million on food waste receiving station, $1.3 million in interconnection fees, and $30 million for the new gas turbine.	$0.4 for FOG and liquid waste receiving station.	$2.02 for microturbines, inline strainer, a mixing pump, a feed pump, and boiler upgrades	$10.4	$2.07 for dual membrane gas storage system, the conditioning and compressor system, microturbines, CNG system
Tipping fees ($/year)	$400,000	$8,000,000	$307,000	$296,800	$10,000	$9,000
Energy derived savings ($/year)	Gas: $396,900 Electric: n/a	Gas & electric: $3,000,000	Gas & electric: $3,000,000	Gas: $296,800 Electric: $366,000	Gas: $30,000 Electric: $50,000	Gas: $28,000 Electric: $224,801 Fuel: $5000
% of electricity demand generated on-site (annual average)	60%	128%	80%–85% (soon to be 100%)	90%	16%–18%	27%

Adapted from US EPA, Sep 2014. Food Waste to Energy: How Six Water Resource Recovery Facilities are Boosting Biogas Production and the Bottom Line. https://www.epa.gov/sites/production/files/2016-07/documents/food_waste_to_energy_-_final.pdf. Retrieved June 2017.

11.9.2 Considerations for Food Waste Digestion and Centralized Digesters

- Complete-mix digesters, such as stirred tank digesters with blade mixing or tank digesters with pumped mixing, show better performance in handling high-solids high organic-content food waste than plug-flow and un-mixed digesters. They also perform better while operating on food waste solely at high daily loading rates and require less retention time in the digester, thereby improving the digester efficiency and operational capacity of digesters.
- Digestion of high strength food waste, such as with higher fat, oil and grease content or with higher sugar content, shows greater biogas generation potential with food waste than with livestock manure waste and wastewater sludge solids. Food waste codigestion at livestock farms and WRRFs has the potential to increase the financial viability of AD systems, which could otherwise struggle to break even.
- AD systems at large and medium industrial food processing facilities may provide excellent opportunities for combined heat and power (CHP) applications at lower technology costs. For example, 100% of the biogas or electricity generation could be consumed at the facility itself; conditioned biogas may be cofired in existing natural gas boilers and separate dedicated boilers for processing heat, steam, and hot water production, thereby reducing capital cost considerations for electrical generation, advanced biogas and biofuel processing, and utility interconnections.
- Some large food plants, which have to manage their own water and wastewater treatment and solid waste disposal in absence of municipal sewer discharge and landfill options, have some experience and understanding of AD system operation. However, other large and medium plants, which usually send out their waste and wastewater with minimal on-site processing, may view owning and operating an AD system on-site as a distraction from their core business with risks of operational disruption in case of AD system failure. Further, many plants do not have the infrastructure to deal with large volumes of digester effluent.
- A preferred model could be centralized digesters, operated by specialized third parties with AD system experience and receiving waste from multiple food waste generators or from one or a few large anchor waste generators.

11.9.3 General Description: Present Waste Handling System

- Most of the process waste water flows to the rotary screen, which separates out the large suspended solids from the liquid stream.
- Only the wastewater from the slicer and slice washer in the potato kitchen flows to a separate starch extraction process line. This wastewater is currently being excluded from the anaerobic digestion scheme. This water is low in energy potential (methane from anaerobic digestion) and the extracted starch is a valued by-product.
- The waste water separated by the rotary screen goes to a clarifier. The clarification process generates a settled sludge and a cleaner waste water stream.
- The waste water from the clarifier is stored in an intermediate holding tank, from where it is discharged to the city POTW.
- The clarifier sludge, along with the rotary screen solids and some process solids, is belt pressed generating a solid waste stream. This is hauled off in trucks.
- Waste water from the belt press is sent back to the clarifier.
- Unsalable products (potato and corn chips) form another dry waste stream.

Table 11.7 summarizes the different final waste stream and characteristics and Fig. 11.3 shows sample laboratory analysis of waste and wastewater streams.

TABLE 11.7 Final Waste Streams & Characteristics

Name	Average Daily Quantity	Average COD (per Unit Weight/ Volume)	Average Volatile Solids (% of Total Solids)	Physical Description	Daily Energy Potential (ft^3 methane)	Daily Value @ $8.3/ MMBtu	Energy Potential (% of Total Potential)
Unsalable products	13,000 lb	1,200,000 mg/kg (1.2 lb./lb)	90%	Potato & corn chips, 1% solids w/ & w/o seasoning, burnt, etc.	90,000 ft^3	$750	50% of total Rank 1
Solid waste	23,000 lb	300,000 mg/kg (0.30 lb./lb)	95%	Belt pressed clarifier sludge & rotary screened solids 25% solids	40,000 ft^3	$330	23% of total Rank 2/3
Dry starch	9,800 lb	110,000 mg/kg (0.11 lb./lb)	95%	Concentrated and dried starch from potato slicer and slice wash-box process wastewater	6,000 ft^3	$50	4% of total Rank 4
Final effluent water	500,000 gal	1,800 mg/L (0.015 lb./gal)	60%	Clarified final waste water discharged to city POTW	40,000 ft^3	$330	23% of total Rank 2/3

- Daily energy potential as ft^3 of methane is calculated as Average Daily Quantity (lb or gal) * Average COD lb. per lb. or gal * 5.6 ft^3 methane per lb. COD. Daily value is calculated using 1,000 Btu/ft^3 of methane and $8.30/MMBtu.
- 100% conversion of COD to methane @ 5.6 ft^3 methane per lb. COD is unrealistic. As a conservative estimate, 90% of COD from both unsalable products and solid waste, and 70% of COD from effluent water can be assumed to be anaerobically digested to yield methane.
- Dry Starch contributes only 4% of the total energy potential and is more valuable as salable starch.
- Annual energy value from unsalable product, solid waste, and final effluent water can be approximated as

[90% of ($750+$330)+70% of $330] * 365 days/year=$440,000

Sample I.D. AE21572 | Location Description: #16 Finished PC
Sample Collector: CLIENT | Collection Date: 7/14/2004 | Collection Time:
Lab Submittal Date: 7/14/2004 | Submittal Time: 16:15 | Received by: DK
Location code: EISC | Validated by: FD | Validation date: 7/27/2004

TEST PARAMETER	RESULT	UNITS	PQL	AN DATE	AN	REF METHOD
SOLIDS, DRY, 104 DEG C	99.7	%	0.01	07/16/04	CF	SM18 2540 G
SOLIDS, VOL. 550 DEG C	89.2	% of DS	0.1	07/16/04	CF	SM17-2540G
BIOCHEMICAL OXYGEN DEMAND	762000	mg/kg	500	07/20/04	WS	SM18 5210 B
CHEMICAL OXYGEN DEMAND	1460000	mg/kg	500	07/15/04	CF	EPA410.4
OIL & GREASE	298000	mg/kg	250	07/23/04	WS	EPA 1664
NITROGEN, KJELDAHL	7850	mg/kg	250	07/27/04	CF	EPA 351.2
NITROGEN, AMMONIA	467	mg/kg	25	07/27/04	SM	350.1
NITROGEN, ORGANIC	7380	mg/kg	10	07/27/04	CF	
PHOSPHORUS	1400	mg/kg	200	07/20/04	CF	SM17-4500B5E

BOD at 10 days: 888000 mg/kg.

Sample I.D. AE21561 | Location Description: #5 Final Effluent
Sample Collector: CLIENT | Collection Date: 7/14/2004 | Collection Time:
Lab Submittal Date: 7/14/2004 | Submittal Time: 16:15 | Received by: DK
Location code: EISC | Validated by: FD | Validation date: 7/26/2004

TEST PARAMETER	RESULT	UNITS	PQL	AN DATE	AN	REF METHOD
BIOCHEMICAL OXYGEN DEMAND	1280	mg/L	4	07/20/04	WS	SM18 5210 B
CHEMICAL OXYGEN DEMAND	2370	mg/L	10	07/15/04	CF	SM 5220 D
OIL & GREASE	9.9	mg/L	5.0	07/15/04	WS	EPA 1664
SOLIDS, TOT. 104 DEG C	2150	mg/L	20	07/19/04	SM	SM18-2540B
SOLIDS, TOT. VOL. 550 DEG C	1330	mg/L	10	07/20/04	SM	SM17-2540E
SOLIDS, SUSP. 104 DEG C	272	mg/L	5	07/16/04	SM	SM18 2540 D
SOLIDS, VOL. SUSP. 550 DEG C	264	mg/L	5	07/16/04	SM	SM18 2540 E
SOLIDS, DISS. 180 DEG C	1590	mg/L	20	07/19/04	SM	SM18 2540 C
NITROGEN, KJELDAHL	36.6	mg/L	0.5	07/23/04	SM	EPA 351.2
NITROGEN, AMMONIA	2.67	mg/L	0.10	07/22/04	SM	EPA 350.1
NITROGEN, ORGANIC	33.9	mg/L	0.5	07/23/04	CF	
CALCIUM, TOTAL	191	mg/L	1.0	07/20/04	AR	200.7/6010B
MAGNESIUM, TOTAL	26.2	mg/L	0.5	07/20/04	AR	200.7/6010B
SODIUM, TOTAL	67.7	mg/L	0.5	07/20/04	AR	200.7/6010B
HARDNESS	585	mg CaCO3/L	2.5	07/21/04	JC	SM 2340 B
PHOSPHORUS	7.28	mg/L	0.04	07/20/04	CF	SM18 4500B5E

BOD at 10 Days: 1490 mg/L.

FIG. 11.3 Sample snack food processing waste and wastewater laboratory analysis for biogas estimation.

11.9.4 Final Effluent Water Surcharge and Gross Economic Value

- Suspended Solids of the final effluent water discharged to the city POTW are mostly below the surcharge limit of 286 mg/L and exceed that occasionally.
- Average COD of the final effluent water is about 1,800 mg/L, well above the surcharge limit of 482 mg/L.
  City wastewater surcharge:
  [Measured COD (mg/l) – 481.898 (mg/L)] * 0.00834 (conversion factor for mg/l to lb./1000 gal) * Flow (1,000 gal) * $0.09530/lb.
  Average annual COD surcharge = $191,000.
- A 50% reduction in the final effluent COD, from 1,800 to 900 mg/L, will save about $130,000 annually.
- $440,000 from biogas and $130,000 from COD surcharge reduction. Total = $570,000 annually.

REFERENCES

Biomass Energy Resource Center (BERC), Aug 2009. Forming, Financing, and Permitting a District Energy Facility in Vermont. http://www.biomasscenter.org/images/stories/District_Energy_Permitting.pdf. Retrieved June 2017.

Burns & McDonnell, 2015. Mixed Waste Processing Economic and Policy Study. Submitted to American Forest and Paper Association. http://www.afandpa.org/docs/default-source/default-document-library/final_mixed-waste-processing-economic-and-policy-study.pdf Accessed June 2017 & July 2018.

CH2MHILL, Jun 2012. Anaerobic Digestion Overview: Feedstocks to Biogas. http://www.ewmce.com/Resources/Documents/Session%204Kraemer%20Anaerobic%20Digestion_UofA.pdf. Retrieved June 2017.

Cornell University, Feb 2008. Anaerobic Digestion at Ridgeline Dairy Farm: Case Study. http://www.manuremanagement.cornell.edu/Pages/General_Docs/Case_Studies/RL_case_study.pdf. Retrieved Sep 2017.

Global Green USA, 2013. Coalition for Resource Recovery. The Business of Organics Recycling in Dense Urban Centers: Updates and Case Studies from New York City. January 29, 2013. Accessed January 2017. Available at: http://compostingcouncil.org/wp/wp-content/uploads/2013/02/Houssaye.pdf.

International Renewable Energy Agency (IRENA), June 2012. Renewable Energy Technologies: Cost Analysis Series—Biomass for Power Generation. https://www.irena.org/DocumentDownloads/Publications/RE_Technologies_Cost_Analysis-BIOMASS.pdf Retrieved June 2017.

Kruger, C., et al., 2008. High-quality fiber and fertilizer as co-products from anaerobic digestion. Journal of Soil and Water Conservation 63 (1), 12A–13A.

Lucas County, Ohio, 2017. Expansions and Improvements: Lucas County WRRF. http://co.lucas.oh.us/405/Expansions-and-Improvements. Accessed September 2017.

MI Department of Agriculture, Anaerobic Digester Case Study: Scenic View Dairy. https://www.michigan.gov/documents/mda/AD_CaseStudy_221950_7.pdf. Retrieved Sep 2017.

Northwest CHP Application Center, Dec 2005. Case Study: Vander Haak Dairy, Lynden, WA. http://www.northwestchptap.org/NwChpDocs/VanderHaakDairyCaseStudy.pdf. Retrieved Sep 2017.

NYSERDA, Matlink Dairy Farm: Project Profile. http://dataint.cdhenergy.com/Fact%20Sheets/Fact%20Sheet%20-%20Matlink%20Farm.pdf.

OSU Extension, Jul 2017. Status of Biorefineries in Ohio. https://ohioline.osu.edu/factsheet/fabe-6602. Accessed September 2017.

Quasar Energy Group, Ohio, 2017. Anaerobic Digester Locations & CNG Fueling Stations. http://quasareg.com/New/locations/. Accessed September 2017.

ReFED (Rethink Food Waste Through Economics and Data), 2016. A Roadmap to Reduce U.S. Food Waste by 20 Percent. https://www.refed.com/downloads/ReFED_Report_2016.pdf. Accessed September 2017.

Regional District of Nanaimo (Canada), May 2016. Residual Waste Management Options for the Regional District of Nanaimo (Canada): Final Report by Morrison Hershfield. http://www.rdn.bc.ca/cms/wpattachments/wpID2663atID7721.pdf. Retrieved June 2017.

Schanbacher, F., Apr 2009. Anaerobic Digestion: Overview & Opportunities. Presented at OSU-OARDC Waste to Energy Workshophttp://www.midwestchptap.org/Archive/pdfs/090407_Ohio/Schanbacher.pdf. Retrieved Sep 2017.

United Nations Environment Program (UNEP)—District Energy in Cities Initiative, 2015. District Energy in Cities: Unlocking the Potential of Energy Efficiency and Renewable Energy. http://www.districtenergyinitiative.org/sites/default/files/publications/districtenergyreportbook-07032017532.pdf. Retrieved June 2017.

US DOE, Federal Business Energy Investment Tax Credit (ITC). https://energy.gov/savings/business-energy-investment-tax-credit-itc. Accessed June 2017.

US DOE, Federal Renewable Electricity Production Tax Credit (PTC). https://energy.gov/savings/renewable-electricity-production-tax-credit-ptc. Accessed June 2017.

US DOE, U.S. Department of Energy—Loan Guarantee Program. https://energy.gov/lpo/loan-programs-office. Accessed June 2017.

US DOE Office of Energy Efficiency and Renewable Energy and North Carolina State University, Database of State Incentives for Renewable Energy (DSIRE). http://www.dsireusa.org/. Accessed June 2017.

US EPA, Sep 2014. Food Waste to Energy: How Six Water Resource Recovery Facilities are Boosting Biogas Production and the Bottom Line. https://www.epa.gov/sites/production/files/2016-07/documents/food_waste_to_energy_-_final.pdf. Retrieved June 2017.

US EPA, Nov 2016. Advancing Sustainable Materials Management: 2014 Fact Sheet. https://www.epa.gov/sites/production/files/2016-11/documents/2014_smmfactsheet_508.pdf. Retrieved June 2017.

US EPA, Aug 2017. AgStar Livestock Anaerobic Digester Database. https://www.epa.gov/agstar/livestock-anaerobic-digester-database. Accessed September 2017.

USDA, USDA Rural Energy for America Program (REAP) Renewable Energy Systems & Energy Efficiency Improvement Loans & Grants. https://www.rd.usda.gov/programs-services/rural-energy-america-program-renewable-energy-systems-energy-efficiency. Accessed June 2017.

USDA, USDA Biorefinery, Renewable Chemical, and Biobased Product Manufacturing Assistance Program. https://www.rd.usda.gov/programs-services/biorefinery-renewable-chemical-and-biobased-product-manufacturing-assistance. Accessed June 2017.

USDA, USDA Rural Energy for America Program Energy Audit & Renewable Energy Development Assistance Grants. https://www.rd.usda.gov/programs-services/rural-energy-america-program-energy-audit-renewable-energy-development-assistance. Accessed June 2017.

USDA National Institute of Food and Agriculture Project Award to OSU, 2004–2009. Anaerobic Digestion of Agricultural and Food Waste Biomass for the Efficient Production of High Quality Biogas. https://portal.nifa.usda.gov/web/crisprojectpages/0200286-anaerobic-digestion-of-agricultural-and-food-waste-biomass-for-the-efficient-production-of-high-quality-biogas.html. Accessed September 2017.

USDA-US EPA-US DOE, Aug 2014. Biogas Opportunities Roadmap. https://www.usda.gov/oce/reports/energy/Biogas_Opportunities_Roadmap_8-1-14.pdf. Retrieved Sep 2017.

Washington State Department of Agriculture, October 2011. Washington Dairies and Digesters. https://agr.wa.gov/fp/pubs/docs/343-washingtondairiesanddigesters-web.pdf. Retrieved Sep 2017.

FURTHER READING

US DOL, U.S. DOL Bureau of Labor Statistics—Consumer Price Index CPI Inflation Calculator. https://www.bls.gov/data/inflation_calculator.htm. Accessed June 2017.

Chapter 12

Policy Landscape and Recommendations to Inform Adoption of Food Waste-to-Energy Technologies

Kirti Richa* and Erinn G. Ryen†

*Energy Systems Division, Argonne National Laboratory, Lemont, IL, United States †Wells College, Aurora, NY, United States

12.1 INTRODUCTION

The current industrial food value chain has been functioning as a "take-make-waste" linear production and consumption system that usually ends with the consumer. To transition toward a more sustainable system, "circular economy" policies are being created to mimic the restorative and circular properties of a natural ecosystem to limit waste and enable the flow of materials (Ghisellini et al., 2016). Globally, food waste represents a significant gap in the food value chain and is a barrier to achieving circular economy goals and policies. Food waste-to-energy is an important technology pathway that can close the resource loop (Brachlianoff, 2013) by diverting a significant amount of waste from landfills, lowering greenhouse gas (GHG) emissions, and providing a secure energy source (USDA et al., 2014; USDA et al., 2015; Edwards et al., 2015). Thus, consistent policies are needed to support infrastructure and technology deployment. In the United States (US), food waste is regulated at the state level and is only guided by voluntary federal goals and a waste management hierarchy established by federal agencies (US Environmental Protection Agency (US EPA) and US Department of Agriculture (USDA)). The variability of state-level policies presents both opportunities and challenges for this technology pathway. At the global level, countries like Japan, South Korea, and members of the European Union (EU), among others, have more mature policies towards food waste-to-energy promotion and offer potential solutions to the limitations posed by US state-level focus.

The goal of this chapter is to understand the food waste-to-energy policy landscape at the US federal and state levels and at the global scale. First, we will define key terms in food waste and explain how this material stream is managed in the US, including benefits and limitations associated with the food waste-to-energy pathway. Then we analyze and compare the current state of US and global policies regulating and managing food waste, including a discussion on the economic, social, and technological barriers and limitations of these policies. Finally, we will identify recommendations and lessons learned to ensure the future development of food waste-to-energy pathways in the US. It is beyond the scope of this chapter to provide a comprehensive review of each US state or every country in the globe in terms of their food waste management and waste-to-energy conversion policies. However, we have handpicked a few states from the US and regions throughout the world that are either leaders in food waste management (such as New England, South Korea, and Japan) or represent a major political and economic unit, such as the EU. Policy approaches to promote food waste management and food waste-to-energy conversion vary by region and are at different levels of maturity. The unique characteristics, as well as common policy aspects of some of these regions can help inform future food waste-to-energy policies in the US, both at a federal and regional level.

12.2 DEFINITIONS

For policy purposes, definitions are very important and often a source of contention and confusion. As such, conflation of terminology is true with the literature on the food waste stream (Thyberg and Tonjes, 2016; Girotto et al., 2015). The United Nations Food and Agriculture Organization (UN FAO), whose goal is to achieve food security in such a way that ensures sustainable management and use of natural resources (UN FAO, 2017), distinguishes between *food loss* and *food waste*. Both terms refer to the amount of food squandered or unused along the food supply chain that is designated as an edible product for people, even if it was diverted for another use (Gustavsson et al., 2011). The UN FAO considers *food loss* as

Sustainable Food Waste-to-Energy Systems. https://doi.org/10.1016/B978-0-12-811157-4.00012-7

discarded food scraps occurring at the beginning of the food supply chain (production, postharvest, and processing stage) and *food waste* at the end of the value chain during retail and final consumption (Gustavsson et al., 2011; Parfitt et al., 2010). Others refer to *food loss* as *surplus food*, which is generated from purchasing and consumption behaviors and practices (Girotto et al., 2015; Mourad, 2016; Eriksson and Strid, 2011; Eriksson et al., 2015; Eriksson et al., 2017). *Food waste* is often defined as scraps, expired, inedible food generated during production, transportation, distribution, and consumption stages (Eriksson et al., 2017).

For the purpose of this book chapter, we will refer to the US EPA terminology of ***wasted food*** and ***food waste*** (US EPA, 2017a). The US EPA defines "*wasted, surplus or excess food*," as "..wholesome, nutritious food that is lost or sent for disposal" such as unsold, edible, food, preparations, or trimmings from retail establishments, cafeterias, or industrial processors (US EPA, 2017a). "*Food waste*" is defined as spoiled or uneaten scraps from served food, cooking oil and fats, and industrial by-products that cannot be consumed by humans (US EPA, 2017a). In general, we will refer to food waste-to-energy as managing any source of wasted food or food waste through energy recovery technologies, such as anaerobic digestion. In discussing policy issues related to recovery of food for human consumption, we will typically use the terms wasted or excess food.

12.3 MANAGING FOOD WASTE

12.3.1 Hierarchy

Food waste and wasted food management priorities in the US and other countries (e.g., South Korea) center on the concepts of prevention (e.g., source reduction), recovery, and recycling (Mourad, 2016; US EPA, 2017b; European Commission, 2008). In the US, the Food Recovery Hierarchy (or the food waste management hierarchy) identifies the type of actions entities and individuals can take to reduce waste and divert wasted food (US EPA, 2017b). Each tier, like the US solid waste management hierarchy, focuses on a specific management strategy. The Food Recovery Hierarchy places the highest priority on preventing food loss, through activities like source reduction or modifying production or transportations practices (Mourad, 2016; US EPA, 2017b). The next level of management practice and policies relate to reuse and diversion but is referred to as *recovery*. The US, unlike its European counterparts, separates out this phase by preferring *wasted food* to be recovered and diverted to *feed people* first and then be used to *feed animal*s (US EPA, 2017b).

Globally, many regulations are focused more on the upper tiers of the Food Recovery Hierarchy (prevention and recovery). Food recovery strategies related to the prevention and recovery tiers involve collecting and redistributing food to donation centers or lower cost food sellers for those (also called economic valorization) who need or want the surplus food (Mourad, 2016). Unlike other countries, the 1996 Bill Emerson "Good Samaritan Act" in the US limits liability and risk for donors and charitable organizations to enable feeding people with surplus food (Priefer et al., 2016). However, there is concern that this law is untested in the courts, preventing many corporations from fully supporting food donations (ReFED, 2016). There are also US federal tax deductions to promote food donations. In December 2015, Congress signed into law a tax break package (Protecting Americans from Tax Hike or PATH Act) with provisions of tax incentives for donating food and allowing businesses to claim the cost of donated food inventories and half the profits of the inventory if sold at fair market value (ReFED, 2016). This new legislation is expected to increase donations from farms and smaller retailers and restaurants (ReFED, 2016). In addition to the federal policies, several states, such as California (10% for CA), Colorado, and Arizona also provide tax credits for wasted food donation, but there is no national-level tax credit that facilitates interstate donations (Lehner, 2013).

The next phase of the management hierarchy, which is central to this chapter, involves *recycling* of the food waste. This includes converting food scraps, byproducts, and/or waste oils into animal feed, soil amendment (compost), or energy. According to the US EPA, in 2014, only 5% of food in the waste stream was composted and 18.6% was combusted for energy recovery (US EPA, 2014). A variety of technologies convert food waste into energy or bio-based fuels including anaerobic digestion, transesterification, pyrolysis and gasification, hydrothermal carbonization, and incineration (Girotto et al., 2015; Pham et al., 2015). *Anaerobic digestion* technology is most often used to convert food waste into biogas energy and consists of a biological process that converts organics (including food waste) into biofuels for energy and digestates that can be applied to the soil as a fertilizer or condition (Chanakya et al., 2007; Guermoud et al., 2009; USDA et al., 2014). In the past, feedstock for anaerobic digestion focused on livestock manure, wastewater sludge, and municipal solid waste (MSW), however, projects are now codigesting traditional feedstocks with commercial food waste and residual waste from food production/processing plants (USDA et al., 2014).

Food waste is good feedstock for anaerobic digestion because of its high degradability and biogas yield (Pham et al., 2015). Wastes like whey, bakery, brewing, and winery residuals, and fats, oils, and greases yield high amounts of biogas while reducing the economic waste management burden from retail establishments and processing plants (USDA et al., 2014). In one study, 1 m^3 of biogas was generated from food waste, which is equivalent to 21 MJ of energy (Murphy et al., 2004). The benefits of this technology pathway include renewable energy generation, waste reduction, and nutrient cycling (Kosseva, 2011). In 2015, the US EPA identified 242 operational anaerobic digester (AD) facilities on livestock farms, of which only approximately 40–50 accepted food scraps (ReFED, 2016; USDA et al., 2014). There are a total of more than 2,000 operational biogas production facilities on farms, wastewater treatment facilities, and landfills, and more than 12,000 sites are available for future development (Serfass, 2012; ReFED, 2016).

Finally, the least preferred, but currently mostly heavily used option in the US is disposition of food waste in the *landfill* (US EPA, 2014). As stated previously, landfill space is limited or on the decline and has adverse environmental implications due to production of methane gas. However, landfill disposal has remained popular in the US because of space availability (as compared to Europe) and because tipping fees are generally low in the US as compared to Europe and other developed countries (Wei et al., 2017; ReFED, 2016).

12.3.2 Potential Benefits and Limitations

Implementing policies to encourage adoption of food waste-to-energy technologies is expected to divert a significant amount of waste from landfills, lower GHG emissions, and has the potential to reduce hauling and disposal costs and create new jobs (USDA et al., 2015; Industrial Economics, Incorporated, 2017; ReFED, 2016). According to the US EPA, if 50% of food waste generated each year was anaerobically digested, it would be equivalent to powering 2.5 million homes for a year (USDA et al., 2015; USDA, 2017b). However, the primary drivers of food waste-to-energy policies are energy security and climate change mitigation, food waste management and recycling, and regional development (Edwards et al., 2015). Food waste has been recently recognized for its under utilized potential to generate energy to meet our growing demands and need for energy security and independence (Pham et al., 2015; Edwards et al., 2015). Biorefineries are seen as a stable and effective option to convert plant-based organic materials into bio-based fuels or products (Girotto et al., 2015). According to Thi et al., (2016), biogas-based electricity production from food waste entails minimum investment costs ($500/kWh) as well as low operation cost ($0.10/kWh) when compared to wind and solar power technologies. Food waste-to-energy technologies could provide the most impact towards diverting waste from the landfill, in comparison to upper levels of the waste management hierarchy (prevention and recovery) that provide the most economic value per ton (ReFED, 2016). However, food waste-to-energy policies provide a multiplier effect on benefits via energy production and reduction of GHG emissions (methane in particular) (ReFED, 2016).

The food waste-to-energy pathway is also appealing as there is an abundance and geographical concentration of food waste stemming from food processors and the agriculture sector (Girotto et al., 2015). Converting food scraps and agriculture waste into biogas and fertilizer allows farms to be more self-sustaining, economical, and environmentally sustainable. These factors are important in ensuring successful symbiotic relationships and circular flow of resources in the value chain (Girotto et al., 2015). Secure, affordable, and local sources of energy are key factors tied to ensuring a strong national economy (Edwards et al., 2015).

To shift the existing trend in the US from sending food to the landfill to food waste-to-energy systems, there are critical challenges to be addressed. Even with the many benefits noted in Section 12.1, the barriers or limitations described later (resource management, social, and financial) are the impetus for policy interventions.

12.3.2.1 Resource Management

One barrier to the food waste-to-energy pathway is the diversity of the food waste material stream. The type of food, location, and season can impact the transportation requirements and effectiveness of the technology (Girotto et al., 2015). Even in Europe, implementation of anaerobic digestion technology has been hampered by efforts to obtain a pure feedstock (Levis et al., 2010). Food waste has a highly variable composition and high moisture content, which can impact the effectiveness of energy conversion process and increase transportation costs (Pham et al., 2015; ReFED, 2016).

While there are many promising benefits for food waste-to-energy pathways, collaboration and coordinated decision-making is required from institutions, governments, and farms. Understanding the concentration and location of food waste is critical to the success of these pathways (Girotto et al., 2015). Entities like New York State Pollution Prevention Institute (NYSP2I, 2017) and the US EPA AgSTAR program (US EPA, 2017c) have conducted research on locating food waste processors in an effort to uncover these relationships (see Tables 12.1 and 12.2 for more information). While food waste may be abundant near farms and food processing facilities, these are concentrated locations, so the successful siting of facilities requires industrial partnerships or "symbiosis" (i.e., waste from one facility is "food" or feedstock for another), as well as financial and technological support (Girotto et al., 2015). The US EPA's AgSTAR program provides resources for companies on the permitting process and technical resources to help analyze and evaluate designs, as well as build, operate, and maintain anaerobic digestion systems (US EPA, 2017c). Finding secure long-term supply of feedstock (20–30 years) for food waste-to-energy facilities is a challenge, municipal contracts are short, and there is competition for other waste management alternatives, which adds additional risk (Edwards et al., 2015). Inventories of food waste generated throughout the supply chain are needed to understand the potential for energy generation and to help reduce resource management barriers (Breunig et al., 2017; LBNL, 2017).

12.3.2.2 Social Attitudes and Cultural Barriers

As with other waste management options like landfills or composting, there are social and cultural barriers to food waste-to-energy implementation. "Not in my backyard" or NIMBY concerns are one potential barrier if facilities are sited near large populations (and large sources of food waste) (Levis et al., 2010). Potential concerns could be associated with increased traffic or odor (ReFED, 2016; Khalid et al., 2011). Strict processing control and optimized operations are needed as operations are sensitive to changes in temperature, pH, and salts (Khalid et al., 2011 in Pham 2015). In the US, facilities are often sited away from large population centers and in more isolated locations, so odor control measures are not needed (Levis et al., 2010). However, because of the scale needed to make these projects economical (50,000 to 250,000 tons of feedstock), facilities need to be sited near energy end users and waste suppliers, presenting a challenge for extremely dense rural or urban areas (ReFED, 2016). While education is very important in the upper tiers of the Food Recovery Hierarchy due to the role of the consumer in preventing food waste, employee training (at restaurants, food processors, etc.) is still needed to help determine which food is edible and can be donated versus being appropriate for energy systems (ReFED, 2016). A change in social attitude, training, and education can enable more food waste being diverted to these two preferred options (donation and energy conversion) and not disposed into landfills.

12.3.2.3 Financial Challenges

The primary barrier to successful implementation of the food waste-to-energy pathway is financial, underscoring the rationale for policy intervention. The financial challenges include project costs, competing waste management fees, and access to secure financing. In Europe, anaerobic digestion has been noted as a successful technology to treat food waste, but implementation has been limited by the capital costs (Levis et al., 2010). The cost-effectiveness of biorefineries and extraction or production of bio-based products or fuels require large, consistent quantities of substrate, stable regional supply chains, and substrate with low heterogeneity to meet large production capacities and economies of scale (Girotto et al., 2015). Centralized anaerobic digestion facilities are capital intensive and require long-term feedstock supply and product offtake agreements (ReFED, 2016). For example, an anaerobic digestion system at an 850-head dairy farm was estimated to cost nearly USD 1.8 million (USDA et al., 2015). High capital costs without subsidies limit anaerobic digestion viability for small to medium size farm operations and currently require high economies of scale. One report identified the average installed and operational costs for anaerobic digestion facilities (accepting food waste) to be $561/ton capacity and $48/ton processed, respectively (Moriarty, 2013). Based on these costs, a pilot project in Louisiana was found to be unviable because of an estimated net present value of −$6.7 million, due to low landfill and energy prices and high capital and operational costs (Moriarty, 2013).

Other cost barriers include permitting, collection, and transportation costs. Permitting can take about 3–5 years, costing nearly $10,000 (ReFED, 2016). Food waste has high moisture content, which contributes to transportation and collection costs (ReFED, 2016). Additional costs are associated with employee training, waste collection, hauling, and transportation (USDA et al., 2014, 2015; Industrial Economics, Incorporated, 2017). Additional uncertainty and costs are associated with the variable material stream (ReFED, 2016). Contaminants in the material stream (plastics, needles, etc.) could present a liability either to worker safety or end product usage and may add additional costs to screen materials (Levis et al., 2010).

As noted previously, low landfill tipping fees present a significant barrier for food waste-to-energy technology. While the average landfill tax increased from 1985 to 1995, from $17.75 to $49.21 per ton (Wei et al., 2017), the average tax in 2013 was only $49.99 per ton (Wei et al., 2017). Landfill tipping fees in New York generally range between $40 and $70 per ton with two extreme cases of $28 to $104 per ton (Industrial Economics, Incorporated, 2017). In contrast, a national survey found the average anaerobic digestion tipping fee in 2010 to be $74.79 (van Haaren et al., 2010). Continued low landfill tipping fees and higher food waste-to-energy facility fees present a financial barrier to ensuring a business case for the pathway.

In addition to high costs, securing access to stable financing and energy market uncertainty are other important challenges (USDA et al., 2015). To overcome the high capital costs, long term, stable purchasing power contracts are required, but they are a challenge to secure due to uncertainty in the government's commitment to bioenergy and tax credit programs (e.g., Federal Production Tax Credit (PTC) program), as well as unfavorable pricing for bio-based renewables in the PTC program and higher and/or uncertain electricity pricing (Navigant Consulting, 2006; Coker, 2016). At the federal level, the tax credit rate for food waste-to-energy projects is nearly half that received by solar and wind projects, or \$0.012/kWh compared to \$0.023/kWh, respectively, (in 2016 dollars) for projects built after January 1, 2017 (Navigant Consulting, 2006; US DOE, 2017a; US EPA, 2017f). Negotiating electricity purchase agreements with utility companies can be challenging, and stable agreements and pricing are vital to the business model and financial success of food waste-to-energy projects (Binkley et al., 2013; USDA et al., 2015). If a project's financing is based on a specified utility rate, which either expires or the conventional electricity rates drop, the economic viability of the installation is threatened (USDA et al., 2015). In one case, a farm in Wisconsin had to alter its waste management practices and stop using an anaerobic digester because the utility rate and revenue stream that the project was based on expired and became less than the regular rate, since the Wisconsin Renewable Energy Portfolio standards had been met resulting in less demand for renewable energy (USDA et al., 2015).

12.4 FOOD WASTE-TO-ENERGY POLICY LANDSCAPE

12.4.1 US Policies

The following sections characterize the current US federal, state, and international policy landscape. As stated previously, food waste-to-energy policies primarily reside at the state level in the US; other countries like the United Kingdom (UK), Denmark, and Germany have national policies in response to an EU framework (Edwards et al., 2015). The federal US policies created during the past administration now face a new level of uncertainty with the administration change in 2017. The international-level policies (EU, Japan, South Korea, etc.) are therefore highlighted to provide a basis for policy benchmarking and to serve as a source of recommendations to help the US incentivize development of this technology pathway.

12.4.1.1 Federal Regulations

At the federal level, food waste management policies are primarily voluntary and programmatic, except for a few related to solid waste disposal, discharge of pollutants into US waters, and renewable fuel transitions. In particular, the Resource Conservation and Recovery Act (RCRA) 40 CFR Part 257 (criteria for classification of solid waste disposal facilities and practices) establishes disposal standards for owners of nonmunicipal nonhazardous wastes, which would also include facilities accepting food wastes (Moriarty, 2013). Indirectly, concentrated animal feeding operations (CAFO) are considered a point source of pollution by the Clean Water Act and require a National Pollutant Discharge Elimination System (NPDES) permit, which regulate the discharge of point source pollutants into US waters (U.S. EPA, 2017d).

Another federal policy that is directly applicable to food waste-to-energy systems is the Renewable Fuel Standard (RFS) program (see Table 12.1), which was originally introduced under the Energy Policy Act of 2005 and was further extended by the Energy Independence and Security Act of 2007 (US DOE, 2017a, 2017b, 2017e, 2017f). The RFS mandates that transportation fuels be blended with renewable fuels with increasing amounts every year, thus leading to lower GHG emissions with respect to the fossil fuel replaced. The goal is to eventually reach 36 billion gallons of renewable transportation fuel by 2022 (US DOE, 2017b). The US EPA is responsible for establishing the volume requirements and annual percentage standards for the four fuel categories (cellulosic biofuel, biomass-based diesel, advanced biofuel, and total renewable fuel) for gasoline and diesel produced and imported, and monitors compliance (US DOE, 2017b). In December 2016, the US EPA promulgated a final rule establishing the annual Renewable Fuel Standard (40 CFR 80) for 2017 and the biomass-based diesel volume for 2018 (Renewable Fuel Standard Program, 2015, 2016). Feedstock used to produce transportation biofuels can be either derived from starch feedstocks (conventional biofuels) or cellulosic or advanced feedstock such as sugarcane or sugar beet-based fuels, and vegetable oil or waste grease for producing biodiesel (advanced biofuels).

Under the RFS program, the US EPA requires a "Separated Food Waste Plan" for companies that are registered under the program to be using either biogenic waste oils, fats or greases *or* noncellulosic portions of separated food waste. Such plans should include information about feedstock suppliers (supplier type, point source such as restaurants or food waste aggregators, supplier name, and address), food waste type (such as yellow grease, animal tallow, etc.), collection region, and additional documentation on waste collection mode (e.g., truck), method for quantification of cellulosic and noncellulosic portions of food waste, and a verification that the waste only consists of food waste (US EPA, 2015).

As of May 30, 2017, a few relevant bills were introduced in the US House and Senate that both support *and* discourage biogas technology and indirectly, food waste-to-energy systems (US Congress, 2017). As shown in Table 12.1, H.R. 119

and H.R. 776 sought to limit or reduce existing renewable fuel standards program, while other bills like S988 encourage the development of biogas energy with tax credits, and HR 2746 establishes renewable electricity standards for electricity and gas suppliers (US Congress, 2017) (see Table 12.1).

12.4.1.2 Federal Voluntary Policies and Programs

At the national level in the US, most food waste-to-energy policies are voluntary and/or provide financial and technical resources. While several policies (Good Samaritan Act and tax incentives) are established to encourage prevention and recovery strategies, public concern in the US as well as the EU has begun to shift from food safety to food waste (Guillier et al., 2016) enabling the recycling of food losses. On June 4, 2013, the US EPA and USDA implemented the US Food Waste Challenge to engage stakeholders and to better understand best practices associated with reducing, recovering, and recycling food waste (USDA, 2017a). In September 2015, the US EPA and USDA announced the first federal voluntary goal, the US 2030 Food Loss and Waste Reduction goal to reduce food waste to 50% by 2030 (USDA, 2017e). Agencies have been collaborating with nonprofits, private sector, and governmental institutions "to improve overall food security and conserve our nation's natural resources" (USDA, 2017a). This goal is being achieved by the USDA Food Waste Challenge program and the US EPA's Food Recovery Challenge under its Sustainable Materials Management Program, working with entities along the food value chain (local, state, and federal government entities, food processors, growers, retailers, NGOs, and trade organizations) (USDA, 2017b). The agencies provide resources, technical guidance, and information on the challenge to help these entities to set goals, create a baseline, and track performance (USDA, 2017b).

Another initiative by the USDA is to educate American consumers and create awareness about food waste through its Center for Nutrition Policy and Promotion (CNPP) (USDA, 2017c) and "Let's Talk Trash" campaign (USDA, 2017d). CNPP is raising awareness about how individuals and families can reduce food waste, in support of larger USDA efforts (USDA, 2017c). According to the USDA website, in 2014, over 4,000 participants were part of the joint US Food Waste Challenge, which exceeded its original milestone of 1,000 entities by 2020 (USDA, 2017a). Educating consumers is critical, as they have the decision-making ability to prevent or reduce food waste by adjusting purchasing and consumption patterns or using technology like sensors and cameras that are being integrated into refrigerators and connected to smartphones to increase awareness and help develop sustainable purchasing patterns (Hebrok and Boks, 2017; Lim et al., 2017). Products are also being designed to recommend actions to mitigate wasteful behaviors or be integrated into a system of products and internet applications/sensors that seek to understand the social context of food and then provide feedback to users on food availability and waste (Lim et al., 2017).

In 2014, the USDA, Department of Energy (US DOE), and US EPA responded to the Obama White House Climate Action Plan to reduce methane emissions and recommended several policy and programmatic actions to encourage the development and use of biogas. These include leveraging existing agency programs' focus on biogas, promoting additional investment in biogas systems, encouraging the development of markets for biogas systems and products, and enhancing communications and awareness (USDA et al., 2014). While traditional feedstock would consist of animal manure, MSW, and wastewater sludge, this report highlighted food waste (edible, postharvest, but not eaten food) and food production residuals as viable feedstocks for biogas. Between 2014 and 2015, the three agencies worked together to create, track, and implement policies and programs to support the development of biogas technology and systems, releasing the *Biogas Opportunities Roadmap Progress Report* that included voluntary measures to increase the use of methane digesters (EESI, 2014; USDA et al., 2014; USDA et al., 2015).

This roadmap provided recommendations to address primary barriers to the development of biogas systems and technology, such as limited understanding of benefits, uncertain market conditions, immature and under-developed markets for by-products, lack of a full environmental value due to high project costs and limited financial recognition of nonenergy services, inconsistent governmental approaches, and limited technical and applied research (USDA et al., 2014). Policies identified to address these challenges and support biogas development consisted of financial incentives (loans, tax credits, and grants) that were used to create mapping and resource tools (e.g., AgSTAR), financial support for pilot projects and technology demonstration projects, and funding for university research (Serfass, 2012; Coker, 2016; USDA et al., 2015; USDA, 2017c, 2017g). Loans and grants have been made possible at the federal level through the Rural Energy for America Program (REAP), Biorefinery, Renewable Chemical, and Bio-based Product Manufacturing Assistance Program, and Rural Utilities Services (RUS) (Serafass, 2012; USDA et al., 2015; USDA, 2017f; Biorefinery, Renewable Chemical, and Biobased Product Manufacturing Assistance Program, 2015; Biomass Research and Development Initiative, 2017). The US EPA was working to update their WARM model to add anaerobic digestion as a waste management practice, which can be used to calculate GHG emissions reduction impacts for alternative management options like food waste-to-energy

(USDA et al., 2015). According to the *Biogas Opportunities Roadmap Progress Report*, currently there is a lack of data on biogas loan performance and efforts to reduce perceived risk for these types of projects, although the North American Product and Industry classification system codes were being established for biogas to help secure financing (USDA et al., 2015).

While bioenergy is seen as an alternative and secure energy source, its success is dependent on subsidies (Edwards et al., 2015). USDA provides financial incentives and collaborates with industry and university partners on research projects to promote food waste-to-energy pathways (USDA, 2017g, 2017f). For example, the USDA's Rural Energy for America (REAP) program awarded $40 million in grants for anaerobic digestion systems (USDA, 2017f). The USDA Agriculture Research Service and University of Maryland have been conducting research on combining food waste and dairy manure to create treatment options for small to medium size farm operations (USDA, 2017b). This program seeks to help dairy farmers in the Chesapeake Bay watershed area find a low-cost treatment option that produces renewable energy and reduces GHG emissions (USDA, 2017b). This area has experienced significant pollution impacts including excess phosphate and nitrogen from spreading animal manure (Layzer, 2015), so the food waste-to-energy pathway may lessen this impact.

With the new administration, the proposed White House 2018 budget sought to cut several agencies' budgets and staffing, which threatened the future of many of these programs. For example, the US EPA's AgSTAR had been identified as a program to be eliminated as part of the White House 2018 Budget proposal (Environmental Protection Network, 2017). Moreover, the proposal also included reducing staff and consolidating regional offices (Environmental Protection Network, 2017), which could indirectly limit the agency's ability to promote the adoption of and spread awareness about food waste-to-energy technologies. Thus the role of the states will continue to be important in promulgating policies to support food waste-to-energy pathways.

TABLE 12.1 US Federal Policies and Programs That Relate to Food Waste Management and Promote Food Waste-to-Energy Pathways

Name of Program, Agency and Program	Description/Current Status	Policy Mechanism
US EPA and USDA US Food Waste Challenge and 2030 Food Loss and Waste Reduction Goal	• Three voluntary recognition programs and a federal goal to reduce food loss and waste by 50% by 2030. Includes a Food Recovery Hierarchy • *Food Recovery Challenge program* launched in 2011 and helps organizations track food waste reduction. US EPA provides technical assistance and reports to participants. Currently has 800 participants. • *US Food Waste Challenge* started in 2013 and creates awareness about food loss and organizations make a one time public disclosure to reduce food waste. Provides information about best practices to reduce, recover, and recycle food loss and waste. Managed by USDA. • *US Food Loss and Waste 2030 Champions program* started in 2016 and publicly recognizes businesses and organizations that have committed to reducing food loss and waste in US by 50% by the year 2030. Managed by USDA.	Education and awareness and voluntary initiative with targets
US EPA Renewable Fuel Standard (RFS)	• Mandates transportation fuels be blended with renewable fuels with increasing amounts every year. Entities covered are oil refiners, and diesel and gasoline importers. • Goal is to reach 36 billion gal of renewable transportation fuel by 2022 with a minimum of 1 billion gal for biomass-based diesel from 2012 to 2022 and 16 billion gal for cellulosic biofuel. Volume assigned to each entity is based on a given percentage of its petroleum product sale. A Renewable Identification Number (RIN) is assigned to each gal of renewable fuel. Entities that have to comply can meet volume targets by either selling required biofuels or purchasing RINs from other entities that exceed their targets. Fine for noncompliance.	Renewable fuel standard

Continued

TABLE 12.1 US Federal Policies and Programs That Relate to Food Waste Management and Promote Food Waste-to-Energy Pathways—contd

Name of Program, Agency and Program	Description/Current Status	Policy Mechanism
US EPA Agstar mapping tool	• Since 1994, program has furthered the development of anaerobic digestion in the agriculture sector and reduced GHG emissions. • Provides a national mapping tool identifying ADs using livestock waste and helps evaluate the potential growth and investment of the technology and compares state incentives, policies, standards, and emissions.	Education and awareness tool and voluntary initiative
US EPA Renewable Energy Production Incentive (REPI)	• As part of the Energy Policy Act of 1992, provides economic incentives for electricity generated and sold by new qualifying renewable energy facilities. Biomass (not part of MSW) and landfill gas qualify • Offers a credit of 1.8 cents per kWh generated for renewable electricity, including closed-loop biomass systems. Lower credit of 0.9 cents per kWh is provided for open-loop biomass and waste incineration (including MSW). Though not directly focusing at food waste-to-energy, can be applied to food waste.	Tax credit
Omnibus Act (H.R. 2029)	• Provided the production tax credit (PTC) and investment tax credit (ITC) for renewable electricity projects. PTCs are offered for biomass-to-electricity projects including ADs (an extension was received for the period of 1st January 2015–31st December 2016). Biomass producers could switch to ITC if construction began during this 2-year period. However, extensions provided for biomass to electricity were very short as compared to other renewable technologies such as wind and solar. • Not generally applicable for municipalities with waste water treatment plants.	Tax credit
USDA programs	• Natural Resources Conservation Service's Environmental Quality Incentive Program (EQUIP) provides technical and financial assistance limited to $20,000 per fiscal year and $80,000 during any 6-year period for persons or legal entities. • Rural Energy for America Program (REAP)—authorized under the Farm Bill and guarantees loan financing and grant funding to agricultural producers and rural small businesses for renewable energy systems. • Bioenergy Program for Advanced Biofuels, and Biorefinery Assistance Program provides loan guarantees up to $250 million for the development, construction, and retrofitting of new and emerging technologies for the development of advanced biofuels. For municipal, industrial, and commercial biogas deployment. • Alternative Fuel Excise Tax Credit—provides incentives for the use of biogas as a transportation fuel. • New Market Tax Credit—incentive for projects in low income areas. • Biofuel Infrastructure Partnership (BIP) provided $100 million in grants to support renewable fuel infrastructure in 20 states. Not currently available.	Tax credit, loan, and grant program
US DOE and USDA National Institute of Food and Agriculture Biomass Research and Development Initiative (BRDI)	• Authorized under the Food, Conservation, and Energy Act of 2008 and the Energy Policy Act of 2005. • Agencies collaborate to provide $9 million in funding through FY 2017 as part of an effort to support the development of bioenergy feedstocks, biofuels, and bio-based products. • Awards amounts are $500,000 to $2 million.	Grant program

TABLE 12.1 US Federal Policies and Programs That Relate to Food Waste Management and Promote Food Waste-to-Energy Pathways—contd

Name of Program, Agency and Program	Description/Current Status	Policy Mechanism
S.988 - Agriculture Environmental Stewardship Act of 2017 (115th Congress (2017–2018)	• Senate *introduced* a bill on April 27, 2017 to amend the Internal Revenue Code to allow energy tax credits through 2021 for investments in qualified biogas property or manure resource recovery property. Also permits new clean renewable energy bonds to be used for such properties. • "Qualified biogas property" is a system that uses ADs or other processes to convert biomass into a gas (at least 52% methane) that can be used as fuel.	Tax credit
H.R. 119 Leave Ethanol Volumes at Existing Levels Act (LEVEL) Act (115th Congress (2017–2018)	• House *introduced* the bill on January 3, 2017 to amend the Clean Air Act and revise the renewable fuel program to *decrease* the volume of renewable fuel that must be contained in gasoline in 2017 through 2022 to 7.5 billion gal for each year and each year thereafter. • Sought to eliminate separate volume requirements for the following renewable fuel categories (advanced biofuels, cellulosic biofuel, and biomass-based diesel) and remove the US EPA's requirement to ensure that renewable fuels emit fewer GHGs than the fuel they replace.	Renewable fuel standard
H.R. 776 (115th Congress (2017–2018)	• House *introduced* a bill on January 31, 2017 to limit the volume of cellulosic biofuel mandated under the renewable fuel program to what is commercially available until a comprehensive study is completed.	Renewable fuel standard
H.R. 2746 (115th Congress (2017–2018)	• House *introduced* a bill on May 25, 2017 to amend Title VI of the Public Utility Regulatory Policies Act to establish a federal renewable electricity standard for retail electricity suppliers and a federal energy efficiency resource standard for electricity and natural gas suppliers	Net metering
S.843 and H.R. 2011 Carbon Capture Improvement Act of 2017	• Senate and House both *introduced* bills (April 5 and 6, 2017) to amend the Internal Revenue Code to authorize the issuance of tax-exempt facility bonds for the financing of qualified carbon dioxide capture facilities. • Included converting gas from biomass or other materials recovered for energy or feedstock value.	Tax credit

12.4.1.3 State-Level Policies and Programs

In the United States, food waste policy and management often occur at the state and local levels, creating a patchwork of policies across the country. Policies consist of cap-and-trade programs, landfill bans, financial incentives (tax credit/loan), and energy standards (e.g., net metering and renewable energy portfolio standards). Table 12.2 lists the US state policies and describes their aims and mechanisms. Other policies that are likely to promote biogas and hence, food waste-to-energy, include net metering, feed-in tariffs (FIT), and loans for technology development (ACORE, 2014).

Several states have GHG reduction programs for the power sector. The Regional Greenhouse Gas Initiative (RGGI) is a market-based regulatory program, employing cap and trade to reduce CO_2 emissions from the electricity sector. RGGI is based on cooperation between nine northeast and mid-Atlantic states: Connecticut, Delaware, Maine, Maryland, Massachusetts, New Hampshire, New York, Rhode Island, and Vermont (RGGI, 2017). Each state has its own CO_2 emission budget trading program enacted through its own state regulation, which is based on the RGGI Model Rule and limits CO_2 emissions from electric power plants and trading allowances (RGGI, 2017). From 2008 to 2014, investment in RGGI resulted in 1.37 billion USD invested in renewable energy and an avoidance of 1.7 million shorts tons of CO_2 emissions

TABLE 12.2 US State Policies and Programs that Relate to Food Waste Management and Promote Food Waste-to-Energy Pathways

Policies	Description	Policy Mechanism
California		
Mandatory Commercial Organics Recycling (MORe) AB 1826 Chesbro (Chapter 727, Statutes of 2014)	• Bans landfill disposal of food waste generated by commercial entities that produce at least 200 tons of organic waste per year. The food waste generated has to be managed by composting or AD. The law phased in from July 2015 and in January 2017, commercial producers of at least 100 ton of organic waste/year were included under this regulation. In 2019, the law will cover commercial generators of total waste of at least 100 tons annually. • Aimed toward reducing GHG emissions and creating opportunities to promote composting and waste-to-energy	Landfill ban
Executive Order S- 06-06	• Mandates 20% of renewable energy within the state to be produced from biomass. Bioenergy Action Plans have been introduced to promote new bio-based energy technology development and facilitate market development for bioenergy. Food waste-to-energy facilities in the state can take advantage of these programs. Lawrence Berkeley National Laboratory is developing a county-level inventory of food waste generated from the entire food supply chain in California, and state-funded research is investigating the energy potential of the waste food stream.	Energy standard
Connecticut		
Connecticut Department of Energy and Environmental Protection, 2015, Public Act 11-217, as amended by Section 4 of Public Act 13-285	• Mandates a landfill ban for commercial entities generating more than 104 tons of food waste annually (decreasing to 52 tons by 2020) with access to a food waste recycling or recovery facility within a 20-mile radius • Requires the entities to source separate organic or food waste materials from other solid waste and send it to a recycling or composting facility that has available capacity and can process such waste.	Landfill ban
Renewable Energy Portfolio Standard	• Requires commercial electricity utility providers to include a minimum percentage (7%) of renewable energy generation, but does not explicitly recognize anaerobic digestion technology as one of the renewable sources • Department of Public Utility Control declared in 2008 that ADs that use food waste as feedstock to produce methane will qualify as Class I renewable energy source and can earn one Renewable Energy Credit for each MW of energy generated • Credits can be sold in the market to other suppliers and utilities who don't meet targets • Food waste can be combined with other farm wastes in the digestion process and as such the state does not mandate a specific manure to food waste ratio	Renewable energy portfolio standard and renewable energy credit

TABLE 12.2 US State Policies and Programs that Relate to Food Waste Management and Promote Food Waste-to-Energy Pathways—contd

Policies	Description	Policy Mechanism
Virtual net Metering	• Allows customers who operate behind-the-meter generation ("Customer Host") to allocate excess production to other metered accounts ("Beneficial Accounts") that are not physically connected to the Customer Host's generator to receive the benefit of the electricity produced • Provides energy credits at retail, not reduced wholesale rates. Capped at upper limit of $10 million annually.	Net metering
Connecticut Green Bank	• Provides capital investment for pilot program AD projects where the biogas generated has to be used by a gas-fired combustion turbine, reciprocating engine, fuel cell, or other commercially available prime mover to generate electricity, where such a system can be no more than 3 MW in size • This pilot program funding opportunity can support up to five anaerobic digestion projects.	Loan program
Investment Tax Credit	• Provides investment tax credit and property tax exemption to anaerobic digestion installations.	Tax credit
Massachusetts		
Code of Massachusetts Regulations 310 CMR 19.000	• Establishes a landfill ban for food waste for businesses and institutions disposing at least one ton per week, with no distance threshold for compliance. Establishes a Renewable Energy Portfolio Standard (REPS) under which a given percentage of the state's electricity is required to be supplied from renewable sources (11% in 2016, with 25% target by 2030). Alternative Energy Portfolio Standard (APS) lists the requirement and incentives for various alternative electricity technologies. The APS includes biomass conversion technologies from organic sources including food waste as a Class I Renewable Energy Source. Includes annual reporting requirements. • MassDEP's Clean Energy Results Program focuses on promoting renewable energy, includes AD of food waste. Developed an Organics Action Plan with a comprehensive set of strategies to meet renewable energy goals.	Landfill ban and renewable energy portfolio standard
New York		
S3418 State Food Waste Prevention and Diversion Act	• Proposes to amend the Environmental Conservation Law to mandate bulk generators of food waste to divert food waste from landfills by 2021. • If passed, the law would require any commercial entity generating more than two tons of food waste per week to either donate the food, and if not suitable for human consumption or animal feed, send it for composting, aerobic digestion, anaerobic digestion, or ethanol production. • Target entities would expand to those generating over 500 pounds per week in 2020. By March of 2022, commercial food waste generators would be required to provide an annual report to the government on the quantity for food waste diverted as well as the pathways, destination, and modes involved.	Proposed landfill ban

Continued

TABLE 12.2 US State Policies and Programs that Relate to Food Waste Management and Promote Food Waste-to-Energy Pathways—contd

Policies	Description	Policy Mechanism
NY Solid Waste Management Facilities Regulations (6 NYCRR Part 360)	• Offers three levels of permits (exempt, registered, permitted) to provide regulatory oversight for organic recycling facilities (composting, anaerobic digestion, landfill application, and other technologies). Exempt facilities have the least impact and require less oversight and permitted require the most oversight as they pose greater environmental impacts • Permitted and registered organic recycling facilities fill out an annual report as part of the regulation	Permitting
NYS Pollution Prevention Institute	• Provides technical resources like regulation overviews and tools that help companies estimate food waste, provides guidelines on self-assessment and tracking, and creates tools to identify and locate organic resources.	
Clean Energy Standard (CES)	• Requires 50% of New York's electricity be obtained from renewable energy sources such as solar and wind by 2030, with a progressive phase-in schedule that started in 2017. • Created two mechanisms: renewable energy standard (RES) and the zero-emissions credit (ZEC) requirements.	Renewable energy portfolio standard and energy credit
Climate Smart Communities Program	• Network of communities collaborating to cut down GHG emissions. Focus includes clean energy development and solid waste reduction. The program is cosponsored by six New York State agencies: Department of Environmental Conservation, Energy Research and Development Authority, Public Service Commission, Department of State, Department of Transportation, and the Department of Health. Developed to promote local initiatives. Provides a "Climate Smart Communities" Certification program, that includes "climate-smart" solid waste management practices.	Voluntary tools and certification
NY Consolidated Laws, Public Service Law -§ 66-j Net Metering	• Originally enacted in 1997 for solar PV, but expanded to include biogas produced from anaerobic digestion of farm waste (including food processing waste). Limited capacity set to 2MW. June 2011, A.B. 6270 allowed eligible farm based generators to engage in 'remote' net metering of farm-based biogas.	Net metering
Oregon		
Oregon 2020 Vision and SB 263	• Passed June 2015, which enables the Department of Environmental Quality (DEQ), local governments, and citizens to make progress under the 2050 Vision. • Established mandatory statewide food waste recovery rate of 25% by 2020.	Recovery rate target

TABLE 12.2 US State Policies and Programs that Relate to Food Waste Management and Promote Food Waste-to-Energy Pathways—contd

Policies	Description	Policy Mechanism
Biomass Producer or Collector Tax Credit	• Provides tax incentives for the production, collection, and transportation of biomass that is used for energy production from feedstock materials obtained from within the state • Other tax credits include credits for renewable energy resource equipment manufacturers, and Property Tax Incentive.	Tax credit
Renewable Portfolio Standard	• Promotes renewable energy with mandates to achieve 25% renewable electricity generation for large utilities by 2025 • For smaller utilities, the renewable generation percentage is mandated to be 5%–10% by 2025	Renewable energy portfolio standard
Renewable Fuels Mandate	• For transportation fuels, requires gasoline retailed in the state to be blended with 10% ethanol and diesel to be blended with 5% biodiesel	Renewable fuel standard
Rhode Island		
Amendments to R.I. Refuse Disposal Laws (Chapter 23-18.9)	• Banned the landfill of commercial food waste in Rhode Island effective January 1, 2016. Targets producers generating more than 104 tons of food waste annually and located within a 15 miles radius of a composting or anaerobic digestion facility. Waiver can be requested if the tipping fee for composting or AD facility is higher than the cost of having the waste taken by the Rhode Island Resource Recovery Corporation. Entities covered are mandated to sort organic waste at source, and arrange for the waste to be transported for composting or anaerobic digestion.	Landfill ban
Solid Waste Regulation No.8- R.I. Waste Composting Facility (250-RICR-140-05-8)	• Regulates the permitting process and standards for anaerobic digestion facilities. Overseen by Rhode Island Department of Environmental Management.	Permitting
The Renewable Energy Growth Program (H.B. 8354)	• Guides and promotes electricity generation from renewable energy sources and includes biogas from AD. It is a tariff-based incentive program that finances distributed renewable electricity projects via a competitive bidding mechanism. Depending on the type and size of project, term length would range from 15 to 20 years. Incentives differ depending on the type of resource, system scale, and the geography. Maximum size for an AD is about 1000kW for a large facility.	Tariff-based incentive program
Vermont		
Universal Recycling and Composting Law [Act 148]	• Bans commercial food waste from landfills, targeting producers generating more than 104 tons of food waste per year with access to recovery facility within 20 miles • Phases in food waste disposal requirements in Vermont from 2014 through 2020. The 2020 target is to cover all waste producers including households under the current food waste disposal ban • Establishes a waste management hierarchy	Landfill ban

Continued

TABLE 12.2 US State Policies and Programs that Relate to Food Waste Management and Promote Food Waste-to-Energy Pathways—contd

Policies	Description	Policy Mechanism
Cow Power Program	• Allows consumers to pay an extra \$0.04 per kWh of electricity produced from biogas, which is then paid to the AD operators	Voluntary tariff
Clean Energy Development Fund	• Established via Act 74 (30 V.S.A. § 8015) in 2005, through Vermont Department of Public Service to promote renewable energy including anaerobic digestion biogas installation	Tax credit
Sustainably Priced Energy Development (SPEED) Program (30 V.S.A. § 8005 and § 8001)	• Enacted in 2005 to ensure that 20% of retail electricity in the state is through renewable energy sources • Anaerobic digestion facilities can be included under the SPEED projects list. Currently two water resource recovery facilities (WRRFs) with ADs in Vermont accept food waste or waste fats, oil and grease (FOG).	Renewable energy portfolio standard
Solid Waste Certifications & Permitting	• Under the Vermont Solid Waste Management Rules (§6-1113), allows digesters to accept any amount of food or nonfarm food processing waste as feedstock on a trial basis without permit provided that the food waste is less than 5% of overall capacity of the anaerobic digestion facility and the trial period does not exceed more than 12 weeks. Permitting requirements simplified with passage of Act 150 on January 1, 2018.	Permitting

(RGGI, 2014). Similar program exists for California, and one was signed, but not implemented in the Midwestern states (Midwestern Greenhouse Gas Reduction Accord). Waste-to-energy is one of the ways through which the regional power utilities can reduce their carbon emissions, and food waste can be one of the key contributors toward these initiatives. As of 2014, 37 states recognized biogas in state renewable energy goals (USDA et al., 2014).

In 2006, California adopted Assembly Bill 32 (AB32), or the California Global Warming Solutions Act, and mandated a statewide reduction of GHG emissions to 1990 levels by 2020 (Air Resources Board, 2017a). The California Cap-and-Trade program, managed by the Air Resources Board, is one measure to achieve this goal (Air Resources Board, 2017b). In this cap and trade program, covered entities in California can use offset credits from anaerobic digestion facilities (Air Resources Board, 2014). In 2014, the program was officially linked with the Québec Cap and Trade System. It was expected that offset prices from the US Livestock Project Protocol would incentivize dairy farmers and waste managers among others to invest in anaerobic digestion technology to reduce GHG emissions and help meet California's statewide reduction goal (Hernandez, 2014).

As stated earlier, low landfill tipping fees present a significant barrier to the food waste-to-energy technology pathway. Twenty states have landfill taxes that can be modified to support food waste-to-energy infrastructure and education (ReFED, 2016). As shown in Table 12.2, four US states (Vermont, Connecticut, Rhode Island, and Massachusetts) have 'organic waste bans' (aka landfill bans) for food waste from certain size generators (Bilsens, 2016; VT DEC, 2016; Industrial Economics, Incorporated, 2017; Mass DEP, 2016; US EPA, 2016). Vermont was the first state to pass such a law (VT DEC, 2016) in 2012 after discovering that recyclables and organic waste contributed to over half of the materials dumped in landfills (Abbey-Lambertz, 2016). California has a waste recycling law requiring food waste from commercial entities of a certain size to be either composted or anaerobically digested. While Oregon does not have a landfill ban, it has set a food waste recovery goal of 25% by 2020 (Oregon DEQ, 2017). A landfill regulatory ban for organic waste can be aimed at different points in the food waste chain, such as manufacturers, distributors, or consumers. Disposal bans may encourage stakeholders to divert food wastes to more productive uses—one of them being conversion of the waste into energy. In addition to statewide bans, there are also several cities that have banned landfill disposal of food waste, including San Francisco, Portland, Seattle (US EPA, 2016), and New York City (NYC Department of Sanitation, 2017). New York City adopted a rule in 2016 that requires large-scale commercial food establishments (e.g., hotels with 150+rooms, arenas with 15,000+ seating, food manufacturers with at least 25,000 square feet floor area, and food wholesalers with at least 20,000 square feet floor area) within a 100 mile radius in New York City to separate organic waste (Industrial Economics, Incorporated, 2017; NYC Department of Sanitation, 2017). While food waste can be safely stored in a title D landfill (Zimlich, 2015), landfill bans are important policy tools as they prevent the use of waste disposal practices and force the use of alternative methods. As seen with the case of cathode ray tubes, effective landfill bans on food waste require adequate regulations and infrastructure and may end up costing the community more (Zimlich, 2015).

New York is in the process of evaluating an organic waste disposal ban. Governor Cuomo announced in his 2016 State of the State address a goal to help legislators pass a bill to ban landfill disposal by large generators of food waste (New York State Senate, 2017; Rosengren, 2017). Approximately 3.9 million tons of food waste is generated as MSW in the state with only 3% diverted from the landfill to waste-to-energy facilities. Within New York, there are about 1,700 large institutional generators of food waste (colleges, hospitals, nursing homes, and correctional facilities), retail, and the service and hospitality sectors, which cumulatively produce over 416,000 tons of waste per year (Industrial Economics, Incorporated, 2017). New York is estimated to have the second highest number of food and beverage manufacturing facilities among all the states in the country. According to the US 2012 Census, New York had 1,990 food manufacturing establishments listed under NAICS code #311 or 8% of the total number (25,619) of establishments (US Census Bureau, 2017). A policy focusing just on large-scale commercial generators has the potential to help the state meet its goal to reduce GHG emissions by 40% by year 2030 (Industrial Economics, Incorporated, 2017).

Several states like New York, Vermont, Connecticut, and California have renewable energy portfolio standards (REPS) to promote the consumption of renewable energy (Goldstein, 2013; NYSERDA, 2017; NYSDPS, 2016; Cammarata, 2016; Breunig, 2017; Oregon Department of Energy, 2018). Some states like Connecticut do not explicitly mention food waste-to-energy, but these facilities can earn a Renewable Energy Credit (REC) per MWh of energy generated. New York Public Service Commission adopted the State Energy Plan (SEP) in 2016 that requires 50% of energy to be from renewable sources by 2030, and created the Clean Energy Standard (CES) to help meet this goal in the state (NYSERDA, 2017). CES consists of a Renewable Energy Standard (RES) requirement and a Zero-Emissions Credit (ZEC) requirement (NYS DPS and NYSERDA, 2016; NYSERDA, 2017). In 2006, California Executive Order S-06-06 mandated that 20% of renewable energy in the state must be generated from biomass (California Energy Commission, 2017). Food processing waste (hulls, shells, pits, and beverage and cheese industry residuals) and organic MSWare sources of inputs used to produce biofuel (Navigant Consulting, 2006). Fuel standards that promote the use of biodiesel are also applicable and support food waste-to-energy technologies in states like New York and Oregon (see Table 12.2).

Another popular policy tool consists of financial support, in the form of renewal energy credits and tax credits (Oregon, Connecticut, and New York), loans for pilot projects (Connecticut) (ACORE, 2014; NYSERDA, 2017; NYS DPS and NYSERDA, 2016; Cammarata, 2016), and tariff incentive program (Rhode Island, Vermont) (DSIRE, 2017; Franchetti and Dellinger, 2014). Vermont has a voluntary tariff program of $0.04 per kWh to support anaerobic digestion on farms (Franchetti and Dellinger, 2014). Additional policy support for food waste-to-energy implementation comes in the form of permitting programs like those in New York, Rhode Island, and Vermont (NYS DEC, 2018; RI DEM, 2018; VT DEC, 2018).

As stated previously, one of the issues associated with the development of food waste-to-energy projects is related to pricing and purchasing agreements with utility companies. There are three types of electricity purchasing agreements: buy all-sell all, surplus sale, or net metering (Binkley et al., 2013). Some utility companies require demand or standby charges in case the process is not running, which adds costs (Binkley et al., 2013). Surplus sale agreements enable only the excess electricity produced to be sold back to the utility company, but the value is often less than the commercial retail price of electricity (Binkley et al., 2013). In a buy all-sell all agreement, all the electricity produced is sold to the utility company at the wholesale rate and any energy demand for the digesters has to be purchased (Binkley et al., 2013). Net metering allows customers to offset electricity consumption and receive credits for surplus generation, but a study comparing net metering purchase agreements found that the standby energy charges result in 20% less revenue (Binkley et al., 2013). Another popular policy is feed-in-tariffs (FIT), a price-based policy that provides guaranteed payment per kilowatt-hour of electricity generated by renewable energy systems and can be used for food waste-to-energy systems (Couture and Cory, 2009; Binkley et al., 2013; Costantini et al., 2015). States like Vermont, Washington, and Wisconsin have FIT policies for biogas (Couture and Cory, 2009). Net metering (e.g., New York and Connecticut), on the other hand, is a demand or load reduction policy (Couture and Cory, 2009; Energize Connecticut, 2018; Findlaw, 2017). While FIT policies do not cover upfront capital costs, this type of policy is intended to help stabilize revenue streams with long-term purchase contracts (Binkley et al., 2013; Couture and Cory, 2009). FIT policies are sensitive to farm sizes, so an effective policy would need to reflect a range of farm sizes (and therefore costs structures) to ensure the recovery of investment costs and some level of profit (Binkley et al., 2013).

12.4.2 Global Food Waste-to-Energy Policies

12.4.2.1 Overview

The United Nations (UN) has agreed on the need to halve per capita food waste in the consumer and retail sectors and reduce losses along production and supply chains by 2030 as part of the Global Sustainable Development Goals. Globally, food waste-to-energy policies have been the focus of many countries and the Food and Agriculture Organization of the United Nations (UN FAO), due to its importance in reducing waste, providing food security, and ensuring energy security (see Table 12.3). In fact, the UN has been financially assisting projects to develop food waste-to-energy systems in developing countries such as India (UNDP, 2006).

TABLE 12.3 Global Policies and Programs That Relate to Food Waste Management and Promote Food Waste-to-Energy Pathways (References Provided in the Main Text)

	Region: EU	
Name of Program/ Policy	**Description/Current Status**	**Type of Policy/ Initiative**
EU Waste Framework Directive (WFD)	• Outlines a waste management hierarchy—can be applied to food waste. • No specific targets for food waste. • Specific guidelines for collection and handling of biowastes to reduce GHG emissions—separate collection of biowaste is directed for composting and AD	Waste management framework; Extended producer responsibility (EPR)
EU Landfill Directive	• Aims to reduce biodegradable waste quantities by 2025 • Requires the diversion of biodegradable municipal waste from landfills	Landfill ban
Circular Economy Package	• Aimed at closing the loop of product life cycles • Food waste reduction across entire supply chain to reduce the waste by 50% in 2030 for member states	Action plan/Strategy for circular economy
EU Renewable Energy Directive (Directive 2009/28/ EC of 23 April 2009)	• Mandatory national renewable energy targets including targets for gross final consumption of energy, and for the share of energy from renewable sources in transport	Renewable fuel standard
	Region: UK	
Name of Program/ Policy	**Description/Current Status**	**Type of Policy/ Initiative**
Courtauld 2025	• Voluntary agreement (years 2015 to 2025) to reduce food waste and GHG emission to at least one-fifth per person • Aims toward 20% reduction in country-wide food and drink waste • Focuses on postfarm wastes	Voluntary initiative
Love Food Hate Waste	• Consumer education through practical advice and effective tools to reduce food waste • Over a five year period it assisted consumers to cut down food waste by 21% (1.1 tons/year)	Education and awareness

Food Waste Recycling Action Plan	• Collaboration of the Waste and Resources Action Programme (WRAP), local authorities and industry associations, with a plan to:
	  ◦ Increase quantity of food waste collected
	  ◦ Provide continuous sustaining feedstocks for ADs and composting
	  ◦ Spread the costs and benefits of food waste management across the supply chain
		Action plan for circular economy

TABLE 12.3 Global Policies and Programs That Relate to Food Waste Management and Promote Food Waste-to-Energy Pathways (References Provided in the Main Text)—contd

Region: Japan		
Name of Program/ Policy	**Description/Current Status**	**Type of Policy/ Initiative**
Food Waste Recycling Law	• Sets recycling rate targets by sectors with annual increments • Mandates reporting of food waste data and recycling strategy for generators across the supply chain	Mandatory recycling targets and reporting
Japan's Feed-in-Tariff Plan	• Power companies encouraged to purchase electricity from renewable sources at a fixed price for a given period of time • Biogas, including that produced from food waste, is covered under the scheme	Feed-in-Tariff
Biomass Nippon Strategy	• National strategy to realize a sustainable society by promoting full biomass utilization • Creation of "Biomass towns"—areas with comprehensive biomass utilization system supported by stakeholder collaboration • Food waste-to-energy included under the plan	Action plan/Strategy for circular economy
Kyoto Protocol Target Achievement Plan	• Meeting of Kyoto Protocol targets by production of 500,000 kL of transportation biofuel to replace fossils fuels • Development of biomass towns • Stimulation of biomass energy conversion technologies. • Aims to cut emissions by 3.8% by 2020 compared with 2005 levels.	Action plan/Strategy for GHG abatement
Basic Act for the Promotion of Biomass Utilization	• Outlines plan for the promotion of biomass utilization at national to municipal levels • Sets up the National Biomass Policy Council	Action plan/Strategy
Basic Energy Plan	• Mandates 10% of renewable energy in the nation's primary energy supply by 2020. • Targets increasing biofuel at a volume equivalent to 3% cut of gasoline demand nationwide by 2020	Renewable fuel standard
Act Concerning Sophisticated Methods of Energy Supply Structure	• Oil refiners to produce a given amount of biofuels (FY2011: 210,000kL to FY2017: 500,000 kL (crude oil equivalent)) • Large electricity retailers to attain 44% of their power from non-fossil sources	Renewable fuel standard
National Plan for the Promotion of Biomass Utilization	• Sets biomass resources utilization target for 2020 to 26 million carbon tons per annum. • 2020 target for food waste-to-energy: utilizing 40% of the viable food waste for energy conversion. • Sets basic policies on development of biomass to energy conversion technologies.	Action plan/Strategy for biomass utilization
Region: South Korea		
Name of Program/ Policy	**Description/Current Status**	**Type of Policy/ Initiative**
Food Waste Reduction Master Plan	• Requires households to separate their food waste for collection • Mandates waste generators to recycle their food wastes	Source separation and recycling
Direct Landfill Ban	• One year after the landfill ban, 94% of food waste was being recycled in some way • Subsequently accompanied by ban on discharge into oceans, of organic waste (including food waste) in 2012, and leachate from food waste treatment processes in 2013	Landfill ban

Continued

TABLE 12.3 Global Policies and Programs That Relate to Food Waste Management and Promote Food Waste-to-Energy Pathways (References Provided in the Main Text)—contd

Region: South Korea		
Name of Program/ Policy	**Description/Current Status**	**Type of Policy/ Initiative**
Volume-based Food Waste Fee System	• Households charged depending on the amount of food waste they generate • Payment chips or sticker system, RFID system, or prepaid standard plastic bags can be used • 142 out of 145 local governments participating in this scheme	Pay-as-you-throw system
Measures for Waste Resource and Biomass to Energy	• Includes a corresponding implementation plan in 2009 • The objective of this plan is to utilize biomass to attain 7% and 10% of the country's primary energy supply by 2030 and 2050 • Waste energy towns developed throughout the country	Action plan/Strategy for waste biomass-to-energy
The National Energy Master Plan	• Develops a roadmap to increase the consumption of renewable fuels every five-years for the next 20 years • Current plan aims to achieve 11% share of renewable energy in the total energy consumption of the country by 2035	Action plan/Strategy for renewable energy promotion
Renewable Portfolio Standard	• Market-based system to promote renewable energy deployment • Targets 13 largest power companies in South Korea • Includes electricity production from biomass, biogas, and waste, among other renewable energy sources • Requires power companies to steadily grow their renewable energy percentage to 10% of total production in 2022	Renewable fuel standard
Basic Plan for the Promotion of the Development, Use and Diffusion of New and Renewable Energy	• Low interest loans to companies to develop renewable energy technologies, processes, and equipment • A 10% investment tax credit provided for energy research, deployment, and development, including renewable energy such a biofuels from food waste	Financial incentives for renewable energy promotion
The Framework Act on Resource Circulation	• Adopts a "zero waste" approach to promote greater recycling and reuse • Promotse the utilization of waste as a resource, with the ultimate goal of raw materials and energy supply security • Can be applied to food waste management	Action plan/Strategy for circular economy

The food waste management policies and related strategies at the global level are guided by a food waste management hierarchy, similar to that proposed by the US EPA. The UN FAO has developed a toolkit that provides three levels at which food waste can be tackled. First, and foremost is to reduce wasted food and food waste by reducing crop wastage, second, reusing the food within the human food chain by means of donation, and finally, if reuse is not possible, considering recycling and recovery via by-product recycling, anaerobic digestion, composting, and incineration with energy recovery. The UN FAO along with the United Nations Environment Programme (UNEP) has published guidance for governments, local authorities, businesses, and other organizations based on this hierarchy of waste management (UNEP, 2014). Similarly, the Directive 2008/98/EC of the European Parliament (or the Waste Framework Directive) that provides a legal framework for the treatment of waste in the EU (European Commission, 2008) has also outlined a waste management hierarchy that can be applied to food waste (see Table 12.3). The five-step waste management hierarchy indicates waste prevention as the most preferred option followed by (in descending order) reuse, recycling, recovery (including energy recovery), and safe disposal. It also outlines the basic definitions, principles, and guidelines for all other EU waste management legislations, and embodies principles such as "polluter pays" and extended producer responsibility. The Directive came into action in December 2010 and governs waste management, including food waste, within member EU countries. For example, waste management policies in the UK have been largely driven by the EU Waste Framework Directive and its underlying principles of the waste management hierarchy. However, with UK's exit from the EU, new policies are likely to be introduced in future.

A world leader in food waste management, South Korea has a similar waste management hierarchy as the US and the EU, with waste reduction at the top of the pyramid, followed by recycling (wherever applicable), anaerobic digestion, aerobic composting, incineration, and ultimately landfill, albeit with a more comprehensive hierarchy of landfill scenarios (Seo, 2013). Specifically for food waste, South Korea has banned direct waste landfill and incineration, thus making it the most progressive country in terms of management of this waste stream (Nguyen et al., 2017).

12.4.2.2 *Landfill Waste Reduction*

The policies for food waste reduction in many countries are centered on landfill waste reduction and GHG emission control. The EU Waste Framework Directive provides specific guidelines for collecting and handling of biowastes in order to reduce GHG emissions. For example, this directive requires member states to ensure that waste oils are collected separately, unless separate collection is not feasible for technical reasons. Under EU waste laws, biowaste is defined as biodegradable garden and park waste; food and kitchen waste from households, restaurants, caterers, and retail premises; and similar waste from food processing plants. Under this directive, there are no specific targets for food waste, but recycling targets for household waste can include food waste (European Commission, 2008). Additionally, the EU Landfill Directive aims to reduce biodegradable waste quantities by 2025 (EU Landfill Directive (99/31/EC)) (see Table 12.3). Under this directive, EU member states are required to set up a national strategy to reduce the quantity of biodegradable waste going into landfill and notify the European Commission of this strategy. This strategy can include measures to achieve the targets primarily via recycling, composting, biogas production, or materials/energy recovery. Despite being at the forefront of environmental sustainability, the EU does not mandate a common landfill waste ban for member countries. However, many EU members like Sweden, Norway, and Austria have independent regulations that prohibit landfill disposal of biodegradable waste or organic waste, which has allowed them to significantly reduce their food waste disposal (Milios, 2013).

With vast economic growth in the country in the 1990s, food waste became a problem in South Korea owing to high resource consumption and waste generation, in parallel with limited landfill space. As shown in Table 12.3, the Korean government established the Food Waste Reduction Master Plan in 1996 that required households to separate their food waste for collection, and mandated waste generators to recycle their food wastes (Innovation Seeds, 2017; Ng, 2013). In 2005, the country banned direct landfill of food waste. Since 2012, discharge of organic waste (including food waste), and from 2013, the leachate from food waste treatment processes into oceans has been banned (Ministry of Environment, 2013). South Korea's improved recycling rates represent the progress made through these policies. One year after the landfill ban took effect, 94% of food waste was being recycled in some way, reflecting an increase of about 40% from five years prior (Kim et al., 2011), thus making the country an international model of waste management. In 2010, a volume-based Food Waste Fee System was introduced in South Korea, under which households are charged depending on the amount of food waste they generate (Ministry of Environment, 2017a). As per the Korean Ministry of Environment, 142 out of 145 local governments are participating in this scheme.

Another international leader in food waste management is Japan (see Table 12.3), which established a law in 2001 (updated in 2007) for Promotion to Recover and Utilize Recyclable Food Resources (also called the Food Waste Recycling Law) (Global Environment Centre Foundation, 2011). This law provides a system for the government to monitor, allocate funding, and provide guidance for food waste recycling across the entire food supply chain in the country. It lays out the reporting requirements for different categories of food waste generators (based on industry groups and generation volumes), and sets annual recycling volume targets for them (Parry et al., 2015). For instance, food-related businesses producing more than 100 tons of waste annually are required to report their food waste data along with a strategy as to how they plan to recycle it. The strategies can include diversion via various routes such as fertilizer, feed, and energy recovery, though priority is given to former two routes. Food waste generators producing less than 100 tons of waste per year are required to fill out a survey, the data from which is added to the national estimates. The country-wide estimates for food waste generation and management are calculated based on parties complying with the law and the sample survey of businesses generating less than 100 tons of waste. Other statistics are also employed for collecting waste specific data from food service industry and the household segment within Japan (Parry et al., 2015).

Apart from regional (EU) and country-specific government regulations, voluntary initiatives can also play a leading role toward food waste reduction and diversion from landfill to energy applications. For example, UK has no mandatory food waste reduction targets, and the country predominantly relies on voluntary food waste reduction initiatives promoted by WRAP. To reduce waste across the entire supply chain, WRAP has launched a 10-year voluntary agreement in the country called Courtauld 2025 (for years 2015 to 2025). With the goal of identifying and targeting hotspots in the resource chain, this program aims to cut down food waste and GHG emission by at least one-fifth per person. The program also has a goal of 20% reduction in food and drink waste arising in the country, focusing on UK food production, manufacturing, distribution, retail, hospitality and food services, and households, and in the future, postfarm gate. Additionally, the country has consumer information campaigns such as "Love Food Waste Food" that began in year 2007 and is aimed at educating consumers through practical advice and effective tools to make decisions to reduce food waste and save money. Over a five-year period, it assisted consumers to cut down food waste by 21% (1.1 tons/year).

The landfill waste reduction policies in various countries can ultimately encourage adoption of recycling routes including food waste-to-energy pathways, such as anaerobic digestion. Due to its strict policies governing food waste disposal, the South Korean government has acknowledged the need for additional routes to manage this waste stream, including energy conversion via AD. The capacity at public recycling facilities around the country has been increased by building biogas and sewage treatment facilities (Innovation Seeds, 2017). However, many countries with less stringent food waste disposal laws or voluntary food waste management initiatives suffer from a major feedstock supply issue for sustaining their AD plants. According to WRAP (2016), only 12% of household food waste collected by local authorities in the UK is recycled, with the remaining 88% ending up in the residual waste stream. On the other hand, food waste processing plant operators consider the shortage of feedstock as one of the primary barriers for financial viability of the plants. Hence, diverting this feedstock from residual streams and landfills appears to be a major focus of the Food Waste Recycling Action Plan in the UK. As per a review study by Thi et al. (2016), food waste-to-energy conversion through the anaerobic digestion pathway can account for a small fraction of total electricity generated in future for some countries such as China (42.9 TWh/year; 0.87% of total electricity supply), Japan (7.04 TWh/year; 0.64% of total electricity supply), and the US (13.3 TWh/year; 0.31% of total electricity generation).

In July 2016, Waste & Resources Action Programme (WRAP) and other collaborators (local authorities, waste treatment operators, private sector waste collectors, and industry bodies such as Association of Anaerobic Digestion Operators and Renewable Energy Association/Organics Recycling Association) launched the Food Waste Recycling Action Plan (WRAP, 2016). The actions under this plan can be summarized in three major objectives. First, to increase the quantity of food waste collected from the household and commercial sectors. Second, to provide long-term sustainable feedstocks for the operators of food waste processing plants, both AD and in-vessel composting. Third, to spread the costs and benefits of collecting and recycling food waste across its whole supply chain. There has been considerable growth in the anaerobic digestion sector in the UK over the last few years. In 2015, the number of dedicated food waste digesters operating in the country rose to 76—up from 52 in 2011 (WRAP, 2016). This trend can continue in the future if barriers to food waste-to-energy conversion are tackled. There still exists a major gap in the country over the total amount of food waste collected and the portion of that being sent for recycling (AD or composting).

12.4.2.3 Renewable Energy Promotion

Similar to the US, several policies internationally that have been formulated to promote renewable energy in the electricity and transportation sectors can also encourage food waste-to-energy pathways. The EU Renewable Energy Directive (Directive 2009/28/EC of 23 April 2009) has established mandatory national targets for gross final consumption of energy and for the share of energy from renewable sources in transport. In November 2016, the directive was revised to reach a minimum target of 27% renewable share for energy consumption in the EU by year 2030. Under this regulation, each member state is mandated to produce 20% of its energy share through renewable sources by year 2020. This directive encourages the application of AD technology to treat food waste. The EU has a target of 10% renewable energy for the transport sector (European Commission, 2017), in which biofuels, including food waste-to-energy, are likely to play a prominent role. To ensure that biofuels are produced in a sustainable manner, companies can demonstrate that they comply with the sustainability criteria through established national systems in individual countries or via voluntary schemes that the EU recognizes.

Many EU member countries have individual plans to promote biogas technology through economic incentives that encourage biofuel development in these countries. For example, Germany is currently the largest producer of biogas in Europe, generating more than 6,000 tons of oil equivalent through decentralized agricultural plants, MSW methanization plants, and centralized codigestion plants in 2013, accounting for 67% of the biogas produced in Europe (Lijó et al., 2017).

South Korea is dependent on imports for 98% of its fossil energy needs (US Energy Information Administration, 2017). Hence, the government is implementing policies and long-term strategic plans to increase the renewable energy share of the country's energy portfolio. The government intends to make bioenergy from organic wastes, including food wastes, a major contributing factor toward these expansion plans, particularly since the country has to deal with increasing amounts of wet food waste after the disposal bans (IEA, 2015). Until a few years back, a meager 2% (160,000 tons/year) of organic waste was being used for energy production. However, the Korean Ministry of Environment has developed measures with the objective to utilize 36% of organic waste including food waste (2,830,000 tons/year) for energy conversion by 2020 (Ministry of Environment, 2017b).

In 2008, the Korean Ministry of Environment outlined the "Measures for Waste Resource and Biomass to Energy" and a corresponding implementation plan in 2009. The objective of this plan is to utilize biomass to attain 7% and 10% of the country's primary energy supply by 2030 and 2050, respectively (Ministry of Environment, 2017b). Under this plan, waste energy towns are being developed throughout the country (Ministry of Strategy and Finance, 2016). Moreover, the government supports the private sectors with 60%–80% of the total investment cost of AD plants converting agricultural raw materials. All biowaste AD facilities in South Korea are constructed and operated by the government (Kang, 2013). Under the Wastes Control Act of 1986 (last amended in 2007), the Ministry of Environment allocates financial and technological support to local governments to expand waste-to-energy conversion facilities, including energy conversion from food waste (Ministry of Environment, 2013). As of 2015, there were 129 biogas plants operating in South Korea, out of which 33 used food waste as the feedstock (IEA, 2015). Additionally, the Korean government has funded research projects to promote biomass, including food waste utilization for energy production. For example, the Ministry of Environment developed a $74 million center to promote organic waste-to-energy research, with three-fourth of the funding from the government and the remaining from the private sector for years 2013–2020 (Kang, 2013). The food waste AD facility developed under this project has a capacity of 1,800 m^3, and analysis is being conducted on issues related to biogas upgrading, operation and maintenance guidelines, and odor control.

Several strategic plans have been established in South Korea to stimulate renewable energy growth that can also promote food waste-to-energy pathways. The National Energy Master Plan has been implemented in South Korea since 2008, which develops a roadmap to increase the consumption of renewable fuels every five years for the next 20 years (IEA, 2017). The second Master Plan was announced in 2014 and aims to achieve 11% share of renewable energy in the total energy consumption of the country by 2035. There are several market-based instruments and financial incentives in South Korea for renewable energy that can benefit the food waste-to-energy sector. In January 2012, the Korean government replaced the existing feed-in-tariff system with a renewable portfolio standard (RPS) to develop a strong market-based system to promote renewable energy deployment in the country. The RPS program targets 13 of the largest power companies in South Korea (with capacity greater than 500 MWh) and includes electricity production from biomass, biogas, and waste among other renewable energy sources (IEA, 2017). It requires that the power companies steadily grow their renewable energy percentage mix in the total power generated, starting with 2% in 2012 and increasing to 10% in 2022. The percentage targets are reviewed and adjusted every three years. The Basic Plan for the Promotion of the Development, Use and Diffusion of New and Renewable Energy of 2001 (last amended in 2014) provides low interest loans to companies toward developing renewable energy technologies, processes, and equipment (EIA, 2017). Additionally, under this plan, a 10% investment tax credit is provided to enterprises investing in an energy research, deployment, and development project, including renewable energy such as waste-derived biofuels.

While Japan has emerged as a leader in food waste management, one of its earlier renewable energy programs, The Kyoto Protocol Target Achievement Plan, was adopted in April 2005 to meet its commitments to reduce GHG emissions by 6% from the 1990 level by 2012. To reach this goal, the country was required to convert biomass energy for various applications, such as transportation fuels, and to develop a nationwide strategy for large-scale adoption of domestically produced biomass by the transportation sector. The target to produce 500,000 kL of transportation biofuel by 2010 (Koizumi, 2011) was achieved through this program. The policy also developed biomass towns and stimulated biomass energy conversion technologies. Now, the country aims to cut emissions by 3.8% by 2020 compared with 2005 levels. Due to Japan's limited capacity for agricultural production, food waste-to-energy is likely to play a big role toward achievement of this goal (Iijima, 2009). The country has more specific renewable energy targets for the power and transportation sector, to which energy produced from food waste can contribute. The Basic Energy Plan of 2010 in Japan mandates 10% of the country's primary energy supply be obtained from renewable energy by 2020 (MAFF, 2013). The policy furthers aims at increasing biofuel at a volume equivalent to 3% displacement of gasoline demand nationwide by 2020 and higher thereon. Other targets include doubling the percentage of electricity generated by renewable sources and nuclear power, and a 30% reduction in energy-related CO_2 emissions by year 2030. Another act "Concerning Sophisticated Methods of Energy Supply Structure" in Japan requires oil refiners to produce a given amount of biofuels (FY2011: 210,000kL and FY2017: 500,000 kL crude oil equivalent). Additionally, for the power sector, it requires electricity retailers producing more than 500 GW annually to attain 44% of their power from non-fossil sources. This target can be achieved independently or jointly with other retailers (IEEJ, 2016).

To promote these policies for the power sector, a feed-in-tariff scheme exists in Japan under which power companies are obligated to purchase electricity from renewable sources at a fixed price for a long, fixed period of time (20 years for bio-based energy). Biogas, including that produced from food waste, is covered under this program (Ushikubo, 2013). Energy plants utilizing food waste and sewage sludge produce biogas and electricity by biogasification, which can be sold to power companies under this scheme. The approach is to attain a closed-loop economy via food waste-to-energy production. Hence, biogas generators such as KAISEI Co. use the residual heat from the process in greenhouses to grow tropical fruits in cold areas, and use the digestive liquid for organic fertilizer to grow rice.

12.4.2.4 Circular Economy-Based Approach

The EU and many countries are gradually moving towards a circular economy-based approach for managing waste. Though this approach has parallels to the food waste management hierarchy, the focus shifts toward closing the loop across the entire food supply chain. The UN FAO has released a toolkit that contains recommendations on how wasted food and food waste can be reduced at every stage of the food supply chain. The toolkit outlines case studies of projects globally that demonstrate how local and national governments, farmers, businesses, and consumers have tackled the problem of food waste (UN FAO, 2013).

Japan was one of the early adopters of such a strategy. The Japanese government promotes biomass utilization at grass-roots levels through the Biomass Nippon Strategy of 2002 (further revised in 2006). Food waste is covered under this policy, which has the primary goals of addressing climate change; creating a "recycling-oriented" society; developing new strategic industries; and activating agriculture, forestry, and fishery. Under this initiative biomass towns have been created—areas where a comprehensive biomass utilization system is established and operated through the cooperation of various stakeholders in the region (MAFF, 2013). In Japan, the Basic Act for the Promotion of Biomass Utilization of 2009 outlines the underlying principles for the promotion of biomass in the country (MAFF, 2013). These principles include diversification of energy sources and mitigation of global warming. The Act advocates full utilization of all types of biomass, including food waste, and advances the country's circular economy principles. This plan sets targets for 2020 by expanding the use of biomass resources equal to 26 million carbon tons per annum in Japan. The target for 2020 for food waste-to-energy is to utilize 40% of the viable food waste for energy conversion (MAFF, 2013). The plan further sets basic policies on the development of biomass to energy conversion technologies.

While South Korea's food waste reduction strategies are already very advanced due to strict regulatory implementation, it plans to gradually adopt a "zero waste" circular economy approach to promote greater recycling and reuse in the country. The Framework Act on Resource Circulation implemented in January 2018 is a step in this direction. This policy aims to promote the utilization of waste as a resource, with the ultimate goal of raw materials and energy supply security in the country (OECD, 2017). This policy can be clearly applied to food waste in South Korea by consolidating different levels of the food waste management hierarchy and corresponding instruments and actions under a common circular economy framework.

Though the EU has been a leader globally in terms of climate change policies, it is still in the process of developing more robust circular economy centric food waste management policies that would include food waste-to-energy as one of the options to avoid landfill disposal. In the EU, there is a call for a proposal for a specific directive for food waste reduction across the entire supply chain, with the goal to reduce the waste by 50% in 2030 for member states. This initiative would be part of the EU Circular Economy Package. The member states are required to report their food waste levels bi-annually. The EU also provides best practices to manage food waste as part of this package. Under this package, the food waste prevention would be the priority (European Parliament, 2016). The Package provides an action plan for circular economy and aims to amend existing EU legislations including the Waste Framework and Landfill Directives. Voluntary initiatives can also promote a circular food waste management system—good examples are the Courtauld 2025 and Food Waste Recycling Action Plan in the UK.

12.5 DISCUSSION AND CONCLUSIONS

It is evident from the analysis presented in this chapter that a wide range of policy instruments exist in the US and abroad that can promote food waste-to-energy conversion. However, it is clear that most of these policies are not directly aimed at food waste-to-energy technology. Rather, their end objectives are renewable energy expansion, GHG reduction, landfill reduction of food waste, and biofuel and biomass utilization. For more sustainable and judicious utilization of food waste-to-energy applications, policies are required in the US that can directly address energy conversion and fuel production from food waste and wasted food. Such policies should be framed with the objective to address the current limitations faced by AD and other technologies that convert food waste to biogas, biodiesel, and ethanol. These limitations, as discussed in Section 12.3.2 can be social, economic, or technological. The pathway for such policies has been outlined with the federal initiatives identified in the *Biogas Opportunities Roadmap and Progress Reports* (USDA et al., 2014; USDA et al., 2015).

Prevention and recovery strategies in the Food Recovery Hierarchy can be aimed directly at consumers and the technology they use because behavior changes are under their control. Industry can play an important role during the prevention phase by more efficient production and distribution practices and by identifying and collaborating with partners to convert wasted food into animal food. In contrast to prevention and recovery, managing food waste requires policy interventions, as the primary decision-makers represent a wide array of government entities, retail establishments, and food processors. For recycling and landfill management strategies, the decision-making ability rests with industry and government, although with small-scale backyard composting, households can also play a small, but critical role in diverting food waste from the landfill.

Several state policies in the US are aimed at reducing food waste currently being sent to landfill from large commercial entities, but do not apply to food waste generated by residential entities or small-scale commercial generators, which, when combined, can comprise a large portion of MSW. Future policies (e.g., residential waste separation and disposal costs, subsidies, grants, or other incentives) should expand to these types of waste generators to maximize food waste-to-energy benefits. Currently, there are very limited regulations in place in the US that strictly require residential sector food consumers to source separate food waste. Because food is one of the largest material streams being sent to the landfill and is primarily managed through this pathway (see Section 12.1), stringent laws are needed to target non-commercial and smaller generators. South Korea provides a compelling example, wherein a federal "pay-as-you-throw" rule is implemented at the municipality levels, targeting residential food waste generators. A large-scale collection network and landfill ban of nonconsumable food waste would enable the material stream to be diverted for either composting or energy generation based on technical and economic viability, thus providing a sustainable feed stream for either application. Stricter and more widespread landfill bans or landfill taxes would enable food waste-to-energy system operators to overcome the economic and technical barriers that are exacerbated by the fragmented waste management policies established at the state and local level.

The federal and state policies in the US that aim to promote renewable energy adoption do include both transportation fuel and power generation via bio-based sources. However, their main focus has been animal farm wastes for biogas production or agricultural feedstock for ethanol and biodiesel production, primarily owing to feedstock availability and compositional uniformity. As a result, though food waste-to-energy is technically covered by these regulations and standards, it has been non-explicitly sidelined. Moreover, in comparison to other forms of renewable energy (solar and wind), food waste-to-energy has received less financial support, so this pathway has to compete with not only conventional waste management strategies (landfill), but also other renewable energy technologies.

Apart from strengthening and extending existing state-level landfill bans for food waste, direct laws and voluntary policies are needed to promote food waste conversion to energy under the circular economy umbrella. Food waste-to-energy is an important technology that can help states meet their renewable energy commitments. State-level landfill bans, landfill reduction targets, renewable energy policies, and targets supporting food waste-to-energy systems are more important than ever as the US may withdraw its national commitment to the Paris Climate Agreement and future federal support is uncertain. A national approach, while unlikely in the near future, would still depend on technical feasibility of regional food waste outflows to be successful and sustainable in the long run.

A major takeaway for the US from the international policy landscape is that federal food waste management efforts need to be further extended to promote grassroot-level initiatives at the regional levels. Since food waste generation, collection, and management are regional problems, high-level federal policies may not be effective if they are not linked to the state-level municipal initiatives and policies. Excellent models for this challenge are Japan and South Korea, where national policies for biomass utilization led to developing biomass villages and waste energy towns all over these countries. For food waste management and associated food waste-to-energy pathways to be a success in the long run, federal and state policies need to be synchronized to result in optimal utilization of resources and technical knowledge.

The analysis also indicates that the US needs to develop food waste management policies that are actionable and circular economy centric. The US EPA food waste management hierarchy and similar frameworks at the state levels provide an extremely high-level snapshot of how food waste should be managed, and where food waste-to-energy fits into that picture. However, actual plans at regional levels, based on past and future food waste generation trends, the percent recoverable, the remaining percent to be recycled, and the composition of the non-consumable food waste stream, are either lacking or are still at a preliminary stage. Encouraging appropriate take back agreements between retail establishments and producers, as seen in Sweden (Eriksson et al., 2017), may provide a stable stream of inputs for food waste-to-energy systems and encourage these establishments to more effectively reduce or prevent waste with technology, sensors, or modified purchasing patterns.

Like many sustainability issues, there is no one-size-fits-all solution to food waste management (Thyberg and Tonjes, 2016), thus a regional approach may be key. Regional policies and programs would enable states to think beyond their own political and physical boundaries and work together as done in the past with watershed and forestry issues. A regional approach could be important in negotiating reduced fees and long-term stable electricity agreements with utility companies, thereby decreasing the perceived and real risks associated with investing in capital-intensive AD technology and encouraging investment for small-to-medium size facilities. A regional perspective could help lower operational costs for food waste-to-energy stakeholders simply by organizing and optimizing transportation routes (ReFED, 2016). Regional economic development programs could encourage and support "biomass towns" in the US, like in Japan, as another policy tool to promote this technology pathway and its benefits while enabling farmers to maintain their way of life. Regional and circular economy policies are needed to identify and support the development of symbiotic relationships between the agriculture, industrial, and waste management sectors to ensure the food waste-to-energy pathway is successful (Girotto et al., 2015). Identifying and maintaining these relationships are important for ensuring steady material inputs and therefore better financing. Educating consumers, commercial entities, and policy makers is critical to ensuring stakeholders are connected with one another and to addressing social barriers such as NIMBY issues related to traffic or odor management.

REFERENCES

Abbey-Lambertz, K., 2016. These 4 states are doing something truly revolutionary with food: ready, set, compost. Huffington Post.com. July 29, 2016. Available from: http://www.huffingtonpost.com/entry/states-food-waste-policies_us_5798a40ce4b0d3568f853698. Accessed 21 July 2017.

Air Resources Board, 2014. Compliance Offset Protocol Livestock Projects. Available from: https://www.arb.ca.gov/regact/2014/capandtrade14/ctlivestockprotocol.pdf. Accessed 21 June 2017.

Air Resources Board, 2017a. Assembly Bill 32 Overview. Available from: https://www.arb.ca.gov/cc/capandtrade/public_info.pdf. Accessed 21 June 2017.

Air Resources Board, 2017b. California's Cap-and-Trade Program. Available from: https://www.arb.ca.gov/cc/capandtrade/public_info.pdf. Accessed 21 June 2017.

American Council on Renewable Energy (ACORE), 2014. Renewable energy in the 50 states: western region. Available from: https://www.acore.org/files/pdfs/states/Oregon.pdf. Accessed 21 March 2017.

Bilsens, L., 2016. Rhode Island—food waste recycling requirements. Institute for Local Self Reliance (ILSR) 2016. Available from: https://ilsr.org/rule/food-scrap-ban/rhode-island-food-waste-recycling. Accessed 8 November 2017.

Binkley, D., Harsh, S., Wolf, C.A., Safferman, S., Kirk, D., 2013. Electricity purchase agreements and distributed energy policies for anaerobic digesters. Energ Policy 53, 341–352.

Biomass Research and Development Initiative, 2017. Available from: https://biomassboard.gov/initiative/initiative.html. Accessed 17 May 2017.

Biorefinery, Renewable Chemical, and Biobased Product Manufacturing Assistance Program, 2015. Interim final rule. Federal Register (24 June 2015), 36409–36455.

Brachlianoff, E., 2013. Turning food waste into energy could be a milestone for the circular economy. The Guardian. November 26. Available from: https://www.theguardian.com/sustainable-business/food-waste-energy-circular-economy. Accessed 27 April 2017.

Breunig, H.M., Jin, L., Robinson, A., Scown, C.D., 2017. Bioenergy potential from food waste in California. Environ. Sci. Technol. 51 (3), 1120–1128.

California Energy Commission, 2017. Waste to energy & biomass in California. http://www.energy.ca.gov/biomass/. Accessed 2 June 2017.

Chanakya, H.N., Ramachandra, T.V., Vijayachamundeeswari, M., 2007. Resource recovery potential from secondary components of segregated municipal solid wastes. Environ. Monit. Assess. 135 (1), 119–127.

Cammarata, 2016. Agricultural Anaerobic Digestion Roadmap for Connecticut. Available from: http://www.ctfarmenergy.org/Pdfs/ADRoadmapCT2016FINAL2.pdf. Accessed 17 May 2017.

Coker, C., 2016. Biomass-to-electricity tax credits extended. BioCycle 57 (1), 60. Available from: https://www.biocycle.net/2016/01/18/biomass-to-electricity-tax-credits-extended. Accessed 5 January 2018.

Costantini, V., Crespi, F., Martini, C., Pennacchio, L., 2015. Demand-pull and technology-push public support for eco-innovation: the case of the biofuels sector. Res. Policy 44 (3), 577–595.

Couture, T., Cory, K., 2009. State clean energy policies analysis project: an analysis of renewable energy feed-in tariffs in the United States. Technical Report NREL/TP-6A2-45551. Available from: https://www.nrel.gov/docs/fy09osti/45551.pdf. Accessed 10 August 2017.

Edwards, J., Othman, M., Burn, S., 2015. A review of policy drivers and barriers for the use of anaerobic digestion in Europe, the United States and Australia. Renew. Sust. Energ. Rev. 52, 815–828.

Energize Connecticut, 2018. Available from: https://www.energizect.com/your-town/solutions-list/virtual-net-metering. Accessed 5 May 2018.

Environmental and Energy Study Institute (EESI), 2014. Administration releases biogas roadmap. August 8. Available from: http://www.eesi.org/articles/view/administration-releases-biogas-roadmap. Accessed 13 July 2017.

Environmental Protection Network, 2017. Analysis of Trump administration proposals for FY2018 budget for the Environmental Protection Agency. March 22, 2017. Available from: http://www.4cleanair.org/sites/default/files/Documents/EPA_Budget_Analysis_EPN_3-22-2017.pdf. Accessed 2 June 2017.

Eriksson, M., Ghosh, R., Mattsson, L., Ismatov, A., 2017. Take-back agreements in the perspective of food waste generation at the supplier-retailer interface. Resour. Conserv. Recycl. 122, 83–93.

Eriksson, M., Strid, I., 2011. Food loss in the retail sector–a study of food waste in six grocery stores. Swedish University of Agricultural Sciences—Department of Energy and Technology, Uppsala, Sweden.

Eriksson, M., Strid, I., Hansson, P.A., 2015. Carbon footprint of food waste management options in the waste hierarchy – a Swedish case study. J. Clean. Prod. 93, 115–125.

European Commission, 2008. Directive 2008/98/EC of the European Parliament and of the Council of 19 November 2008 on Waste (Waste Framework Directive). Official Journal of the European Union, Strasbourg.

European Commission, 2017. Renewable Energy. Moving towards a Low Carbon Economy. Available from: http://ec.europa.eu/energy/en/topics/renewable-energy. Accessed 10 August 2017.

European Parliament, 2016. Circular Economy Package. Four Legislative Proposals on Waste. Members' Research Service. Available from: http://www.europarl.europa.eu/EPRS/EPRS-Briefing-573936-Circular-economy-package-FINAL.pdf. Accessed 10 August 2017.

Findlaw, 2017. Available at http://codes.findlaw.com/ny/public-service-law/pbs-sect-66-j.html. Accessed 15 July 2017.

Ghisellini, P., Cialani, C., Ulgiati, S., 2016. A review on circular economy: the expected transition to a balanced interplay of environmental and economic systems. J. Clean. Prod. 114, 11–32.

Girotto, F., Alibardi, L., Cossu, R., 2015. Food waste generation and industrial uses: a review. Waste Manag. 45, 32–41.

Global Environment Centre Foundation, 2011. Law for Promotion to Recover and Utilize Recyclable Food Resources (Food Recycling Law). Available from: http://nett21.gec.jp/Ecotowns/data/et_c-08.html. Accessed 10 August 2017.

Goldstein, N., 2013. Farm digester evolution in Vermont. Biocycle. 54(2).

Guermoud, N., Ouadjnia, F., Abdelmalek, F., Taleb, F., 2009. Municipal solid waste in Mostaganem city (Western Algeria). Waste Manag. 29 (2), 896–902.

Guillier, L., Duret, S., Hoang, H.M., Flick, D., Laguerre, O., 2016. Is food safety compatible with food waste prevention and sustainability of the food chain? Proc. Food Sci. 7, 125–128.

Gustavsson, J., Cederberg, C., Sonesson, U., van Otterdijk, R., Meybeck, A., 2011. Global Food Losses and Food Waste: Extent, Causes and Prevention. FAO, Rome.

Hebrok, M., Boks, C., 2017. Household food waste: drivers and potential intervention points for design – an extensive review. J. Clean. Prod. 151, 380–392.

Hernandez, S., 2014. The Business Case for Carbon Offsets from Waste Diversion: Waste Digestion and Composting. Waste Advantagemag.com. (September 21, 2014) Available from: https://wasteadvantagemag.com/business-case-carbon-offsets-waste-diversion-waste-digestion-composting/. Accessed 21 June 2017.

IEA, 2015. Electricity from Biomass: from small to large scale Summary and Conclusions from the IEA Bioenergy ExCo72 Workshop. Available from: http://www.ieabioenergy.com/wp-content/uploads/2015/04/ExCo72-Electricity-fromBiomass-Summary-and-Conclusions-13.04.15.pdf. Accessed 10 August 2017.

IEA, 2017. IEA/IRENA Joint Policies and Measures Database. Available from: http://www.iea.org/policiesandmeasures/renewableenergy/?country=Korea. Accessed 10 August 2017.

IEEJ, 2016. Mapping the energy future. Available from: http://eneken.ieej.or.jp/en/jeb/160316.pdf. Accessed 5 April 2017.

Iijima, M., 2009. Japan to focus on next generation biofuels. USDA Foreign Agricultural Service. GAIN Report Number JA9044.

Industrial Economics, Incorporated, 2017. Benefit-Cost Analysis of Potential Food Waste Diversion Legislation. Prepared for New York State Energy Research and Development Authority, Cambridge, MA. NYSERDA Report 17-06. Available from: https://www.nyserda.ny.gov/-/media/Files/Publications/Research/Environmental/Benefit-Cost-Analysis-of-Potential-Food-Waste-Diversion-Legislation.pdf. Accessed 8 June 2017.

Innovation Seeds, 2017. South Korea's food waste reduction policies. Available from: http://www.innovationseeds.eu/policy-library/core-articles/south-koreas-food-waste-reduction-policies.kl. Accessed 10 August 2017.

Kang, H., 2013. South Korea country report. IEA Bioenergy Task 37. In: IEA Bioenergy Task 37 country reports. Available from: http://www.iea-biogas.net/country-reports.html. Accessed 10 August 2017.

Khalid, A., Arshad, M., Anjum, M., Mahmood, T., Dawson, L., 2011. The anaerobic digestion of solid organic waste. Waste Manag. 31 (8), 1737–1744.

Kim, M.H., Song, Y.E., Song, H.B., Kim, J.W., Hwang, S.J., 2011. Evaluation of food waste disposal options by LCC analysis from the perspective of global warming: Jungnang case, South Korea. Waste Manag. 31 (9), 2112–2120.

Koizumi, T., 2011. Biofuel programs in East Asia: developments, perspectives, and sustainability. In: Environmental Impact of Biofuels. InTech., Rijeka, Croatia 2011.

Kosseva, M.R., 2011. Management and processing of food wastes. In: Reference Module in Earth Systems and Environmental Sciences Comprehensive Biotechnology. 2nd ed. Elsevier B.V., pp. 557–593.

Layzer, J.A., 2015. The Environmental Case: Translating Values Into Policy. CQ Press.

Lawrence Berkeley National Lab (LBNL), 2017. Food Waste-to-Energy to Help California Meet Energy, Zero-Waste, Climate Goals. Available from: https://eaei.lbl.gov/news/article/food-waste-energy-help-california. Accessed 24 September 2017.

Lehner, P., 2013. What if government took food waste seriously? April 3, 2016. Available from: https://www.nrdc.org/experts/peter-lehner/what-if-government-took-food-waste-seriously. Accessed 6 April 2017.

Levis, J.W., Barlaz, M.A., Themelis, N.J., Ulloa, P., 2010. Assessment of the state of food waste treatment in the United States and Canada. Waste Manag. 30 (8), 1486–1494.

Lijó, L., González-García, S., Bacenetti, J., Moreira, M.T., 2017. The environmental effect of substituting energy crops for food waste as feedstock for biogas production. Energy.

Lim, V., Funk, M., Marcenaro, L., Regazzoni, C., Rauterberg, M., 2017. Designing for action: an evaluation of social recipes in reducing food waste. Int. J. Hum. Comput. Stud. 100, 18–32.

Mass DEP, 2016. Commercial food waste disposal ban. Available from: http://www.mass.gov/eea/agencies/massdep/recycle/reduce/food-waste-ban.html. Accessed 8 June 2017.

Milios, L., 2013. Municipal Waste Management in Sweden. In: A Report Prepared for the European Environmental Agency and European Topic Centre on Sustainable Consumption and Production. Available from: https://www.eea.europa.eu/publications/managing-municipal-solid-waste/sweden-municipal-waste-management/view. Accessed 12 August 2017.

Ministry of Agriculture, Forestry and Fisheries of Japan (MAFF), 2013. Biomass policies and assistance measures in Japan. Available from: http://www.maff.go.jp/e/pdf/reference6-8.pdf. Accessed 10 August 2017.

Ministry of Environment, 2013. Special Management Plan for Prohibition of Food Waste Leachate Dumping in the Sea. Ministry of Environment, Korea.

Ministry of Environment, 2017a. Volume-Based Food Waste Fee System. Ministry of Environment, Korea. Available from: http://eng.me.go.kr/eng/web/index.do?menuId=387. Accessed 10 August 2017.

Ministry of Environment, 2017b. Waste-to-energy. Ministry of Environment, Korea.http://eng.me.go.kr/eng/web/index.do?menuId=390. Accessed 10 August 2017.

Ministry of Strategy and Finance, 2016. 2016 Modularization of Korea's Development Experience: Waste Resources Management and Utilization Policies of Korea. Republic of Korea, Ministry of Strategy and Finance. Available from: https://seoulsolution.kr/sites/default/files/gettoknowus/%5BKSP%20Modularization%5D%20Waste%20Resources%20Management%20and%20Utilization%20Policies%20of%20Korea_2016.pdf. Accessed 17 August 2017.

Moriarty, K., 2013. Feasibility study of anaerobic digestion of food waste in St. Bernard, Louisiana. Technical Report NREL/TP-7A30-57082.

Mourad, M., 2016. Recycling, recovering and preventing "food waste": competing solutions for food systems sustainability in the United States and France. J. Clean. Prod. 126, 461–477.

Murphy, J.D., McKeogh, E., Kiely, G., 2004. Technical/economic/environmental analysis of biogas utilisation. Appl. Energy 77 (4), 407–427.

Navigant Consulting, 2006. Recommendations for a Bioenergy Action Plan for California. Report for the Bioenergy Interagency Working Group. Available from: http://www.energy.ca.gov/2006publications/CEC-600-2006-004/CEC-600-2006-004-F.PDF. Accessed 21 July 2017.

New York State Department of Environmental Conservation (NYS DEC), 2018. Regulations for organic recycling facilities.Available from: http://www.dec.ny.gov/chemical/97488.html. Accessed 5 June 2018.

New York State Department of Public Service (NYS DPS) and New York State Energy Research and Development Authority (NSYERDA), 2016. The Clean Energy Standard: Informational Webinar. September 13, 2016. Available from: http://www3.dps.ny.gov/W/PSCWeb.nsf/96f0fec0b45a3c6485257688006a701a/56c58a580d2cf2e185257fd4006b90ce/$FILE/Sept%2014%20Webinar%20Slides%20-9-13-16%20340%20pm.pdf. Accessed 21 July 2017.

New York State Senate, 2017. Senate Bill S3418.Available from: https://www.nysenate.gov/legislation/bills/2017/S3418. Accessed 30 May 2017.

Ng, T., 2013. Legislative Council Secretariat. Information Note, South Korea's waste management policies. Available from: http://www.legco.gov.hk/yr12-13/english/sec/library/1213inc04-e.pdf. Accessed 10 August 2017.

Nguyen, D.D., Yeop, J.S., Choi, J., Kim, S., Chang, S.W., Jeon, B.H., Guo, W., Ngo, H.H., 2017. A new approach for concurrently improving performance of South Korean food waste valorization and renewable energy recovery via dry anaerobic digestion under mesophilic and thermophilic conditions. Waste Manag. 66, 161–168.

NYC Department of Sanitation, 2017. Food Scraps + Yard Waste for Businesses. Available here http://www1.nyc.gov/assets/dsny/zerowaste/businesses/food-scraps-and-yard-waste.shtml. Accessed 21 July 2017.

NYS Pollution Prevention Institute (NYSP2I), 2017. NYS Food System Sustainability Clearinghouse. Available from: https://www.rit.edu/affiliate/nysp2i/food/regulations-laws-and-related-activities. Accessed 11 August 2017.

NYSERDA, 2017. Clean energy standard.Available from: https://www.nyserda.ny.gov/All-Programs/Programs/Clean-Energy-Standard. Accessed 21 July 2017.

OECD, 2017. Environmental Performance Reviews. Korea Highlight. 2017, Retrieved from https://www.oecd.org/env/country-reviews/OECD_EPR_Korea_Highlights.pdf. Accessed 15 September 2017.

Parfitt, J., Barthel, M., Macnaughton, S., 2010. Food waste within food supply chains: quantification and potential for change to 2050. Philos. Trans. Roy. Soc. Lond. B: Biol. Sci. 365 (1554), 3065–3081.

Parry, A., Bleazard, P., Okawa, K., 2015. Preventing Food Waste: Case Studies of Japan and the United Kingdom. OECD Food, Agriculture and Fisheries Papers, No. 76. OECD Publishing, Paris, France.https://doi.org/10.1787/5js4w29cf0f7-en. Accessed 17 August 2017.

Pham, T.P.T., Kaushik, R., Parshetti, G.K., Mahmood, R., Balasubramanian, R., 2015. Food waste-to-energy conversion technologies: current status and future directions. Waste Manag. 38, 399–408.

Priefer, C., Jörissen, J., Bräutigam, K.R., 2016. Food waste prevention in Europe – a cause-driven approach to identify the most relevant leverage points for action. Resour. Conserv. Recycl. 109, 155–165.

ReFED, 2016. A roadmap to reduce U.S. food waste by 20 percent. March, 2016. Available from: http://www.refed.com/?sort=economic-value-per-ton. Accessed 15 March 2017.

Regional Greenhouse Gas Initiative (RGGI), 2014. Fact Sheet: The Investment of RGGI Proceeds Through 2014. Available from: https://www.rggi.org/docs/ProceedsReport/RGGI_Proceeds_FactSheet_2014.pdf. Accessed 5 June 2017.

Renewable Fuel Standard Program, 2015. Standards for 2014, 2015, and 2016 and Biomass-Based Diesel Volume for 2017, 2015. 40 CFR Part 80, No. 239, December 14, 2015. [EPA–HQ–OAR–2015–0111; FRL–9939–72– OAR] RIN 2060–AS22. Available from: https://www.gpo.gov/fdsys/pkg/FR-2015-12-14/pdf/2015-30893.pdf. Accessed 2 June 2017.

Renewable Fuel Standard Program, 2016. Standards for 2017 and Biomass- Based Diesel Volume for 2018, 2016. 40 CFR Part 80, No. 238, December 12, 2016. [EPA–HQ–OAR–2016–0004; FRL–9955–84– OAR] RIN 2060–AS72. Available from: https://www.gpo.gov/fdsys/pkg/FR-2016-12-12/pdf/2016-28879.pdf. Accessed 2 June 2017.

RGGI, 2017. Welcome. Available from: https://www.rggi.org. Accessed 5 June 2017.

Rhode Island Department of Environmental Management (RI DEM), 2018. Waste Management.Available from: http://www.dem.ri.gov/programs/wastemanagement/. Accessed 5 June 2018.

Rosengren, C., 2017. New York may be the next state to pass a commercial organics law. WasteDive. 2017 (January 26) Available from: https://www.wastedive.com/news/new-york-may-be-the-next-state-to-pass-a-commercial-organics-law/434795. Accessed 30 May 2017.

Seo, Y., 2013. Current MSW Management and Waste-To-Energy Status in the Republic of Korea. M.S. Thesis, Columbia University, New York, NY.

Serfass, P., 2012. Biogas Markets and Federal Policy. American Biogas Council Presentation. Available from: https://energy.gov/sites/prod/files/2014/03/f10/june2012_biogas_workshop_serfass.pdf. Accessed 21 April 2017.

Thi, N.B.D., Lin, C.Y., Kumar, G., 2016. Electricity generation comparison of food waste-based bioenergy with wind and solar powers: A mini review. Sustain. Environ. Res. 26 (5), 197–202.

Thyberg, K.L., Tonjes, D.J., 2016. Drivers of food waste and their implications for sustainable policy development. Resour. Conserv. Recycl. 106, 110–123.

United Nations Development Programme (UNDP), 2006. Green Energy from Wastes. Ministry of Non-conventional Energy Sources, Government of India and UNDP.

United Nations Environmental Programme (UNEP), 2014. Prevention and reduction of food and drink waste in businesses and households – Guidance for governments, local authorities, businesses and other organisations. Version 1.0. Available from: http://www.fao.org/fileadmin/user_upload/save-food/PDF/Guidance-content.pdf. Accessed 5 April 2017.

United Nations Food and Agriculture Organization (UN FAO), 2013. Food waste harms climate, water, land and biodiversity – new FAO report. September 11, 2013. Available from: http://www.fao.org/news/story/en/item/196220/. Accessed 21 July 2017.

United Nations Food and Agriculture Organization (UN FAO), Food loss and food waste. Available from: http://www.fao.org/food-loss-and-food-waste/en/, Accessed 28 March 2017.

US Census Bureau, 2017. Industry snapshot: food manufacturing (NAICS 311). Available from: http://thedataweb.rm.census.gov/TheDataWeb_HotReport2/econsnapshot/2012/snapshot.hrml?STATE=ALL&COUNTY=ALL&IND=%3DCOMP%28%28C4*C4%29%2FC4%29&x=23&y=11&NAICS=311. Accessed 23 June 2017.

US Congress, 2017. Current legislative activities.Available from: https://www.congress.gov. Accessed 30 May 2017 Available from:.

US Department of Agriculture (USDA), US Environmental Protection Agency (US EPA), US Department of Energy (US DOE), 2014. Biogas opportunities road map: voluntary actions to reduce methane emissions and increase energy independence. Available from: https://www.usda.gov/oce/reports/energy/Biogas_Opportunities_Roadmap_8-1-14.pdf. Accessed 2 June 2017.

US Department of Energy (US DOE), 2017a. Renewable electricity production tax credit (PTC). Available from: https://energy.gov/savings/renewable-electricity-production-tax-credit-ptc. Accessed 9 July 2017.

US Department of Energy (US DOE), 2017b. Alternative fuel data center: renewable fuels standard. Available from: http://www.afdc.energy.gov/laws/RFS.html. Accessed 2 June 2017.

US Energy Information Administration, 2017. Country analysis brief: South Korea. Available from: http://www.marcon.com/library/country_briefs/SouthKorea/south_korea.pdf. Accessed 10 August 2017.

US Environmental Protection Agency (US EPA), 2015. RFS registration for renewable fuel production: separated food waste guidance. Available from: https://www.epa.gov/sites/production/files/2015-09/documents/rfs-sfwp-pres-2015-07.pdf. Accessed 15 June 2017.

US EPA, 2017a. Sustainable management of food. Available from: https://www.epa.gov/sustainable-management-food/sustainable-management-food-basics#whatiswastedfood. Accessed 10 August 2017.

US EPA, 2017b. Food recovery hierarchy. Available from: https://www.epa.gov/sustainable-management-food/food-recovery-hierarchy. Accessed 15 June 2017.

US EPA, 2017c. AgSTAR: Biogas recovery in the agriculture sector. Available from: https://www.epa.gov/agstar. Accessed 10 August 2017.

US EPA, 2017d. National Pollutant Discharge Elimination System (NPDES): Animal feeding operations (AFOs). Available from: https://www.epa.gov/npdes/animal-feeding-operations-afos. Accessed 2 June 2017.

US EPA, 2017e. Renewable fuel standard program. https://www.epa.gov/renewable-fuel-standard-program. Accessed 15 June 2017.

US EPA, 2017f. Renewable energy production incentives. Available from: https://www.archive.epa.gov/epawaste/hazard/wastemin/web/html/. Accessed 15 June 2017.

US EPA, Office of Land and Emergency Management, 2014. Advancing sustainable materials management: 2014 fact sheet – assessing trends in material generation, recycling, composting, combustion with energy recovery and landfilling in the United States. US EPA, Office of Land and Emergency Management, Washington, DC, EPA 530-R-17-01.

US EPA, Office of Resource Conservation and Recovery, 2016. Food waste management in the United States, 2014. Available from: https://www.epa.gov/sites/production/files/2016-12/documents/food_waste_management_2014_12082016_508.pdf. Accessed 21 July 2017.

USDA, 2017a. US food waste challenge FAQs. Available from: https://www.usda.gov/oce/foodwaste/faqs.htm. Accessed 12 June 2017.

USDA, 2017b. Recycle. Available from: https://www.usda.gov/oce/foodwaste/resources/recycle.htm. Accessed 12 June 2017.

USDA, 2017c. Center for nutrition policy and promotion. Available from: https://www.cnpp.usda.gov. Accessed 11 August 2017.

USDA, 2017d. Let's talk trash. https://www.choosemyplate.gov/lets-talk-trash. Accessed 11 August 2017.

USDA, 2017e. US food loss and waste 2030 champions.Available from: https://www.usda.gov/oce/foodwaste/index.htm. Accessed 6 December 2017.

USDA, 2017f. Rural energy for America program renewable energy systems&energy efficiency improvement loans & grants.Available from: https://www.rd.usda.gov/programs-services/rural-energy-america-program-renewable-energy-systems-energy-efficiency. Accessed 13 July 2017.

USDA, 2017g. Biofuel infrastructure partnership.Available from: https://www.fsa.usda.gov/programs-and-services/energy-programs/bip/index. Accessed 13 July 2017.

USDA, US EPA, US DOE, 2015. Biogas opportunities roadmap progress report. Available from: https://energy.gov/sites/prod/files/2015/12/f27/biogas_opportunites_roadmap_progress_report_0.pdf. Accessed 2 June 2017.

Ushikubo, A., 2013. Recycling Food Waste in Japan. Available from: https://www.oecd.org/site/agrfcn/Session%204_Akikuni%20Ushikubo.pdf. Accessed 5 April 2017.

van Haaren, R., Themelis, N., Nora Goldstein, N., 2010. The state of garbage in America: 17th nationwide survey of MSW management in the U.S. BioCycle. October 2010. Available here: https://www.biocycle.net/images/art/1010/bc101016_s.pdf. Accessed 5 April 2017.

Vermont Department of Environmental Conservation (VT DEC), 2016. Vermont Universal Recycling Law: Status Report. Available from: http://dec.vermont.gov/sites/dec/files/wmp/SolidWaste/Documents/Universal.Recycling.Status.Report.Dec_.2016.pdf. Accessed 21 July 2017.

Waste and Resources Action Programme (WRAP), 2016. Food Waste Recycling Action Plan for England. Available from: http://www.wrap.org.uk/content/food-waste-recycling-action-plan. Accessed 5 April 2017.

Wei, Y., Li, J., Shi, D., Liu, G., Zhao, Y., Shimaoka, T., 2017. Environmental challenges impeding the composting of biodegradable municipal solid waste: a critical review. Resour. Conserv. Recycl. 122, 51–65.

Zimlich, R., 2015. Food waste bans can be problematic for landfills. Waste360. July 7, 2015. Available from: https://www.waste360.com/operations/food-waste-bans-can-be-problematic-landfills. Accessed 11 August 2017.

FURTHER READING

IEA Bioenergy, 2016. IEA Bioenergy Countries' Report. Bioenergy Policies and Status of Implementation. Available from: http://www.ieabioenergy.com/wp-content/uploads/2016/09/iea-bioenergy-countries-report-13-01-2017.pdf. Accessed 10 August 2017.

Chapter 13

Challenges and Innovations in Food Waste-to-Energy Management and Logistics

William R. Armington, Roger B. Chen and Callie W. Babbitt
Golisano Institute for Sustainability, Rochester Institute of Technology, Rochester, NY, United States

13.1 ATTRIBUTES OF FOOD WASTE MANAGEMENT SYSTEMS

The food waste recovery system is comprised of diverse actors, each with personal goals, influence, and decisions that affect system-level performance. Decisions and interactions between these actors influence how the food waste recovery system functions. The interactions between food waste generators, transporters, and recovery facilities comprise the core of this recovery system, whereas regulations or policies externally influence core operations (Fig. 13.1).

13.1.1 Influence of Food Waste Legislation

Properly designed and operated management systems are necessary to minimize landfilling and maximize energy recovery from food waste. Even though a reasonable number of current food waste management systems have evolved as profit-seeking businesses without government finance and support, policies arise that facilitate the development of these systems. For example, as reviewed in Chapter 12, state-level food waste landfill disposal bans, which depart from more common regulatory approaches to nonhazardous waste management, have been driving forces for improved management and logistics. The aim of these regulations is to reduce greenhouse gas (GHG) emissions from food waste degrading in landfills under anaerobic conditions. These regulations have a number of impacts on the ultimate design and operation of logistics systems and decisions made by key stakeholders within these systems.

Consider, for example, food waste landfill disposal bans put in place in the States of Massachusetts and Connecticut (Connecticut Department of Energy and Environmental Protection, 2017; Department of Environmental Protection, 2014). These regulations target large food waste generators (i.e., those producing more than one or two tons per week). Facilities that meet the generation quantity threshold for bans are required to send their food waste to a recovery facility if one exists within a distance given by policy, usually a 40–50 miles radius from the source. Additionally, from a regulatory standpoint, food waste is still considered a waste and must be treated similarly to other waste with respect to collection, transportation, and disposal.

In conventional solid waste management, collected waste is deposited at the least expensive facility as determined by the transportation costs and the tipping fee, or the cost to unload waste at a facility, such as a landfill. Typically, waste is unloaded at the closest landfill or at a transfer station at the end of a collection route to save on transportation costs. Separate food waste hauling will likely encounter similar unloading behaviors unless partnerships emerge that make unloading at better waste-to-energy facilities less costly to the transporter. In states without policies that specifically target food waste, such collection and transportation activities are voluntary, but businesses may participate to meet corporate sustainability goals or in anticipation of future food waste disposal bans. For voluntary systems, recovery facilities may have the flexibility to choose generators from whom to receive food waste. These decisions can optimize the throughput or volume of food waste treated for an operating scale or optimize the proportions of food waste types that allow for optimal energy recovery, given a specific waste-to-energy technology. In either case, design of logistics systems will face two challenges:

- Some food waste may be ill-suited for energy recovery but will still require treatment at a recovery facility due to food waste legislation.

Sustainable Food Waste-to-Energy Systems. https://doi.org/10.1016/B978-0-12-811157-4.00013-9

- Organizations subject to a landfill disposal ban, but not currently participating in a collection service due to infeasible distance from a recovery facility, may now face increased costs from collection.

Therefore, effective logistics solutions are critical for proactively responding to food waste policies and creating systems that allow both large and small producers of food waste to participate economically.

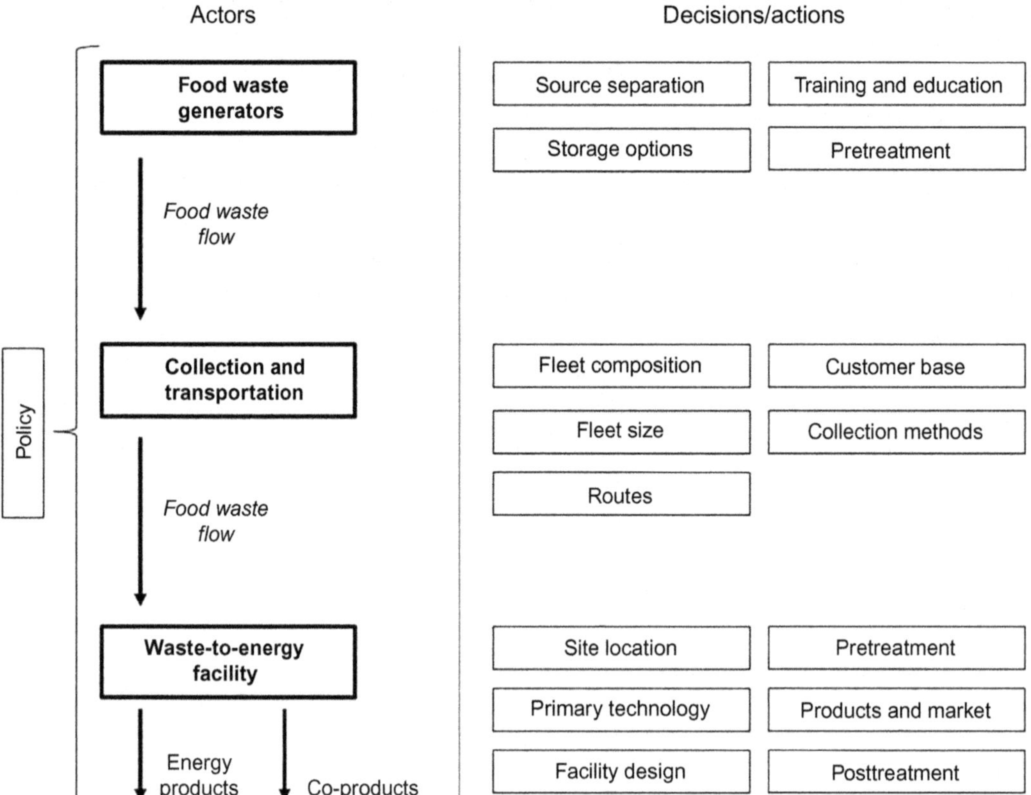

FIG. 13.1 An overview of the actors and decisions in the food waste management system. Policy is a main driver of decision making by each actor within their space of influence.

13.1.2 Actor Interactions

From the perspective of the food waste generator, the food waste management system functions similarly to conventional waste management. We can gain insight from leveraging existing and analogous municipal solid waste (MSW) collection practices. For conventional MSW practice, waste generators produce waste and contract with a waste hauler (or similar service) to collect this waste according to a fixed schedule. Depending on the frequency and quantity of waste produced, collection could range from one to five times per week. Waste haulers gain a service fee based on the contract terms, obligating the hauler to collect that generator's waste. There are, however, key differences between MSW and food waste management related to business practice:

- Employees require training for proper disposal of food waste in specific containers located in the business and for determining which foods are suitable for separate disposal.
- Logistic decisions must be addressed with regard to using collection bins within businesses, such as placement of bins and frequency of disposal.

At the residential level, food waste management would most likely mirror conventional solid waste management services; collection programs should be structured similarly. Motivation for residents to participate depends on their intrinsic environmentalism, which can motivate their dedication to properly separating food as well as their willingness to pay for the extra collection service (Neff et al., 2015; Qi and Roe, 2016). Although this service is typically offered in addition to conventional MSW and recycling services, there has been a steadily growing number of municipalities that are implementing successful residential food waste collection programs (Yepsen, 2015). For example, Denver, Colorado offers collection services to 4500 households; San Antonio, Texas offers the program to all 350,000 residents and has 19,000 subscribers as of 2014; and Princeton, New Jersey offers a collection service to all 9500 households within the municipality.

From the perspective of the energy recovery facility operator, collected food waste becomes the input in a supply chain that feeds into waste-to-energy operations to make products, including electricity, biogas, and other waste-derived materials. Food waste, unlike other materials, is constantly undergoing uncontrolled degradation during transport, potentially reducing the amount of energy recoverable through technological processing (Nilsson Påledal et al., 2018). Energy recovery facilities must coordinate with food waste generators and collection companies to source material quickly to help maintain facility operation efficiency. As such, educational and training material specific to the requirements of the partner recovery technology is needed to educate generators on food separation to reduce complications in the transportation of food waste for energy recovery.

Collection companies are responsible for meeting the needs of their customers, who demand reliable collection of waste from their locations, and for supplying waste-to-energy facilities, who require consistent, suitable feedstock. Performing both tasks requires coordination and some understanding of both the food waste generation and energy recovery processes. Many food waste collection companies operate their service like traditional MSW management. To keep operation costs down, collection firms may have agreements with multiple food waste accepting facilities to deliver collected waste via the shortest route or to the nearest facility. Even though transportation costs and emissions are minimized, composting or incineration facilities may be chosen for managing food waste instead of waste-to-energy, resulting in reduced energy recovery. Food waste logistics should make collection of waste from small generators and residential sources viable to maximize the material collected, and to ensure that transport of food waste to waste-to-energy facilities is preferred.

13.1.3 Insights from Analogous Systems

While food waste-to-energy is a more recent challenge, a significant body of literature exists that addresses solutions and insights from potentially analogous recovery systems such as MSW management and reverse logistics. Table 13.1 highlights these recovery systems and provides example representative literature in the field.

Presently, the research available specific to food waste logistics is limited; however, research concepts from other logistics systems with similar characteristics can guide the design and operation of food waste management. For example, challenges encountered in conventional MSW management may parallel those encountered with food waste. Waste type, vehicle use, disposal methods, system-level objectives, and vehicle routing methods considered in other systems may have analogous counterparts in food waste logistics systems. Alternatively, the supply chain literature may be useful for understanding the coordination of material inputs for waste-to-energy facilities, where analytical, economic, and simulation as well as performance measures used in conventional material supply chains might be useful indicators for food waste resources.

TABLE 13.1 Logistics Systems With Similar Properties to Food Waste Management

System Type	Description	Representative Literature Examples
Municipal solid waste	Collection and disposal of solid waste generated from residential and commercial sources	(Beliën et al., 2012)
Supply chains	Distribution of materials and resources in conventional manufacturing systems to create products for sale	(Beamon, 1998; Hesse and Rodrigue, 2004; Srivastava, 2007)
Reverse logistics	Logistics system setup to collect products from end users for recycling or refurbishing at recycling facilities, commonly for electronics	(Fleischmann et al., 1997; Srivastava, 2008)
Biomass-to-energy logistics	Biomass recovery systems consist of collection and transporting agricultural crops, residue, or woody biomass to energy recovery facilities	(Gold and Seuring, 2011; Iakovou et al., 2010; Pan et al., 2015)
Conventional recycling	Recycling of glass, paper, and plastic separate from municipal solid waste collection via materials recovery facilities usually colocated with landfills to take advantage of the existing logistics systems	(Pohlen and Theodore Farris, 1992)

Similarly, reverse logistics involves the collection, transportation, remanufacturing, and redistribution of end user products that are replaced or no longer wanted (Fleischmann et al., 1997; Srivastava, 2008). This systems perspective envisions waste materials as part of a closed-loop production cycle, similar to the goal of converting food waste into energy and value-added products. Research in reverse logistics for product recovery, particularly involving remanufacturing of electronic devices, illustrates models for similar systems with varying materials, unknown operational costs, location of remanufacturing facilities, as well as consideration of material supply and demand (Beullens, 2004). However, as food waste management systems become more prevalent, these systems may begin to mirror methods used by traditional reverse supply chains such as plastic, paper, and glass recycling systems common since the 1980s. In this case, waste is collected from commercial and residential sources, transported to a materials recovery facility (MRF) where sorting, cleaning, and repackaging occurs, and then sold on the market as a raw material. Food waste literature has yet to explore the potential of a MRF analog as part of the food waste management system; it may be a useful concept, particularly from the standpoint of waste-to-energy facilities that require consistent quantity and quality of food waste materials. Many of these logistics concepts could parallel food waste management strategies as recovery systems expand.

13.2 FOOD WASTE MANAGEMENT AS A SYSTEM

Sustainability of food waste-to-energy technologies depends on the broader system in which these technologies exist, spanning from separating and collecting waste as it is generated, to transportation, and ultimately to the treatment facility where distribution of recovered energy and other products to markets occurs. Each of the steps in this process is overviewed in Table 13.2, which also provides representative literature examples from the wide array of research on this topic.

Step 1: Source Separation

Source separation of food waste is the first, and arguably most important step in food management systems. Control of the organic materials entering the waste stream allows for the generation of homogenous, contaminant-free feedstocks suitable for specific types of energy recovery technologies. There are two main reasons for source separation. First, source separation is relatively cheaper compared to separation at other points in the logistics chain where heterogeneous waste materials are combined. Second, source separation is the main point of information collection and exchange between the generator and recovery technology operator to ensure that the feedstock is suitable for waste-to-energy processes. Generators can use source separation to assess the characteristics or types of food waste generated and pass that information

TABLE 13.2 Steps in the Food Waste Management System

Step	Description	Representative Literature Examples
Source separation	Separation and temporary storage of food wastes like produce, meats, grains, packaged foods, grease, and process wastewater	(Dahlén and Lagerkvist, 2010; Leanpath, 2016; New York State Pollution Prevention Institute, n.d.; Tompkins County Department of Recycling and Materials Management, n.d.; United States Environmental Protection Agency, 2014)
Collection	Activities performed when transferring food wastes from generation sources into collection vehicles	(Bernstad and la Cour Jansen, 2012)
Transportation	Conveyance, routing, and delivery of food wastes between generation sources and food waste-to-energy facilities	(Sandhu et al., 2015)
Pretreatment	Process to prepare food waste for conversion via primary energy recovery technology. Pretreatment may be necessary for primary technology or performed to increase energy recovery efficiency	(Ariunbaatar et al., 2014; Patinvoh et al., 2017; Sullivan, 2012)
Energy recovery operations	Conversion of food waste to energy through biological, chemical, thermal, and mechanical pathways	Chapters 4–9 of this book
Product distribution	Transfer and sales to end users of the energy products and coproducts produced at a food waste-to-energy facility	(No references)

along to the recovery technologies. Recovery technology operators can potentially evaluate the feedstock and either give the generator further separation instructions or alter their operations to effectively process the incoming material.

Understanding the characteristics of food waste is necessary to ensure the feedstock meets the requirements of the energy recovery technology. A commercial food waste generator, such as a restaurant or supermarket, should begin by performing a food waste audit or assessment to inventory the potential feedstock materials produced. The results of the assessment can inform downstream collection and management decisions. For residential programs, a food waste assessment is impractical since households produce significantly less food waste than commercial sources; therefore communicating educational information on food waste to include in the collection bins is more practical for achieving suitable feedstock when collecting from an entire community of households.

Containers are needed to temporarily store separated food waste before being collected. A variety of storage containers exist for collecting food waste from commercial and residential sources. At a commercial scale, 32- and 64-gal bins with wheels are common for establishments producing hundreds of pounds of food waste daily. These containers are usually small enough to bring inside a facility and have the mobility to be moved by a single person. Bins can be wheeled to the conventional location where MSW is collected or another convenient location so that employees do not have to substantially alter their routines. Commercial entities are often concerned with appearance, odor, safety, and pests associated with food waste collection, which potentially adds unique design constraints not found in the case of typical MSW separation.

At the residential scale, smaller 4- or 5-gal containers with handles and covers are adequately sized for the generation rate of an average family over the course of a week and are small enough to be handled by one person. If the residential collection program combines food waste and yard waste, 64- and 96-gal wheeled bins could be used. Residents then put their containers on the curb to be collected in the same manner as conventional MSW or recycling. Alternatively, centralized community drop-off locations can be organized where residents can bring their food waste. This approach reduces or eliminates the need for curbside collection and allows residents to choose when they wish to remove food waste from their household instead of adhering to a collection schedule. However, added separation and handling steps can also limit household willingness to participate.

Step 2: Collecting and Loading Food Waste

Collection and loading refer to the activities performed transferring the food waste from collection storage into the collection truck. Food waste can be classified as either solid or liquid for the purposes of collection methods.

For solid food waste, collection bins can be loaded onto a waste collection truck by a mechanical lift, commonly on the side and rear of the truck. One consideration is the height of storage compared to the collection vehicle. At commercial facilities, it is common to set up storage options on the loading dock, whereas storage will be at ground level for residential collection. Solid food wastes such as vegetables or bread products are commonly collected by liquid-tight waste collection vehicles that could be fully enclosed or top loading.

Liquid food wastes can be transported in liquid tanker trucks to their disposal sites but are often sent to waste water treatment plants via sewer systems. If a waste water treatment plant has a food waste recovery technology like anaerobic digestion installed, then sewer conveyance of other food waste types would serve as a cost-effective method for feedstock transportation (Bernstad and la Cour Jansen, 2012). Grease trap waste is commonly collected via liquid tight vacuum trucks that are specially designed for grease collection.

An important consideration is cleaning or exchanging containers for customers so that odors from food waste degradation are minimized. Commercial waste collection trucks may have high pressure washers so that after the contents of a bin are emptied, the bin can be cleaned before returning. For residential collection, it may be easier to have a supply of cleaned containers and exchange them as households are visited.

Step 3: Transportation

Transportation refers to the movement of collected food waste from generators to energy recovery facilities, including vehicle routing and waste unloading. There are many types of collection vehicles suitable for commercial and residential collection including, but not limited to:

- 14–20 cubic yard open top, liquid tight waste collection truck;
- specialized collection truck with completely encapsulated storage compartment;
- pickup truck modified to collect and hold food waste;
- tanker truck for collection of bulk liquids; and
- vacuum trucks.

Transportation of commercial food waste can leverage the large quantities generated at each facility to perform more than one round of collection and delivery per day. Residential transportation solutions are different than transportation of food waste from larger generators. The small amount of food waste generated by single-family residences may not fill a large collection truck within a single day, causing reluctance among drivers to transport the load to a distant waste-to-energy facility. One option is to collect the food waste over a period of multiple days, allowing the food waste to sit until the truck is full. However, food waste will begin degrading in a short amount of time, losing energy potential. A second option is to use a collection vehicle with smaller capacity, which might have the added benefit of better fuel economy, but would necessitate multiple trips to service a wide area.

Specific commercial collection routes may be determined by drivers that are familiar with the pickup locations. However, as smaller generators are added to the service and the complexity of the system increases, software may be necessary for developing routing solutions to increase efficiency. Voluntary residential collection programs might require the use of routing software such as ArcGIS Network Analyst, eRoute Logistics, or Roadnet Transportation Suite to solve for the collection routes between participants (Hall and Partyka, 2016). As more residents participate in the program or residential food waste collection becomes mandatory for homeowners, collection paths will approach the routes used by MSW and recycling since every residence will be participating.

Step 4: Pretreatment

Pretreatment is the process where food waste feedstock is treated physically, thermally, chemically, or biologically to prepare it for the primary energy recovery technology. Pretreatment can occur at many places in the management system; however, currently it is most commonly performed at the waste-to-energy location. In the future, however, pretreatment systems deployed at a commercial food waste generator may enable volume or mass reduction (e.g., dehydration) that minimizes storage volume, pickup frequency, or overall cost to the generator. This concept is discussed later in this chapter because of the challenges and solutions that accompany operation.

Step 5: Energy Recovery Operations

In this step, food waste is converted into energy products for use or sale. Refer to Chapters 4–9 where the technology and conversion processes are extensively discussed.

Step 6: Product Distribution

The final step in food waste management logistics is the distribution of products from the energy recovery technology. While no longer focused on food waste per se, this step is critical in the logistics system and cannot be overlooked. Many types of products can be produced from the range of waste-to-energy technologies discussed in this book. For logistics considerations, the physical state of the product (electricity, solid, liquid, and gas) will be the biggest determinant of the transportation method from the energy recovery facility to the product buyer. Product distribution presents many challenges of its own and is further discussed later in this chapter.

13.3 CHALLENGES AND RECOMMENDATIONS FOR FOOD WASTE LOGISTICS

Across the food waste management system, a series of challenges emerge that impact individual decisions and processes taking place at each of the steps described previously, as well as the overall performance and sustainability of the system. Some of the key challenges are overviewed in Table 13.3, with representative examples of literature addressing these issues.

13.3.1 Compositional Variation

The composition of food waste related to physical and chemical characteristics can be highly variable across generators and time periods. The waste-to-energy technologies detailed in this book rely on consistency in the feedstock supply, just as any other industrial process. Variability in feedstock composition may impact the effectiveness and efficiency of these technologies; therefore maintaining consistency is important and has implications for material sourcing.

In any given area of generators participating in a food waste recovery program, the food waste produced will be variable in terms of quantity and composition. Food waste hauling companies will have to balance a volume of material that maximizes the efficiency of their collection and hauling route with the quality of material desired by downstream management facilities. For example, transportation routes that separate waste based on moisture content may allow for directing drier

TABLE 13.3 Challenges and Opportunities Across the Food Waste Management System

Challenge	Description	Representative Literature
Compositional variation	Variability in the chemical and physical attributes of the food waste itself, which can impact the effectiveness of energy recovery technologies. Variability may arise from types of food waste generated or from the separation practices put in place at the point of generation	(Fisgativa et al., 2016; Lopez et al., 2016; Zhang et al., 2007)
Spatial variation	Fluctuation in quantity of food waste generated from sources across a geographic region. Important contributor to collection and transportation cost	(Business for Social Responsibility (BSR), 2013; New York State Pollution Prevention Institute, 2017a)
Temporal variation	Fluctuation in the quantity of food waste generated from sources over time. Little data available and logistics effects are speculative	(Eriksson et al., 2012; Lebersorger and Schneider, 2014)
Infrastructure deployment	Decisions for the development and operations of food waste logistics systems will vary depending on key actors within a region; their individual priorities; and any collective policy, economic, or technology constraints	(Iakovou et al., 2010; Lopez et al., 2016; Ma et al., 2005; Srivastava, 2007; Thompson et al., 2013)

waste to a pyrolysis facility and wetter waste to an anaerobic digester. Haulers may also consider chemical oxygen demand (COD), pH, and other nutrient characteristics for feedstock compositions (Fisgativa et al., 2016; Lopez et al., 2016).

Contamination in feedstocks is a challenge that must also be considered, referring to materials that are not biodegradable, like plastics, metals, or undesired organic materials that reduce the efficiency of energy recovery. While contamination may not affect collection route logistics, the frequency of contamination, in general, may influence food waste handling and the types of pretreatment technologies that are implemented. For instance, if food usually sold in packages is identified as desirable for an energy recovery facility, but the generator cannot afford to manually separate the waste streams, then installing depackaging equipment at the recovery facility may be a feasible solution. Composition compatibility challenges are likely the most important, but most difficult to overcome due to the general lack of data recorded by generators on their food waste streams.

13.3.2 Spatial Variation

The generation rate of food waste can range from a few pounds per week for households (Van Garde and Woodburn, 1987) to thousands of pounds per week for businesses (New York State Pollution Prevention Institute, 2017b). However, food waste generators can be loosely divided into two categories for the purposes of collection and management:

- Point-source generation: A single location that generates a large quantity of food waste. For example, a restaurant that generates 2000 pounds a week.
- Diffuse generation: Multiple locations that generate small quantities of food waste that add up when combined. For example, 200 households that produce 10 pounds a week.

The spatial variation in food waste generation is a considerable logistics challenge to overcome to achieve the goal of maximizing food waste collection for energy recovery. Utilizing food waste from commercial generators offers the advantage of point-source generation, which may increase efficiency because collection vehicles can collect more food waste from fewer locations, thus reducing the time spent per ton of food waste collected. Fortunately, many businesses that produce food waste are often located near each other, from the natural development of commerce hubs intended for customer convenience; therefore collection routes may further benefit from the reduced distance between businesses (Fig. 13.2).

Of course, efficiency also depends on the spatial distribution of pickup locations in the system. Collecting food waste from organizations that are far apart is less efficient per ton of food waste than collecting from businesses that are clustered together. Collection and transportation efforts will need to coordinate with generators to identify affordable collection options. Options could include altering conventional collection schedules to allow more waste to accumulate between pickups or offering collection services to additional establishments to defray some of the operational costs. Ultimately, rural commercial food waste generation will present the hardest challenges for efficient logistics systems, particularly under food waste disposal bans.

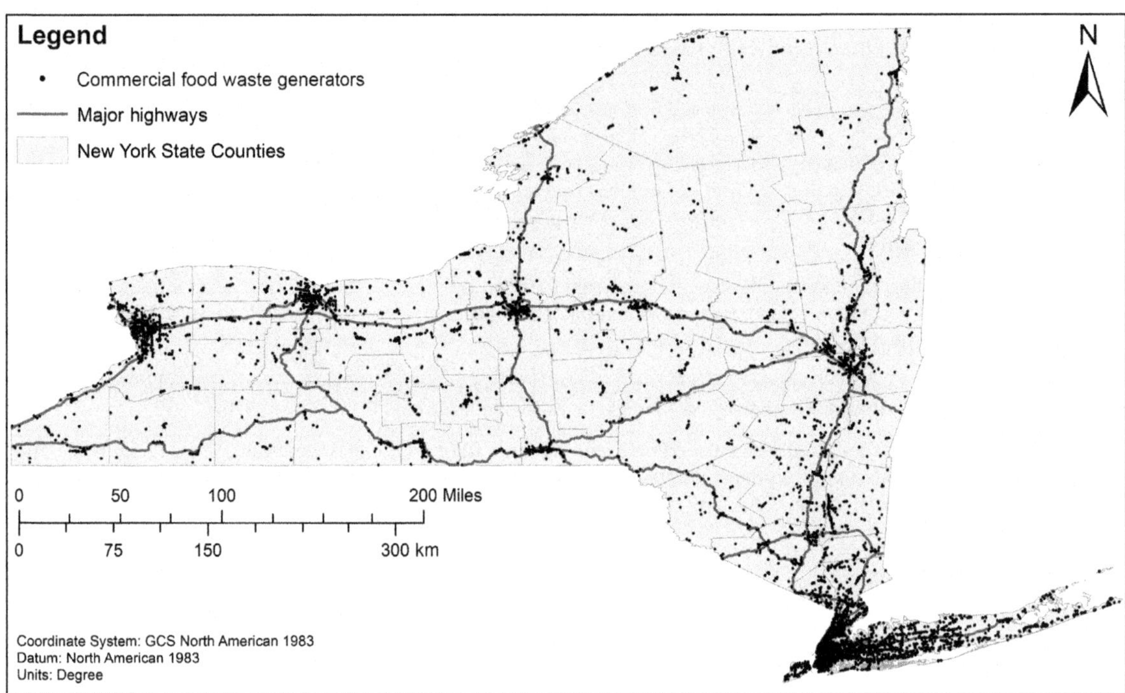

FIG. 13.2 Locations of commercial food waste generators in New York State. Population centers and major roads affect the density and distribution of commercial generators.

The other major source of food waste generation is from households and residences, estimated to be equivalent to the quantity generated by businesses; however, the generation is more diffuse than its commercial counterparts. In the United States, an average person may only generate a few pounds of food waste per week, but cumulatively, consumers produce over 81 billion pounds of food waste per year (Buzby and Hyman, 2012; Kantor et al., 1997). Although the total potential for energy recovery is similar to that of commercial food waste streams, curbside collection will not be as efficient in terms of energy recovered relative to energy invested in collection and hauling.

Altering collection schedules to allow for more material to accumulate at smaller generators will reduce the number of routes and vehicles required to collect food waste. However, odor and pest concerns increase the longer food waste degrades, which is a major barrier for many household participants. For residential collection, utilizing community drop-off locations can centralize the collection and mitigate issues that might arise from degrading organics. Provided that consumers are willing to participate in drop-off systems, the accumulation of residential food waste at these central locations will likely rival volumes generated by businesses, thus making collection more efficient for the service provider (Seven Generations Ahead, 2015).

13.3.3 Temporal Variation

Temporal variation in food waste generation is an additional challenge for waste-to-energy technologies, particularly for waste streams from larger commercial and industrial sources. The rate of food waste generation is typically reported annually (Buzby and Hyman, 2012; New York State Pollution Prevention Institute, 2017b; Panera Bread, 2016) and then converted to average generation per week. This approach is convenient for regulatory purposes because the current laws that exist refer to average weekly or yearly generation. However, in reality, food waste generation will vary over time, often influenced by seasonality, special events, and holidays.

Industrial generators will often produce food waste in episodes coinciding with seasonal material availability or product or process changes. For example, a yogurt manufacturer changes its production lines from light yogurt to Greek yogurt, resulting in washing the storage vats and product lines. This transition could produce hundreds of thousands of gallons of waste water, which may be sent to an anaerobic digester as a cost effective alternative to surcharges applied to sewer disposal of waste water with high organic content. A food waste-to-energy facility may receive little waste from a manufacturer for a period of time and then receive hundreds of thousands of gallons of waste rapidly. These surges of food waste from both commercial and industrial sources can be challenging to compensate for when the recovery facility has limited on-site storage space and most of its input requirements are met by a more consistent source of material.

Similar trends might be prevalent in residential generation. Just as supermarkets will create more food waste from unsold inventory during major events such as national holidays or sports games, residences may also generate more food waste during seasonal events or holidays associated with large meals. Increased generation might disrupt logistics to the extent of requiring additional collection routes to be scheduled to handle the increase in food waste.

13.3.4 Infrastructure Deployment

Infrastructure design presents further challenges to food waste recovery logistics, particularly with respect to siting centralized waste-to-energy facilities. Siting these plants requires considering many factors, such as regulations that apply to waste disposal and treatment, resource supply, and product distribution. Food waste treatment and disposal must comply with all waste relevant regulations, including environmental regulations pertaining to the siting of waste disposal facilities. Environmental issues such as wetland displacement, odor control, and runoff water treatment must be considered. Additionally, community perception and stigma should also be considered, as operating an energy recovery facility near commercial or residential area may create unforeseen negative externalities, such as noise and odor generated by operations and truck traffic. Alternatively, locating in industrial areas or rural locations may minimize these impacts, but may increase the transportation cost for material delivery. It is also important to consider the proximity of other organic waste or MSW management facilities in the area. Tipping fees, especially at landfills, vary regionally (Manson, 2017) and across states in the United States; the economic viability of a waste-to-energy recovery facility may be regionally dependent on the competitiveness of these facilities compared to incumbent disposal options. This concept is discussed in more detail in Chapter 11.

Access to a supply of food waste resources is an important and controllable factor when siting recovery facilities. Depending on the technology, we need to consider a feasible area that provides suitable food waste resources that satisfy the technology requirements. Proximity to a freeway or arterial road is an important consideration for collection vehicles access. Ease of facility access should be balanced with spatial availability of resources. Intuitively, more food waste will be available in more densely populated areas; however, ease of access might be restricted due to congestion effects of increased population. As the potential facility site moves away from population centers, ease of access will increase but resource availability will decrease.

For example, anaerobic digesters already operating in Western New York State currently face challenges related to resource scarcity. Initially, these digesters were constructed alongside farms with hundreds of dairy cows to take advantage of the close proximity of cow manure with which to codigest food waste for greater process stability. However, the dairy farms are located relatively far away from population centers (more than 20 miles) and have limited access to major roads. This spatial variability has led to many digesters struggling to stay economically viable because of the limited access to food waste resources and high cost of transportation from generators to the dairy farms. The trade-off between resource availability and a suitable site is one of the larger system challenges prospective waste-to-energy facilities face.

Transportation of products off-site is another key component to effective siting. Each recovery technology mentioned previously converts food waste into some sort of product, be it solid, liquid, gas, or by extension, electricity. Part of waste-to-energy facility operations is ensuring that avenues exist to move products to markets off-site. Siting and development of recovery technologies with many potential product choices will complicate the site selection process. Road access is generally important for moving any solid, liquid, or compressed gas outputs by truck. If electricity is produced, then access to power lines that can handle transmission should be readily available. For technologies producing natural gas for pipeline injection, the proximity and cost to connect to the pipeline network should be considered in the planning process. Particularly with respect to biogas produced via anaerobic digestion, the economic benefits of the system can depend heavily on utilization pathways (i.e., vehicle fuel, heating fuel, or electricity), access to infrastructure, and economic incentives for producing and utilizing the biogas (Lantz et al., 2007).

13.4 INNOVATIONS IN FOOD WASTE MANAGEMENT SYSTEM DESIGN

Many of the challenges discussed in this chapter stem from the time, cost, labor, and environmental impacts associated with collecting and transporting heavy food waste streams over significant distances. While recovering energy from this waste stream offers the promise of recouping these costs and impacts, a net benefit is not assured. Major barriers to consistent energy recovery benefits include impurities and inconsistencies in food waste composition, waste-to-energy facility processing inefficiencies, and/or lack of markets or infrastructure for utilizing the generated energy and/or by-products (De Clercq et al., 2016; Nghiem et al., 2017). These challenges give rise to opportunities for innovating the overall design of the waste management system. Three specific emerging solutions are discussed in greater detail here.

13.4.1 Pretreatment and Posttreatment Technologies

Characteristics of food waste when generated may not always meet the operational requirements of waste-to-energy technologies; thus, pretreatment can play a critical role in efficient operations. Pretreatment processes can be operated anywhere along the logistic chain depending on the specific needs of generators and energy recovery facilities and may alter how food waste is transported. For instance, if the system feedstock needs to be converted into a slurry, the required grinder or shredder could be located at either the waste generator or the recovery facility. If located at the recovery facility, the food waste could be transported as produced in a conventional solid waste collection truck and then pulped at the recovery facility. If the grinder or shredder was operated at the waste generator, the food waste can be pumped into a tanker truck and delivered to the recovery facility requiring less pretreatment. Table 13.4 lists types of pretreatment processes currently used and in development for waste-to-energy pretreatment.

From a systems perspective, installing and operating pretreatment at generators may not produce benefits, unless they bring reduced transportation costs, seamless transfer, or compatibility with other secondary operations. Consumer-facing businesses generally lack waste management as part of their business model and produce insufficient food waste for pretreatment process operations at an economically viable scale. Additionally, recovery facilities act as centralized locations for storage and pretreatment where operators control properties of the feedstock added to the treatment technology. To relieve recovery facilities from these responsibilities, each food waste generator in the system would have to pretreat their waste under similar standards appropriate for input. Furthermore, the recovery facility would have to trust generators in performing pretreatment correctly. This level of coordination within the system is unlikely and may reduce the flexibility for logistics choices.

Posttreatment technologies are also important, not just for energy products, but other coproducts associated with facility operations. An anaerobic digester that produces biogas may choose to clean and compress it for pipeline injection, rather than combust it to generate electricity, each choice requiring drastically different technologies (Lantz et al., 2007). That same digester facility may also choose to separate solids and liquids from the residual slurry using a screw press or similar equipment, selling or finding onsite use for the solids as animal bedding and the liquids as fertilizer.

TABLE 13.4 Lists and Descriptions of Pretreatment Technologies

Type	Examples	Description	Representative Literature Example
Physical	Depackaging Vacuum	Contamination removal	(Sullivan, 2012)
	Screw press Bead mill Sonication Maceration Lysis centrifuge	Particle size reduction for to allow for higher efficiency of primary energy recovery	(Ariunbaatar et al., 2014; Patinvoh et al., 2017)
Thermal	Steam heating Electric heating Microwave Supercritical liquid	Assists the physical and chemical breakdown of materials, potentially breaking down long chain molecules or creating new compounds	(Ariunbaatar et al., 2014; Patinvoh et al., 2017)
Chemical	Alkali addition Acid addition Oxidant addition Organic Solvent Addition	Changes the chemical composition of the feedstock material, usually to make the material environment more receptive to biological treatments	(Ariunbaatar et al., 2014; Patinvoh et al., 2017)
Biological	Aerobic Anaerobic Enzyme addition Fungi growth	Decomposition and conversion of nutrients via biological processes to increase effectiveness of primary recovery process	(Ariunbaatar et al., 2014; Patinvoh et al., 2017)

13.4.2 Intermediate Facilities

Various types of logistics systems such as material supply chains, product recovery, and similar biomass-to-energy systems include intermediate facilities to aggregate materials and stabilize inventory (Fleischmann et al., 2000; Gold and Seuring, 2011; Iakovou et al., 2010; Pohlen and Theodore Farris, 1992), a concept that has not been extensively researched for food waste management. These facilities could act as central hubs within waste supply chains for multiple organic recovery facilities, including food waste that requires sorting, treatment, and distribution. The concept would require generators to drop-off organic wastes ignoring the composition during collection. Within the facility, managers may inventory the food wastes and sort it into specific categories. Then, the wastes could be combined and treated to form a feedstock with specific attributes depending on the composition desired by the recovery technology client.

By centralizing the storage and distribution effort, the burden of identifying and addressing heterogeneity in food waste may be mitigated by technical capabilities that leverage economies of scale and reduce operating costs. Incorporating the centralized storage concept into food waste transportation would significantly decrease the logistics and infrastructure complexity faced by food waste collection companies and recovery facility operations.

13.4.3 Food Waste Biorefineries

As described previously, centralizing food waste collection and distribution operations can leverage economies of scale and minimize natural fluctuations in waste stream composition and volume that may impact waste-to-energy facility operation. However, an even greater benefit may be realized by a similar integration of technologies, whereby food waste streams can be converted into a wide array of value-added bioenergy and bioproducts. This concept, known as a "food waste biorefinery," is akin to oil refineries in that incoming feedstocks (biomass instead petroleum) are converted to multiple coproducts, which may include electrical or thermal energy, liquid fuels, compost, fertilizer, specialty chemicals, or solvent-grade alcohols (Hegde et al., 2018), which can make the operator more competitive, particularly in response to fluctuating demand for and prices of these bioproducts (Cherubini, 2010; Lohrasbi et al., 2010).

The biorefinery concept will require integrating many of the technologies discussed earlier in this book, for example, combining anaerobic digestion and pyrolysis to convert certain organic waste streams to both biogas and biochar (Barber and Trabold, 2017; Hegde et al., 2018). Other technology integrations may lead to enhanced economic and environmental performance, particularly if they can be colocated as an integrated biorefinery "ecosystem." For example, anaerobic digesters may be deployed in close proximity to address waste management requirements, like manure management regulations for dairy farms or commercial food waste disposal bans. The resulting effluent (digestate) can be used as a liquid fertilizer, but nutrient (nitrogen, phosphorus, potassium) content depends on the waste feedstocks, which change over time. For integrated biorefineries to displace synthetic fertilizers in a meaningful way, a secondary process is needed to blend digestate streams into bio-based fertilizers to meet specifications of the receiving firm. This arrangement depends on characterization of feedstock and effluent properties, spatial organization of digestate production and demand, and economic and policy constraints in the area. Nevertheless, the biorefinery concept offers the potential for significant technological innovation and for rethinking food waste as a bio-based feedstock for the circular economy.

13.5 CONCLUSIONS

As food waste-to-energy technologies continue to emerge, their role within the broader waste management system requires careful consideration from logistics perspectives. Food waste supply encompasses the activities of separating and collecting food waste from a generator and then transporting it to intermediate and final treatment facilities. Energy recovery relates to operations, matching the technology correctly with a highly variable input stream, and to ultimately utilizing the resulting bioenergy and other products. The food waste management system exists within the broader context of policy initiatives (e.g., landfill bans), economic constraints (e.g., biogas incentives or regional tipping fees), and regional infrastructure systems (e.g., roads, facility siting, and bioenergy integration into energy infrastructure).

The overall sustainability of a food waste management system can be assessed at a direct level, in terms of net resources diverted from disposal, emissions avoided by removing food waste from landfill, or replacement of fossil energy sources with waste-derived bioenergy. More broadly, sustainability of food waste-to-energy systems cross-cuts the entire nexus of food, energy, and water, and includes economic, environmental, and social ramifications. Therefore, assessment of food waste-to-energy must also comprehend any systems-level impacts, such as the downstream disposal of liquid residues or the social implications of siting treatment facilities. Sustainable waste-to-energy systems will require continuous innovation, such as novel pretreatment technologies that can make waste streams easier to transport, treat, and utilize for energy recovery. To reduce the complexity of food waste management systems and to maximize waste diversion and energy recovery, innovative logistics methods and system designs are required. Food waste biorefineries that centralize waste inputs with integrated treatment technologies are capable of generating a wide array of bioenergy, biomaterial, and specialty chemical products as part of the growing circular economy.

REFERENCES

Ariunbaatar, J., Panico, A., Esposito, G., Pirozzi, F., Lens, P.N.L., 2014. Pretreatment methods to enhance anaerobic digestion of organic solid waste. Appl. Energy 123, 143–156.

Barber, S.T., Trabold, T.A., 2017. In: Waste paper biochar as a sustainable carbon black pigment replacement in ink jet printer inks.6th World Sustainability Forum, Cape Town, p. 55.

Beamon, B.M., 1998. Supply chain design and analysis: Int. J. Prod. Econ. 55, 281–294.

Beliën, J., De Boeck, L., Van Ackere, J., 2012. Municipal solid waste collection and management problems: a literature review. Transp. Sci. 48, 78–102.

Bernstad, A., la Cour Jansen, J., 2012. Separate collection of household food waste for anaerobic degradation—comparison of different techniques from a systems perspective. Waste Manag. 32, 806–815.

Beullens, P., 2004. Reverse logistics in effective recovery of products from waste materials. Rev. Environ. Sci. Bio/Technol. 3, 283–306.

Business for Social Responsibility (BSR), 2013. Analysis of U.S. Food Waste Among Food Manufacturers, Retailers, and Wholesalers. Prepared for the Food Waste Reduction Alliance. pp. 1–24.

Buzby, J.C., Hyman, J., 2012. Total and per capita value of food loss in the United States. Food Policy 37, 561–570.

Cherubini, F., 2010. The biorefinery concept: using biomass instead of oil for producing energy and chemicals. Energy Convers. Manag. 51, 1412–1421.

Connecticut Department of Energy & Environmental Protection, 2017. DEEP: Commercial Organics Recycling Law. Public Act 11-217. http://www.ct.gov/deep/cwp/view.asp?a=2718&q=552676&deepNav_GID=1645 (accessed 10.20.17).

Dahlén, L., Lagerkvist, A., 2010. Pay as you throw. Strengths and weaknesses of weight-based billing in household waste collection systems in Sweden. Waste Manag. 30, 23–31.

De Clercq, D., Wen, Z., Fan, F., Caicedo, L., 2016. Biomethane production potential from restaurant food waste in megacities and project level-bottlenecks: a case study in Beijing. Renew. Sust. Energ. Rev. 59, 1676–1685.

Department of Environmental Protection, 2014. Commercial Food Waste Disposal Ban | MassDEP. Energy Environ. Aff. http://www.mass.gov/eea/agencies/massdep/recycle/reduce/food-waste-ban.html.

Eriksson, M., Strid, I., Hansson, P.A., 2012. Food losses in six Swedish retail stores: wastage of fruit and vegetables in relation to quantities delivered. Resour. Conserv. Recycl. 68, 14–20.

Fisgativa, H., Tremier, A., Dabert, P., 2016. Characterizing the variability of food waste quality: a need for efficient valorisation through anaerobic digestion. Waste Manag. 50, 264–274.

Fleischmann, M., Bloemhof-Ruwaard, J.M., Dekker, R., van der Laan, E., van Nunen, J.A.E.E., Van Wassenhove, L.N., 1997. Quantitative models for reverse logistics: a review. Eur. J. Oper. Res. 103, 1–17.

Fleischmann, M., Krikke, H.R., Dekker, R., Flapper, S.D.P., 2000. A characterisation of logistics networks for product recovery. Omega 28, 653–666.

Gold, S., Seuring, S., 2011. Supply chain and logistics issues of bio-energy production. J. Clean. Prod. 19, 32–42.

Hall, R.W., Partyka, J., 2016. Vehicle routing software survey: higher expectations drive transformation. OR/MS Today. 43.

Hegde, S., Lodge, J.S., Trabold, T.A., 2018. Characteristics of food processing wastes and their use in sustainable alcohol production. Renew. Sust. Energ. Rev. 81, 510–523.

Hesse, M., Rodrigue, J.-P., 2004. The transport geography of logistics and freight distribution. J. Transp. Geogr. 12, 171–184.

Iakovou, E., Karagiannidis, A., Vlachos, D., Toka, A., Malamakis, A., 2010. Waste biomass-to-energy supply chain management: A critical synthesis. Waste Manag.—Anaerobic Digest. (AD) Solid Waste 30, 1860–1870.

Kantor, L.S., Lipton, K., Manchester, A., Oliveira, V., others, 1997. Estimating and addressing America's food losses. Food Rev. 20, 2–12.

Lantz, M., Svensson, M., Björnsson, L., Börjesson, P., 2007. The prospects for an expansion of biogas systems in Sweden—incentives, barriers and potentials. Energy Policy 35, 1819–1829.

Leanpath, 2016. How to Conduct a Food Waste Audit. http://info.leanpath.com/waste-audit-guide.

Lebersorger, S., Schneider, F., 2014. Food loss rates at the food retail, influencing factors and reasons as a basis for waste prevention measures. Waste Manag. 34, 1911–1919.

Lohrasbi, M., Pourbafrani, M., Niklasson, C., Taherzadeh, M.J., 2010. Process design and economic analysis of a citrus waste biorefinery with biofuels and limonene as products. Bioresour. Technol. 101, 7382–7388.

Lopez, V.M., De la Cruz, F.B., Barlaz, M.A., 2016. Chemical composition and methane potential of commercial food wastes. Waste Manag. 56, 477–490.

Ma, J., Scott, N.R., DeGloria, S.D., Lembo, A.J., 2005. Siting analysis of farm-based centralized anaerobic digester systems for distributed generation using GIS. Biomass Bioenergy 28, 591–600.

Manson, C., 2017. Benefit-Cost Analysis of Potential Food Waste Diversion Legislation. https://www.nyserda.ny.gov/About/Newsroom/2017-Announcements/2017-03-16-NYSERDA-Diverting-Food-Scraps-From-Landfills-Produce-Net-Benefit-22M-Annually.

Neff, R.A., Spiker, M.L., Truant, P.L., 2015. Wasted food: U.S. consumers' reported awareness, attitudes, and behaviors. PLoS One. 10.

New York State Pollution Prevention Institute, 2017a. Organic Resource Locator | NYS Food System Sustainability Clearinghouse | NYSP2I. Org. Resour. Locator. https://www.rit.edu/affiliate/nysp2i/food/organic-resource-locator.

New York State Pollution Prevention Institute, 2017b. Organic Resource Locator | NYSP2I. https://www.rit.edu/affiliate/nysp2i/organic_resource_locator (accessed 4.25.16).

New York State Pollution Prevention Institute, n.d. NYSP2I Self-assessment Toolbox. https://www.rit.edu/affiliate/nysp2i/food/nysp2i-self-assessment-toolbox (accessed 11.11.17).

Nghiem, L.D., Koch, K., Bolzonella, D., Drewes, J.E., 2017. Full scale co-digestion of wastewater sludge and food waste: bottlenecks and possibilities. Renew. Sust. Energ. Rev. 72, 354–362.

Nilsson Påledal, S., Hellman, E., Moestedt, J., 2018. The effect of temperature, storage time and collection method on biomethane potential of source separated household food waste. Waste Manag. 71, 636–643.

Pan, S.-Y., Du, M.A., Huang, I.-T., Liu, I.-H., Chang, E.-E., Chiang, P.-C., 2015. Strategies on implementation of waste-to-energy (WTE) supply chain for circular economy system: a review. J. Clean. Prod. 108, 409–421.

Panera Bread, 2016. Panera Responsibility Report. https://www.panerabread.com/foundation/documents/press/2017/panera-bread-csr-2015-2016.pdf (accessed 11.20.17).

Patinvoh, R.J., Osadolor, O.A., Chandolias, K., Sárvári Horváth, I., Taherzadeh, M.J., 2017. Innovative pretreatment strategies for biogas production. Bioresour. Technol. 224, 13–24.

Pohlen, T.L., Theodore Farris, M., 1992. Reverse logistics in plastics recycling. Int. J. Phys. Distrib. Logist. Manag. 22, 35–47.

Qi, D., Roe, B.E., 2016. Household food waste: multivariate regression and principal components analyses of awareness and attitudes among U.S. consumers. PLoS One. 11.

Sandhu, G.S., Frey, H.C., Bartelt-Hunt, S., Jones, E., 2015. In-use activity, fuel use, and emissions of heavy-duty diesel roll-off refuse trucks. J. Air Waste Manage. Assoc. 65, 306–323.

Seven Generations Ahead, 2015. Food Scrap Composting Challenges and Solutions in Illinois Report. http://illinoiscomposts.org/files/IFSC-FoodScrapReportFINAL-Jan2015.pdf (accessed 11.29.17).

Srivastava, S.K., 2007. Green supply-chain management: a state-of-the-art literature review. Int. J. Manag. Rev. 9, 53–80.

Srivastava, S.K., 2008. Network design for reverse logistics. Omega (Spec. Issue Logist. N. Perspect. Challenges) 36, 535–548.

Sullivan, D., 2012. Depackaging organics to produce energy. Biocycle 53, 42.

Thompson, E., Wang, Q., Li, M., 2013. Anaerobic digester systems (ADS) for multiple dairy farms: a GIS analysis for optimal site selection. Energy Policy 61, 114–124.

Tompkins County Department of Recycling and Materials Management, n.d., Food Scraps Recycling. https://recycletompkins.org/Recycling/Food-Scraps-Recycling (accessed 11.14.17).

United States Environmental Protection Agency, 2014. A Guide to Conducting and Analyzing a Food Waste Assessment. https://www.epa.gov/sites/production/files/2015-08/documents/r5_fd_wste_guidebk_020615.pdf.

Van Garde, S.J., Woodburn, M.J., 1987. Food discard practices of householders. J. Am. Diet. Assoc. 87, 322–329.

Yepsen, R., 2015. BioCycle Nationwide Survey: residential food waste collection in the U.S. Biocycle 56, 53.

Zhang, R., El-Mashad, H.M., Hartman, K., Wang, F., Liu, G., Choate, C., Gamble, P., 2007. Characterization of food waste as feedstock for anaerobic digestion. Bioresour. Technol. 98, 929–935.

Index

Note: Page numbers followed by *f* indicate figures, and *t* indicate tables.

D

E

F

W

Y

Z

Printed in the United States
By Bookmasters